Mastering iOS 12 Programming *Third Edition*

Build professional-grade iOS applications with Swift and Xcode 10

Donny Wals

BIRMINGHAM - MUMBAI

Mastering iOS 12 Programming
Third Edition

Commissioning Editor: Amarabha Banerjee
Acquisition Editor: Trusha Shriyan
Content Development Editor: Onkar Wani
Technical Editor: Ralph Rosario
Copy Editor: Safis Editing
Project Coordinator: Sheejal Shah
Proofreader: Safis Editing
Indexer: Tejal Daruwale Soni
Graphics: Alishon Mendonsa
Production Coordinator: Aparna Bhagat

First published: December 2016
Second edition: October 2017
Third edition: October 2018

Production reference: 1311018

Published by Packt Publishing Ltd.
Livery Place
35 Livery Street
Birmingham
B3 2PB, UK.

ISBN 978-1-78913-320-2

www.packtpub.com

This book is dedicated to Dorien for always believing in me and pushing me to stay on top of my game.

– Donny Wals

`mapt.io`

Mapt is an online digital library that gives you full access to over 5,000 books and videos, as well as industry leading tools to help you plan your personal development and advance your career. For more information, please visit our website.

Why subscribe?

- Spend less time learning and more time coding with practical eBooks and Videos from over 4,000 industry professionals
- Improve your learning with Skill Plans built especially for you
- Get a free eBook or video every month
- Mapt is fully searchable
- Copy and paste, print, and bookmark content

Packt.com

Did you know that Packt offers eBook versions of every book published, with PDF and ePub files available? You can upgrade to the eBook version at `www.packt.com` and as a print book customer, you are entitled to a discount on the eBook copy. Get in touch with us at `customercare@packtpub.com` for more details.

At `www.packt.com`, you can also read a collection of free technical articles, sign up for a range of free newsletters, and receive exclusive discounts and offers on Packt books and eBooks.

Contributors

About the author

Donny Wals is a passionate, curious, iOS developer from The Netherlands. With several years of experience in building apps and sharing knowledge under his belt, Donny is a respected member of the iOS development community. Donny enjoys delivering talks on smaller and larger scales to share his knowledge and experiences with his peers.

In addition to sharing knowledge, Donny loves learning more about iOS, Apple's frameworks and development in general. This eagerness to learn has made him into a versatile iOS developer with knowledge of a significant number of Apple's frameworks and tools. During WWDC you will often find Donny binge-watching the talks that Apple engineers deliver to introduce new features and frameworks.

This book goes out to the entire iOS development community as a whole. Without such a great community I would not be where I am today (and there would be nobody to read this book). I would like to thank Ties Baltissen in particular for being my guinea pig during the process of writing and refining the content for this book.

About the reviewers

Cecil Costa, also known as Eduardo Campos in Latin countries, is a Euro-Brazilian freelance developer. He has been giving onsite courses for companies such as Ericsson, Roche, TVE (a Spanish TV channel), and others. He has also worked for different companies, including IBM, Qualcomm, Spanish Lottery, and Dia. Heis also the author of Swift Cookbook, Swift 2 Blueprints, Reactive Programming with Swift, and a video course called Building iOS 10 Applications with Swift, by Packt Publishing.

Nikola Brežnjak is an engineer at heart and a jack of all trades kind of guy. Currently, he's the director of mobile engineering at Teltech and is responsible for the management, mentoring and coaching of mobile app developers.
He loves his job!

He wrote books on the Ionic framework and the MEAN stack and was a technical reviewer for few of Packt's books.

He likes to help out on StackOverflow where he's a top contributor.
He records a podcast called DevThink with his friend Shawn Milochik and runs a local meetup called MeCoDe.

I want to thank my wife for supporting me in all my geeky endeavors and my parents for teaching me the power of hard and consistent work.

Shaun Rowe is a software engineer with over 20 years of professional experience. He wrote his first line of code at the age of 11 and began his career at the height of the dot com boom. He has been developing for Apple platforms for over 8 years, as soon as Swift was officially supported by XCode he became an early adopter. During his 20 years in the industry he has worked in many fields and many disciplines, gaining knowledge and experience in all aspects of software development, from the web to mobile. He ran a successful mobile development agency for a number of years before moving to Amsterdam for a new and exciting opportunity at the end of 2017.

I would like to thank the author for offering me the opportunity to review this excellent book.

Packt is searching for authors like you

If you're interested in becoming an author for Packt, please visit `authors.packtpub.com` and apply today. We have worked with thousands of developers and tech professionals, just like you, to help them share their insight with the global tech community. You can make a general application, apply for a specific hot topic that we are recruiting an author for, or submit your own idea.

Table of Contents

Preface

Mastering iOS 12 is the third book in the *Mastering iOS* series that started in 2016 when iOS 10 was released. Since the release of the first version, this book has grown in size to make sure all of the new and interesting bits of are iOS covered and to teach you how to make sure your iOS applications meet Apple's latest standards. You will learn everything you need to know to build great applications that are packed with great features.

This book follows a learn-by-doing style of teaching. Every newly introduced topic in every chapter is accompanied by a sample application that shows you how to put the new theory you read into practice. This means that this book is a very practical guide to becoming an iOS master. In addition to teaching you a lot of Apple's new and existing frameworks, a lot of focus is placed on Apple's own documentation. Since not every single iOS framework can be covered in one book, it is crucial that you train yourself in scanning and understanding Apple's documentation.

It doesn't matter whether you haven't managed to build an app before, or if you're an experienced iOS developer. This book will help you become a better iOS developer, and it will teach you everything new that iOS 12 has to offer. You will step beyond the basics and you will learn everything you need to know to build high-quality iOS applications.

Who this book is for

This book is for developers who have some experience with iOS programming but want to take their skills to the next level by unlocking the full potential of the latest version of iOS with Swift to build impressive applications. It is assumed that you have some basic knowledge of iOS development throughout the book.

What this book covers

`Chapter 1`, *UITableView Touch-up*, covers UITableView, which is the core of many iOS applications. Most developers that attempt to learn iOS will start by implementing a table view. This chapter is intended to provide a warm-up for the next chapters while also providing in-depth information about UITableView and its internals.

`Chapter 2`, *A Better Layout with UICollectionView*, explores UITableView's super-powered sibling, UICollectionView. This chapter covers implementing a collection view and writing a custom layout object for it.

`Chapter 3`, *Creating a Detail Page*, explains that when you build apps, you'll need to make them work on many different screen sizes. This chapter will show you how to use AutoLayout to create beautiful, adaptive layouts.

`Chapter 4`, *Immersing Your Users with Animation*, shows how the best apps set themselves apart with beautiful and subtle animations. In this chapter, you will learn the best animation techniques, and to top it off, you'll learn how to create custom transitions between view controllers.

`Chapter 5`, *Understanding the Swift Type System*, focuses on the different data types Swift uses, how they relate to each other, and how to decide what type is best for certain use cases.

`Chapter 6`, *Writing Flexible Code with Protocols and Generics*, moves on from learning about the type system's basic concrete types; it's also important to learn about abstractions that make your code more flexible, robust, and future-proof. This chapter will show you how you can improve your Swift code with the right abstractions and principles.

`Chapter 7`, *Improving Application Structure*, teaches you how to apply the lessons learned from the previous two chapters to improve an app you've already worked on.

`Chapter 8`, *Adding Core Data to Your App*, explains that many apps need to store data in a database. This chapter introduces Apple's Core Data framework as a way to include a database for user data in your app.

`Chapter 9`, *Fetching and Displaying Data from the Network*, shows you how you can access data and resources over a network connection and display the results to the user.

`Chapter 10`, *Being Proactive with Background Fetch*, explains that iOS allows apps to refresh and load data in the background. You will learn how to implement Background Fetch and you will briefly be introduced to Dispatch Groups.

`Chapter 11`, *Syncing Data with CloudKit*, covers how to store data in the cloud with CloudKit. It also covers a scenario where CloudKit is used as online storage, while Core Data is used to persist data locally.

`Chapter 12`, *Using Augmented Reality*, demonstrates how you can use Apple's groundbreaking ARKit framework to build an augmented reality experience.

`Chapter 13`, *Improving Apps with Location Services*, covers several ways that apps can implement location tracking to enhance and improve a user's experience.

`Chapter 14`, *Making Smarter Apps with CoreML*, teaches everything you should know about the CoreML framework. Readers will implement a machine learning model that recognizes dominant objects in a scene. This chapter also covers CreateML to augment existing models with more specific training data.

`Chapter 15`, *Tracking Activity Using HealthKit*, explains how to access a user's health information and how you can implement an app that tracks a user's workouts and stores them in the Health app.

`Chapter 16`, *Streamlining Experiences with Siri*, shows you how to integrate the SiriKit APIs in your own applications. This enables you to integrate your app deeply into the iOS platform. It also covers the new Siri shortcuts that allow users to quickly perform actions in certain apps.

`Chapter 17`, *Using Media in Your App*, explains how to play back audio and video, take photos, and apply interesting filters to images that users take with your app.

`Chapter 18`, *Implementing Rich Notifications*, is a walk-through of everything you need to know about providing a great notification experience for your users. It covers both the UI extension and the content extension.

`Chapter 19`, *Instant Information with a Today Extension*, shows that developers can add widgets to the Notification Center to disclose quick information to users; this chapter will teach you how.

`Chapter 20`, *Exchanging Data with Drag and Drop*, takes you through how to augment an app to allow users to drag contents from your app to other apps and vice versa.

`Chapter 21`, *Improved Discoverability with Spotlight and Universal Links*, explores how you can make an iOS index for your app's contents to make it available through the powerful Spotlight search index.

`Chapter 22`, *Extending iMessage*, shows how to build a simple sticker pack and app for iMessage.

`Chapter 23`, *Ensuring App Quality with Tests*, focuses on testing, which is an often-overlooked aspect of developing an app. You will learn how to set up tests for your application.

`Chapter 24`, *Discovering Bottlenecks with Instruments*, explains how to profile your app's performance with instruments. You also learn how to implement and use your own instrument tools.

`Chapter 25`, *Offloading Tasks with Operations and GCD*, covers apps that perform increasingly complex tasks. You will learn how to ensure that complex or slow tasks don't freeze your user interface.

`Chapter 26`, *Submitting Your App to the App Store*, explains how to distribute your app to beta testers through TestFlight and how to submit your app for review in order to publish it to the App Store.

To get the most out of this book

All sample code in this book was written in Swift 4.2 with Xcode 10.0 on a Mac running macOS Mojave. To follow along with all the examples in this book you must have at least Xcode 10.0 installed on your machine. It is recommended that you also have at least macOS Mojave installed on your Mac because not all code samples are compatible with older versions of macOS.

This book assumes that you are somewhat familiar with Swift and iOS development. If you have no experience with Swift at all, it is recommended that you skim through Apple's Swift manual and go over the basics of iOS development. You don't have to be an expert on iOS development yet, but a solid foundation won't hurt you since the pacing of the book is aimed at somewhat experienced developers.

Download the example code files

You can download the example code files for this book from your account at `www.packt.com`. If you purchased this book elsewhere, you can visit `www.packt.com/support` and register to have the files emailed directly to you.

You can download the code files by following these steps:

1. Log in or register at `www.packt.com`.
2. Select the **SUPPORT** tab.
3. Click on **Code Downloads & Errata**.
4. Enter the name of the book in the **Search** box and follow the onscreen instructions.

Once the file is downloaded, please make sure that you unzip or extract the folder using the latest version of:

- WinRAR/7-Zip for Windows
- Zipeg/iZip/UnRarX for Mac
- 7-Zip/PeaZip for Linux

The code bundle for the book is also hosted on GitHub at `https://github.com/PacktPublishing/Mastering-iOS-12-Programming-Third-Edition`. In case there's an update to the code, it will be updated on the existing GitHub repository.

We also have other code bundles from our rich catalog of books and videos available at `https://github.com/PacktPublishing/`. Check them out!

Download the color images

We also provide a PDF file that has color images of the screenshots/diagrams used in this book. You can download it here: `https://www.packtpub.com/sites/default/files/downloads/9781789133202_ColorImages.pdf`.

Conventions used

There are a number of text conventions used throughout this book.

`CodeInText`: Indicates code words in text, database table names, folder names, filenames, file extensions, pathnames, dummy URLs, user input, and Twitter handles. Here is an example: "Mount the downloaded `WebStorm-10*.dmg` disk image file as another disk in your system."

A block of code is set as follows:

```
func doSomething(completionHandler: (Int) -> Void) {
  // perform some actions
  var result = theResultOfSomeAction
  completionHandler(result)
}
```

Bold: Indicates a new term, an important word, or words that you see onscreen. For example, words in menus or dialog boxes appear in the text like this. Here is an example: "Select **System info** from the **Administration** panel."

Warnings or important notes appear like this.

Tips and tricks appear like this.

Get in touch

Feedback from our readers is always welcome.

General feedback: If you have questions about any aspect of this book, mention the book title in the subject of your message and email us at `customercare@packtpub.com`.

Errata: Although we have taken every care to ensure the accuracy of our content, mistakes do happen. If you have found a mistake in this book, we would be grateful if you would report this to us. Please visit `www.packt.com/submit-errata`, selecting your book, clicking on the Errata Submission Form link, and entering the details.

Piracy: If you come across any illegal copies of our works in any form on the Internet, we would be grateful if you would provide us with the location address or website name. Please contact us at `copyright@packt.com` with a link to the material.

If you are interested in becoming an author: If there is a topic that you have expertise in and you are interested in either writing or contributing to a book, please visit `authors.packtpub.com`.

Reviews

Please leave a review. Once you have read and used this book, why not leave a review on the site that you purchased it from? Potential readers can then see and use your unbiased opinion to make purchase decisions, we at Packt can understand what you think about our products, and our authors can see your feedback on their book. Thank you!

For more information about Packt, please visit `packt.com`.

1
UITableView Touch-up

There's a good chance that you have built a simple app before, or maybe you have tried but didn't quite succeed. If this is the case, it is likely that you have used `UITableView` or `UITableViewController`. `UITableView` is a core component of many iOS applications. If an app shows a list of things, it was likely built using `UITableView`. Because `UITableView` is such an essential component on iOS, I want to make sure that we cover it right away in this book. It doesn't matter whether you have looked at `UITableView` before. This chapter will ensure that you are up to speed with the ins and outs of `UITableView` and understand how Apple has made sure that every table view scrolls with a smooth speed of 60 frames per second, which we all strive for when developing apps.

In addition to covering the basics of `UITableView`, such as how it uses delegation to obtain information about the contents it should display, you'll also learn the basics about accessing a user's data, in this case, their contacts. The application that you will end up building displays a list of the user's contacts in a table view.

Every `UITableView` uses cells to render each item it displays. In this chapter, you will create your own `UITableViewCell` that uses *Auto Layout*. Auto Layout is a layout technique that is used throughout this book due to its essential part of every iOS developer's toolkit. If you haven't used *Auto Layout* before, or if you haven't heard of it, that's OK. This chapter starts with the basics, and your knowledge will expand as you go through this book.

In summary, this chapter covers:

- Configuring and displaying `UITableView`
- Fetching a user's contacts through `Contacts.framework`
- The delegate and data source of `UITableView`
- Creating a custom `UITableViewCell`
- `UITableView` performance characteristics

Setting up the user interface

Every time you start a new project in Xcode, you have the option to pick a template for your application. Every template contains a small amount of code to get you started. Sometimes, a basic layout will even be set up for you. Throughout this book, you should default to using the **Single View Application** template. Don't be fooled by its name; you can add as many views to your app as you would like, this template only provides you with only one simple view. Using this template allows you to build your application from scratch, giving you the freedom to set up all the components as you like.

In this chapter, you will create an app that is called **Hello-Contacts**. This app renders your user's contact list in a `UITableView` that you will set up. Let's create a project for this app right now. Select **File | New | Project** in the menu bar. Next, select **Single View Application** from the list of project templates. When prompted to give your project a name, call it **Hello-Contacts**. Make sure that **Swift** is selected as the programming language for your app and uncheck all the Core Data and testing-related checkboxes; we won't need those right now.

Your configuration should resemble the following screenshot:

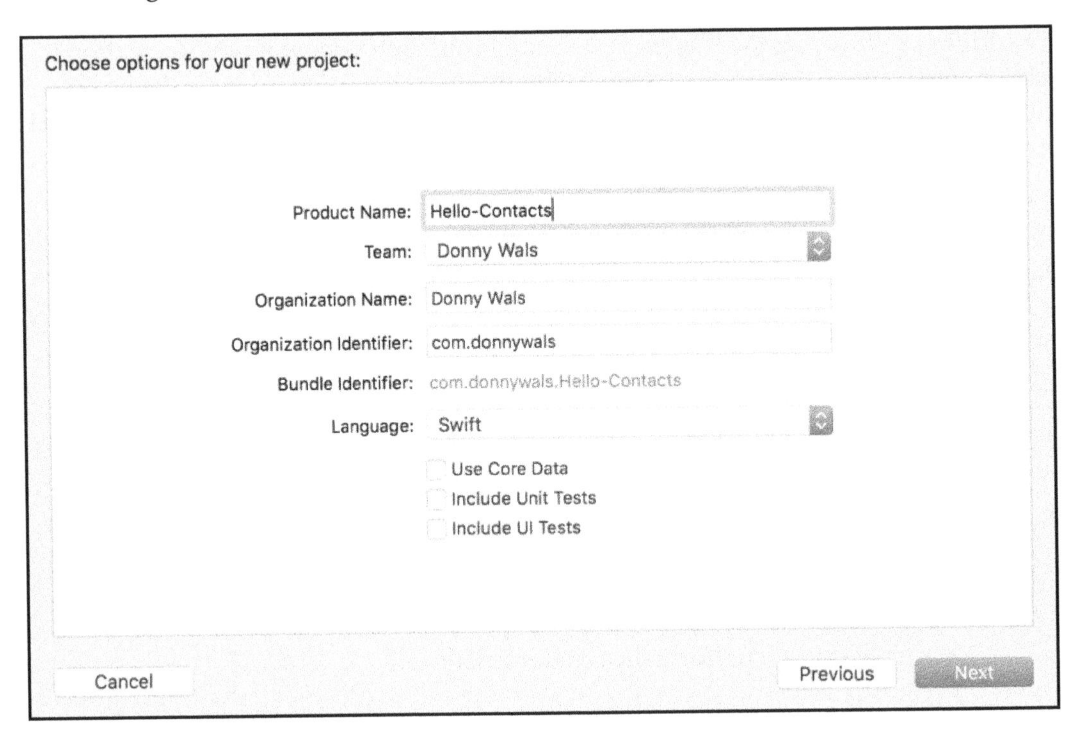

Once your application is configured, open the file named `Main.storyboard`. The storyboard file is used to lay out all of your application's views and to connect them to the code you write. The editor you use to manipulate your storyboard is called **Interface Builder**. Storyboards are a great way to edit your files and see the results of your actions immediately.

If you have used `UITableView` in the past, you may have used `UITableViewController`. The `UITableViewController` class is a subclass of a regular `UIViewController`. The difference is that `UITableViewController` contains a lot of setup that you would otherwise have to perform on your own. To fully understand how `UITableView` is configured and set up, you will not use `UITableViewController` now.

Take a look at the top bar in the **Interface Builder** window. There is a button there that has a circle with a square in it. This button opens the **Object Library**. The following screenshot shows you the **Object Library** and the button you can click to access it:

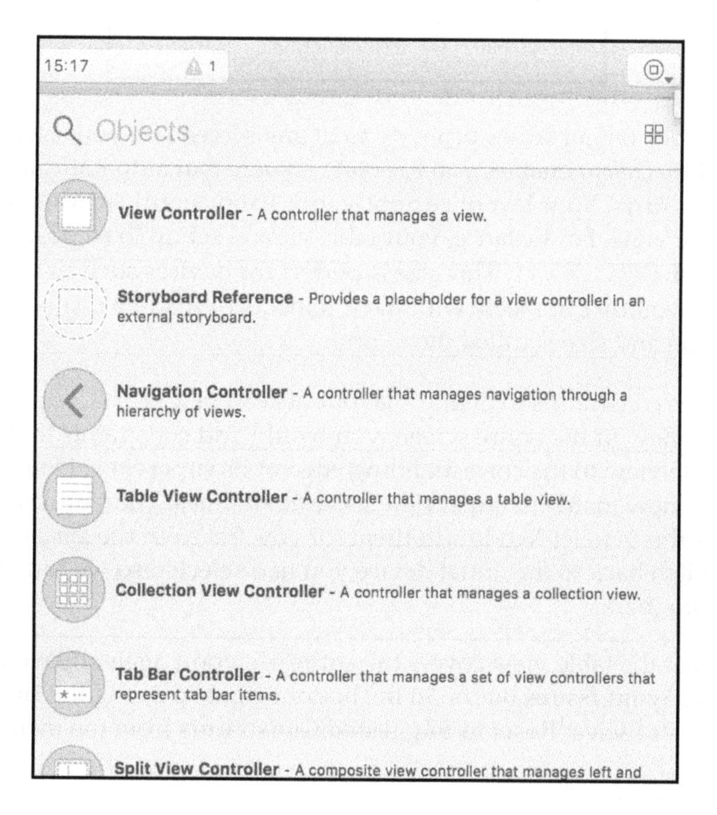

With the **Object Library** opened, look for `UITableView`. If you begin typing the name of the component you're looking for in the **Object Library**, all potential matches should automatically be filtered. Once you find the table view, drag it to the app's view. After doing this, use the white squares at the corners of the table view to make sure that the table covers the entire view.

If you look at the bottom of the window, you can see the dynamic viewport inspector. If you don't see it, try clicking on the name of the current preview device. In the inspector, select a device that either has a larger or a smaller screen than the current device. When you have done this, you will notice that the table view doesn't cover the viewport as nicely anymore. On smaller screens, you'll see that the table view has become larger than the view and on larger screens, the table view isn't large enough to cover the viewport:

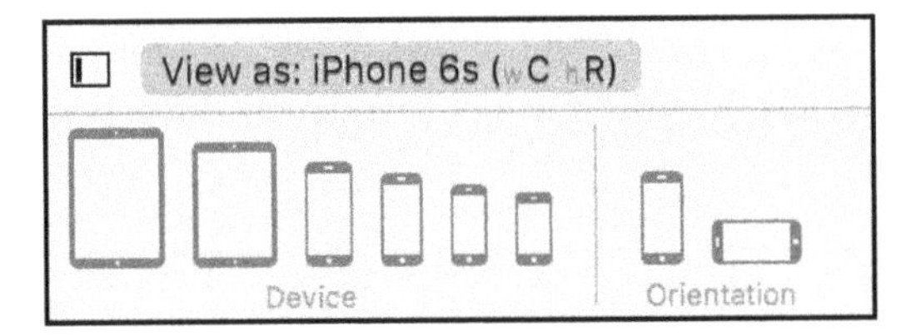

To make sure your layout scales properly to fit any screen size you select, you use *Auto Layout*. *Auto Layout* enables you to create layouts that automatically adapt to any screen size that exists. Your layout currently uses fixed coordinates and dimensions to lay out the table view. For instance, your table view is set up to be positioned at `(0, 0)` with a size of `(375, 667)`. This size is perfect for devices such as the iPhone 6, 6S, 7, and 8. But it wouldn't fit nicely with the iPhone Xs or iPad Pro. This combination of a view's position and size is called the `frame`.

Auto Layout uses constraints to define a layout instead of a `frame`. For instance, to make the table view fit the entire screen, you would add constraints that pin every edge of the table view to the corresponding edge of its superview. Doing so would make the table view match its superview's size at all times. The simplest way to set up constraints for this is to let Xcode add them for you. Let's use the *dynamic viewport inspector* to switch back to the initial device you had selected so we can add constraints from there.

First, ensure that the table view covers the entire viewport again, then click on the **Resolve Auto Layout Issues** button in the bottom-right corner of the **Interface Builder** screen and select **Reset to Suggested Constraints** from the menu that pops up:

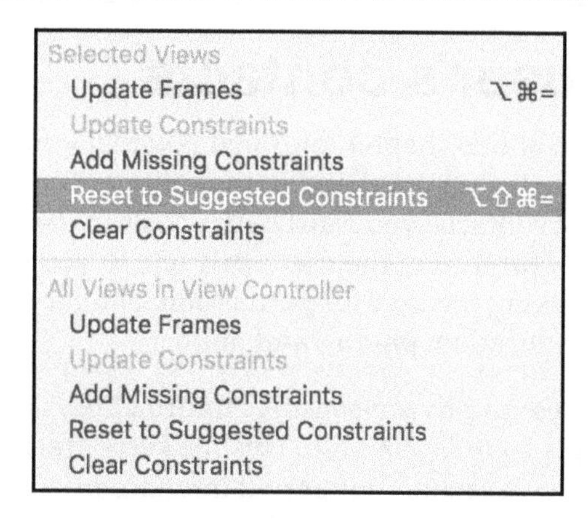

Selecting this option automatically adds the constraints that Xcode considers required for your view. The constraints that were added by Xcode pin all of the table view's edges to its superview edges, which is precisely what you wanted to happen. You can manually inspect these constraints in the **Document Outline** on the left-hand side of the **Interface Builder** window:

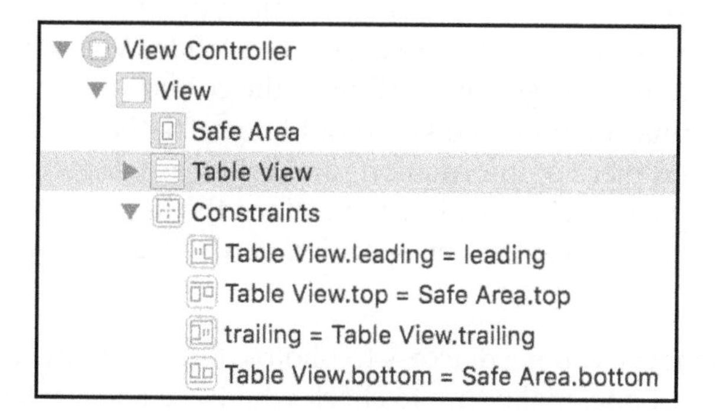

Each constraint that got added describes how the table view should behave relative to another view. In this case, the other view is the superview. If you change the preview device in the *dynamic viewport inspector*, you should see that the table view always covers the entire screen. Try picking a couple of different screen sizes to make sure this works.

Now that you have a table view added to your view and its layout is exactly what you need, it is time to provide the table with some contents to display. To do this, you're going to write some code that uses Apple's `Contacts.framework` to fetch the app user's contacts list from their address book.

Fetching a user's contacts

The introductory section of this chapter informed you that you would use `Contacts.framework` to retrieve that app user's contacts and show this in a table view. To display a list of contacts, you must have access to the user's address book. Apple does a great job of protecting the user's privacy, so you can't read any of their contacts' data without asking the user for permission. Similar restrictions apply to access the user's camera, location, photos, and more.

Whenever you need access to privacy-sensitive information, you are required to specify this in your app's `Info.plist` file. This file keeps track of many of your app's properties, such as its display name, supported interface orientations, and, in the case of accessing a user's contacts, `Info.plist` also contains information about why you need access to the user's contacts.

To add this information to `Info.plist`, open it from the list of files in the **Project Navigator** on the left. Once opened, hover over the word **Information Property List** at the top of the file. A plus icon should appear, clicking it adds a new empty item with a search field to the list. When you begin typing `Privacy - contacts`, Xcode will filter out options for you until there is only one left for you to pick. This option, called **Privacy – Contacts Usage Description**, is the correct option to choose for this case. The value for this newly-added key should describe the reason that you need access to the specified piece of information for. In this case, *reads contacts and shows them in a list* should be sufficient explanation. When the user is asked for permission to access their contacts, the reason you specified here will be shown, so make sure you add an informative message.

Whenever you need access to photos, Bluetooth, the camera, and the microphone, make sure to check whether a privacy key in `Info.plist` is required. If you don't provide this key, your app will crash and will not make it past Apple's review process.

Now that you have configured your app so it specifies that it wants to access contact data, let's get down to writing some code. Before you can read contacts, you must make sure that the user has given the appropriate permissions for you to access contact data. To do this, the code must first read the current permission status. Once this is done, the user must either be prompted for permission to access contacts, or the contacts must be fetched. Add the following code to `ViewController.swift`, we'll cover the details for this code after you have implemented it:

```
import UIKit
// 1
import Contacts

class ViewController: UIViewController {

  override func viewDidLoad() {
    super.viewDidLoad()

    let store = CNContactStore()
    let authorizationStatus = CNContactStore.authorizationStatus(for:
.contacts)

    // 2
    if authorizationStatus == .notDetermined {
      // 3
      store.requestAccess(for: .contacts) { [weak self] didAuthorize,
      error in
        if didAuthorize {
          self?.retrieveContacts(from: store)
        }
      }
    } else if authorizationStatus == .authorized {
        retrieveContacts(from: store)
    }
  }

  func retrieveContacts(from store: CNContactStore) {
    let containerId = store.defaultContainerIdentifier()
    let predicate =
CNContact.predicateForContactsInContainer(withIdentifier: containerId)
    // 4
    let keysToFetch = [CNContactGivenNameKey as CNKeyDescriptor,
                       CNContactFamilyNameKey as CNKeyDescriptor,
                       CNContactImageDataAvailableKey as
                       CNKeyDescriptor,
                       CNContactImageDataKey as CNKeyDescriptor]

    let contacts = try! store.unifiedContacts(matching: predicate,
keysToFetch: keysToFetch)

    // 5
    print(contacts)
  }
}
```

In the preceding code, the first step is to import the `Contacts` framework into the current file. If you don't do this, the compiler won't be able to understand `CNContactStore` or `CNContact` because these classes are part of the `Contacts` framework.

The second step is to check the value of the current authorization status. For this example, only the `notDetermined` and `authorized` statuses are relevant. However, the user can also deny access to their address book. In that case, the authorization status would be `denied`. If the status has not been determined yet, the user is asked for permission. When the app already has access, the contacts are fetched right away.

In the third step, permission is asked to access the user's contacts. The request access method takes a completion-handler as its last argument. In asynchronous programming, completion-handlers are used often. It allows your app to perform some work in the background and then call the completion-handler when the work is completed. You will find completion-handlers throughout `Foundation`, `UIKit`, and many other frameworks. If you implement a very simple function of your own that takes a callback, it might look as follows:

```
func doSomething(completionHandler: (Int) -> Void) {
  // perform some actions
  var result = theResultOfSomeAction
  completionHandler(result)
}
```

Calling a completion-handler looks just like calling a function. The reason for this is that a completion-handler is a block of code, called a closure. **Closures** are a lot like functions because they both contain a potentially reusable block of code that is expected to be executed when called. You will find plenty of examples of closures and completion-handlers in this book because they are ubiquitous in iOS, and programming in general.

Step four in the big chunk of code you added created a list of keys that you'll need to render a list of contacts. Since these keys are of the `String` type and you must provide a list of `CNKeyDesriptor` later, you can use `as CNKeyDescriptor` to convert these `String` values to `CNKeyDescriptor` values. Note that this won't always work because not every type is convertible to a specific other type. For example, you wouldn't be able to do this type of conversion with `UIViewController`.

Finally, when the contacts are fetched, they are printed to the console. Of course, you'll want to update this so that the contacts aren't printed in the console, but rendered in the table view. You might notice the `try!` keyword before fetching the contacts. This is done because fetching contacts could fail and throw an error.

In Swift, you are expected to make the right decision in regards to handling errors. By using `try!`, you inform the compiler that you are 100%, sure that this fetch call will never fail. This is fine for now so you can focus on the essential bits, but when you're writing an app that is expected to make it to production, you might want to handle errors more gracefully. A good practice is to use a `do {} catch {}` block for code that could throw an error. The following code shows a basic example of such a construct:

```
do {
  let contacts = try store.unifiedContacts(matching: predicate,
keysToFetch: keysToFetch)
  print(contacts)
} catch {
  // something went wrong
  print(error) // there always is a "free" error variable inside of a
catch block
}
```

If you run the app with the code you added, the app will immediately ask for permission to access contacts. If you allow access, you will see a list of contacts printed to the console, as shown in the following screenshot:

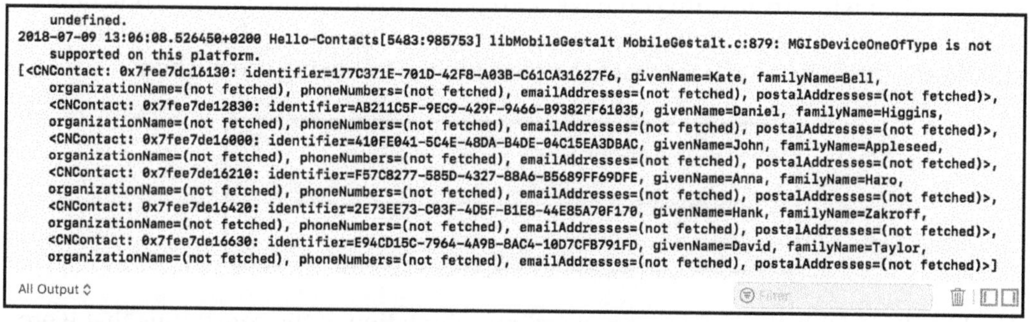

```
    undefined.
2018-07-09 13:06:08.526450+0200 Hello-Contacts[5483:985753] libMobileGestalt MobileGestalt.c:879: MGIsDeviceOneOfType is not
    supported on this platform.
[<CNContact: 0x7fee7dc16130: identifier=177C371E-701D-42F8-A03B-C61CA31627F6, givenName=Kate, familyName=Bell,
    organizationName=(not fetched), phoneNumbers=(not fetched), emailAddresses=(not fetched), postalAddresses=(not fetched)>,
    <CNContact: 0x7fee7de12830: identifier=AB211C5F-9EC9-429F-9466-B9382FF61035, givenName=Daniel, familyName=Higgins,
    organizationName=(not fetched), phoneNumbers=(not fetched), emailAddresses=(not fetched), postalAddresses=(not fetched)>,
    <CNContact: 0x7fee7de16000: identifier=410FE041-5C4E-48DA-B4DE-04C15EA3DBAC, givenName=John, familyName=Appleseed,
    organizationName=(not fetched), phoneNumbers=(not fetched), emailAddresses=(not fetched), postalAddresses=(not fetched)>,
    <CNContact: 0x7fee7de16210: identifier=F57C8277-585D-4327-88A6-B5689FF69DFE, givenName=Anna, familyName=Haro,
    organizationName=(not fetched), phoneNumbers=(not fetched), emailAddresses=(not fetched), postalAddresses=(not fetched)>,
    <CNContact: 0x7fee7de16420: identifier=2E73EE73-C03F-4D5F-B1E8-44E85A70F170, givenName=Hank, familyName=Zakroff,
    organizationName=(not fetched), phoneNumbers=(not fetched), emailAddresses=(not fetched), postalAddresses=(not fetched)>,
    <CNContact: 0x7fee7de16630: identifier=E94CD15C-7964-4A9B-8AC4-10D7CFB791FD, givenName=David, familyName=Taylor,
    organizationName=(not fetched), phoneNumbers=(not fetched), emailAddresses=(not fetched), postalAddresses=(not fetched)>]
All Output
```

Now that you have the user's contact list, let's see how you can make the contacts appear in your table view!

Creating a custom UITableViewCell to show contacts

To display contacts in your table view, you must set up a few more things. First, you are going to need a table-view cell that displays contact information. All code for a custom table view cell should live in a `UITableViewCell` subclass. The design for your custom cell can be made in **Interface Builder**. When you make a design in **Interface Builder**, you can connect your code and the design using `@IBOutlet`. `@IBOutlet` is a connection between an object in the visual layout and a variable in your code.

Designing a table-view cell

Open your app's storyboard in **Interface Builder** and look for `UITableViewCell` in the **Object Library**. When you drag it into the table view that you have already added, your new cell is added as a **prototype cell**. A prototype cell functions as a blueprint for all cells the table view is going to display. That's right; you only need to set up a single cell to display many. You'll see how this works when you implement the code for your table-view cell. First, let's focus on the layout.

After dragging `UITableViewCell` to the table view, find and drag out `UILabel` and `UIImageView`. Both views should be added to the *prototype cell*. Arrange the label and image as shown in the following screenshot. After doing this, use the **reset to suggested constraints** feature you have used before to add Auto Layout constraints to the label and image. When you select both views after adding the constraints, you should see the same blue lines that are present in the following screenshot:

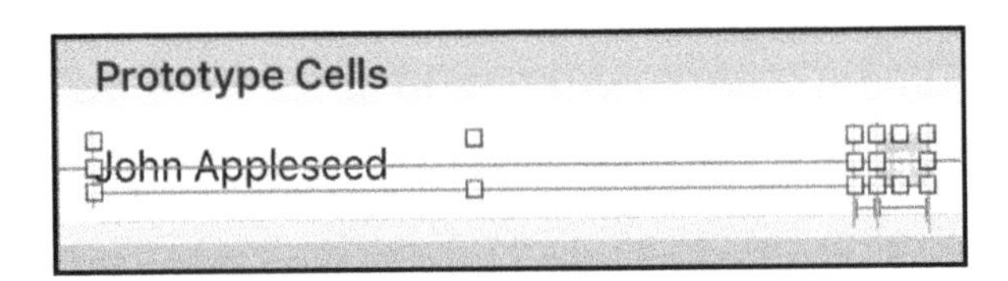

The blue lines from the image are a visual representation of the constraints that were added to lay out your label and image. In the image, you can see a constraint that offsets the label from the left side of the cell. Between the label and the image, you can see a constraint that defines the spacing between these two views. The line that runs through the cell horizontally shows that the label and image are centered on the vertical axis.

You can use **Document Outline** on the left side of **Interface Builder** to explore these constraints. The table-view cell design is now complete, it's time to implement the `UITableViewCell` subclass and create some `@IBOutlets` to connect design and code.

Creating the table-view cell subclass

To create a new `UITableViewCell` subclass, you need to create a new file (**File | New | File**) and choose a `Cocoa Touch` file. Name the file `ContactTableViewCell` and select `UITableViewCell` as the superclass for your file, as shown in the following screenshot:

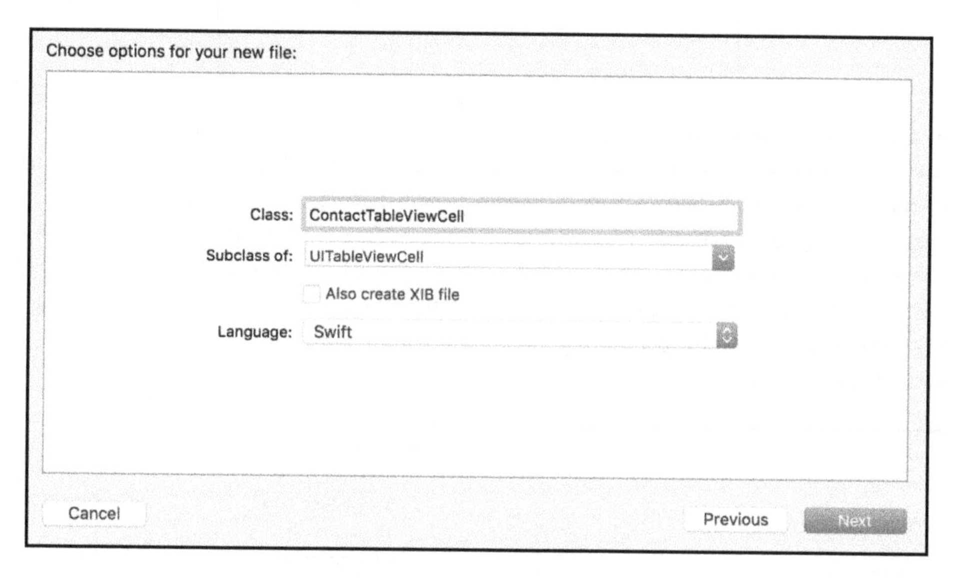

When you open the newly-created file, you'll see that two methods were added to the `ContactTableViewCell` for you. These methods are `awakeFromNib()` and `setSelected(_:animated:)`. The `awakeFromNib()` method is called the very first time an instance of your class is created. This method is the perfect place to do some initial setup that should only be performed once for your cell.

The second method in the template is `setSelected(_:animated:)`, you can use this method to perform some customizations for the cell when a user taps on it. You could, for instance, change the text or background color for a cell there. For now, delete both methods from the class and replace its contents with the following code:

```
@IBOutlet var nameLabel: UILabel!
@IBOutlet var contactImage: UIImageView!
```

The preceding code should be the entire body for the `ContactTableViewCell` class. The variables in the class are annotated with `@IBOutlet`; this means that those variables can be connected with your prototype cell in **Interface Builder**. To do this, open `Main. storyboard`, select your prototype cell and look for the **Identity Inspector** in the sidebar on the right. Set the **class** property for your cell to `ContactTableViewCell`, as shown in the following screenshot. Setting this makes sure that your layout and code are correctly connected:

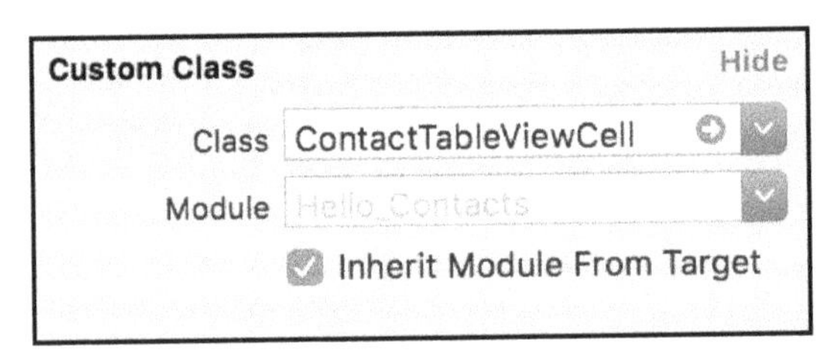

Next, select the table view cell that you added to the table view. Open the **Connections Inspector** in the right sidebar. Under the **Outlets** header, you'll find a list of names. Among those names, you can find the `nameLabel` and `contactImage` you added to `ContactTableViewCell`. Drag from the circle next to the `nameLabel` towards the label inside of your cell. By doing this, you connect the `@IBOutlet` that was created in code to its counterpart in the layout. Perform the same steps outlined in the preceding paragraph for the image, as shown in the following screenshot:

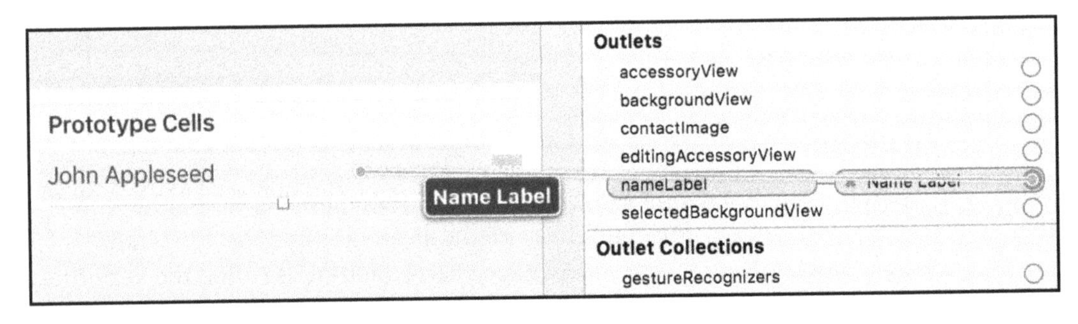

The last step is to provide a reuse-identifier for your cell. The table view uses the reuse-identifier so it can reuse instances of table-view cells. Cell-reuse is an optimization feature that will be cover in depth later in this chapter.

To set the reuse-identifier, open the **Attributes inspector** after selecting your cell. In the **Attributes inspector**, you'll find, and an input field labeled **Identifier**. Set this field to the **ContactTableViewCell** value.

With your layout fully set up, we need to take a couple more steps to make the table view show a list of contacts by assigning it a data source and delegate.

Displaying the list of contacts

One easily-overlooked fact about the table view is that no matter how simple it might seem to use one in your app, it's one of the more complex components of `UIKit`. Some of the complexity is exposed when you add a table view to a regular view controller instead of using `UITableViewController`. For instance, you had to manually set up the layout, so your table view covered the viewport. Then, you had to manually set up a prototype cell to display data in.

The next step toward displaying contacts to your user is providing the table view with information about the contents it should show. To do this, you must implement the data source and delegate for the table view. These properties use advanced concepts that you may have seen before, you probably just weren't aware of them yet. Let's make sure you know exactly what is going on.

Protocols and delegation

Throughout the iOS SDK and the `Foundation` framework, a design pattern named *delegation* is used. Delegation allows an object to have another object perform work on its behalf. When implemented correctly, it's a great way to separate concerns and decouple code within your app. The following figure illustrates how `UITableView` uses delegation for its data source using `UITableViewDataSource`:

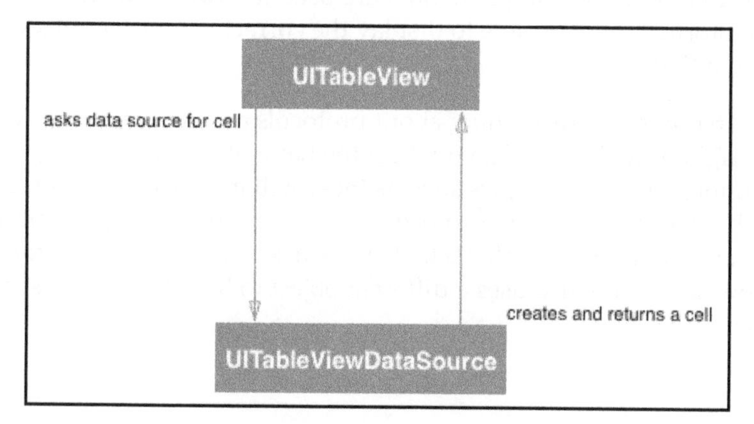

The table view uses the help of two objects to function correctly. One is `delegate`, and the other is `dataSource`. Any time you use a table view, you must configure these two objects yourself. When the time comes for the table view to render its contents, it asks `dataSource` for information about the data to display. `delegate` comes into play when a user interacts with the items in the table view.

If you look at the documentation for `UITableView`, you can find the `delegate` property. The type for `delegate` is `UITableViewDelegate?`. This tells you two things about `delegate`. First of all, `UITableViewDelegate` is a protocol. This means that any object can act as a delegate for a table view, as long as it implements the `UITableViewDelegate` protocol. Second, the question mark behind the type name tells you that the delegate is an `Optional` property. An `Optional` property either has a value of the specified type, or it is `nil`. The table view's delegate is `Optional` because you do not *have to* set it to create a functioning table view.

A protocol, such as `UITableViewDelegate`, defines a set of properties and methods that must be implemented by any type that wants to conform to the protocol. Not all methods must be explicitly implemented by conforming objects. Sometimes, a protocol extension provides a reasonable default implementation. You can read more about this in `Chapter 6`, *Writing Flexible Code With Protocols And Generics*.

In addition to `delegate`, `UITableView` has a `dataSource` property. The data source's type is `UITableViewDataSource?`, and just like `UITableViewDelegate`, `UITableViewDataSource` is a protocol. However, `UITableViewDelegate` only has optional methods, meaning you don't need to implement any methods to conform to `UITableViewDelegate`. `UITableViewDataSource` does have required methods. The methods that need to be implemented are used to provide the table view with just enough information to be able to display the correct amount of cells with the right content in them.

If this is the first time you're learning about protocols and delegation, you might feel a little bit lost right now. That's OK; you'll get the hang of it soon. Throughout this book, your understanding of topics such as these will improve bit by bit. You will even learn about a concept called protocol-oriented programming! For now, it's important that you understand that a table view asks a different object for the data it needs to show and that it also uses a different object to handle certain user interactions.

We can break the flow of displaying contents in a table view down into a couple of steps:

1. The table view needs to reload the data
2. The table view checks whether a `dataSource` is set, and asks it for the number of sections it should render
3. Once the number of sections is passed back to the table view, the `dataSource` is asked for the number of items for each section
4. With knowledge about the number of sections and items that need to be shown, the table view asks its `dataSource` for the cells it should display
5. After receiving all of the configured cells, the table view can finally render these cells to the screen

These steps should give you a little bit more insight into how a table view uses another object to figure out the contents it should render. This pattern is compelling because it makes the table view an extremely flexible component. Let's put some of this newfound knowledge to use!

Conforming to the UITableViewDataSource and UITableViewDelegate protocols

To set up the table view's delegate and data source, you need to create an `@IBOutlet` for the table view in `ViewController.swift`. Add the following line to your `ViewController` class, just before `viewDidLoad()`:

```
@IBOutlet var tableView: UITableView!
```

Now, using the same technique as you used before when connecting outlets for your table view cell, select the table view in `Main.storyboard` and use the **Connections Inspector** to connect the outlet to the table view.

To make `ViewController` both the delegate and the data source for its table view, it will have to conform to both protocols. It is a best practice to create an `extension` whenever you make an object conform to a protocol. Ideally, you make one `extension` for each protocol you want to implement. Doing this helps to keep your code clean and maintainable.

Add the following two extensions to `ViewController.swift`:

```
extension ViewController: UITableViewDataSource {
  // extension implementation
```

```
}
extension ViewController: UITableViewDelegate {
  // extension implementation
}
```

After doing this, your code contains an error. That's because none of the required methods from `UITableViewDataSource` have been implemented yet. There are two methods you need to implement to conform to `UITableViewDataSource`. These methods are `tableView(_:numberOfRowsInSection:)` and `tableView(_:cellForRowAt:)`.

Let's go ahead and fix the error Xcode is showing by adjusting the code a little bit. This is also a great time to refactor the contact-fetching code a little bit. You will want to access the fetched contacts in multiple places, so the list should be an instance variable on the view controller. Also, if you're adding code to create cells anyway, you might as well make them display the correct information.

Add the following updates to `ViewController.swift`:

```
class ViewController: UIViewController {

  var contacts = [CNContact]()

  // viewDidLoad
  // retrieveContacts
}

extension ViewController: UITableViewDataSource {
  // 1
  func tableView(_ tableView: UITableView, numberOfRowsInSection
section: Int) -> Int {
    return contacts.count
  }

  // 2
  func tableView(_ tableView: UITableView, cellForRowAt indexPath:
IndexPath) -> UITableViewCell {
    // 3
    let cell = tableView.dequeueReusableCell(withIdentifier:
"ContactTableViewCell") as! ContactTableViewCell
    let contact = contacts[indexPath.row]

    cell.nameLabel.text = "\(contact.givenName) \(contact.familyName)"

    // 4
    if contact.imageDataAvailable == true, let imageData =
contact.imageData {
```

```
            cell.contactImage.image = UIImage(data: imageData)
        }

        return cell
    }
}
```

The preceding code completes the implementation of `UITableViewDataSource`. Let's go over the commented sections of code to clarify them a little bit:

- Since this table view only has a single section, the number of contacts as returned for the number of items in every section. This is OK because we know that there will always be just a single section. When you build an app that shows a table view with multiple sections, you would have to implement the `numberOfSections` property to inform the table view about the number of sections it needs to render.
- This method is responsible for creating and configuring one of the `ContactTableViewCell` cells you created earlier.
- Earlier in this chapter, you learned that cells are reused, that's why you had to set a reuse-identifier in the storyboard. Here, the reuse-identifier is used to ask for a cell to display a fetched contact in. Reusing cells that have been scrolled off screen is a performance optimization that enables a table view to display vast amounts of items without choppy scrolling or consuming tons of memory. The `dequeueReusableCell(withIdentifier:)` method has `UITableViewCell` as its return type. Therefore, you need to cast the result of that method to be the cell you set up in **Interface Builder** earlier. In this case, that is `ContactTableViewCell`.
- The last step safely extracts image data from the contact if it's available. If it is, the image data is used to set up an image for the cell.

This doesn't wrap up the refactoring of fetching contacts just yet. Contacts are being fetched, but the array of contacts you added to `ViewController` earlier is not set up correctly yet, the fetched contacts are not attached to this array. In its current state, the last couple of lines in `retrieveContacts` look as follows:

```
let contacts = try! store.unifiedContacts(matching: predicate,
keysToFetch: keysToFetch)
print(contacts)
```

Change these lines to the following code:

```
contacts = try! store.unifiedContacts(matching: predicate,
keysToFetch: keysToFetch)
DispatchQueue.main.async { [weak self] in
```

```
        self?.tableView.reloadData()
    }
```

With this update, the result of fetching contacts is assigned to the variable you created earlier. Also, the table view is instructed to reload its data. Note that this is wrapped in a `DispatchQueue.main.async` call. Doing this ensures that the UI is updated on the main thread. Since iOS 11, your app will crash if you don't perform UI work on the main thread. If you want to learn more about this, have a look at Chapter 25, *Offloading Tasks with Operations and GCD*, it covers threading in more depth.

There is one more step before you're done. The table view is not aware of its `dataSource` and `delegate` yet. You should update the `viewDidLoad()` method to assign the table view's `dataSource` and `delegate` properties. Add the following lines to the end of `viewDidLoad()`:

```
tableView.delegate = self
tableView.dataSource = self
```

Now go ahead and run your app. If you're running your app on the simulator, or you don't have any images assigned to your contacts, you won't see any images. You can add images to contacts in the simulator by dragging images from your Mac onto the simulator and saving them in the photo library. From there, you can add pictures to contacts the same way you would do it on a real device. If you have a lot of contacts on your device, but don't have an image for everybody, you might encounter an issue when scrolling. Sometimes, you might see a picture of a different contact than the one you expect! This is a performance optimization that is biting you. Let's see what's going on and how to fix this bug.

Under-the-hood performance of UITableView

Earlier in this chapter, you learned about cell-reuse in table views. You assigned a reuse-identifier to a table-view cell so that the table view would know which cell it should use to display contacts in. Cell-reuse is a concept that is applied so a table view can reuse cells that it has already created. This means that the only cells that are in memory are either on the screen or barely off the screen. The alternative would be to keep all cells in memory, which could potentially mean that hundreds or thousands of cells are held in memory at any given time. For a visualization of what cell reuse looks like, have a look at the following diagram:

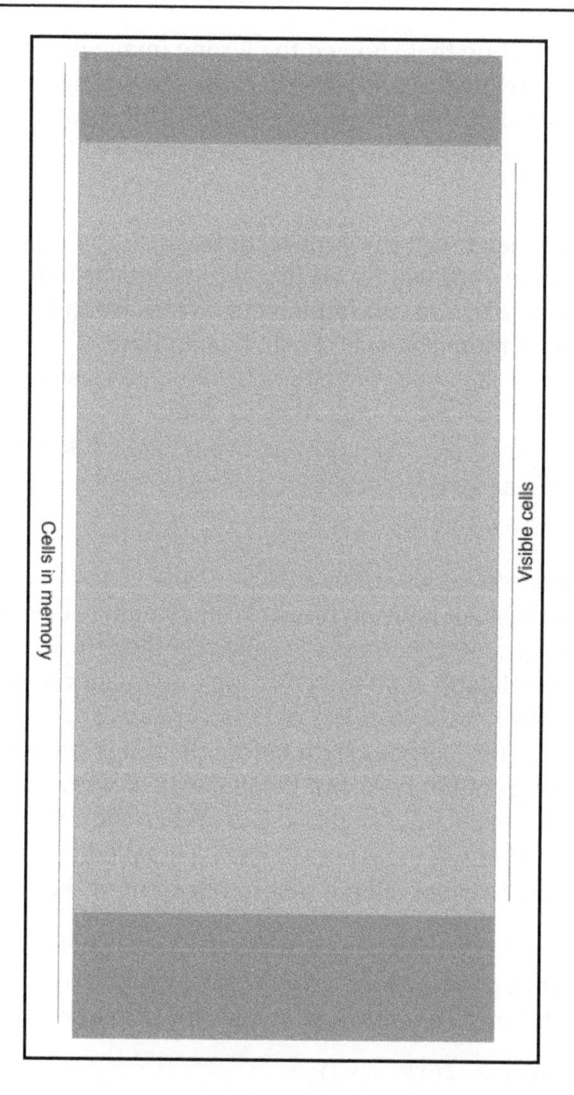

As you can see, there are just a few cells in the picture that are not on the visible screen. This roughly equals the number of cells that a table view might keep in memory. This means that regardless of the total amount of rows you want to show, the table view has a roughly constant pressure on your app's memory usage.

Earlier, you witnessed a bug that showed the wrong image next to a contact in the table view. This bug is related to cell-reuse because the wrong image is actually only visible for contacts that don't have their own image. This means that the image from the contact that was previously shown in that particular cell is now shown for a different contact.

If you haven't seen this bug occur because you don't have that many contacts in your list, try adding more contacts in the contacts app. Alternatively, you can implement a workaround to pretend that you have a lot more contacts. To do this, update `tableView(_:numberOfRowsInSection:)` so it returns `contacts.count * 10`. Also, update `tableView(_:cellForRowAtIndexPath:)` so the contact is retrieved as `let contact = contacts[indexPath.row % contacts.count]`.

A cell is first created when `dequeueReusableCell(withIdentifier:)` is called on the table view and it does not have an unused cell available. Once the cell is either reused or created, `prepareForReuse()` is called on the cell. This is a great spot to reset your cells to their default state by removing any images or setting labels back to their default values. Next, `tableView(_:willDisplay:forRowAt:)` is called on the table views's delegate. This happens right before the cell is shown. You can perform some last-minute configuration here, but the majority of work should already be done in `tableView(_:cellForRowAtIndexPath:)`. When the cell scrolls offscreen, `tableView(_:didEndDisplaying:forRowAt:)` is called on the delegate. This signals that a previously-visible cell has just scrolled out of the view's bounds.

With all this cell life cycle information in mind, the best way to fix the image-reuse bug is by implementing `prepareForReuse()` on `ContactTableViewCell`. Add the following implementation to remove any images that have previously been set:

```
override func prepareForReuse() {
  super.prepareForReuse()

  contactImage.image = nil
}
```

Quite an easy fix for a pesky bug, don't you think? Let's have a look at another performance optimization that table views have, called prefetching.

Improving performance with prefetching

In addition to `UITableViewDelegate` and `UITableViewDataSource`, a third protocol exists that you can implement to improve your table view's performance. It's called `UITableViewDataSourcePrefetching` and you can use it to enhance your data source. If your data source performs some complex task, such as retrieving and decoding an image, it could slow down the performance of your table view if this task is performed at the moment the table view wants to retrieve a cell. Performing this operation a little bit sooner than that can positively impact your app in those cases.

Since **Hello-Contacts** currently decodes contact images for its cells, it makes sense to implement prefetching to make sure the scrolling performance remains smooth at all times. The current implementation performs the decoding in `tableView(_:cellForRowAt:)`. To move this logic to `UITableViewDataSourcePrefetching`, there is one method that needs to be implemented, it's called `tableView(_:prefetchRowsAt:)`. Add the following extension to `ViewController.swift` to create a nice starting point for implementing prefetching:

```
extension ViewController: UITableViewDataSourcePrefetching {
  func tableView(_ tableView: UITableView, prefetchRowsAt indexPaths:
[IndexPath]) {
    for indexPath in indexPaths {
      // Prefetching will be implemented here soon
    }
  }
}
```

Instead of receiving just a single `IndexPath`, `tableView(_:prefetchRowsAt:)` receives a list of index path for which you should perform a prefetch. Before implementing the prefetching, take a step back to come up with a good strategy to implement prefetching. For instance, it would be ideal if each image only has to be decoded once to prevent duplicate work from being done. Also, a mechanism is needed to decode images in cases where the image wasn't prefetched. And also in that case, only having to decode once would be great. This can be achieved by creating a class that wraps `CNContact` and has some helper methods to make prefetching and decoding nice and smooth.

First, create a new file (**File** | **New** | **File...**) and select the **Swift file** template. Name this file `Contact.swift`. Add the following code to this file:

```
import UIKit
import Contacts
```

```
class Contact {
  private let contact: CNContact
  var image: UIImage?

  // 1
  var givenName: String {
    return contact.givenName
  }

  var familyName: String {
    return contact.familyName
  }

  init(contact: CNContact) {
    self.contact = contact
  }

  //2
  func fetchImageIfNeeded(completion: @escaping ((UIImage?) -> Void) =
{_ in }) {
    guard contact.imageDataAvailable == true, let imageData =
contact.imageData else {
      completion?(nil)
      return
    }

    if let image = self.image {
      completion?(image)
      return
    }

    DispatchQueue.global(qos: .userInitiated).async { [weak self] in
      self?.image = UIImage(data: imageData)
      DispatchQueue.main.async {
        completion?(self?.image)
      }
    }
  }
}
```

The first thing to note about this code is the use of so-called *computed variables*. These variables act as a proxy for properties from the private `CNContact` instance that `Contact` wraps. It's good practice to set up proxies such as these because they prevent exposing too many details to other objects. Imagine having to switch from `CNContact` to a different type of contact internally. That becomes a lot easier when as few places as possible know about `CNContact`.

The second segment of code you should pay extra attention to is the image-fetching part that ensures we fetch images as efficiently as possible. First, the code checks whether an image is present on the contact at all. If it is, a check is done to see whether a decoded image already exists. And if it does, the completion closure is called with the decoded image. If no image exists yet, it is decoded on the global dispatch queue. By executing code on the global dispatch queue, it is automatically executed off the main thread. This means that no matter how slow or lengthy the image decoding gets, the table view will never freeze up because of it, since the main thread is not doing the heavy lifting. Because this code is asynchronous, a `completion` closure is used to call back with the decoded images. Calling back is done on the main thread since that is where the image should be used eventually. Note that the completion closure has a default value in the signature for `fetchImageIfNeeded(completion:)`. Sometimes, the result of prefetching isn't needed so no completion handler will be given. Again, if this dispatch stuff makes you dizzy, don't worry. Or skip ahead to `Chapter 25`, *Offloading Tasks with Operations and GCD*, if you can't wait to learn more.

There are only a couple more changes that must be made to `ViewController` to make it uses the new `Contact` class and you're good to go. The following code snippet shows all the updates you will need to incorporate:

```
class ViewController: UIViewController {
  var contacts = [Contact]()

  //...

  func retrieveContacts(from store: CNContactStore) {
    //...

    // 1
    contacts = try! store.unifiedContacts(matching: predicate,
keysToFetch: keysToFetch)
      .map { Contact(contact: $0) }

    // ...
  }
}

extension ViewController: UITableViewDataSource {
  func tableView(_ tableView: UITableView, numberOfRowsInSection
section: Int) -> Int {
    return contacts.count
  }

  func tableView(_ tableView: UITableView, cellForRowAt indexPath:
```

```
IndexPath) -> UITableViewCell {
    let cell = tableView.dequeueReusableCell(withIdentifier:
"ContactTableViewCell") as! ContactTableViewCell
    let contact = contacts[indexPath.row]

    cell.nameLabel.text = "\(contact.givenName) \(contact.familyName)"

    // 2
    contact.fetchImageIfNeeded { image in
      cell.contactImage.image = image
    }

    return cell
  }
}

extension ViewController: UITableViewDelegate {
  // extension implementation
}

extension ViewController: UITableViewDataSourcePrefetching {
  func tableView(_ tableView: UITableView, prefetchRowsAt indexPaths:
[IndexPath]) {
    for indexPath in indexPaths {
      // 3
      let contact = contacts[indexPath.row]
      contact.fetchImageIfNeeded()
    }
  }
}
```

The first addition is to make use of `map` to transform the list of `CNContact` instances to `Contact` instances. The second update uses the `fetchImageIfNeeded(completion:)` method to obtain an image. This method can be used because it has been set up to return either the existing decoded image or a freshly-decoded one if the prefetching wasn't able to decode the image in time.

The last change is to prefetch images as needed. Because `fetchImageIfNeeded(completion:)` has a default implementation for its `completion` argument, it can be called without a `completion` closure. The result isn't immediately relevant, so not having to provide a closure is convenient in this case. The prefetching is fully implemented now; you might not immediately notice any improvements when you run the app, but rest assured that proper use of prefetching can greatly benefit your apps.

UITableViewDelegate and interactions

So far, the `ViewController` class has implemented the `UITableViewDelegate` protocol but none of the delegate methods have been implemented yet. Any time certain interactions occur on a table view, such as tapping a cell or swiping on a cell, the table view will attempt to call the corresponding delegate methods to inform the delegate about the action that occurred.

The `UITableViewDelegate` protocol does not specify any required methods, which is why it has been possible to conform to this protocol without actually doing work. Usually, you will want to implement at least one method from `UITableViewDelegate` because simply displaying a list without responding to any interactions is quite boring. Let's go ahead and explore some of the methods from `UITableViewDelegate` to create a better experience for **Hello-Contacts**. If you take a look at the documentation for `UITableViewDelegate`, you'll notice that it has a large collection of delegate methods that you can implement in your app.

You can hold the *Alt* key when clicking on a class, struct, enum, or protocol name to navigate to the documentation for the type you clicked on.

The documentation for `UITableViewDelegate` lists methods for configuring cell height, content-indentation level, cell-selection, and more. There are also methods that you can implement to get notified when the table view is about to display a cell or is about to stop displaying one. You can handle cell-selection, cell-highlighting, reordering cells, and even deleting them. One of the more common methods to implement is `tableView(_:didSelectRowAt:)`. After you have implemented this method, you'll also implement cell-reordering and -removal.

Responding to cell-selection

Cell-selection refers to a user tapping on a cell. In order to respond to cell-selection, the `UITableViewDelegate` method called `tableView(_:didSelectRowAt:)` should be implemented. In **Hello-Contacts**, `ViewController` already has an extension implemented to make it conform to `UITableViewDelegate` so all you need to do is implement `tableView(_:didSelectRowAt:)` in the extension.

Since a table view will call methods on its delegate whenever they are implemented, you don't need to tell the table view that you want to respond to cell-selection. This automatically works if the table view has a delegate, and if the delegate implements `tableView(_:cellForRowAt:)`. The implementation you'll add to **Hello-Contacts**, for now, is a very simple one. When the user taps a cell, the app displays an alert. In Chapter 3, *Creating a Detail Page*, you will learn how to perform more meaningful actions such as displaying a detail page. Add the following code to the `UITableViewDelegate` extension in `ViewController.swift`:

```
func tableView(_ tableView: UITableView, didSelectRowAt indexPath:
IndexPath) {
  let contact = contacts[indexPath.row]
  let alertController = UIAlertController(title: "Contact tapped",
                                          message: "You tapped
\(contact.givenName)",
                                          preferredStyle: .alert)

  let dismissAction = UIAlertAction(title: "Ok", style: .default,
handler: { action in
    tableView.deselectRow(at: indexPath, animated: true)
  })

  alertController.addAction(dismissAction)
  present(alertController, animated: true, completion: nil)
}
```

The `tableView(_:cellForRowAt:)` method receives two arguments, the first is the table view that called this delegate method. The second argument is the index path at which the selection occurred. The implementation you wrote for this method uses the index path to retrieve the contact that corresponds with the tapped cell so the contact name can be shown in an alert. You could also retrieve the contact's name from the tapped cell. However, this is not considered good practice because your cells and the underlying data should be as loosely-coupled as possible. When the user taps the **Ok** button in the alert, the table view is told to deselect the selected row. If you don't deselect the selected row, the last tapped cell will always remain highlighted. Note that the alert is displayed by calling `present(_:animated:completion:)` on the view controller. Any time you want to make a view controller display another view controller, such as an alert controller, you use this method.

Even though this setup is not extremely complex, there is some interesting stuff going on. The delegation pattern is a very powerful one when implemented correctly. Especially in the case of a table view, you can add a lot of functionality simply by implementing the delegate methods that correspond to the desired behavior. Because the table view's delegate could be any object that conforms to `UITableViewDelegate`, you could split up `ViewController` and `UITableViewDelegate` entirely. Doing so would enable you to reuse your delegate implementation across multiple table views. For now, I'll leave it as an exercise for you to do this. Attempting such a refactor will certainly help you to increase your understanding of delegation and its powers.

Try to extract your delegate and/or data source for the table view to a separate class or struct. This will allow you to reuse your code, and you will gain a deeper understanding of what delegation is and how it works.

Implementing cell-deletion

Now that you know how to respond to cell-selection, let's have a look at a slightly more advanced topic – cell-deletion. Deleting data from a table view is a feature that many apps implement. If you have ever used the mail app on iOS, you might have noticed that several actions appear when a user swipes either right or left on a table-view cell. These swipe actions are a great feature to implement for **Hello-Contacts** so users can swipe over contacts and easily delete them. Of course, we won't be actually deleting contacts from a user's address book, but it would be possible to implement this if you truly wanted to.

In this example, you'll learn how to delete contacts from the array of contacts that is used to populate the table view. To implement support for cell-deletion, you need to implement another method from `UITableViewDelegate`. The method you need to implement is `tableView(_:trailingSwipeActionsConfigurationForRowAt:)`. This delegate method is called when a user swipes over a table view cell and returns the actions should be shown when the cell moves sidewards. A good example of this is found in the mail app on iOS.

Add the following implementation of `tableView(_:trailingSwipeActionsConfigurationForRowAt:)` to the `UITableViewDelegate` extension for `ViewController`:

```
func tableView(_ tableView: UITableView,
               trailingSwipeActionsConfigurationForRowAt indexPath:
IndexPath) -> UISwipeActionsConfiguration? {

  // 1
  let deleteHandler: UIContextualActionHandler = { [weak self] action,
view, callback in
    self?.contacts.remove(at: indexPath.row)`
    callback(true)
  }

  // 2
  let deleteAction = UIContextualAction(style: .destructive,
                                        title: "Delete", handler:
deleteHandler)

  // 3
  return UISwipeActionsConfiguration(actions: [deleteAction])
}
```

If you run the app now and swipe from right to left over a table view cell, a delete button will appear from underneath the cell. In the code snippet, the first step is to set up a delete-handler that takes care of the actual deletion of the contact. This handler is a closure that is passed to the `UIContextualAction` instance that is created in step two. You have seen closures passed directly to method calls already, for instance as completion-handlers. However, you can also store a closure in a variable. This allows you to reuse your closure in several places and can make your code more readable. The third and last step is to create an instance on `UISwipeActionsConfiguration` and pass it the actions that you want to display. Since you can pass an array of actions, it is possible to show multiple actions when the user swipes over a cell. In this case, only a single action is added – the delete action.

Currently, tapping the delete button doesn't do much. While the contact is removed from the underlying data source, the table view itself doesn't update. Table views don't automatically stay in sync with their data sources. Add the following `deleteHandler` implementation to make sure the table view updates when the user taps the **delete** button:

```
let deleteHandler: UIContextualActionHandler = { [weak self] action,
view, callback in
  self?.contacts.remove(at: indexPath.row)
```

```
        self?.tableView.beginUpdates()
        self?.tableView.deleteRows(at: [indexPath], with: .fade)
        self?.tableView.endUpdates()
        callback(true)
    }
```

This new version of `deleteHandler` ensures that the table view updates itself by removing the row that the user has decided to remove. Note that the contacts array is updated before updating the table view. When you update the table view like this, it will verify that it is in sync with the data source, which is the contacts array in this case. If the data source does not contain the expected amount of sections or rows, your app will crash. So whenever you update a table view, make sure to update the data source first. Also, note the calls to `beginUpdates` and `endUpdates`. These methods make sure that the table view doesn't reload in the middle of being manipulated. This is especially useful if you're performing a lot of complex updates, such as moving cells, inserting new ones, and removing old ones all at the same time.

With cell-deletion out of the way, let's have a look at reordering cells.

Allowing the user to reorder cells

In some applications, it makes sense for users to be able to reorder cells that are shown in a table view, such as in a to-do list application or a grocery list. Implementing cell-reordering takes a couple of steps. First, a table view needs to be put in editing mode. When a table view is in editing mode, the user can begin sorting cells visually. Typically, a button in the top right or left corner of the screen is used to enter and exit the editing mode. The easiest way to make space for a button is by wrapping your `ViewController` in a `UINavigationController`. Doing this makes a navigation bar appear at the top of the screen that has space for a title, back button, and also for custom buttons such as the **Edit/Done** button we need to make the table view enter and exit the editing mode.

Placing the table view in editing mode is actually really simple if you know how. Every `UIViewController` instance has a `setEditing(_:animated:)` method. If you override this method, you can use it as an entry point to call `setEditing(_:animated:)` on the table view so it enters edit mode. Once this is implemented, you need to implement `tableView(_:moveRowAt:to:)` from `UITableViewDelegate` to commit the reordered cells to your data source by updating the contacts array.

First, open `Main.storyboard` so you can wrap the view controller in a navigation controller. When you have selected the view controller in your storyboard, click **Editor | Embed In | Navigation Controller** in the menu bar at the top of the screen. This will automatically embed the view controller in a navigation controller and configure everything as needed. To add the **Edit/Done** button, open `ViewController.swift` and add the following code to `viewDidLoad`:

```
navigationItem.rightBarButtonItem = editButtonItem
```

This line adds a `UIBarButtonItem` that automatically toggles itself and calls `setEditing(_:animated:)` on the view controller. Since it's set as `rightBarButtonItem`, it will appear on the right side of the navigation bar. If you go ahead and build the app now, you'll see that you have a button that toggles between a label that says **Edit** and **Done**. To put the table view in edit mode, you must override the `setEditing(_:animated:)` method in `ViewController.swift`, as follows:

```
override func setEditing(_ editing: Bool, animated: Bool) {
  super.setEditing(editing, animated: animated)

  tableView.setEditing(editing, animated: animated)
}
```

What this method does should be self-explanatory. Go ahead and run the app now. If you tap the **Edit** button, every cell suddenly displays a red circle – while this is interesting, it's not quite what's needed. Cells don't show reorder controls when in edit mode by default. Open `Main.storyboard` again and select your prototype cell. In the **Attributes inspector**, you'll find a checkbox named **Shows Re-order controls**. You want to make sure this checkbox is checked so you can reorder cells.

The final step to implementing this feature is adding `tableView(_:moveRowAt:to:)` in the `UITableViewDelegate` extension in `ViewController.swift`. This method will make sure that the contacts array is updated in the same way the cells are, ensuring that the table view and data source remain nicely in sync. Add the following code to the `UITableViewDelegate` extension in `ViewController.swift`:

```
func tableView(_ tableView: UITableView, moveRowAt sourceIndexPath:
IndexPath, to destinationIndexPath: IndexPath) {
  let contact = contacts.remove(at: sourceIndexPath.row)
  contacts.insert(contact, at: destinationIndexPath.row)
}
```

Even though it's only two lines, this code updates the data source by moving a contact from its old position in the array to its new position. You now have everything in place to correctly reorder cells in a table view. Go ahead and try it out!

Summary

The **Hello-Contacts** app is complete for now. The next few chapters will focus on improving it with a new layout, a detail page, and a couple more changes. You've covered a lot of ground on the way toward mastering iOS. You've used Auto Layout and the Contacts framework, you learned about delegation and custom table view cells, and you implemented several delegate methods to implement various features on your table view.

If you want to learn more about `UITableView`, I don't blame you! The table view is a very powerful and versatile component in the iOS developer's toolkit. Make sure to explore Apple's documentation because there is a lot more to learn and study. One of the most important patterns you learned about is delegation. You'll find implementations of the delegate pattern throughout this book and `UIKit`. Next up? Converting the `UITableView` to its even more powerful and interesting sibling, `UICollectionView`.

Questions

1. What happens if you don't provide a reason for wanting to access a user's contacts?

   a) Nothing
   b) An empty alert is shown
   c) The app crashes

2. What is a reuse-identifier on a table-view cell used for?

   a) It helps you identify cells by their name
   b) It is useful for accessibility
   c) It is used by the table view to optimize performance

3. Where does a table view obtain information about the cells it displays from?

    a) `CellProvider`
    b) `UITableViewDataSource`
    c) `UITableViewDelegate`

4. How does a table view make sure to keep its memory footprint as small as possible?

    a) It loads all cells at once
    b) It reuses cells that were displayed before
    c) It limits the number of cells it can handle

5. What is a placeholder cell called in Interface Builder?

    a) Prototype cell
    b) Placeholder cell
    c) Designer cell

6. What is a connection between an item Interface Builder and a variable in code called?

    a) `@IBConnection`
    b) `@IBInlet`
    c) `@IBOutlet`

7. Where is the best place to reset a table-view cell?

    a) In `awakeFromNib()`
    b) In `reset()`
    c) In `prepareForReuse()`

Further reading

- Apple's documentation on `UITableView`: `https://developer.apple.com/documentation/uikit/uitableview?preferredLanguage=occ`

2
A Better Layout with UICollectionView

The previous chapter showed you how to build a list of items using a table view. In this chapter, you'll learn how to display the same items using a more flexible component. This component is called `UICollectionView`, and it has a very similar interface to `UITableView`. The most significant difference is that you can define a much more flexible layout using a collection view. An example of this is the grid layout that `UICollectionView` provides out of the box. Of course, you are free to implement any layout you want for a collection view. This is probably the most significant advantage a collection view has to offer.

You will build on top of **Hello-Contacts** by refactoring the existing application to switch from a table view to a collection view. You are going to replace all table-view-related configuration code with its collection view counterparts. You will also make use of a grid layout to show contacts. Lastly, you will create a custom collection view cell, just like you did for the table view.

To take the grid layout one step further than the default implementation, you will also explore `UICollectionViewFlowLayout` to implement a cool custom version of a grid that is a bit more playful than the default version. Since collection views and table views are so similar, you will also implement all of the features that existed for the table view version of the app.

This chapter covers the following topics:

- Converting from a table view to a collection view
- Creating a custom collection view cell
- Using `UICollectionViewFlowLayout`
- Creating a custom collection view layout
- User interactions for `UICollectionView`

Converting from a table view to a collection view

Showing a user's contacts in the form of a list is a fine idea. It's functional, looks all right, and people are used to seeing data displayed in the form of a list. However, wouldn't it be nice if there was a more interesting way to display the same contact data? For instance, using bigger images for the contact photos. And instead of showing everything in a list, maybe a grid is a nice alternative.

Exciting and compelling layouts drive user engagement. Users enjoy interacting with well-designed apps, and they certainly notice when you put in a bit of extra effort. Of course, merely implementing a grid layout won't automatically make your app great but when used well, users can certainly feel more at home with your app than they would if you just stick to a table view. Every app benefits from a different kind of design and layout, so collection views are no magical solution for all apps, but they indeed are a powerful tool to master.

To show contacts in a grid layout, some of the existing code has to be cleaned up. All code and designs related to the table view must be removed. That's what you should do first. When you've cleaned up the project, you're left with a great starting point to begin implementing a collection view. The conversion process has three steps:

1. Delete all table view code.
2. Delete the table view from the storyboard and replace it with a collection view.
3. Add code for the collection view.

Let's start off by deleting all the code in `ViewController.swift` that is related to your old table view implementation. This includes removing all extensions for conformance to `UITableView` related protocols. After doing so, you should be left with the following implementation of `ViewController.swift`:

```
import UIKit
import Contacts

class ViewController: UIViewController {
  var contacts = [Contact]()

  override func viewDidLoad() {
    super.viewDidLoad()

    let store = CNContactStore()
    let authorizationStatus = CNContactStore.authorizationStatus(for:
```

```
.contacts)

    if authorizationStatus == .notDetermined {
      store.requestAccess(for: .contacts) { [weak self] didAuthorize,
error in
        if didAuthorize {
          self?.retrieveContacts(from: store)
        }
      }
    } else if authorizationStatus == .authorized  {
      retrieveContacts(from: store)
    }
  }

  func retrieveContacts(from store: CNContactStore) {
    let containerId = store.defaultContainerIdentifier()
    let predicate =
CNContact.predicateForContactsInContainer(withIdentifier: containerId)
    let keysToFetch = [CNContactGivenNameKey as CNKeyDescriptor,
                       CNContactFamilyNameKey as CNKeyDescriptor,
                       CNContactImageDataAvailableKey as
                       CNKeyDescriptor,
                       CNContactImageDataKey as CNKeyDescriptor]

    contacts = try! store.unifiedContacts(matching: predicate,
keysToFetch: keysToFetch)
      .map { Contact(contact: $0) }
  }
}
```

You might be tempted also to remove the custom table view cell that you created in the previous chapter. You don't have to remove that for now since you can reuse the code for the collection view cell you will create later.

Now that the code is cleaned up, the user interface should also be cleaned up. Open `Main.storyboard` and delete the table view. You should now be left with an empty view controller that has no child views. Drag `UICollectionView` from the **Object Library** over to the view controller's view. Make sure this collection view covers the entire view, including the navigation bar that is visible at the top of the view. Next, use **Reset to Suggested Constraints** in the **Resolve Auto Layout Issues** menu to add the proper layout constraints for the collection view, just like you did before with the table view in `Chapter 1`, *UITableView Touch-Up*. This will pin the collection view to the top, right, bottom, and left edges of the view controller's view.

With the collection view added to the user interface, the time has come to connect the interface and the code again. Add the following `@IBOutlet` to `ViewController.swift` so you can connect the outlet and the interface using **Interface Builder**:

```
@IBOutlet var collectionView: UICollectionView!
```

Back in the storyboard file, connect the new `@IBOutlet` to the collection view by selecting the view controller, opening the **Connections Inspector**, and dragging a referencing outlet from the **collectionView** property to the collection view you just added.

You have now successfully removed the old table view implementation, and you have swapped it out for a collection view. Of course, there is more work to be done to conclude the conversion, such as creating the custom collection view cell that will hold all of the contact information.

Creating and implementing a custom collection view cell

When you implemented the table view cell in the previous chapter, you designed a custom cell. This cell was a special kind of view that the table view reused for every contact that was visible on the screen. A collection view uses cells in a very similar way, but you can't use table view cells in a collection view and vice versa. However, both cells share a lot of functionality. For instance, they both have a `prepareForReuse` and an `awakeFromNib` method. This is why you haven't removed any code for the table view cell earlier; you can reuse the implementation code.

When you dragged the collection view onto the storyboard, it came with a default cell. This cell is a lot more flexible than a table view cell. For instance, try resizing it in your storyboard. Doing this is not possible with a table view cell, but a collection view cell will happily resize itself for you.

If you look at the **Document Outline** on the left-hand side of the window, you can see an object called **Collection View Flow Layout**. This object is responsible for the collection view's layout, and you will learn a lot more about it later in the chapter when you implement your custom layout object. For now, click it so it is selected and look at the **Size Inspector** on the right-hand side of the window. Set the item height property on the layout object to 90 and set the width to 110. The cell in the storyboard should automatically resize accordingly.

Now that the cell size is configured, drag an image and a label into the cell. Try to position your views as shown in the following screenshot:

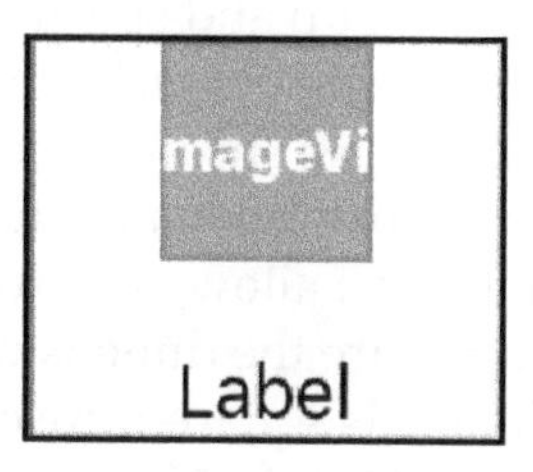

When you have manually placed these two views, you should add Auto Layout constraints to make sure they are nicely positioned at all times. Use the technique you have used before to make Xcode add the constraints for you. You'll notice that Xcode didn't do a great job this time around. Instead of centering the image and the label in the cell, they are offset from the left edge of the cell. This might be fine if you don't intend on ever changing the cell size but otherwise you'll want to improve the added constraints manually. Undo this step by using *Cmd* + *Z* or by selecting **Edit** | **Undo** from the menu.

When adding constraints, it's important that every view can figure out its position on the screen and its size. Views always use constraints for the *x* and *y* values. For a view's size, either constraints or the intrinsic content size of the view are used. The intrinsic content size can be calculated by a view depending in its context. UILabel does this to determine its size based on its text contents.

When you apply the rules from the information block when adding constraints to the image view in your custom cell, you should conclude that you want to add the following constraints:

- Center the image horizontally in the cell (*x* position).
- Stick the image to the top of the cell (*y* position).
- Make the image 50 points wide and 50 points tall (width and height).

For the label, you need the following constraints:

- Center the label horizontally in the cell (*x* position).
- Stick the label to the bottom of the cell (*y* position).

There is no need to set up width and height constraints for the label because these values are calculated using the intrinsic content size.

To add the required constraints to your image view, select it in the storyboard and click the **Align** button in the bottom-right corner of the window. This button opens a popup that allows you to manually set up and specify constraints that align the current view to other views. For your current task, you need the **Horizontally in Container** constraint to center the image. The other constraints you need are added using the **Pin** button. This button is positioned next to the **Align** button. The dialog that opens when clicking the **Pin** button allows you to pin the selected view to edges on its superview. You can also configure the dimensions for the selected view. Set the **Width** and **Height** to `50` so the image is squared. Also, click the red constraint marker that represents the top anchor and set it to `0`. This pins the top of the image to the top of your cell. Refer to the following screenshot to make sure you set everything up correctly:

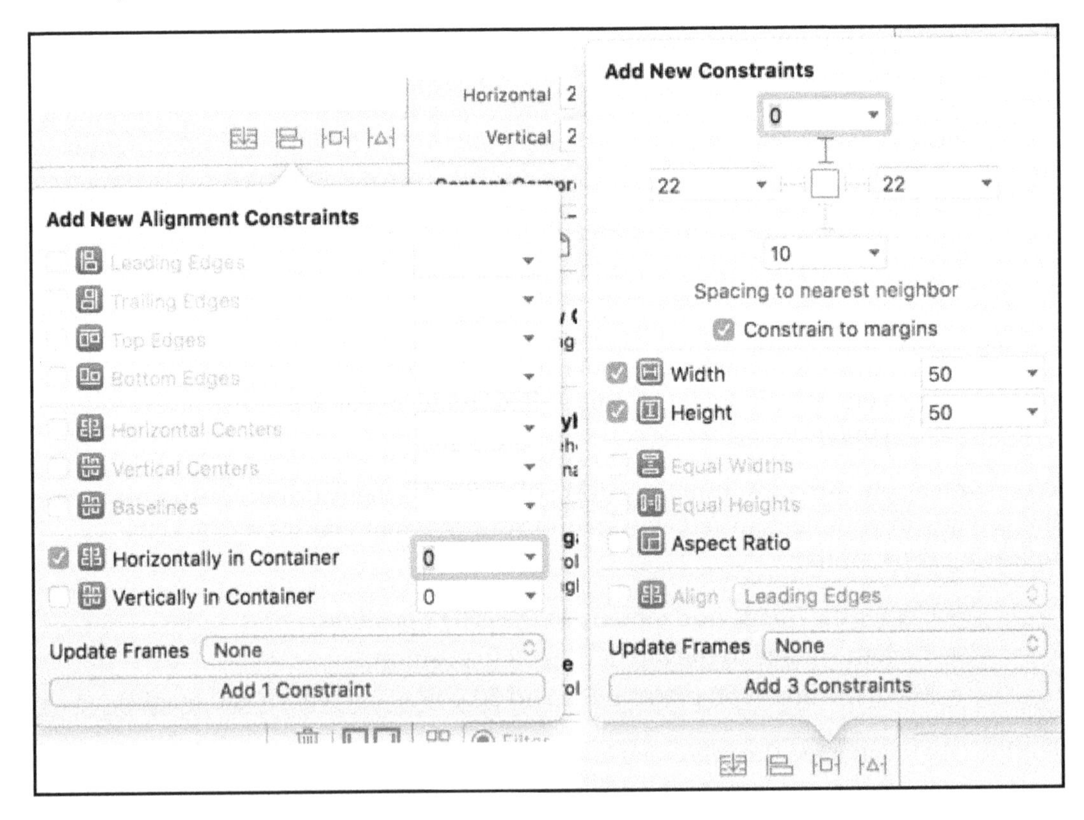

For the label, you should repeat the horizontal centering step. Next, pin the label to the bottom of the cell using the same technique as you used before. Don't set up constraints for the width and height because the label will use its intrinsic content size for this. Centering the label and positioning it at the bottom of the cell provides Auto Layout with enough information to render your cell's layout.

Since you have kept the code for your old table view cell around, you can now use this code and turn it into a collection view cell. The collection view cell displays the same data as the table view cell did before, and they have very similar methods available. For instance, `prepareForReuse()` is available on both cells. This means that all you need to do to switch from a collection view cell to a table view cell is change its superclass from `UITableViewCell` to a `UICollectionViewCell`. However, that would leave you with a collection view cell that is called `ContactTableViewCell`, so it's a good idea also to rename the class itself. Adjust your code as follows:

```
class ContactCollectionViewCell: UICollectionViewCell
```

You can now go ahead and use this class in **Interface Builder** just like you did before. Before setting up all the `@IBOutlet` connections, take a moment and rename the file for `ContactCollectionViewCell`, so the filename matches the class name.

To rename a file, select it in the **File Inspector** on the left-hand side and press *Enter*. You can now rename the file and press *Enter* again to complete the renaming action.

With the file renamed, switch back to the storyboard and assign `ContactCollectionViewCell` as the class for the collection view cell. You can now also hook up the image and label to their `@IBOutlet` counterparts using the **Connections Inspector.** Lastly, make sure to set a reuse identifier on your cell in the **Attributes Inspector**.

Collection views use delegation to retrieve cells, just like table views do. Currently, your collection view does not have a data source set up yet so if you run your app, you won't see any cells yet. If you have followed along in the previous chapter, you should be able to figure out how to implement a collection view data source. Doing this on your own would be a great exercise in navigating Apple's documentation, but you would also train your ability to apply knowledge from other, related topics to new subjects.

When you're done, refer to the following code snippet to ensure you implemented everything correctly:

```
import UIKit
import Contacts

class ViewController: UIViewController {
  var contacts = [Contact]()

  @IBOutlet var collectionView: UICollectionView!

  override func viewDidLoad() {
    super.viewDidLoad()

    collectionView.dataSource = self

    let store = CNContactStore()
    let authorizationStatus = CNContactStore.authorizationStatus(for:
.contacts)

    if authorizationStatus == .notDetermined {
      store.requestAccess(for: .contacts) { [weak self] didAuthorize,
      error in
        if didAuthorize {
          self?.retrieveContacts(from: store)
        }
      }
    } else if authorizationStatus == .authorized  {
      retrieveContacts(from: store)
    }
  }

  func retrieveContacts(from store: CNContactStore) {
    let containerId = store.defaultContainerIdentifier()
    let predicate =
CNContact.predicateForContactsInContainer(withIdentifier: containerId)
    let keysToFetch = [CNContactGivenNameKey as CNKeyDescriptor,
                       CNContactFamilyNameKey as CNKeyDescriptor,
                       CNContactImageDataAvailableKey as
                       CNKeyDescriptor,
                       CNContactImageDataKey as CNKeyDescriptor]

    contacts = try! store.unifiedContacts(matching: predicate,
    keysToFetch: keysToFetch)
      .map { Contact(contact: $0) }

    DispatchQueue.main.async { [weak self] in
      self?.collectionView.reloadData()
```

```
        }
      }
    }

    extension ViewController: UICollectionViewDataSource {
      func collectionView(_ collectionView: UICollectionView,
                          numberOfItemsInSection section: Int) -> Int {
        return contacts.count
      }

      func collectionView(_ collectionView: UICollectionView,
                          cellForItemAt indexPath: IndexPath) ->
    UICollectionViewCell {
        let cell = collectionView.dequeueReusableCell(withReuseIdentifier:
          "contactCell", for: indexPath) as! ContactCollectionViewCell
        let contact = contacts[indexPath.row]
        cell.nameLabel.text = "\(contact.givenName) \(contact.familyName)"
        contact.fetchImageIfNeeded { image in
          cell.contactImage.image = image
        }

        return cell
      }
    }
```

`Chapter 1`, *UITableView Touch-Up* covers the details regarding the preceding code. It implements data source methods for cell count and cell creation. It also fetches contacts and reloads the collection view, very similar to how the table view was reloaded.

If you run the app now, it doesn't look amazing. Images are a little distorted and using plain squares doesn't help the design either. You can fix this by configuring the image view in a slightly different way to prevent it from distorting the image and to make it scale it proportionally to cover the available space.

Open your Storyboard and select the image view in the cell. Look for the option called **Content Mode** in the **Identity Inspector**. This option describes how images should be rendered in the image view. The default value for this property is **Scale to Fill**. This setting scales and distorts the image to make it cover the viewport. A better option is **Aspect Fill**. This scales the image to cover the available space while maintaining the correct aspect ratio. Try playing with the other available options as well to see what happens.

Next, look for the **Background** option for the image view. Set this property to a light gray color. That will function as a nice placeholder for the real image. The final step is to set rounded corners on the image view. This option isn't easily found in **Interface Builder**, so you'll add it through code. Before you do this though, make sure that **Clips to Bounds** for the image view is enabled. This will make sure the rounded corner effect is visible.

In addition to the technique demonstrated below, you can set a **User defined runtime attributes** panel on a view using the **Identity Inspector**. You can use this panel to configure properties that are normally not exposed to **Interface Builder**. To set a corner radius, you can use the following attributes: Key Path: `layer.cornerRadius`, `Type: Number`, and `Value: 25`.

The best place to set the corner radius for the collection view cell is in `awakeFromNib()`. This method is only called once for each cell which makes it the perfect place to do some extra view configuration. Add the following piece of code to `ContactCollectionViewCell.swift` to implement the rounded corner effect:

```
override func awakeFromNib() {
  super.awakeFromNib()

  contactImage.layer.cornerRadius = 25
}
```

The preceding snippet only sets the corner radius for the image's layer to 25. Every view has a layer that is used for animations and rendering. When you set a corner radius on an image, you must do this on the layer instead of the view. You can also use the view's layer to add drop shadows, borders, and more.

All you need to do is assign a radius to the layer and run the app. All the other required work, such as making sure the view clips to its bounds, has already been done in **Interface Builder.** Now that you know how to improve the look of your cells, it's time to explore the collection view's layout object.

Understanding UICollectionViewFlowLayout and its delegate

When you use a standard collection view, it uses a grid layout. In this grid layout, all items are evenly spaced, and all items have the same size. You can visualize this quite easily by giving the collection view cell a background color in your Storyboard and running your app. This standard grid layout makes adding a grid to your app a breeze. However, the standard implementation of the grid layout is not quite perfect. If you look at the grid you have set up, it should look good on, for instance, an iPhone 8 but when switching to a smaller screen such as the iPhone SE, the layout looks nothing like the beautiful grid layout you saw before. Viewing the grid on a larger screen such as an iPhone 8 Plus doesn't make it any better either. The grid would be significantly improved if every cell was sized based on the size of the screen. Let's see how this can be achieved.

In your storyboard, select the **Collection View Flow Layout** object in the **Document Outline**. In the **Attributes Inspector**, you can change several properties for the collection view's layout. For instance, you can change its scroll axis from vertical to horizontal. This is something that can't be achieved using a table view since table views only allow scrolling on their vertical axis. If you switch to the **Size Inspector** for the layout object, you will find several available options to change the cell's spacing and size. As you may or may not know, the iPhone SE has a display width of 320 points, so try updating the layout's item size from 110 points wide to a width of 106 points. That should allow the grid to lay out all items with just a single pixel in between every cell. Also, update the minimum spacing for cells and lines to 1 point. This value indicates the minimum spacing for cells and lines. Note that when the layout is actually calculated, the spacing could be larger than you specified to ensure everything fits the screen nicely. Because it's a minimum spacing, cells will never be placed closer to each other than you have specified.

Try running your app on an iPhone SE now. You will notice that the layout looks great for this device. However, on larger screens the spacing between cells is a lot larger than you would like it to be, so we haven't achieved much yet. Luckily, you can adjust the cell size for your collection view items in your code as well. This means that you can use the currently available space for the collection view to determine precisely what size every cell should be to look great on all devices.

The standard layout that is used by a collection view is an instance of `UICollectionViewFlowLayout`. This object has a delegate property that conforms to the `UICollectionViewDelegateFlowLayout` protocol. This protocol specifies several methods that you can implement to take fine-grained control over your collection view's layout, allowing you to dynamically calculate cell spacing, line spacing, and cell size. For this example, dynamically calculating the cell size is the most interesting one. Since the spacing between cells should always be at least 1 point regardless of screen size, this property doesn't have to be calculated dynamically.

The requirement for the cell size can be summed up as follows:

- Three cells should fit on a single row in the grid.
- The cell's sizing should allow for approximately 1 pixel of room between every cell in a row.

To meet these requirements, the method from `UICollectionViewDelegateFlowLayout` that needs to be implemented is `collectionView(_:layout:sizeForItemAt:)`. This method will be used to return a size that is roughly one-third or less of the collection view's width, and that has a height of 90 points. To do this, add the following extension to your `ViewController` class:

```
extension ViewController: UICollectionViewDelegateFlowLayout {
  func collectionView(_ collectionView: UICollectionView, layout
collectionViewLayout: UICollectionViewLayout,
                      sizeForItemAt indexPath: IndexPath) -> CGSize {

    let width = floor((collectionView.bounds.width - 2) / 3)
    return CGSize(width: width, height: 90)
  }
}
```

The method in this extension is called for every cell in the collection view. This means that you can dynamically calculate the size for each collection view cell. The implementation you just added uses the `floor` function to make sure the width for each item is always one-third or less of the collection view width. To account for the 1 point spacing between each cell, 2 is subtracted from the collection view's width before dividing it by 3.

If you run your app like this, you will find that the spacing between items is nice and tight. However, if you look at this layout on an iPhone 8, you might notice that the spacing between cells on the horizontal axis does not really match the spacing between the cells on the vertical axis. You might have to look real close, but once you notice this it is hard to not see it. You can fix this by dynamically calculating the spacing between each cell and then using this value as the new minimum line spacing. The following code snippet takes care of this calculation. Add it to your `UICollectionViewDelegateFlowLayout` extension in `ViewController.swift`:

```
func collectionView(_ collectionView: UICollectionView, layout
collectionViewLayout: UICollectionViewLayout,
                    minimumLineSpacingForSectionAt section: Int) ->
CGFloat {

  let availableWidthForCells = collectionView.bounds.width - 2
  let totalGutterSpace =
availableWidthForCells.truncatingRemainder(dividingBy: 3)
  let cellSpacing = totalGutterSpace / 2

  return 1 + cellSpacing
}
```

The calculation to come up with the appropriate value is somewhat complex so breaking it down step by step helps to make sense of it. Since the spacing will never be less than 1, and there are two spacing gutters per row, these gutters are subtracted from the available width. This available width is the same width that is available for the collection view cells. Next, `truncatingRemainder(dividingBy:)` is used to find out how much space is left over after dividing it between the collection view cells. This remaining space is then divided by 2 since there are two gutters in the collection view. And lastly, because the standard spacing will always be 1, the calculated gutter width is incremented by 1.

The fact that you can take fine-grained control over a collection view's grid layout by implementing methods from `UICollectionViewDelegateFlowLayout` makes it even more powerful and flexible than its out-of-the-box implementation. You can create all kinds of grids with custom spacing, cells that have different heights or widths, and more. Sometimes all this power and flexibility isn't quite enough. Imagine, for instance, you want your grid to be a little bit more playful. Something where cells are a little bit more scattered across the screen, for instance. You can achieve this by creating your `UICollectionViewLayout` subclass.

Creating a custom UICollectionViewLayout

Implementing a large and complex feature such as a custom collection view layout might seem like a huge challenge for most people. Creating your layout involves calculating the position for every cell that your collection view will display. You must ensure that these calculations are performed as quickly and efficiently as possible because your layout calculations directly influence the performance of your collection view. A poor layout implementation will lead to slow scrolling and a lousy user experience eventually. Luckily, the documentation that has been provided for creating a collection view layout is pretty good as a reference to figure out if you're on the right track.

If you take a look at Apple's documentation on `UICollectionViewLayout`, you can read about its role in a collection view. The available information shows that a custom layout requires you to handle layout for cells, supplementary views, and decoration views. Supplementary views are header and footer views. The **Hello-Contacts** app does not use headers and footers in its collection view so these views can be skipped. Decoration views are views that don't correspond with any data in the collection view's data source, but they are part of the collection view's layout. The purpose of these views is purely decorative, just like the name *decoration view* suggests. Since these views are also not used by **Hello-Contacts,** the focus will be on the collection view cells only. This makes sure that the implementation remains simple and understandable.

In addition to listing the kinds of views you should layout, the documentation also mentions several methods that should be implemented for any custom layout. Note that not all of these methods are mandatory. For instance, any method that relates to supplementary and decoration views can be omitted if you don't need them. The methods and properties that you need to implement for your custom grid of collection view cells are the following:

- `collectionViewContentSize`
- `layoutAttributesForElements(in:)`
- `layoutAttributesForItem(at:)`
- `shouldInvalidateLayouts(forBoundsChange:)`
- `prepare()`

Further down in the documentation page for `UICollectionViewLayout`, you can find some information about updating the layout. For now, you won't bother implementing this, but it's most certainly worth it to familiarize yourself with all of the available documentation. The `prepare()` method is a very interesting one. It provides a great opportunity for developers to pre-calculate a collection view's layout. Since all cells in **Hello-Contacts** are the same size, the entire layout for the collection view can be calculated in `prepare()`. In this case, it's a great choice to do this due to the extremely predictable nature of the layout.

A custom collection view layout always is a subclass of `UICollectionViewLayout`. To make your layout object, create a new Cocoa Touch class and name it `ContactsCollectionViewLayout`. Make sure to have this class inherit from `UICollectionViewLayout`. The design you are going to implement in your custom layout object is shown in the following screenshot. It looks a lot like a regular horizontally scrolling layout except that the cells are distributed quite playfully, with every other row slightly offset to the right:

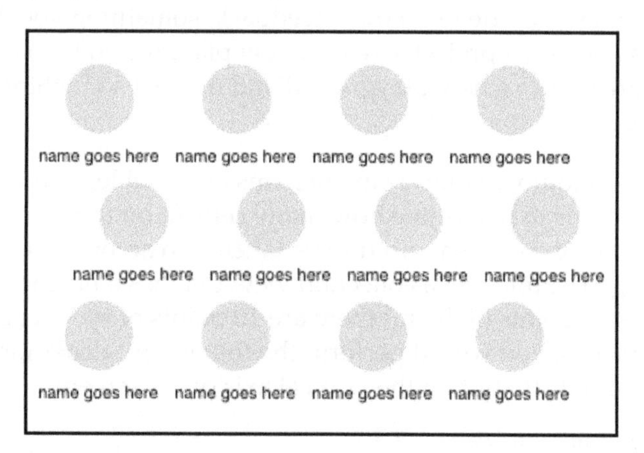

Since all elements in this layout have a predictable size that doesn't depend on any external factors such as the bounds of the collection view this layout belongs to, all heavy lifting can be done in the `prepare()` method. This is where all layout attributes for cells will be calculated, so they are available long before the collection view starts asking for sizes for specific cells that are about to be shown on screen.

The steps required to implement the entire custom layout are the following:

1. Pre-calculate the layout in the `prepare()` method.
2. Implement `collectionViewContentSize` to provide the collection view with enough information to configure scrolling.

3. Implement `layoutAttributesForElements(in:)` to provide layout attributes for items in a certain area.
4. Implement `layoutAttributesForItem(at:)` to provide information about the layout of a specific item.
5. Implement `shouldInvalidateLayout(forBoundsChange:)` to determine whether a certain change in a collection view's size should trigger a recalculation of the layout.
6. Assign the custom layout to the collection view.

Pre-calculating the layout

To figure out how to calculate the layout for your collection view, you can use a playground to quickly test and iterate over your attempts at calculating a layout, without the overhead of having to run an app first. Playgrounds are a fantastic way to test, explore, and validate ideas you have quickly. Since playgrounds don't need an entire app to run, they provide immediate feedback, something you don't get when you usually build and run a project. Create a new playground by navigating to **File** | **New** | **Playground**. Create a new playground and name it something like `LayoutExploration`.

The collection view should display as many items as possible, so the first step when calculating the layout is to determine how many cells fit on the vertical axis of the screen. The horizontal axis doesn't need to be taken into account since the horizontal axis is the scrolling axis. Imagine a collection view that has a height of 667 points, where every cell is 90 points high and there are 10 points of spacing between every cell. Using these values, you would perform the following calculation to come up with the number of cells that fit on the vertical axis of the screen:

```
(667 + 10) / 100 = 6.77
```

This means that you would fit 6.77 cells on the vertical axis of the collection view. Note that 10 is added to the height of the collection view and that the resulting value is divided by 100. This is done to properly compensate for the spacing between cells. Once you know the number of cells that fit on the vertical axis, you can also determine how many cells will exist on the horizontal axis. All you need to know is the total number of contacts that the collection view is going to display. When the collection view will show 60 cells in total, you can use the following calculation to determine the number of cells per row:

```
60 / 6 = 10
```

Note that the number of cells is not divided by 6.77 but by 6. The number of cells on the vertical axis is rounded down because cells should always fit on the screen completely, not partially. The vertical axis only fits six complete cells, so dividing by 6 is the correct way to perform the calculation.

With this information in mind, you can begin setting up the playground by defining a couple of variables. Open the playground you created earlier and add the following code to it:

```
import UIKit

let collectionViewHeight = 667
let itemHeight = 90
let itemWidth = 100
let itemSpacing = 10
let numberOfItems = 60
let numberOfRows = (collectionViewHeight + itemSpacing) / (itemHeight
+ itemSpacing)
let numberOfColumns = numberOfItems / numberOfRows
```

This snippet sets up all the variables that you have seen earlier. Note that `numberOfRows` does not need to be rounded down. All the input numbers have the type `Int`. Since `Int` doesn't support floating point numbers, all calculations done with `Int` are automatically rounded to the correct value. You can now use these values to write a loop that calculates the exact location for each of the 60 cells that you want to display. The following code snippet does just that:

```
let allFrames: [CGRect] = (0..<numberOfItems).map { itemIndex in
  let row = itemIndex % numberOfRows
  let column = itemIndex / numberOfRows

  var xPosition = column * (itemWidth + itemSpacing)
  if row % 2 == 1 {
    xPosition += itemWidth / 2
  }

  let yPosition = row * (itemHeight + itemSpacing)

  return CGRect(x: xPosition, y: yPosition, width: itemWidth, height:
itemHeight)
}

print(allFrames)
```

The preceding code creates an array of `CGRect` objects by mapping over a range. Whenever you have a list of something - in this case, a list of indexes that is created using `(0..<numberOfItems)` - you can then use `map` to transform each item in that list into something else. In this case, `map` turns an `itemIndex` object that corresponds to one of the cells that the collection will display into a `CGRect` object that could be used as the `frame` object for a cell. This loop also takes care of that playful offset every other row should have. The `%` operator is used to determine whether the item belongs on an even or odd row and then offsets the item accordingly.

After the array of frames is created, it is printed to make sure the layout contains all the expected values. You can now run your playground by hitting *Shift* + *Enter* or clicking the play button in the bottom-left corner of the editor. After running your playground, you should see the following output in the console:

```
[(0.0, 0.0, 100.0, 90.0), (50.0, 100.0, 100.0, 90.0), (0.0, 200.0,
100.0, 90.0), (50.0, 300.0, 100.0, 90.0), (0.0, 400.0, 100.0, 90.0),
(50.0, 500.0, 100.0, 90.0), (110.0, 0.0, 100.0, 90.0), (160.0, 100.0,
100.0, 90.0), (110.0, 200.0, 100.0, 90.0), (160.0, 300.0, 100.0,
90.0), (110.0, 400.0, 100.0, 90.0), (160.0, 500.0, 100.0, 90.0),
(220.0, 0.0, 100.0, 90.0), (270.0, 100.0, 100.0, 90.0), (220.0, 200.0,
100.0, 90.0), (270.0, 300.0, 100.0, 90.0), (220.0, 400.0, 100.0,
90.0), (270.0, 500.0, 100.0, 90.0), (330.0, 0.0, 100.0, 90.0), (380.0,
100.0, 100.0, 90.0), (330.0, 200.0, 100.0, 90.0), (380.0, 300.0,
100.0, 90.0), (330.0, 400.0, 100.0, 90.0), (380.0, 500.0, 100.0,
90.0), (440.0, 0.0, 100.0, 90.0), (490.0, 100.0, 100.0, 90.0), (440.0,
200.0, 100.0, 90.0), (490.0, 300.0, 100.0, 90.0), (440.0, 400.0,
100.0, 90.0), (490.0, 500.0, 100.0, 90.0), (550.0, 0.0, 100.0, 90.0),
(600.0, 100.0, 100.0, 90.0), (550.0, 200.0, 100.0, 90.0), (600.0,
300.0, 100.0, 90.0), (550.0, 400.0, 100.0, 90.0), (600.0, 500.0,
100.0, 90.0), (660.0, 0.0, 100.0, 90.0), (710.0, 100.0, 100.0, 90.0),
(660.0, 200.0, 100.0, 90.0), (710.0, 300.0, 100.0, 90.0), (660.0,
400.0, 100.0, 90.0), (710.0, 500.0, 100.0, 90.0), (770.0, 0.0, 100.0,
90.0), (820.0, 100.0, 100.0, 90.0), (770.0, 200.0, 100.0, 90.0),
(820.0, 300.0, 100.0, 90.0), (770.0, 400.0, 100.0, 90.0), (820.0,
500.0, 100.0, 90.0), (880.0, 0.0, 100.0, 90.0), (930.0, 100.0, 100.0,
90.0), (880.0, 200.0, 100.0, 90.0), (930.0, 300.0, 100.0, 90.0),
(880.0, 400.0, 100.0, 90.0), (930.0, 500.0, 100.0, 90.0), (990.0, 0.0,
100.0, 90.0), (1040.0, 100.0, 100.0, 90.0), (990.0, 200.0, 100.0,
90.0), (1040.0, 300.0, 100.0, 90.0), (990.0, 400.0, 100.0, 90.0),
(1040.0, 500.0, 100.0, 90.0)]
```

If you want to run your complete playground again, click the stop button in the bottom-left corner to reset the playground and then hit play again to run it. The output you produced is not very easy to read but if you examine it closely, you should be able to see that this layout is exactly what you were looking for. All items have the correct size, the indentation works, and the items are created from top to bottom and from left to right. You have already created a placeholder file called `ContactsCollectionViewLayout`; now add the following skeleton code to it as a first step toward implementing your custom layout:

```
import UIKit

class ContactsCollectionViewLayout: UICollectionViewLayout {
  private let itemSize = CGSize(width: 100, height: 90)
  private let itemSpacing: CGFloat = 10

  private var layoutAttributes = [UICollectionViewLayoutAttributes]()

  override var collectionViewContentSize: CGSize {
    return .zero
  }

  override func prepare() {

  }

  override func shouldInvalidateLayout(forBoundsChange newBounds:
CGRect) -> Bool {
    return false
  }

  override func layoutAttributesForElements(in rect: CGRect) ->
[UICollectionViewLayoutAttributes]? {
    return nil
  }

  override func layoutAttributesForItem(at indexPath: IndexPath) ->
UICollectionViewLayoutAttributes? {
    return nil
  }
}
```

All this code does is provide some placeholders that you will gradually replace with proper implementations. Note that `itemSize` and `itemSpacing` are defined as constants here. If you were to create your own layout and you want it to behave a bit more like `UICollectionViewFlowLayout` by allowing developers to customize the item sizes and spacing, you should make these constants variables instead.

In this case, the item size and spacing won't be modified externally so having them as constants makes sense. Because external sources don't need to modify or even access any of the defined properties, they are marked as `private`. It is considered good practice to mark properties and methods as `private` if other classes do not need to access them.

Also note that the layout is not defined as an array of `CGRect` instances, but instead it is defined as an array of `UICollectionViewLayoutAttributes`. This is the type of object that a collection view uses to describe the layout attributes of its items. You can now add the following implementation for `prepare()`:

```
override func prepare() {
  // 1
  guard let collectionView = self.collectionView
    else { return }

  let availableHeight = Int(collectionView.bounds.height +
itemSpacing)
  let itemHeight = Int(itemSize.height + itemSpacing)

  numberOfItems = collectionView.numberOfItems(inSection: 0)
  numberOfRows = availableHeight / itemHeight
  numberOfColumns = Int(ceil(CGFloat(numberOfItems) /
CGFloat(numberOfRows)))

  layoutAttributes.removeAll()

  // 2
  layoutAttributes = (0..<numberOfItems).map { itemIndex in
    let row = itemIndex % numberOfRows
    let column = itemIndex / numberOfRows

    var xPosition = column * Int(itemSize.width + itemSpacing)
    if row % 2 == 1 {
      xPosition += Int(itemSize.width / 2)
    }

    let yPosition = row * Int(itemSize.height + itemSpacing)

    // 3
    let indexPath = IndexPath(row: itemIndex, section: 0)
    let attributes = UICollectionViewLayoutAttributes(forCellWith:
indexPath)
    attributes.frame = CGRect(x: CGFloat(xPosition), y:
CGFloat(yPosition),
 width: itemSize.width, height: itemSize.height)
```

```
        return attributes
    }
}
```

This implementation is very similar to the one you wrote in the playground. The first comment highlights a key difference though. Instead of using hardcoded values for the available space, the layout's `collectionView` property is used to determine the available space.

The second comment shows the code that calculates the position of every item. Just like before, `map` is used to configure the list of attributes. Note that many conversions between `Int` and `CGFloat` take place to make sure all types are used correctly. Since iOS expresses positions in a layout using `CGFloat` but the layout calculations rely on the rounding properties of `Int`, it is important to convert between types properly.

The third and last comment highlights the creation of the `UICollectionViewLayoutAttributes` instance for each item. In addition to a `CGRect` instance for size and positioning, `UICollectionViewLayoutAttributes` uses an `IndexPath` object to tie itself to the position of a certain cell. This wraps up the preparation of the layout. Next up, we will calculate `collectionViewContentSize`.

Implementing collectionViewContentSize

A collection view uses the `collectionViewContentSize` property from its layout to figure out the size of its contents. This property is especially important because it is used to configure and display the scrolling indicators for the collection view. It also provides the collection view with information about the direction in which scrolling should be enabled.

Implementing this property uses the number of rows and columns in the collection view. It also takes the item size and item spacing into account to come up with the size of all of its contents together. Add the following implementation for `collectionViewContentSize`:

```
override var collectionViewContentSize: CGSize {
  let width = CGFloat(numberOfColumns) * itemSize.width +
CGFloat(numberOfColumns - 1) * itemSpacing
  let height = CGFloat(numberOfRows) * itemSize.height +
CGFloat(numberOfRows - 1) * itemSpacing

  return CGSize(width: width, height: height)
}
```

Determining the entire size of the collection view is quite simple now because the `prepare()` method has already performed all the hard work to set up the layout.

Implementing layoutAttributesForElements(in:)

More complex than `collectionViewContentSize` is `layoutAttributesForElements(in:)`. This method is responsible for providing a collection view with the layout attributes for several elements at once. The collection view always provides a rectangle, for which it needs layout attributes. The layout is responsible for providing these attributes to the collection view as fast as possible. It is essential that the implementation of this method is as efficient as you can get it to be; your scroll performance depends on it.

Even though there is only a small number of cells visible at a time, the collection view has a lot more content outside of its current viewport. Sometimes it is asked to jump to a particular cell, or the user scrolls extremely fast. There are many cases for the collection view to ask for all layout attributes for several cells at once. When this happens, the layout object can help the cell determine which cells should be visible for a particular rectangle. This is possible because the layout attributes do not only contain the rectangle in which a cell should be rendered, it also knows the `IndexPath` object that corresponds with that specific cell.

This is pretty complicated matter, and it's okay if you find this to be a little bit confusing. As long as you understand that a collection view can ask its layout which cells are present in a certain `CGRect` instance and how they should be rendered, you understand what `layoutAttributesForElements(in:)` does. The most straightforward implementation we can come up with for `layoutAttributesForElements(in:)` is the following:

```
override func layoutAttributesForElements(in rect: CGRect) ->
[UICollectionViewLayoutAttributes]? {
  return layoutAttributes.filter { attributes in
    return attributes.frame.intersects(rect)
  }
}
```

This code filters the list of attributes and returns a new list that only contains attributes that are within the bounds of the supplied rectangle. Note that this implementation might not be the best possible implementation for your custom layout. As the number of items in the collection grows, the time to run the filter grows too. You might be better off implementing a different method to look up the correct items. This depends on your exact use case, layout, and bottlenecks.

Implementing layoutAttributesForItem(at:)

Another way a collection view can request layout attributes for its layout is by requesting the attributes for a single item. Because the collection view does so by supplying an index path, this method is quite simple to implement. The layout you implemented assumes that only a single section exists in the collection view and the layout attributes array is sorted by index path because that's the order in which all items were inserted into the array. This means that you can use the supplied index path's row to fetch the correct layout attributes from the layout attributes array:

```
override func layoutAttributesForItem(at indexPath: IndexPath) ->
UICollectionViewLayoutAttributes? {
  return layoutAttributes[indexPath.row]
}
```

The last remaining method to implement is `shouldInvalidateLayout(forBoundsChange:)`.

Implementing shouldInvalidateLayout(forBoundsChange:)

Getting the implementation for `shouldInvalidateLayout(forBoundsChange:)` is crucial to having a great collection view layout that has amazing performance. If you implement this method incorrectly, you could constantly be invalidating the layout, meaning you will need to recalculate all the time. It's also possible that the collection view will never update its layout at all, even when it should. The collection view will call this method any time its size changes. For instance, when the user rotates their device or when your app runs on an iPad, the user opens another app in multitasking mode.

The layout you have created only need to update when the number of items that fit on the vertical axis changes. The following diagram shows a scenario where invalidating the bounds isn't required because the number of items that fit on the vertical axis does not change:

As you can see, it would not make sense to recalculate the entire layout if the layout of the cells doesn't have to change. The best implementation for this specific layout is to calculate the new number of items on the vertical axis, checking whether this new number is different and then invalidating the layout when it is. Add the following implementation to your layout class:

```
override func shouldInvalidateLayout(forBoundsChange newBounds:
CGRect) -> Bool {
  guard let collectionView = self.collectionView
    else { return true }

  let availableHeight = newBounds.height -
```

```
collectionView.contentInset.top - collectionView.contentInset.bottom
  let possibleRows = Int(availableHeight + itemSpacing) /
Int(itemSize.height + itemSpacing)

  return possibleRows != numberOfRows
}
```

This method uses a similar technique as you have seen before to determine the maximum amount of items on the vertical axis. When this new amount is different from the old amount, `possibleRows != numberOfRows` evaluates to true, and the layout will be invalidated.

This wraps up the work you have to do to set up your layout object; the time has come to see how you can use this custom layout in your project.

Assigning a custom layout to your collection view

The final step to using your custom layout is telling your collection view to use your layout. You have already seen that you can assign a custom class to the layout for a collection view in **Interface Builder**. However, this only works when your layout inherits from `UICollectionViewFlowLayout`, which your layout does not inherit from. Luckily, you can also set your collection view's layout in code. Update your `viewDidLoad` method in `ViewController.swift` by adding the following line to it:

```
collectionView.collectionViewLayout = ContactsCollectionViewLayout()
```

This line sets your new layout as the current layout. You can now remove the `UICollectionViewDelegateFlowLayout` extension from `ViewController.swift` as it is not needed anymore.

Final words on the custom layout

The custom layout you created in this subsection is an excellent example of how powerful collection views are. However, with great power comes great responsibility. For instance, you will need to make sure your layout is as performant as you can make it. Slower implementations will quickly lead to lousy scrolling performance, which you want to avoid at all costs. However, don't go overboard with optimizing your code right away. Think about the use case for your collection view. Maybe you won't ever display more than a certain amount of items, and you can get away with a slightly slower implementation. Or perhaps you want to reuse your layout, meaning you should make sure it works great no matter how many items you throw at it.

Being able to create fantastic collection view layouts is a skill you will most certainly want to have as it enables you to implement amazing, performant layouts that will delight your users. You know now that creating your collection view layout does not have to be extremely complex. With some careful planning and experimentation in playgrounds, you will build amazing layouts for your users in no time!

Reader exercise: The layout you created is aligned to the top of the collection view and fills the screen up as much as possible. This leaves empty space at the bottom of the screen. Try to adapt the layout so that it's centered vertically, making sure the space up top matches the space at the bottom. Implementing this will involve updating your layout preparation and invalidation code. Good luck!

UICollectionView performance

You have probably already noticed that, apart from the ability to add a custom layout, a collection view is very similar to a table view. When you look at what collection view does under the hood to maintain excellent scrolling performance, you will find even more similarities. The collection view is optimized to show as many cells on screen as quickly as it possibly can while keeping its memory footprint as small as possible. These optimizations are important for table views but they are even more important for collection views because a collection view might show a lot more cells on screen at a time than a table view does. The following diagram visualizes this:

The fact that collection views show so many cells at once has made it so that Apple added an extra optimization to it compared to table views. A table view always loads one or two items that are near the user's current scroll position. This is fine, especially considering that table views only one new cell to the screen at once. Collection views sometimes have to visualize three or more new cells all at once. This means that instead of configuring a single cell, you are now supposed to configure and lay out multiple cells in the same time window as before.

If the layout of your cells is complex or the collection view contains many cells in a single row, your scrolling performance can quickly degrade if you're not careful. Bad scrolling performance is something that should always be avoided at all costs.

When you aim for smooth 60-frames-per-second scrolling, you only get about 16 milliseconds to perform all operations and calculations needed to configure your cells. Once this time window is up, the layout engine will want to render a new frame on screen. If you miss this 16-millisecond window, the frame is considered dropped because the layout engine could not render it in time, and your users will notice your performance is not what it should be.

Luckily, Apple has optimized collection views in a way that prevents the need to deliver all cells that are about to be shown all at once. While a table view only has very few extra cells in memory, collection views request new cells a bit earlier to make sure that all cells that are about to be rendered are ready to go. The following diagram visualizes what this would look like when shown in a similar way to before:

This optimization divides the work that needs to be done by the collection view's data source and delegates more evenly so that the workload is quite constant. This is much better because, instead of having to set up a bunch of cells at once just before they are shown, small bits of work are now being done more often.

Just like table views, collection views have prefetching. In the previous chapter, you implemented prefetching to decode and prepare images for the table view. Using this knowledge, you should be able to add the same functionality to the collection view. Implementing prefetching on a collection view will have a more significant impact than implementing it on a table view. This is because collection views typically display more cells at once than a table view, so there is a lot more work to be done at any given time.

If you need some guidance for implementing prefetching on the collection view, refer to this book's code bundle to see how to implement it.

Implementing user interactions for the collection view

In the previous chapter, you saw how to implement several interactions with a table view by implementing several of its delegate methods. Any time the user interacts with a cell, the table view calls a delegate method to see what exactly should happen as a response to this interaction. Collection views have incredibly similar capabilities, except for a few details. For instance, implementing reordering on a collection view requires a little bit more work than it does for a table view, as you will see shortly. Also, collection view cells cannot be swiped on to reveal actions to, for instance, delete a cell like you can with a table view cell. Similar functionality can be added to collection views though, and this section will show you exactly how.

You will learn how to implement the following interactions:

- Cell selection
- Cell deletion
- Cell reordering

Cell selection is the simplest feature to implement. The delegate method for this is called when the user taps on a cell. This is the first interaction you will learn about.

Enabling cell selection for your collection view

Implementing cell selection for a collection view is almost identical to implementing it on a table view. With the knowledge you have gained in the previous chapter, you should be able to implement basic cell selection on your own. To keep this section interesting, you won't implement the same alert as you did in the previous chapter. Because users like to see some feedback when they tap something that they can interact with, it's a great idea to add a little animation to your collection view cells to make them bounce a little bit when they are tapped.

Add the following `UICollectionViewDelegate` extension to `ViewController.swift`. Don't forget to set your view controller as the delegate for your collection view in `viewDidLoad()`:

```
extension ViewController: UICollectionViewDelegate {
  func collectionView(_ collectionView: UICollectionView,
didSelectItemAt indexPath: IndexPath) {
    guard let cell = collectionView.cellForItem(at: indexPath) as?
ContactCollectionViewCell
      else { return }

    UIView.animate(withDuration: 0.1, delay: 0, options:
[.curveEaseOut], animations: {
      cell.contactImage.transform = CGAffineTransform(scaleX: 0.9, y:
0.9)
    }, completion: { _ in
      UIView.animate(withDuration: 0.1, delay: 0, options:
[.curveEaseIn], animations: {
        cell.contactImage.transform = .identity
      }, completion: nil)
    })
  }
}
```

This snippet implements a simple animation using `UIView.animate`. In Chapter 4, Immersing your users with animation, you will learn more about animations, and you'll see other, more powerful ways to enable rich interactions to your app.

Note that the collection view is asked for the tapped cell using an index path. Since this method returns a `UICollectionViewCell?` object, you need to make sure that a cell was found and you must also cast it to a `ContactCollectionViewCell` object to access the contact image. If no cell exists at the requested index path or if the obtained cell is not a `ContactCollectionViewCell` object, the method is aborted.

If the cell is retrieved successfully, the animation is performed. The `animate` method takes several parameters that you will learn more about later. Chapter 4, Immersing your users with animation covers animation in depth, and you will be up to speed with what exactly is going on in this bounce animation in no time. For now, you might want to go ahead and run the app to see your bounce animation in action. If your animation doesn't work, double-check that you have set `ViewController` as the collection view's delegate.

Allowing users to delete cells

Any good contacts app enables users to remove contacts. The previous version of **Hello-Contacts** used a swipe gesture on a table view cell to delete a cell. Swiping a cell made a button appear that the user could tap and the corresponding `UITableViewDelegate` method was called.

Unfortunately, `UICollectionViewDelegate` does not specify a similar delegate method for deleting cells. This means that you'll need to do a little more work to implement cell deletion for a collection view. A very simple implementation would be to have the user long-press on a cell, ask them whether they want to delete the contact, and reload the collection view if needed. This would work but the deleted contact would quickly disappear, and the whole thing would look quite choppy.

Luckily, `UICollectionView` defines several methods that you can use to update the collection view's contents in a very nice way. For instance, when you delete a cell from the collection view, you can have the remaining cells animate to their new positions. This looks way better than simply reloading the entire collection view without an animation. Good iOS developers will always make sure that they go the extra mile to find a nice, smooth way to transition between interface states. So animating cells when deleting one is a great thing to have.

If you look at Apple's documentation on `UICollectionView`, you'll find that there is a lot of information available about collection views. If you scroll all the way down to the **Symbols** section, there is a subsection named **Inserting, Moving and Deleting items**. Perfect, this is precisely the kind of information you need to implement cell deletion. More specifically, the `deleteItems(_:)` method looks like it's exactly what you need to get the job done.

The requirements for the cell deletion feature are the following:

1. The user long-presses on a cell.
2. An action sheet appears to verify whether the user wants to delete this cell.
3. If the user confirms the deletion, the cell is removed, and the layout animates to its new state. The contact is also removed from the contacts array.

To detect certain user interactions such as double-tapping, swiping, pinching, and long-pressing, you make use of gesture recognizers. A gesture recognizer is a special object provided by `UIKit` that can detect certain gestures. When such a gesture occurs, it calls a selector (method) on a target object. This method can then perform a specific task in response to the gesture.

To keep things simple, you should add a gesture recognizer to the collection view as a whole. Adding the gesture recognizer to the whole collection view rather than its cells is a lot simpler because you can reuse a single recognizer, and figuring out information about the tapped cell and responding to the long-press is easier to do from the view controller than the tapped cell itself.

In a moment, you'll see how to find the tapped cell in the collection view. First, set up the recognizer by adding the following lines to `viewDidLoad()` in `ViewController.swift`:

```
let longPressRecognizer = UILongPressGestureRecognizer(target: self,
action: #selector(self.userDidLongPress(_:)))
collectionView.addGestureRecognizer(longPressRecognizer)
```

The first line sets up the long-press gesture recognizer. The target given to the gesture recognizer is `self`. This means that the current instance of `ViewController` is used to call the action on. The action is the second argument for the gesture recognizer. The selector passed to it refers to the method that is called when the user performs the gesture.

The second line adds the gesture recognizer to the collection view. This means that the gesture recognizer will only detect long-presses that occur within the collection view. When a long-press occurs on the collection view, the gesture recognizer will inform the `ViewController` of this event by calling `userDidLongPress(_:)` on it.

Now that your gesture recognizer is set up and you know a bit about how it works, add the following implementation of `userDidLongPress(_:)` to `ViewController.swift`:

```
@objc func userDidLongPress(_ gestureRecognizer:
UILongPressGestureRecognizer) {
  // 1
  let tappedPoint = gestureRecognizer.location(in: collectionView)
  guard let indexPath = collectionView.indexPathForItem(at:
tappedPoint),
    let tappedCell = collectionView.cellForItem(at: indexPath)
    else { return }

  // 2
  let confirmationDialog = UIAlertController(title: "Delete contact?",
message: "Are you sure you want to delete this contact?",
preferredStyle: .actionSheet)

  let deleteAction = UIAlertAction(title: "Yes", style: .destructive,
handler: { [weak self] _ in
    // 3
    self?.contacts.remove(at: indexPath.row)
    self?.collectionView.deleteItems(at: [indexPath])
  })

  let cancelAction = UIAlertAction(title: "No", style: .default,
handler: nil)

  confirmationDialog.addAction(deleteAction)
  confirmationDialog.addAction(cancelAction)

  // 4
  if let popOver = confirmationDialog.popoverPresentationController {
    popOver.sourceView = tappedCell
  }

  present(confirmationDialog, animated: true, completion: nil)
}
```

Note that this method is prefixed with `@objc`. This is required because selectors are a dynamic feature that originated in Objective-C. The `userDidLongPress(_:)` method must be exposed to the Objective-C runtime by prefixing it with `@objc`.

The first step in the implementation of this method is to obtain some information about the cell that was tapped. By taking the location at which the long-press occurred inside of the collection view, it is possible to determine the index path that corresponds with this long press. The index path can then be used to obtain a reference to the cell that the user was long-pressing on. Note that the user can also long-press outside of a cell. If this happens, the `guard` will cause the method to return early because there will be no index path that corresponds to the long-press location. If everything is fine, the code continues to the second step in this method.

To display an action sheet to the user, an instance of `UIAlertController` is created. You have already seen this object in action in the previous chapter when it was used to show an alert when the user tapped on a table view cell. The main difference between the alert implementation and this implementation is `preferredStyle`. Since this alert should show as an action sheet, the `.actionSheet` style is passed as the preferred style. This style will make an action sheet pop up from the bottom of the screen.

In the delete action for this action sheet, the contact is removed from the contacts array that is used as a data source. After updating the data source, the cell is removed from the collection view using `deleteItems(at:)`. When you update the items in a collection view, it is imperative to make sure that you update the data source first. If you don't do this, the app is likely to crash due to internal inconsistency exceptions. To see this crash occur, reverse the order of commands in the delete action.

Whenever you update a collection view in a way that moves or deletes its contacts, *always* make sure to update the underlying data source *first*. Not doing this will crash your app with an internal inconsistency error.

The fourth and last step before presenting the action sheet is some defensive programming. Larger screens, such as an iPad screen, display action sheets as popover views instead of action sheets. You can detect whether an action sheet will be shown as a popover by checking the `popoverPresentationController` object on the alert controller instance you have created. If a `popoverPresentationController` object exists, the action sheet will be presented as a popover and requires a `sourceView` object to be set. Not setting the `sourceView` object on a popover crashes your app, so it's better to provide a `sourceView` object than have your app crash when something unexpected happens.

When you display an action sheet, make sure to check whether a `popoverPresentationController` object exists and if it does, make sure to set a `sourceView` or `sourceRect` object. Devices with larger screens present action sheets as popovers and not setting a source for the popover crashes your app.

This wraps up implementing cell deletion for the collection view. You can try long-pressing on a contact cell now to see your action sheet appear. Deleting a contact should nicely animate the update. Even though it took a little bit more effort to implement cell deletion for collection view than it did for table view, it wasn't too bad. Next up is cell reordering.

Reordering cells in a collection view

Since collection views don't have the same awesome reordering API that table views have, you'll need to do a little bit of extra work again to get this to work. Doing extra work doesn't mean that implementing reordering is extremely complex, quite the opposite. There are some nice delegate methods available that aid greatly in implementing reordering for your collection view.

It will take a couple of steps to implement reordering for your collection view. Apple's documentation for `UICollectionView` lists several methods that relate to collection view reordering. Every method has its place in the reorder interaction, and you are expected to call each one at the right time on the collection view. Two methods from this list are of special interest:

- `endInteractiveMovement()`
- `beginInteractiveMovementForItem(at:)`

These two methods are interesting because when you call these on your collection view, the collection view will then call methods on its data source. When you end the interactive movement, the data source is asked to update its underlying dataset by moving an item from its old index path to a new index path. When you call `beginInteractiveMovementForItem(at:)`, the data source is asked whether the currently selected item is allowed to be reordered.

Collection views do not keep track of moving cells around on its own; you must implement this yourself. You can do this by adding a pan gesture recognizer but, coincidentally, the long-press gesture recognizer can also track movements the user makes with their finger.

To reuse the existing long-press gesture recognizer, you will need to refactor it a little bit. To differentiate between wanting to delete or move a cell, an edit button should be added to the collection view. The user can then tap this button to toggle between edit mode and normal mode. When the collection view is in edit mode, it allows cell reordering and otherwise it allows users to delete items.

To implement reordering, you will perform the following steps:

1. Refactor the long-press handler to differentiate between reordering and deleting cells.
2. Implement the sequence of method calls for cell reordering based on the gesture recognizer's state.
3. Implement the data source methods that allow cell reordering.
4. Add the edit button to the navigation bar.

Refactoring the long-press handler

Since the existing long-press handler will now have two different actions depending on the view controller's `isEditing` state, it's a wise idea to split the long-press handler into several methods that will be called depending on the desired action. The handler itself will still ensure that a valid cell and index path exist. But after that, it will forward to another method to perform further actions. Update your code as shown to refactor the long-press handler:

```
@objc func userDidLongPress(_ gestureRecognizer:
UILongPressGestureRecognizer) {
  let tappedPoint = gestureRecognizer.location(in: collectionView)
  guard let indexPath = collectionView.indexPathForItem(at:
tappedPoint),
    let tappedCell = collectionView.cellForItem(at: indexPath)
    else { return }

  if isEditing {
    beginReorderingForCell(tappedCell, atIndexPath: indexPath,
gestureRecognizer: gestureRecognizer)
  } else {
    deleteContactForCell(tappedCell, atIndexPath: indexPath)
  }
}

func beginReorderingForCell(_ cell: UICollectionViewCell, atIndexPath
indexPath: IndexPath, gestureRecognizer: UILongPressGestureRecognizer)
{
```

```
}

func deleteContactForCell(_ tappedCell: UICollectionViewCell,
atIndexPath indexPath: IndexPath) {
 // The existing cell deletion code goes here
}
```

First, the `isEditing` property on the view controller is used to determine what should happen when the user long-presses. If this property is `true`, `beginReorderingForCell(_:atIndexPath:gestureRecognizer:)` is called. You will implement this method later to support reordering. If `isEditing` is `false`, `deleteContactForCell(_:atIndexPath:)` is called to perform the delete action you have already implemented.

Implementing the reordering method calls

The second step to implement cell reordering is to call the correct `UICollectionView` methods at the right times. To do so, the long-press gesture's state is tracked and used to inform the collection view about the current state of the reorder life cycle it should be in. Add the following implementation for `beginReorderingForCell(_:atIndexPath:gestureRecognizer:)`:

```
func beginReorderingForCell(_ cell: UICollectionViewCell, atIndexPath
indexPath: IndexPath, gestureRecognizer: UILongPressGestureRecognizer)
{
  switch gestureRecognizer.state {
  case .began:
    collectionView.beginInteractiveMovementForItem(at: indexPath)
    UIView.animate(withDuration: 0.2, delay: 0, options:
[.curveEaseOut], animations: {
      cell.transform = CGAffineTransform(scaleX: 1.1, y: 1.1)
    }, completion: nil)
  case .changed:
    let position = gestureRecognizer.location(in: collectionView)
    collectionView.updateInteractiveMovementTargetPosition(position)
  case .ended:
    collectionView.endInteractiveMovement()
  default:
    collectionView.endInteractiveMovement()
  }
}
```

You can use a gesture recognizer's `state` property to find out more about the state it is in and take action based on this state. In this case, when the gesture begins, the collection view should enable its interactive movement mode for the pressed index path. To inform the user that the cell is being dragged, it is also made slightly larger using an animation.

When the recognizer updates, and it's in the `changed` state, the collection view should update accordingly. The collection view is informed about the position of the user's finger so it can update the visible cells to make room for the cell that is now hovering at a new location.

If the gesture is `ended`, or in another state, the interactive movement is ended. This means that the collection view will ask the data source to update its underlying data and the interface is animated to its new state. Since the collection view communicates with its data source for cell reordering, the next logical step is to update the data source, so it allows items to be reordered, and to make it persist the changes made by the user into the underlying data storage.

Implementing the data source methods

To support cell reordering, you must implement two `UICollectionViewDataSource` methods. The first method is `collectionView(_:moveItemAt:)`. This method is called to determine whether the currently selected item can be moved around. The second method you must implement is `collectionView(_:moveItemAt:to:)`. This method is called to tell the data source that it must commit the changes made by the user to the underlying data store. Let's jump straight into the implementation code. The following methods should be added to your `UICollectionViewDataSource` extension in `ViewController.swift`:

```
func collectionView(_ collectionView: UICollectionView, canMoveItemAt
indexPath: IndexPath) -> Bool {
  return true
}

func collectionView(_ collectionView: UICollectionView, moveItemAt
sourceIndexPath: IndexPath, to destinationIndexPath: IndexPath) {
  let movedContact = contacts.remove(at: sourceIndexPath.row)
  contacts.insert(movedContact, at: destinationIndexPath.row)
}
```

The first method always returns true because every item in the collection view can be reordered. The second method moves the reordered contact from its old position in the dataset to the new position. This is very similar to what you have done before to implement table view reordering. The last step to enable cell reordering is implementing the edit button in the navigation bar. When the user taps this button, the collection view enters edit mode, and the contacts can be dragged around after long-pressing them.

Adding the edit button

Adding an edit button to a navigation bar is fairly straightforward. Add the following code to the `viewDidLoad()` method of `ViewController`:

```
navigationItem.rightBarButtonItem = editButtonItem
```

If you build and run your app now, you can immediately begin dragging contacts around to reorder them. However, it's not really obvious that cells are in a different state after entering edit mode. The home screen on an iPhone does a great job at indicating when it's in edit mode. The icons begin to wiggle so the user knows they can now be moved around freely.

For now, an animation like that is a little bit too advanced so, for now, the cells will just be given a different background color to indicate that they are now in a different state. Add the following code to `ViewController` to update the cell's background color:

```
override func setEditing(_ editing: Bool, animated: Bool) {
  super.setEditing(editing, animated: animated)

  for cell in collectionView.visibleCells {
    UIView.animate(withDuration: 0.2, delay: 0, options:
[.curveEaseOut], animations: {
      if editing {
        cell.backgroundColor = UIColor(red: 0.9, green: 0.9, blue:
0.9, alpha: 1)
      } else {
        cell.backgroundColor = .clear
      }
    }, completion: nil)
  }
}
```

This updates all currently visible cells. However, because the collection view proactively keeps some off-screen cells in memory, you need to make sure to update those as well. You can implement `collectionView(_:willDisplay:forItemAt:)` to configure a cell right before it becomes visible to the user. This method is great for updating cells that are loaded but haven't become visible yet. Implement the following method on your `UICollectionViewDelegate` extension:

```
func collectionView(_ collectionView: UICollectionView, willDisplay
cell: UICollectionViewCell, forItemAt indexPath: IndexPath) {
  if isEditing {
    cell.backgroundColor = UIColor(red: 0.9, green: 0.9, blue: 0.9,
alpha: 1)
  } else {
    cell.backgroundColor = .clear
  }
}
```

That's it! Try running your app now to see all your hard work come together. Even though not every feature was as simple to implement on a collection view as it was on a table view, they are truly similar and the benefits of being able to create your own custom layout make the collection view one of the most important components to master in iOS.

Summary

This chapter has taught you how to harness the powers given to you by `UICollectionView` and `UICollectionViewLayout` in your apps. You saw that collection view come with a nice grid layout out of the box and that you can easily customize this grid layout by implementing a couple of delegate methods. Next, you created your layout to make your contacts app look more interesting and engaging. You also learned about the collection view's performance characteristics and how to implement great features such as contact selection, deletion, and reordering.

You have learned some fairly complex matter in this chapter, and you have implemented some pretty advanced features, such as your custom layout, gesture recognition, and using animations to provide feedback to users. All of these techniques are important parts of your toolbox as an iOS developer because they are applicable in many applications you will build in the future. The next chapter covers another important aspect of iOS: navigation. You'll see how you can navigate from one screen to another using the storyboard in **Interface Builder**. You'll also learn a bit more about Auto Layout and `UIStackView`, a powerful component that makes certain layouts a lot easier to create.

Questions

1. `UICollectionView` is very similar to another component on iOS. Which component is this?

   a) `UIImageView`
   b) `UIScrollView`
   c) `UITableView`

2. Which class did you have to subclass to create your own collection view layout?

   a) `UICollectionViewFlowLayout`
   b) `UICollectionViewLayout`
   c) `UICollectionLayout`

3. Which technique is applied by collection views to ensure great performance?

   a) Fetching cells a couple of rows ahead of the user's current position
   b) Fetching all cells at once
   c) Fetching cells just in time

4. Where in **Interface Builder** can you find the collection view's layout object?

   a) The **Attributes Inspector**
   b) The **Document Outline**
   c) The **Object Library**

5. Which feature that can be implemented for collection views is not available for table views?

   a) Reordering cells
   b) Deleting cells
   c) Horizontal scrolling

6. What is a gesture recognizer?

    a) An object that responds to taps only
    b) An object that predicts whether a user will scroll
    c) An object that responds to several user interactions, such as tapping, swiping, pinching, and more

7. What is the type of the button that you added to the navigation bar to toggle the collection view's editing state?

    a) `UIEditButton`
    b) `UINavigationItem`
    c) `UIBarButtonItem`

Further reading

- Apple's documentation on `UICollectionView`: `https://developer.apple.com/documentation/uikit/uicollectionview`
- Apple's documentation `UICollectionViewLayout`: `https://developer.apple.com/documentation/uikit/uicollectionviewlayout`

3
Creating a Detail Page

So far, you have managed to build an app that shows a set of contacts on a custom grid in a collection view. This is pretty impressive, but not very useful. Typically, a user will expect to be able to see more information when tapping on an item in an overview. In this case, they would likely expect to see more details about the tapped contact, for instance, their email address and phone number. In this chapter, you will see how to do just that.

You will learn how to use the storyboard to set up a connection between the collection view cells and a detail page. In addition to this, you will learn how to pass some data along when navigating between view controllers to make populating your detail pages with data a breeze. You won't stop there though. Currently, the contacts app is built with the iPhone in mind, but a great app will work on any device. So you're going to make use of **Auto Layout** and **size classes** to ensure your layouts look great on all screens, even the iPad!

Setting up a good page with a layout where many objects are placed side by side or above each other can be a tedious job using Auto Layout constraints as you have done so far. In this chapter, you're going to use a new component to make specific layouts a lot easier to set up. This tool is called `UIStackView`, and it is incredible. You're going to end this content-packed chapter by implementing a feature that allows users to preview the content they're about to navigate to when they 3D Touch on one of their contacts.

To sum them all up at once, these are the topics you will learn about in this chapter:

- Universal applications
- Segues
- Advanced Auto Layout
- `UIStackView`
- Passing data between view controllers
- 3D Touch

Building universal applications

It is not uncommon for people to own more than one iOS device. A lot of people who own an iPhone are likely to have an iPad laying around somewhere as well. Users who own multiple devices expect their apps to work well on any screen they have nearby. Ever since Apple launched the iPad, it has encouraged developers to make their apps universal to make sure they run on both the iPhone and the iPad.

A lot has changed since then. Currently, there are more screen sizes to keep in mind than ever before and the tools that you have at your disposal to make beautiful layouts have gotten better and better over the years. You can use a single storyboard file to set up your layout for any screen size that is available to users. The best part is that you don't have to start from scratch setting up your layout for every screen size. You can use a single layout and tweak it as needed for certain screen sizes.

If you don't make your apps adaptive, your users will have a sub-par experience when they view your app on a different device than you may have expected. Imagine having designed your app for the iPhone SE form factor, but one of your users runs it on a 12.9" iPad Pro. You can probably imagine how terrible your layout would look.

Users tend to love apps that work on any device they own, and if you don't make your app compatible with iPad if you have developed it for an iPhone, then Apple will make it available on iPads anyway. Your app will run in a scaled-up fake iPhone resolution that looks awful. It typically only takes a small effort to make your app work well on an iPad, and the rewards are great. To make your app look great on all screen sizes, you use size classes.

A size class is a property that belongs to the `UITraitCollection` class. A trait collection describes the features that make up the environment in which a certain view controller exists. Some of these traits describe the available screen space. But they also tell you whether a screen supports a wide color gamut or whether force touch capabilities are present. For now, you'll focus on size classes. At the time of writing, two size classes exist: **regular** and **compact**. The following list shows several available device form factors and their corresponding size classes in all orientations:

- Plus-sized iPhones, iPhone X(s) and iPhone Xs max:
    - Portrait orientation: compact width x regular height
    - Landscape orientation: regular width x compact height
- All other iPhones:
    - Portrait orientation: compact width x regular height
    - Landscape orientation: compact width x compact height

- All iPad sizes:
    - Portrait orientation: regular width x regular height
    - Landscape orientation: regular width x regular height

The preceding list is not entirely exhaustive for all possible size class combinations. On the iPad, an app can appear in multitasking view, potentially giving it a compact width size class.

You should not use size classes as an indicator for the device the user is holding. An iPad could look very similar to an iPhone if you base your assumptions on the size class. Typically, this is good: you should optimize your app for capabilities, not a specific device. However, if you need to know what type of device the user is holding,, you can use `userInterfaceIdiom` on `UITraitCollection` to find out the name of the current device.

Whenever you create a new project in Xcode, the app is configured to be universal by default. This means that your app will work on all devices and when you publish your app to the App Store, it will be natively available on the iPad and iPhone. If you need to change this setting, you can do so in your project settings by changing the **Devices** option from **Universal** to something else, for instance, **iPhone**. The following screenshot shows this option in the project settings window:

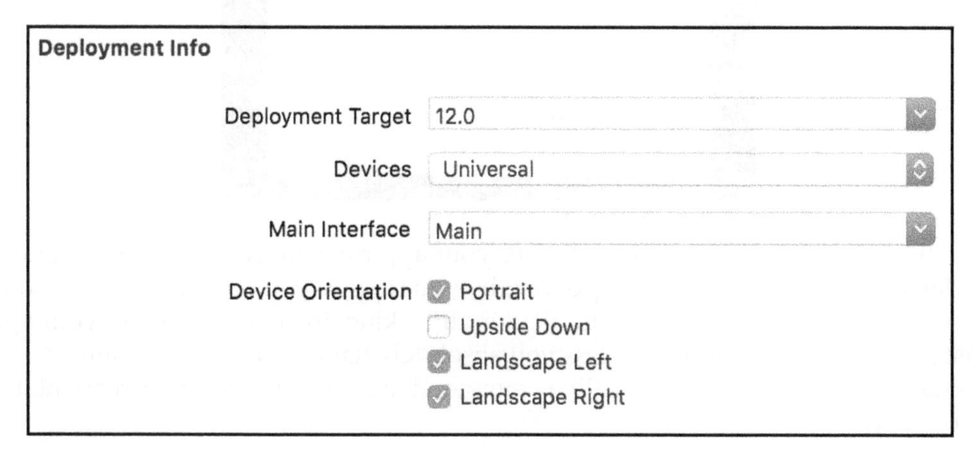

Typically, you will prefer to build your app universally, but sometimes you want to target a single device. Be aware that selecting **iPhone** does not make it impossible for your app to run on an iPad. Your users can download the iPhone version of your app on their iPad, and the iPad will then render the app in a scaled window. The following screenshot shows how this would look for the **Hello-Contacts** app:

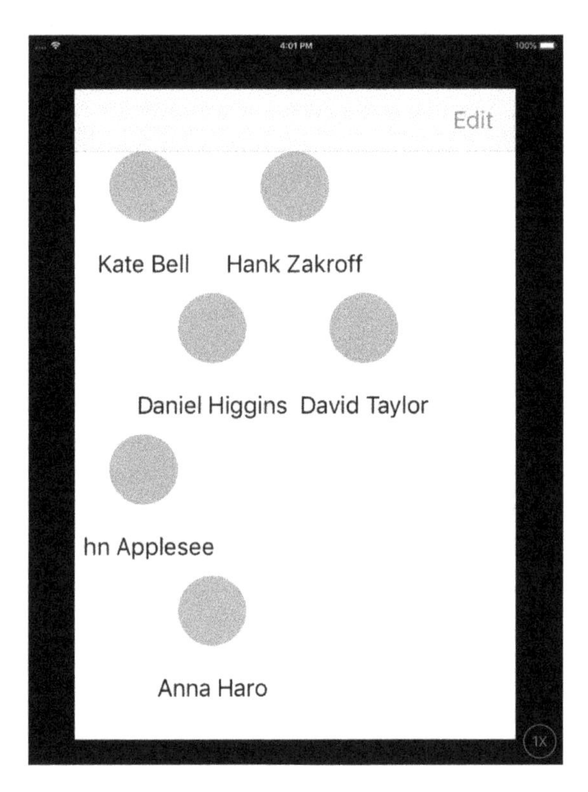

In addition to selecting the type of devices your app runs on, you can also configure the device orientations that are supported by your app. If your app is universal and you wish to make it compatible with iPad multitasking, the iPad version of your app must support all orientations. Unless you absolutely have to stick to a certain orientation on the iPad, it is typically recommended to support any screen orientation in your app.

On the iPhone, it is typical for apps to only work in portrait mode. This is okay, and you can support the landscape orientation whenever it fits your app. The only orientation that is a little bit strange to support on an iPhone is upside down. It is typically discouraged to support this orientation in an iPhone app.

A single app can support different screen orientations for iPhone and iPad. This allows you to make an app portrait only on iPhone and support any orientation on iPad. By default, Xcode sets up your project so that it supports all iPad orientations, and only the ones you check in your project settings are supported for the iPhone. Open your `Info.plist` file to see the general and iPad-specific orientation configurations under the **Supported interface orientations** keys. If you change the configuration in your project settings, your `Info.plist` will update as well.

Remember that you implemented some code to add support for showing your action sheet on the iPad? Go ahead and run **Hello-Contacts** on an iPad simulator and long-press on one of the contacts to see the delete popover appear. The popover should be presented as shown in the following screenshot:

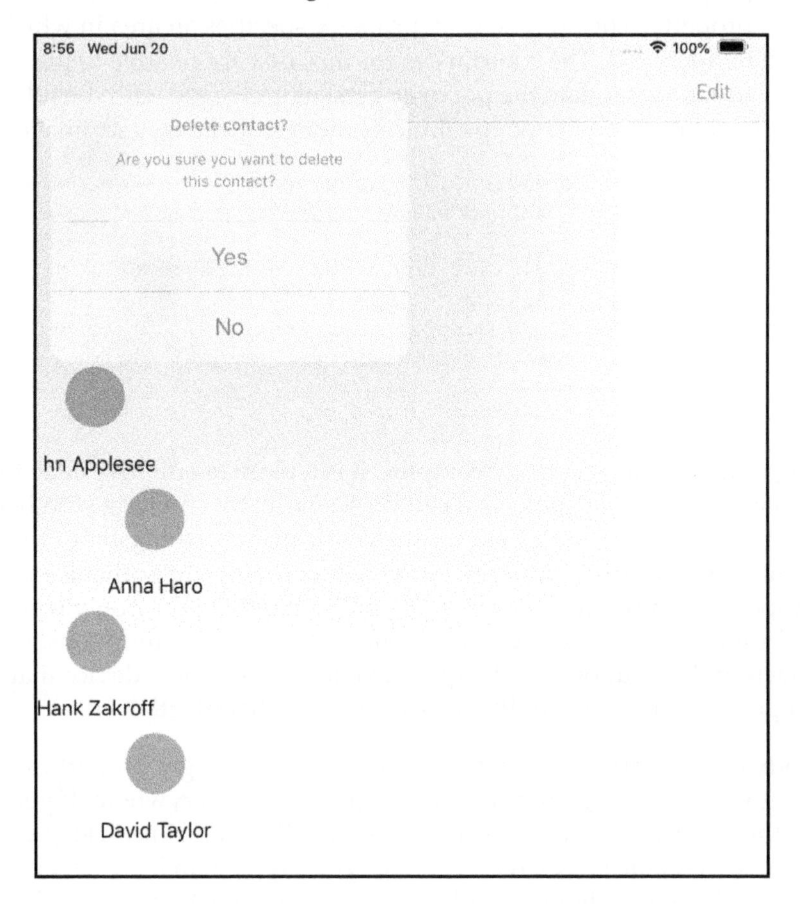

You'll notice that the popover isn't presented in the best position. Users will likely expect to see the popover appear alongside the contact image instead of seeing positioned entirely over the contact cell. The current implementation for showing the popover looks like this:

```
if let popOver = confirmationDialog.popoverPresentationController {
  popOver.sourceView = tappedCell
}
```

The preceding code sets a `sourceView` property on the popover. The popover uses the `sourceView` property as a container for the popover window. This means that the popover is added as a subview of the `sourceView` property. In addition to setting the `sourceView` property, you can also provide the popover with a `sourceRect` property. The `sourceRect` property specifies an area in which the popover will be anchored. The coordinates for this area are relative to the `sourceView` property. Update the popover presentation code with the following implementation to position the popover nicely alongside the contact image view:

```
if let popOver = confirmationDialog.popoverPresentationController {
  popOver.sourceView = tappedCell

  if let cell = tappedCell as? ContactCollectionViewCell {
    let imageCenter = cell.contactImage.center
    popOver.sourceRect = CGRect(x: imageCenter.x,  y: imageCenter.y,
                                width: 0,  height: 0)
  }
}
```

Even though this update isn't crazy complex, it is a piece of adaptive code that optimizes the way your app looks in a certain scenario. By checking whether a `popoverPresentationController` exists, you learn something about the environment that your app runs in. Since this property is only available on specific devices in specific conditions, you want to configure the popover only when it makes sense. The beauty of truly adaptive code is that it makes no assumptions. It simply reads the facts and acts upon that. If Apple would release a new device that also shows a popover, your code would already be able to handle it!

It has already been mentioned, but it's critical to check the right properties of your environment when building apps that should look great everywhere. If you are doing something that is specific to a device, especially an iPad, it is time to step back and reconsider. The available screen real estate for your app on an iPad can be extremely narrow and tall if the user holds their iPad in portrait and uses multitasking. The more flexible you make your code and the fewer assumptions you make, the better your apps will adapt to their environment.

If you have never seen an app running in multitasking mode on an iPad, now is the time to build and run your app on an iPad simulator. When the app is running, swipe up from the bottom of the screen to reveal the Dock and drag an app from the dock upwards, so it appears next to the **Hello-Contacts** app. You can resize each app by dragging the spacer in the middle of the screen and rotate the device. The app layout will always adapt to the available space because the entire app was set up with adaptivity in mind.

It's pretty cool that you have this single screen working for all sizes. How about adding a second screen and some navigation next?

Implementing navigation with segues

Most good applications have more than a single screen. I bet that most app ideas you have in your head involve at least a couple of different screens. Maybe you would like to display a table view or a collection view that links to a detail page. Or perhaps you want your user to drill down into your app's contents in a different way. Maybe you don't have detail views, but you would like to display a couple of modal screens for data input instead.

Every time your user moves from one screen in your app to another, they are navigating. Navigation is an essential aspect of building an app, and it's vital that you understand the possibilities and patterns for building good navigation on the iOS platform. The easiest way to gain insight into navigation is by using a storyboard to explore the available options.

Up until now, you have used your storyboard to create the layout for a single screen. However, the name *storyboard* implies that you can do a lot more than laying out a single screen. The purpose of using a storyboard is the ability to lay out all screens of your application in a single place so you can easily see how screens and sections of your app are related and how users navigate between them. In this section, you will add a second view controller to your storyboard that functions as a detail page when a user taps on a contact.

Open the `Main.storyboard` file and drag out a view controller from the **Object Library** by clicking the round icon with a square in the middle in the top bar of **Interface Builder**. Drop it next to the existing view controller. Next, look for a label in the **Object Library** and add it to the new view controller you just added to the storyboard.

Place the label in the center of the view and add some text to it so you can easily recognize it as your second view controller later. Then, add two constraints to center the label in its parent view. Do this by clicking the **Align** button in **Interface Builder**'s bottom-right corner and selecting the **Horizontal center** and **Vertical center** constraints.

Before you add all the content for the contact detail page to the second view controller, it's a good idea to configure the navigation from the overview page to the detail page. To do this, you're going to create a **selection segue**. A segue is a transition from one screen to the next. Not all segues are animated; sometimes you might need to present the next screen without performing a beautiful animation. Both animated and static transitions can be set up with segues.

Any time you connect one screen to the next to perform navigation, you are creating a segue. Some segues are performed when the user taps a button; these are called **action segues**. Segues that are only triggered through code are called **manual segues**. The selection segue you're going to use in this sample is a segue that is set up by connecting a table view cell or collection view cell to a next screen. The segue is performed when the user taps on a cell.

To set up your selection segue, select the prototype collection view cell you have created for the contacts overview page. Next, press and hold the *Ctrl* key while you drag from the cell to the second view controller. When you release the mouse over the second view controller, a list of options is shown. This list of possibilities describes how the detail view will be presented to the user. For instance, you can choose a modal presentation style. This will display the detail page with an upward animation from the bottom of the screen. Not quite the kind of animation you're looking for in this instance but interesting to explore either way.

A better way to show the contact is by adding it to the navigation stack. Doing this will make a back button appear in the navigation bar, and the user will immediately understand that they are looking at a detail page due to the animation that moves the new view controller in from the right-hand side of the screen. To set this up, you need to select the **show** segue. This segue pushes the newly presented view controller onto the existing navigation controller's navigation stack. Go ahead and run your app; you can navigate back and forth between the overview and the detail page.

In the previous chapter, you implemented a nice bounce animation when a user taps the cell. Unfortunately, the transition to the detail page takes place before the animation has a chance to execute. So even though the show segue works precisely as needed, it's better to use a manual segue in this case. This allows you to perform the animation and trigger the segue manually once the animation is finished.

Go back to the storyboard and click on the line that connects the collection view cell to the view controller. Press *Backspace* to delete the segue. Drag from the yellow circle at the top of the first view controller window to the second view controller. This is how you create a manual segue. When the dialog to determine how the segue is performed appears, select **show** again because you don't want to use a different animation. Click on the connecting line to inspect the segue and set the value for the **Identifier** field to `detailViewSegue` in the **Attributes Inspector**. Similar to how you set a reuse identifier on table view cells and collection view cells, segues also use a string as their identifier.

To trigger the segue after the animation, you must manually do so from your code. Open `ViewController.swift` and update the implementation of `collectionView(_:didSelectItemAt:)` as shown in the following snippet. Pay special attention to the last completion closure because that's where the segue is performed:

```
func collectionView(_ collectionView: UICollectionView,
didSelectItemAt indexPath: IndexPath) {
  guard let cell = collectionView.cellForItem(at: indexPath) as?
ContactCollectionViewCell
    else { return }

  UIView.animate(withDuration: 0.1, delay: 0, options:
[UIView.AnimationOptions.curveEaseOut], animations: {
    cell.contactImage.transform = CGAffineTransform(scaleX: 0.9, y:
0.9)
  }, completion: { _ in
    UIView.animate(withDuration: 0.1, delay: 0, options:
[UIView.AnimationOptions.curveEaseIn], animations: {
      cell.contactImage.transform = .identity
    }, completion: { [weak self] _ in
      self?.performSegue(withIdentifier: "detailViewSegue", sender:
self)
    })
  })
}
```

This change makes sure the animation is finished before moving to the detail page. Build and run your app on any device to see your manual segue in action. The next step is to learn a bit more about layouts that react to changes in their environment by setting up the contact detail page.

Creating adaptive layouts with Auto Layout

At the beginning of this chapter, you learned what adaptive layouts are and why they are essential. You learned a little bit about size classes and the `traitCollection` property. In this section, you'll deep dive into these topics by implementing an adaptive contact details page. You'll learn some best practices to implement a layout that is tailored to work well with different size classes. You will learn how to do this in code as well as storyboards because it's not always possible to define your entire layout in a storyboard. Refer to the following diagram to see what the layout is going to look like:

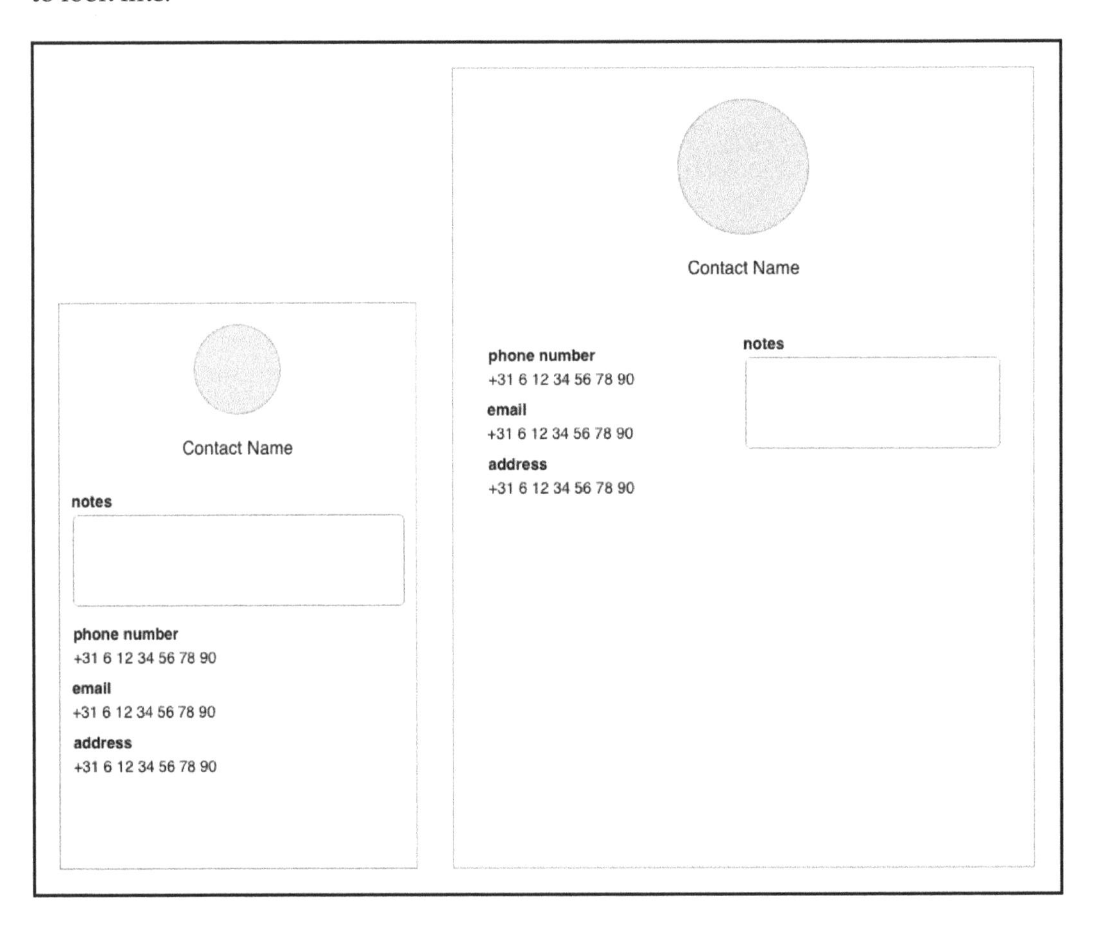

The detail page you're going to set up contains some information about the contact the user has tapped, and the user can add some notes for this contact. On small screens, the layout will be a simple, one-column layout. On larger screens, the layout will adapt to make good use of the screen by splitting into a two-column layout. First, you'll create this layout in **Interface Builder**. Once you're done, you will make some tweaks to the layout in code.

Auto Layout with Interface Builder

A good strategy for implementing a layout that is adaptive is to begin by setting up Auto Layout constraints that apply to all different versions of the layout. We'll call these **common constraints**. These are constraints that remain active for any size class. By implementing common constraints first, you typically end up with fewer and clearer constraints than you would if you take a different approach.

When you add constraints in **Interface Builder**, they apply to all size classes by default. You can add constraints for a specific size class by activating the **Vary for Traits** option by clicking on the corresponding button in the bottom-right corner of the device preview section. Before pressing this button, make sure to select a device that has the size classes you want to work with. For instance, if you want to define constraints for a regular width and regular height, you can select any iPad.

Before you start working on the contact detail page, try to get a feel for how the **Vary for Traits** feature works. Make sure to turn the feature off before you continue and remove the label you added earlier; you'll start from scratch to implement the new layout.

The contact detail page contains a text area where the user can add some notes. When the user taps this text area, the keyboard will appear, potentially covering the notes field itself on smaller screens. This isn't great because if the keyboard covers the text area, the user cannot see what they are typing. To fix that, you can wrap the entire detail page in a scroll view, allowing the user to scroll the page whenever the keyboard covers the notes field so they can make it visible again.

Implementing a scroll view while using Auto Layout has proven to be non-trivial. Once you understand how it works and what it takes to implement a scroll view with Auto Layout properly, it all makes sense. However, figuring it out for the first time is tough. So the first thing you'll learn is how to implement the scroll view.

You will follow the following steps to implement the full detail page:

1. Implement the scroll view that will contain the entire layout.
2. Add the big contact image and name label.
3. Implement variation for the contact image on large screens.
4. Add the bottom section of the detail page.
5. Adjust the bottom section of the detail page for small screens.

Implementing a scroll view with Auto Layout

A scroll view is a convenient tool you can use to prevent a keyboard from obscuring contents in your app. Using a scroll view, the user can quickly decide which parts of the interface they would like to see when the keyboard overlays a portion of the view. However, a scroll view is not always trivial to configure. It must know how large its contents are to determine whether scrolling needs to be enabled and to know how far down a user can scroll. Setting this up in Auto Layout requires a little bit of practice.

When you first add the scroll view to your view in **Interface Builder**, you must make sure that you add constraints that pin the scroll view to the edges of the view controller's view. This makes sure the scroll view takes up the entire available space on the screen. Next, you should add a view to the scroll view. This view will act as a container view for all the content and will be able to inform the scroll view about its content size.

The content view should have a width that is equal to the view controller's view. This will make sure the content width is never larger than the view controller so the scroll view won't scroll on the horizontal axis. Next, the content view should have all its edges pinned to the scroll view edges. Let's go over everything you have read so far again and implement it for the detail page right away. Before you begin, make sure you uncheck **Adjust Scroll View Insets** in the **Attributes Inspector** for the entire contact detail view controller.

Now drag a scroll view from the **Object Library** and add it to the view. Resize it so it covers the entire available view; don't make it cover the navigation bar. Add constraints to the scroll view using the **Add New Constraints** menu, as shown in the following screenshot:

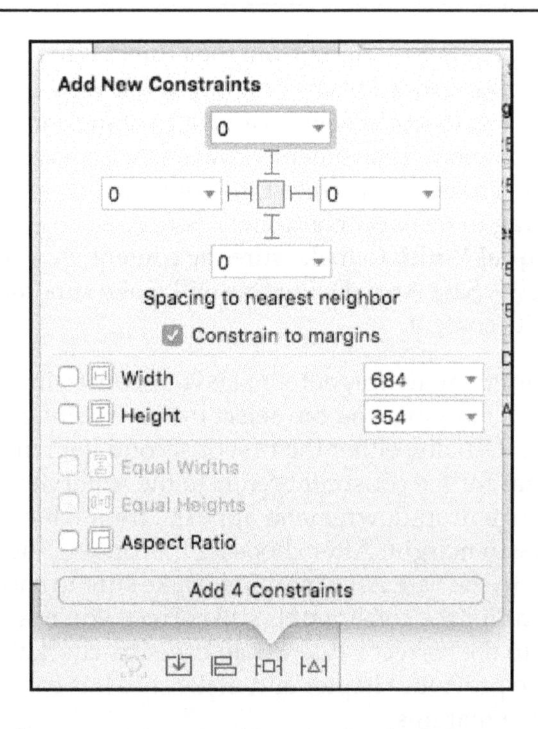

Note that the **Constrain to margins** checkbox is checked. By constraining your scroll view relative to the margins of its container, you make sure that your view will not exceed the Safe Area bounds. This is especially important for devices like the iPhone X and Xs, where part of the view is used for the home indicator and the camera notch. The following diagram visualizes the Safe Area bounds on a landscape-oriented iPhone Xs:

The next step is to drag a `UIView` object from the **Object Library** and add it to the scroll view. Manually size it so it covers the entire scroll view. Add the same constraints you added to the scroll view, so all edges of the content view are pinned to the edges of the scroll view. The content view must also have an equal width to the view controller's view. To set this up, select the content view and press *Ctrl*. While holding down *Ctrl*, drag to the view controller's Safe Area object. When you let go, you can now select **Equal Width** to make sure the content view always has an equal width to the main view's Safe Area. Doing this will make sure the scroll view knows the correct width for its content.

If you accidentally forget to make your constraints relative to the Safe Area margins, you can select the constraint in the **Document Outline**. Usually, either the first or second item in the **Size Inspector** for the constraint refers to the superview. If you click on this item, a drop-down menu appears, from which you can select **Relative to margin**. After choosing this option, the constraint will be relative to the Safe Area margins. Make sure to check whether the **Constant** for the constraint is still correct; you might have to update it back to the correct value. As an exercise, check the collection view from the previous chapter and make sure it is configured to use the Safe Area margins.

After following these steps, you should have now encountered an Auto Layout issue. This issue has appeared because the scroll view is unable to determine its height since the content view also cannot determine its height. You could say that the scroll view now has an ambiguous scroll height. Once you add some children to the container view, this problem will resolve itself. Because the container view will be able to calculate its height based on its child views, the scroll view can then use the container view's height to figure out the scroll height.

Laying out the image and name label

The next step is to add the contact's image and its name label to the content container view. Start by dragging a `UIImageView` and `UILabel` object from the **Object Library**. Position them so that they both are horizontally centered in the view. Also, position them so they have about 10 points of spacing on the vertical axis. Don't forget to apply a nice light gray background color to the image. You've already seen the following two ways to add constraints:

- Have Xcode add constraints for you.
- Add constraints through the menu in the bottom-right corner.

You'll add a third technique to that list: *Ctrl* + dragging. You already used *Ctrl* + dragging while setting up the scroll view, but it also works when dragging from one view to another in your design preview.

If you select the `UIImageView` object and drag upward while pressing *Ctrl*, you can add constraints that relate to the view you started dragging from, the view you let go in, and the direction you were dragging in. Try this now; a menu such as the one in the following screenshot should appear. Press and hold *Shift* and select **Vertical Spacing to Top Layout Guide** and **Center Horizontally in Container**. An **Add Constraints** option should have appeared at the bottom of the list after you selected the constraints you want to add. Click on that to add the selected constraints. Moving forward, this will be the preferred way to add constraints:

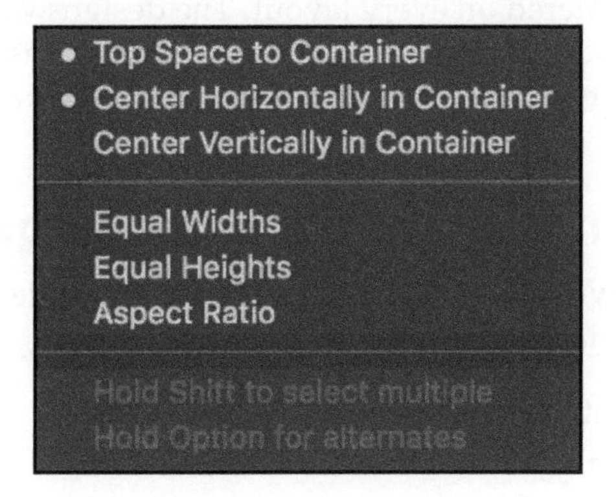

To set an appropriate width and height for the `UIImageView` object, press *Ctrl* and drag on the image itself. If you drag and release in a horizontal direction, you can add a width constraint, and dragging vertically allows you to add a height constraint. You can't set the value though; the value will be set to whatever the current size is.

If you want to change the value of a constraint, go to the **Size Inspector** on the right-hand side of the screen and double-click on a constraint. This allows you to change the **Constant** for a constraint. When you set a size constraint, the constant will refer to that size. When setting spacing, the constant will refer to the current amount of spacing. Click on the image and correct the width and height constraints so the image will be 60 x 60 points. Update the preview by clicking on the **Update Frames** menu option at the bottom of the **Interface Builder** window located in the bottom-right area of the window.

With the image view set up, it is time to add some constraints to layout out the name label. Position the name label approximately 10 points beneath the image and make sure that it's centered. Now, press *Ctrl* and drag from the label to the image view. The dialog you saw before should appear; press *Shift* and add **Vertical Spacing** and **Center Horizontally** constraints. This will set up constraints that horizontally center the label relative to the image. It also adds a constraint that vertically spaces the label from the image. Finally, drag down from the label to set its vertical spacing to the container view. Modify this constraint, so the label has a spacing of 8 points to the container's bottom. This will make sure that the content view has a height, and this resolves the Auto Layout issue you saw earlier.

Change the preview to a couple of different devices, and the image and label should always appear nicely centered on every layout. The designs we saw earlier had a larger image on the iPad. This means that separate sizing constraints should be used for the iPad's larger screen. This is what the next step is about.

Adjusting the image and label for large screens

When you click the **Vary for Traits** button in **Interface Builder**, you have three options to base your variation on:

- Vary on height.
- Vary on width.
- Vary on width and height.

If you select width, any new constraints you add will apply to all trait collections that match the size class axis you have chosen to vary on. So if you select the iPad view and then select to vary for width only, all Xcode will present several options that your variation would apply to. The following screenshot shows an example of this:

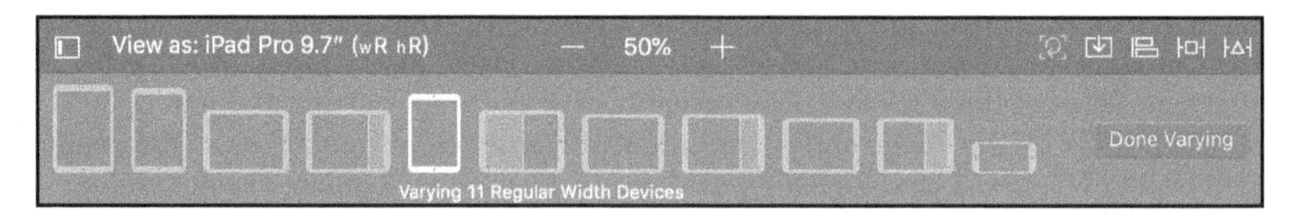

To make the larger image look good, you should vary both the width and height. If the larger image appears on an iPhone 6 Plus in landscape mode, it would cover too much of the screen, so the bigger image should only be shown on devices that can accommodate it on both the horizontal and vertical axis. Click on **Vary for Traits** again, and this time check the width and height boxes. You'll notice that the iPhone isn't in the list of devices anymore because a large screen iPhone has a regular width and a compact height. On the left-hand side of the **Interface Builder** window, you can find the document outline. In this outline, look for your image view. You can see the constraints that are added to the image view listed underneath the image view. Expand the constraints and select the width constraint. At the bottom of **Attributes Inspector** on the right-hand side of the screen, you'll find a plus icon, and a checkbox labeled **Installed**. Click on the plus symbol and navigate to the **Regular width | Regular height | Any gamut** (current) option as shown in the following screenshot:

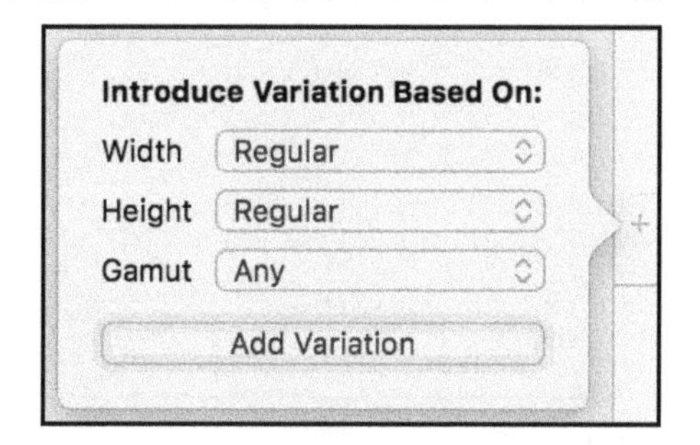

After doing this, an extra checkbox will appear. Uncheck it so the width constraint is not installed for the current variation. Do the same for the image's height constraint; after doing this, the image might not be visible anymore because Auto Layout doesn't know its dimensions. With the image view selected, go to the **Size Inspector** and set the image's size to 120 x 120. This will make the image reappear. Now, use the *Ctrl* + drag technique to add width and height constraints, as you have done before. If you inspect these constraints, you'll see that they have a similar setup for their installation checkboxes as before, except that the general one is unchecked and the specific one is checked. Great! This means you've successfully set up the constraint variations for your app.

It is possible that the created configuration installs the constraint only based on the width size class and ignores the height size class. If this happens, click the plus icon to manually set up the correct configuration and delete the wrong one.

Go ahead and exit the variation mode by clicking on the **Done Varying** button and check out your layout on several screens. If the image size doesn't update automatically, select the entire view and update the frames from the **Resolve Auto Layout Issues** menu. This wraps up the layout of the top section of the detail page.

Laying out the bottom section

The bottom section for this page looks simple, but you need to think very carefully about the way it is set up. On small screens, all of the elements should be displayed in a single column, while larger screens have two columns. The simplest way to do this is to wrap both of these sections in a container view as shown in the following screenshot:

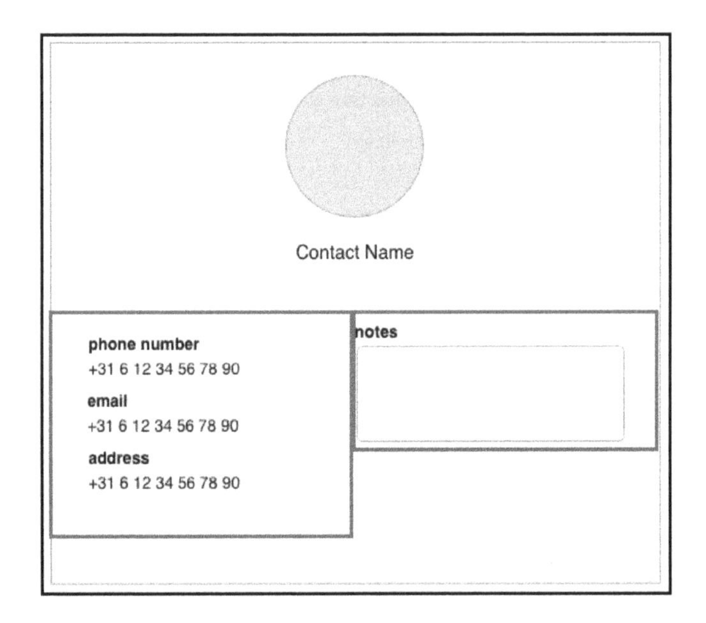

The biggest challenge you face is to make sure the scroll view's content view knows how to calculate the height of its contents. Previously, you added a constraint that pinned the label to the bottom of the content view. You should remove this constraint because it isn't needed anymore. Instead, the left column of the layout will be pinned to the bottom of the content view since it's the tallest view of the two. Remove the label's bottom constraints by selecting the label in the **Document Outline** and navigating to the **Size Inspector**. Click the **Bottom Space to:** constraint and press *Backspace* to remove this constraint. Now you can manually make the content view a bit taller, so you have some space to work with.

Drag two regular views from the **Object Library** into the content view and size them so they are roughly the same size and so they meet in the middle. The views shouldn't overlap, but they also shouldn't have any space between them. Drag the left edge of the view on the left until you see the blue helper lines. Do the same for the right view but drag the right side of the view to the right side of the window. Position both views so they are positioned roughly 40 points below the label.

Now, select the view on the left and *Ctrl* + drag to the name label and add a vertical spacing constraint. Then, *Ctrl* + drag to the left side to pin this view's left side to the content view's leading edge. Now, select the right view and do the same, except instead of dragging left, drag right to pin the right side of the view to the content view's trailing edge. Finally, *Ctrl* + drag from the left view to the right view and select the equal width constraint and vertical spacing constraint. This should leave you with two equally sized views that don't have a height yet. Lastly, *Ctrl* + drag from the left view downward into the container view to pin its bottom 10 points from the bottom of the container view.

Drag six labels into the left view. Align them as they are in the design. Use the blue lines as guides for the margins, and make the labels wide enough to cover the view. Stop at the blue lines again; they help you by providing some nice margins. Press *Ctrl* and drag from the first label upward to pin it to the top side of the view. Press *Ctrl* and drag left into the left view to pin the label to the left side of the view, and press *Ctrl* and drag right to pin it to the right side. Then, select the second label and *Ctrl* + drag upward to the first label. While holding the *Shift* key, select the leading, trailing, and vertical spacing constraints. Repeat this for all other labels. The final label should be pinned to the bottom of the left view.

Now, drag a label and a text field to the right side's view. Align the elements by using the blue lines again. Pin the label to the top, right, and left sides of its containing view. Drag the text field to the label and select the leading, trailing, and vertical spacing constraints. Finally, drag upward inside the text field to add a height constraint. Modify this constraint so the text field has a height of 80. Finally, press *Ctrl* + drag downward to pin the bottom of the text field to the bottom of the view.

The step to create the layout is to press *Ctrl* and drag the left view down into the content view. Select the vertical spacing constraint and modify the constant so the spacing is eight points. Now use the **Attributes Inspector** to make all header labels use a bold font and provide the appropriate text values for them. Refer to the design to see what the correct label values should be.

Well, that was a lot of instructions, but you should have successfully completed the layout for regular x regular devices. Try to run your app on an iPad; looks pretty good, right?

Adjusting the bottom section for small screens

The layout you have so far looks pretty good on the iPad. However, it's just too cramped on the iPhone in portrait mode. Let's create some room by changing it to a single-column layout. First, select iPhone view and click on the **Vary for Traits** button. This time, make sure that you only vary for the width. The easiest way to pull off a change this big is to rearrange the view first without modifying the constraints yet. Rearrange the views so they look as shown in the following screenshot:

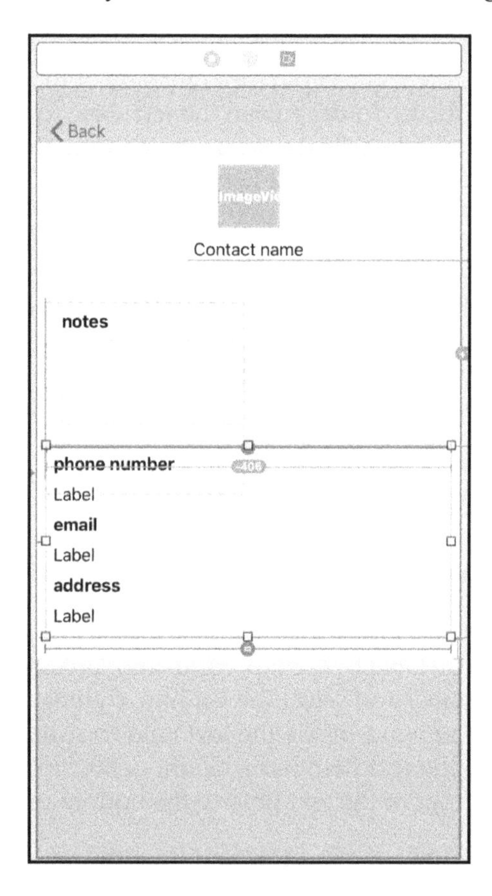

Once you have rearranged the views, it's a good idea to add constraints that you know are missing. This isn't always straightforward but with some practice, you should get the hang of this pretty soon. Always make sure that each view can figure out its *x* and *y* position and its width and height.

In this case, the following constraints are missing:

- Top view with notes to the left side of the content view
- Vertical spacing between the top and bottom view
- Bottom view to the right side of the content view

Add the constraints after you have started your **Vary for Traits** by pressing *Ctrl* and dragging as you have done before. You'll end up with a lot of errors and red lines. This is because you haven't disabled the constraints that were created for the two-column layout. To fix the errors, the following constraints that only apply to the two-column layout must be removed for the compact width variation:

- Vertical spacing between the view with the contact information and the contact name label
- Horizontal spacing between the former left and right view
- Equal width constraint for the two bottom views

To remove the mentioned constraints, select the top view and look for the **Leading Space to: View** constraint. Click this constraint and delete it to disable it for the **Compact Width** version of your layout. Make sure you are still in **Vary for Traits** mode when you do this. Otherwise, the constraint will be deleted for all size variations. Do the same for the other two constraints. You can find these constraints by selecting either of the views related to the constraint you're looking for. Finally, select the vertical spacing constraint for the top and bottom view and modify the constant to `8`; this will provide nice spacing between the two sections. You're done modifying the constraints now; no conflicts should remain, and your layout should now look good on all devices. Check out the plus-sized iPhone in particular; it will switch between the single- and two-column layouts! Pretty cool, isn't it? If you run the app in landscape on other devices, you'll notice that even though the layout is taller than the screen, the scroll view makes sure all content is accessible.

Before you deep-dive into Auto Layout on the code side, there's one last thing that needs to be fixed. Whenever the keyboard appears, the scroll view should resize so a user can still see the entire page. To do this, you need to create an `@IBOutlet` object for the bottom constraint of the scroll view, and then you will need to programmatically set the constant for the bottom constraint to the height of the keyboard. To get started, create a new `UIViewController` subclass (**File** | **New** | **File...**) and name it `ContactDetailViewController`. Assign this class to the contact details screen's view controller in **Interface Builder**.

Define the following `@IBOutlet` object in `ContactDetailViewController.swift` and connect it by dragging from the **Connections Inspector** to the scroll view's bottom constraint in the **Document Outline**:

```
@IBOutlet var scrollViewBottomConstraint: NSLayoutConstraint!
```

If an item is hard to find or access in the visual view hierarchy, you can look for the view or constraint in the **Document Outline**. Once you have found the thing you wish to connect to an `@IBOutlet` object, you can *Ctrl* + drag from the outlet in the **Connections Inspector** to the item in the **Document Outline** to configure the connection.

Next, add the following code to `ContactDetailViewController.swift`:

```
override func viewDidLoad() {
  super.viewDidLoad()

  NotificationCenter.default.addObserver(self, selector:
#selector(keyboardWillAppear),
                                         name:
UIApplication.keyboardWillShowNotification,
                                         object: nil)

  NotificationCenter.default.addObserver(self, selector:
#selector(keyboardWillHide),
                                         name:
UIApplication.keyboardWillHideNotification,
                                         object: nil)

}
```

The preceding code subscribes the details page to keyboard notifications. Whenever the keyboard appears, the system posts a notification to an object called `NotificationCenter` that any object can listen to. Whenever the notifications are fired, a selector is executed on the view controller. This selector points to a function in the view controller. The code as it won't compile just yet because you still need to implement the `keyboardWillAppear` and `keyboardWillHide` methods. Add the following implementation for these methods:

```
@objc func keyboardWillAppear(_ notification: Notification) {
  guard let userInfo = notification.userInfo,
    // 1
    let keyboardFrame =
userInfo[UIResponder.keyboardFrameEndUserInfoKey] as? NSValue,
    let animationDuration =
```

```
userInfo[UIResponder.keyboardAnimationDurationUserInfoKey] as? Double
    else { return }

  // 2
  scrollViewBottomConstraint.constant =
keyboardFrame.cgRectValue.size.height
  UIView.animate(withDuration: TimeInterval(animationDuration),
animations: { [weak self ] in
    // 3
    self?.view.layoutIfNeeded()
  })
}

@objc func keyboardWillHide(_ notification: Notification) {
  guard let userInfo = notification.userInfo,
    let animationDuration =
userInfo[UIResponder.keyboardAnimationDurationUserInfoKey] as? Double
    else { return }

  scrollViewBottomConstraint.constant = 0
  UIView.animate(withDuration: TimeInterval(animationDuration),
animations: { [weak self ] in
    self?.view.layoutIfNeeded()
  })
}
```

The preceding code snippet first reads some values from the `userInfo` dictionary on the `notification` object. The keyboard's final frame and the animation duration are extracted from this dictionary. Next, the scroll view's bottom constraint is updated so it is pushed upward by the keyboard's height. To animate this change, `layoutIfNeeded()` is called on the view controller's view inside of an animation block. Calling this method inside of an animation block ensures that the Auto Layout updates are performed with an animation.

When the keyboard hides, a similar flow is used except the scroll view's bottom constant is reset back to `0`.

Using Auto Layout in code

Auto Layout is a technique that can be used in **Interface Builder** as well as in code. You have created the contacts page entirely in **Interface Builder**, and that's very convenient. However, there are times when you might not be able to create the layout before the app runs. Let's call the moment of designing in **Interface Builder** the design time. At design time, it's possible that not all variables in your design are known. This means that you will need to dynamically add, remove, or update constraints when the app is running. When you do something while the app is running, you do it at runtime.

Setting up a layout visually with Auto Layout is very well supported, and **Interface Builder** even has some tools that make setting constraints programmatically fairly straightforward. If you have a very dynamic layout, you don't always know every constraint in advance. Especially if you are dealing with unpredictable data from external sources, a layout might need to change dynamically based on the available contents.

To help you with this, **Interface Builder** allows you to mark constraints as placeholder constraints in the **Attributes Inspector**. Marking a constraint as a placeholder means that it was used to create a valid layout for **Interface Builder** at design time, but you'll replace that constraint at runtime.

When you use Auto Layout in code, you can dynamically update constraints by changing the constant for a constraint at runtime. It's also possible to activate or deactivate certain constraints and to add or remove them. It doesn't matter if you created the constraints that you want to modify in code or **Interface Builder**. Any constraints that affect a view are accessible through code.

The best way to explore this is to get started with some coding. For the sake of this exercise, you will recreate the top part of the contact details page in code. This means you're going to set the constraints affecting the contact image and the contact name label as placeholder constraints in **Interface Builder**. The constraints for these views will then be added to the view controller, and you'll toggle some constraints' active states based on the current size class. The final code will also watch for changes in the size class and update the constraints accordingly if needed.

To get started, open `ContactDetailViewController.swift` and add the following two outlets to the class:

```
@IBOutlet var contactImage: UIImageView!
@IBOutlet var contactNameLabel: UILabel!
```

Now, open the `Main.storyboard` file, select your view controller, and connect the outlets in the **Connections Inspector** to the correct views. After you have connected the outlets, you should change the existing constraints that position the image and label so they become placeholder constraints instead of regular constraints. The constraints that need to be changed are the following:

- Width constraints for the image (both regular and compact)
- Height constraints for the image (both regular and compact)
- Top spacing constraint for the image
- Horizontal center constraint for the image
- Spacing constraint between the label and the image
- Horizontal center constraint for the label

The vertical spacing constraint between the label and the bottom views should remain intact. To change a constraint to a placeholder, you must select it and check the placeholder checkbox in the **Attributes Inspector**. If you build and run your app after doing this, you'll end up with a mostly white view. That's okay; you have just removed some essential constraints that need to be re-added in code.

To get everything up and running again, you'll write some code to implement the eight constraints that were removed. One of the tools you can use to do so is **Visual Format Language (VFL)**. This language is an expressive, declarative way to describe Auto Layout constraints. You can describe multiple constraints that affect multiple views at once, which makes this language very powerful.

A visual format string contains the following information:

- **Information regarding the axis on which the constraint should work:** This is either horizontal (H), vertical (V), or not specified. The default is horizontal.
- **Leading space to superview:** This is optional.
- **Affected view:** This is required.
- **Connection to another view:** This is optional.
- **Trailing space to superview:** This is optional.

An example of a visual format string looks like this:

```
V:|-[contactImageView(60)]-[contactNameLabel]
```

If you take this string apart piece by piece, it contains the following information:

- `V:`: This specifies that this format string applies to the vertical axis.
- `|`: This represents the superview.
- `-`: This applies a standard spacing of about 8 points. This is equivalent to spacing a view in **Interface Builder** using the blue guidelines.
- `[contactImageView(60)]`: This places the `contactImageView` and gives it a height of 60 points. The placement will be about 8 points from the top of the superview.
- `-`: This applies another standard spacing.
- `[contactNameLabel]`: This places `contactNameLabel` with about 8 points of spacing from `contactImageView`.

This way of describing layouts takes some getting used to, but it's a really powerful way to describe layouts. Once you get the hang of all definitions and possibilities, you'll find that a visual format string is a very descriptive representation of the layout you're trying to create.

Time to dive right in and take a look at how to implement the entire layout for the top section of the contact details page. You'll only create the constraints for the compact size class for now and you'll add the regular constraints later.

Implementing the compact size layout

To implement the compact size layout, you'll combine VFL and anchors. Anchors are fairly straightforward; the best way for you to understand how they work is by using them.

First, add the following two variables that will be used later to activate and deactivate the compact size constraints:

```
var compactWidthConstraint: NSLayoutConstraint!
var compactHeightConstraint: NSLayoutConstraint!
```

These variables will be set in `viewDidLoad`, and we're using implicitly unwrapped optionals for them. This means that you must set these variables before attempting to use them. If you fail to do this, the app will crash due to an unexpected `nil` value.

You often want to avoid implicitly unwrapping optionals. Using optionals without implicit unwrapping enforces safety because you need to unwrap these values before attempting to use them. However, in some cases, you want your program to crash if a value isn't set; for instance, when there's no sensible way to recover from such a missing value. Scenarios like these are very rare, and you should use implicit unwrapping with great caution. In this example, it's used for brevity.

The following code should be added to the `viewDidLoad` method:

```
// 1
let views: [String: Any] = ["contactImage": contactImage,
"contactNameLabel": contactNameLabel]

var allConstraints = [NSLayoutConstraint]()

// 2
compactWidthConstraint =
contactImage.widthAnchor.constraint(equalToConstant: 60)
compactHeightConstraint =
contactImage.heightAnchor.constraint(equalToConstant: 60)

// 3
let verticalPositioningConstraints = NSLayoutConstraint.constraints(
  withVisualFormat: "V:|-[contactImage]-[contactNameLabel]",
  options: [.alignAllCenterX], metrics: nil, views: views)

allConstraints += verticalPositioningConstraints

// 4
let centerXConstraint = contactImage.centerXAnchor.constraint(equalTo:
self.view.centerXAnchor)

// 5
allConstraints.append(centerXConstraint)
allConstraints.append(compactWidthConstraint)
allConstraints.append(compactHeightConstraint)

//6
NSLayoutConstraint.activate(allConstraints)
```

There is a lot going on in the preceding snippet so take some time to study it closely. First, a dictionary of views is created. This dictionary is used by VFL later in step 3 to figure out which views correspond to the views mentioned in the format string. In step 2, the width and height for the image view are being set up. These constraints are configured with anchors. Every attribute that you can use for constraints can be represented with an anchor, in this case, width and height anchor, but there are also left, right, top, and bottom anchors that you can use to specify a layout with. Step 3 defines a visual format string to define the spacing between the `contactImage` object and the `contactNameLabel` object. By passing `.alignAllCenterX` to the options list for this format string, all views that participate in this string are aligned horizontally. Step 4 defines a constraint that places the image view in the middle of the view controller's view. Step 5 adds all anchor constraints to the `allConstraints` array. The final step is to call `NSLayoutConstraint.activate(_:)` to activate all constraints in the `allConstraints` array.

Try running your app on a device with a small screen; your layout should look just like it did before. The next step is to adapt this layout to larger screens by implementing the regular size layout.

Implementing the regular size layout

In order to implement the layout for regular-sized devices, the `traitCollection` property of the details view controller is used. As mentioned earlier, the `traitCollection` property contains information about the current environment your app is running in. All `UIView` instances, `UIViewControllers`, `UIWindows`, `UIPresentationControllers`, and `UIScreens` conform to the `UITraitEnvironment` protocol.

This protocol dictates that all objects that conform to `UITraitEnvironment` must have a `traitCollection` attribute. They must also have a `traitCollectionDidChange(_:)` method. This method is called whenever the `traitCollection` attribute changes. This could happen if a user rotates their device or when a multitasking window on the iPad changes its size. You'll use `traitCollection` and `traitCollectionDidChange(_:)` to correctly adapt the layout.

First, you must update `viewDidLoad` so it applies the correct layout for the current `traitCollection` attribute as soon as possible. Then, the code should watch for changes in the `traitCollection` attribute and update the constraints accordingly. Start by adding the following two variables to `ContactDetailController.swift`:

```
var regularWidthConstraint: NSLayoutConstraint!
var regularHeightConstraint: NSLayoutConstraint!
```

These variables will hold the larger width and height constraints for `contactImage`. Now, update `viewDidLoad` as follows; only areas where you should update the code are included in this snippet:

```
// unchanged implementation

compactWidthConstraint =
contactImage.widthAnchor.constraint(equalToConstant: 60)
compactHeightConstraint =
contactImage.heightAnchor.constraint(equalToConstant: 60)

// 1
regularWidthConstraint =
contactImage.widthAnchor.constraint(equalToConstant: 120)
regularHeightConstraint =
contactImage.heightAnchor.constraint(equalToConstant: 120)

// unchanged implementation

allConstraints.append(centerXConstraint)

// 2
if traitCollection.horizontalSizeClass == .compact &&
traitCollection.verticalSizeClass == .regular {
  allConstraints.append(regularWidthConstraint)
  allConstraints.append(regularHeightConstraint)
} else {
  allConstraints.append(compactWidthConstraint)
  allConstraints.append(compactHeightConstraint)
}

NSLayoutConstraint.activate(allConstraints)
```

The first modification is to create two new anchor-based constraints. These constraints represent the image's size for larger screens. The second step is to check the current traits and make sure that both the horizontal and the vertical size classes are regular. Size classes in code work the same as they do in **Interface Builder**. When you implemented this layout in **Interface Builder**, you only wanted the image view to be bigger on devices that were of regular width and regular height, so this still applies. By selectively appending these constraints to all the constraints array, you can immediately apply the correct layout.

When a user is using an iPad, an app can suddenly change from a regular x regular environment to a compact x regular environment when multitasking is used. To adapt the layout accordingly, you need to implement the `traitCollectionDidChange(_:)` method. By implementing this method, you can check the new and old traits and decide whether to activate or deactivate certain constraints. Add the following method to `ContactDetailViewController` to implement this behavior:

```
override func traitCollectionDidChange(_ previousTraitCollection:
UITraitCollection?) {
  super.traitCollectionDidChange(previousTraitCollection)

  // 1
  guard let previousTraitCollection = previousTraitCollection,
     (previousTraitCollection.horizontalSizeClass !=
traitCollection.horizontalSizeClass ||
      previousTraitCollection.verticalSizeClass !=
traitCollection.verticalSizeClass)
    else { return}

  // 2
  if traitCollection.horizontalSizeClass == .regular &&
traitCollection.verticalSizeClass == .regular {
    NSLayoutConstraint.deactivate([compactHeightConstraint,
compactWidthConstraint])
    NSLayoutConstraint.activate([regularHeightConstraint,
regularWidthConstraint])
  } else {
    NSLayoutConstraint.deactivate([regularHeightConstraint,
regularWidthConstraint])
    NSLayoutConstraint.activate([compactHeightConstraint,
compactWidthConstraint])
  }
}
```

The first step in this code is to check whether an old trait collection exists since the method receives an optional `UITraitCollection` object. This means that the argument must first be unwrapped using either a `guard let` statement of an `if let` statement. In this case, `guard let` is used. In the same `guard`, the new size classes and the old ones are compared to make sure something related to the size classes has changed.

The second step is to check the new traits to see whether the app is running in a regular x regular environment. When this is the case, the compact constraints are deactivated and the regular constraints are activated. If the app is not running in a regular x regular environment, the regular constraints are deactivated and the compact constraints are activated.

Implementing an adaptive layout in code requires more work than in **Interface Builder**. However, the tools make it fairly straightforward. Most of the time, you'll find that **Interface Builder** works perfectly fine, but when you find that you need more control, you can drop down to the code level and take it from there. The layout you have implemented right now uses quite some constraints. Luckily, you can refactor the detail page to have fewer constraints by using a `UIStackView` object. The next section will show you how to do this.

Improving layouts with UIStackView

You have learned a lot by manually configuring the entire layout for your contact detail page because you had to make some decisions about how you have set up certain constraints. Luckily, a lot of this work can be done by a powerful component called `UIStackView`. Stack views are able to lay out views that are added next to each other or on top off each other all by itself. This saves you adding constraints for the vertical spacing between labels like you have done for the contact detail information.

Since stack views can also layout objects that are next to each other and stack view can be nested, it could also take care of the two-column layout that you implemented for screens with a regular width size class. And to top it all off, you can swap the direction in which a stack view lays out its items at runtime, meaning you can change it from horizontal to vertical depending on the available space.

This will simplify a lot of the work you have already done and it makes your layout a lot more maintainable as well. To use a stack view, all you need to do is add one to your storyboard and begin adding items to it.

Containing labels in a stack view

Open your `Main.storyboard` file and select a device that has a compact width, such as an iPhone Xs, for instance. Select the six labels in the contact info view and embed them in a stack view by using the **Embed In** menu as shown in the following screenshot.

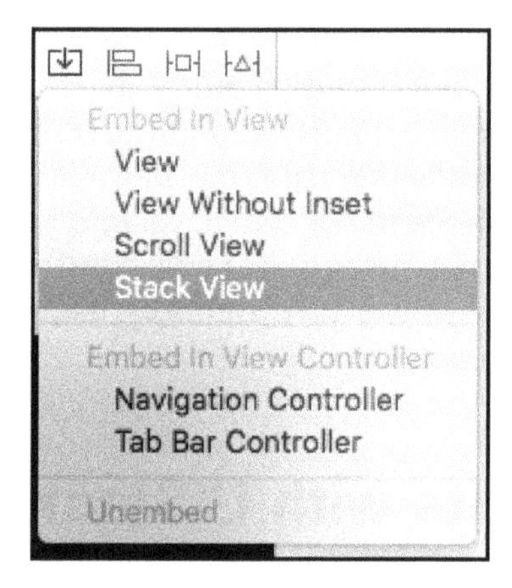

Now find the stack view in the **Document Outline** and drag it up so that it ends up above the view that currently contains it. Once you have done this, the stack view is no longer a child view of that view so you can remove it. Move the stack view to roughly to the position that the old view was in.

The next step is to select the notes label and the text field and embed them in a stack view too using the **Embed In** menu. Drag this stack view out from its parent view as well so it ends up on the same level in the **Document Editor** as the other stack view you just created. Remove the remaining, empty view as well and position the new stack view roughly where the old view was positioned.

Select both stack views and use the **Embed In** menu once more to embed the stack views in another stack view so both stack views are placed on top of each other. You are now ready to add some constraints that position the newly created stack view. You will need to add the following constraints:

- The stack view's top should be positioned 40 points below the name label's bottom.
- The stack view's left edge should be positioned 8 points from the superview's left edge.
- The stack view's right edge should be positioned 8 points from the superview's right edge.
- The stack view's bottom should be positioned 8 points from the superview's bottom.

Add these constraints using the *Ctrl* + drag technique. You can either drag within the views itself or you can use the **Document Outline** for easy access by dragging from the stack view to its superview.

After you have added the constraints, your layout should resemble the following screenshot:

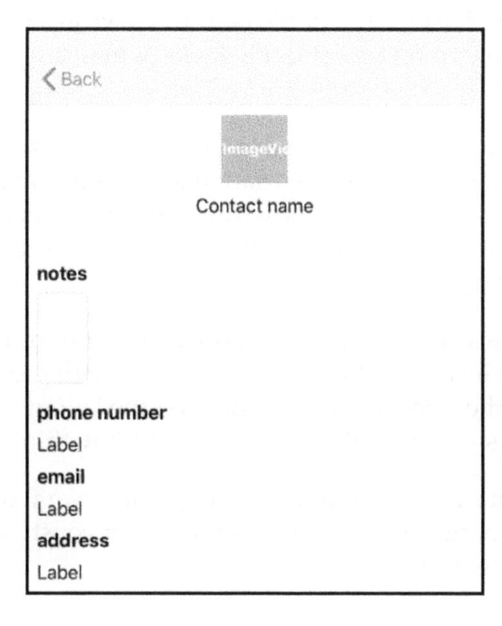

This is quite close to the layout you implemented before but it's not quite the same yet. To achieve the same effect, you're going to have to tweak some settings on the stack view. Select the stack view and look at the **Attributes Inspector**. You'll find that there are a couple of settings that you can configure. The settings you should look at for now are **Spacing**, **Distribution**, and **Alignment**. First, update the **Spacing** and set it to 16. This value controls the amount of space between each of the stack view's children. In this case, the only two children are the stack views that contain the contact's details and the notes area.

After adjusting the spacing, update the **Distribution** property so it is set to **Equal Spacing**; this will make sure that the stack view does everything it can to make sure the spacing between its children is always the same. In this case, you won't find much difference but you'll find that this property can be crucial to set in some other cases.

Lastly, set the **Alignment** to be **Fill**. This setting makes sure that items are positioned a certain way within the stack view. In this case, **Leading** makes the items stick to the left side of the stack view. Setting this value to **Center** would align them in the middle and **Fill** ensures that the stack's children are all stretched out to fill the entire width.

Once you have adjusted the stack view's configuration, you'll see that the layout looks exactly as planned for the compact size class. The regular size class still needs some work though because it doesn't lay out the bottom views in a two-column layout anymore with this setup. To fix this, select a device with a large screen and begin a **Vary for Traits** session.

Select the outer stack view and go to the **Attributes Inspector**. Using the plus icon next to the **Axis** setting, you can add a variation for the current size class configuration (regular width). Instead of setting the **Axis** to **Vertical**, you should set it to **Horizontal** and the stack view will update the view on large screens so the column is a two-column layout again.

To make sure both views are the same size, add a variation for **Distribution** as well and select **Fill Equally** instead of **Equal Spacing**. This setting ensures that all children of the stack view take the same amount of space on the layout axis. In this case, this means both child views will have the exact same width within the stack view.

You can now click **Done varying** and run your app on several devices. You'll find that the layout adapts perfectly with significantly fewer constraints in the bottom section of the app. Quite neat, right?

Passing data between view controllers

The final bridge to cross for the **HelloContacts** app is to display some actual information about a selected contact. In order to do this, a couple of new outlets should be added to `ContactDetailViewController`. The code that fetches contact information also needs to be expanded a little bit so a contact's phone number, email address, and postal address are fetched.

Finally, the contact data needs to be passed from the overview to the details page so the details page is able to actually display the data. The steps involved in this process are the following:

1. Update the data loading and model.
2. Pass the model to the details page.
3. Implement new outlets and display data.

Updating the data loading and model

Currently, the code in `ViewController.swift` specifies that just the given name, family name, image data, and image availability for a contact should be fetched. This needs to be expanded to the email address, postal address, and phone number are fetched as well. Update the `retrieveContacts(store:)` method with the following code for `keysToFetch`:

```
let keysToFetch = [CNContactGivenNameKey as CNKeyDescriptor,
                   CNContactFamilyNameKey as CNKeyDescriptor,
                   CNContactImageDataAvailableKey as CNKeyDescriptor,
                   CNContactImageDataKey as CNKeyDescriptor,
                   CNContactEmailAddressesKey as CNKeyDescriptor,
                   CNContactPhoneNumbersKey as CNKeyDescriptor,
                   CNContactPostalAddressesKey as CNKeyDescriptor]
```

Now that the code will fetch these new attributes for the contacts, the `Contact` class must be updated so it allows these new properties to be used. Add the following variable declarations to `Contact.swift`:

```
var emailAddress: String {
  // 1
  return String(contact.emailAddresses.first?.value ?? "--")
}

var phoneNumber: String {
  // 2
```

```
        return contact.phoneNumbers.first?.value.stringValue ?? "--"
    }

    var address: String {
        // 3
        let street = contact.postalAddresses.first?.value.street ?? "--"
        let city = contact.postalAddresses.first?.value.city ?? "--"

        return "\(street) \(city)"
    }
```

Each of the three properties you have added is annotated with a comment. This is because these properties are not as straightforward to pass on as the others. Let's go over these comments one by one:

- Because `emailAddresses` is an array and the value of an email address is an `NSString?` object instead of a `String` object, the first item in the list of email addresses is retrieved using optional chaining. If there is no first email address, the string `--` is returned instead. This is then passed to the `String` initializer to convert the resulting `NSString` object to the `String` object that is expected to be returned. Being able to implement logic like this on the `Contact` model is convenient because it makes using the email address a lot more convenient for anybody who wants to display it.
- Retrieving the string value for the phone number is a bit simpler. Optional chaining is used again to try and retrieve the first item in a list. If a phone number is found, the value's `stringValue` is used to turn the phone number into a `String` object. If no phone numbers are found, `--` is returned here too.
- The address is built by retrieving the first available address and extracting the `street` and `city` objects from it. These values are then combined in a string and returned. Again, missing values are replaced with `--`.

Now that the model is prepared, the time has come to pass the model to the contact details page so it can display real contact data.

Passing the model to the details page

The transition from the overview page to the details page is implemented with a segue. The segue is triggered when the user taps a contact, putting the detail page on the screen. Because this transition uses a segue, there is a special method that can be implemented to pass data from the first view controller to the second view controller. This special method is called `prepare(for:sender:)`.

This method is called on the source view controller right before a segue is performed and it provides access to the destination view controller. The segue's destination is used to configure data on the view controller that is about to be presented. Let's implement this right now so you can pass the tapped contact to the detail page. Add the following extension to `ViewController.swift`:

```
extension ViewController {
  override func prepare(for segue: UIStoryboardSegue, sender: Any?) {
    if let contactDetailVC = segue.destination as?
ContactDetailViewController,
      segue.identifier == "detailViewSegue",
      let selectedIndex =
collectionView.indexPathsForSelectedItems?.first {
        contactDetailVC.contact = contacts[selectedIndex.row]
    }
  }
}
```

This implementation first verifies that the destination of the segue has the correct type. Then, it also makes sure that the segue's identifier matches the identifier that was added for this segue in **Interface Builder**. Finally, the first (and only) selected index path is read from the collection view. This information is then used to assign the correct contact from the contacts array to a contact property on the destination view controller. This property does not exist yet, but you will add it to `ContactDetailViewController` in a moment.

Implementing the new outlets and displaying data

The initial view controller is now fully prepared to pass data along to the contact detail view controller. To display a contact, you will need to add a variable that holds on to the selected contact and a couple of outlets so you can update the view with the appropriate contact information. Add the following code to `ContactDetailViewController` to prepare it for displaying a contact:

```
@IBOutlet var contactPhoneLabel: UILabel!
@IBOutlet var contactEmailLabel: UILabel!
@IBOutlet var contactAddressLabel: UILabel!

var contact: Contact?

override func viewDidLoad() {
  // existing implementation...
```

```
    if let contact = self.contact {
      // 1
      contact.fetchImageIfNeeded { [weak self] image in
        self?.contactImage.image = image
      }

      contactNameLabel.text = "\(contact.givenName)
\(contact.familyName)"
      contactPhoneLabel.text = contact.phoneNumber
      contactEmailLabel.text = contact.emailAddress
      contactAddressLabel.text = contact.address
    }
  }
```

The preceding snippet configures the view controller so it uses contact data to populate its labels and image view. The comment in this snippet highlights the `fetchImageIfNeeded(completion:)` method that is used to retrieve and use the image.

The final step is to go to the storyboard, select the detail view controller, and connect the outlets in the **Connections Inspector** to the user interface elements. After doing this, build and run your app to see the details page in its full glory. Now, let's add some icing to this cake by implementing peek and pop using 3D Touch!

Enhancing the user experience with 3D Touch

One of iOS's lesser-used features is 3D Touch. 3D Touch allows users to perform special interactions with apps by pressing a little bit more firmly on the screen than usual. The iPhone 6s and newer devices support this functionality and it allows for some pretty neat interactions. One of these interactions is called peek and pop.

With peek and pop, a user can 3D Touch an element on the screen and they'll see a preview of the detail page they would see if they had performed a regular tap on the UI element. The following screenshot shows an example of such a preview:

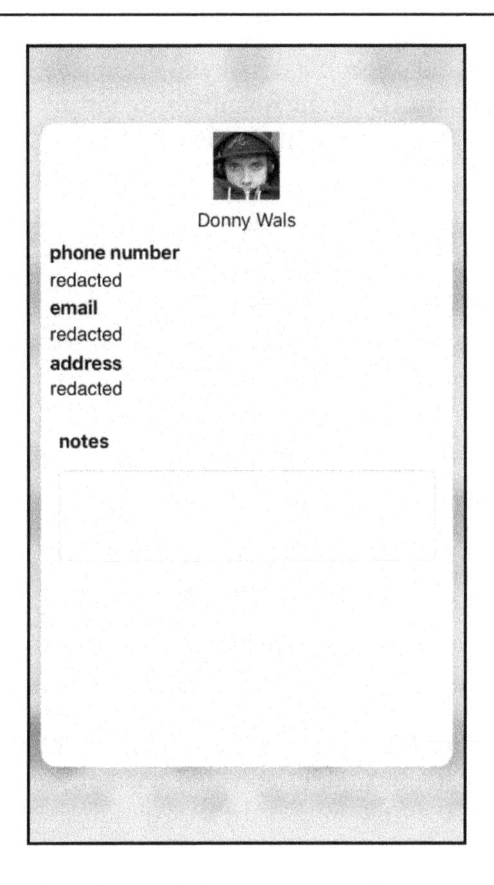

If the user sees a preview like this and they press on their screen a little bit harder, the user commits to seeing this view and they will be taken to the detail page as if they had normally tapped on the contact. Implementing this feature requires only a small amount of effort.

To implement peek and pop, the source view controller must register its intent to display previews. The source view controller must also conform to `UIViewControllerPreviewingDelegate` so it can provide and commit to the preview view controller.

Before you get to writing the code, there is one change you'll need to make to the storyboard. You currently don't have access to the detail page view controller from code. However, to provide it as a preview view controller, you must be able to provide an instance of `ContactDetailViewController`.

Open the storyboard file and select the detail view controller. Use the **Identity Inspector** to set a **Storyboard ID** on the detail view controller. Use `ContactDetailViewController` as the value for the **Storyboard ID** as shown in the following screenshot:

You can use the **Storyboard ID** to create instances of the detail view controller from code. You'll see how in just a second. First, you'll register `ViewController` for previewing content. Add the following code to `viewDidLoad()` on the `ViewController`:

```
if traitCollection.forceTouchCapability == .available {
  registerForPreviewing(with: self, sourceView: collectionView)
}
```

When the current trait collection supports 3D Touch, the view controller is registered for previewing using itself as a delegate. The collection view is used as a source for this interaction since it's the component for which you want to provide previews. The next step is to make `ViewController` conform to the `UIViewControllerPreviewingDelegate` protocol. Add the following extension to `ViewController.swift` to add this conformance:

```
extension ViewController: UIViewControllerPreviewingDelegate {
  func previewingContext(_ previewingContext:
UIViewControllerPreviewing, viewControllerForLocation location:
CGPoint) -> UIViewController? {
    guard let tappedIndexPath = collectionView.indexPathForItem(at:
location)
      else { return nil }
```

```
        let contact = contacts[tappedIndexPath.row]

        guard let viewController =
storyboard?.instantiateViewController(withIdentifier:"ContactDetailVie
wController") as? ContactDetailViewController
            else { return nil }

        viewController.contact = contact
        return viewController
    }

    func previewingContext(_ previewingContext:
UIViewControllerPreviewing,
                            commit viewControllerToCommit:
UIViewController) {

        navigationController?.show(viewControllerToCommit, sender: self)
    }
}
```

This extension implements two methods:
`previewingContext(_:viewControllerForLocation)` and
`previewingContext(_:commit)`. The first method is responsible for providing the previewed view controller. To provide the correct view controller with the correct contact, this code checks whether the 3D Touch occurred on a collection view cell. Next, a view controller is obtained through the storyboard by using the identifier you configured earlier. Finally, the view controller is assigned a contact and returned for previewing.

The second method is used to commit to seeing the details page. This implementation simply tells the navigation controller to present the view controller that was previewed. When you implement different preview view controllers than the ones you actually want to use when the user commits to navigating to the detail page, you can use the commit method to configure a real detail view controller instead of the preview.

If you run the project on a device that supports 3D Touch, you should be able to apply some force while tapping on a contact to see their previews appear.

Summary

Congratulations, you have successfully created an application that runs on all iOS devices and screen sizes. You took an application that had just a single page and turned it into a simple contacts application with a detail view. To achieve this, you made use of Auto Layout and `traitCollections`. You learned about size classes and what they tell you about the available screen's real estate for your application. You also learned how to make use of Auto Layout through code and how to respond to changes in your app's environment in real time. Finally, you learned how to simplify a lot of the Auto Layout work you've done by implementing `UIStackView`. To top it all off, you saw how `prepare(for:sender)` allows you to pass data from one view controller to another.

The lessons you've learned in this chapter are extremely valuable. The increase in possible screen sizes over the past few years has made Auto Layout an invaluable tool for developers and not using it will make your job much harder. If Auto Layout is still a bit hard for you, or it doesn't fully make sense, don't worry. We'll keep using Auto Layout in this book so you will get the hang of it. Going over this chapter again if you feel lost should help you as well; it's a lot to take in, so there's no shame in reading it again. In the next chapter, you'll add some finishing touches to **Hello-Contacts**. The best applications feature clever and useful animations. In the previous chapters, you have already used little bits of animation. In the next chapter, you are going to dive deep into more complex animations and custom transitions between view controllers.

Questions

1. Where can you configure the supported Interface Orientations for an iPad in the Hello-Contacts project?

   a) Project Settings
   b) `Info.plist`
   c) The storyboard

2. How many different size classes exist today?

   a) 2
   b) 3
   c) 4

3. What object contains information about the current environment your app is running in?

    a) `UIEnvironmentTraits`
    b) `UISizeClass`
    c) `UITraitCollection`

4. What is the best way to lay out several objects on top of each other?

    a) Auto Layout
    b) Using `UIStackView`
    c) Using `UIScrollView`

5. How do you animate an update to layout constraints?

    a) By calling `view.layoutSubviews()` in an animation
    b) By updating the constraints inside an animation

6. What is the correct way to create an instance of a view controller from a storyboard?

    a) `let myViewController = MyViewController()`
    b) `let myViewController = storyboard?.instantiateViewController(withIdentifier:"MyViewController")`
    c) `let myViewController = storyboard?.MyViewController()`

7. How many methods did you have to implement to support peek and pop?

    a) 3
    b) 2
    c) 1

Further reading

- *Auto Layout Guide*: `https://developer.apple.com/library/archive/documentation/UserExperience/Conceptual/AutolayoutPG/index.html`
- *Building Adaptive User Interfaces*: `https://developer.apple.com/design/adaptivity/`

4
Immersing Your Users with Animation

The **Hello-Contacts** app is shaping up quite well. You have already covered a lot of ground by implementing a custom overview page and a contact detail page that both work great on any screen size that exists today. However, you can still make some considerable improvements to this app by implementing awesome animations. You have already used some very basic animations. In this chapter, you're going to learn about advanced animation techniques.

The UIKit framework provides some very powerful APIs that you can utilize in your apps to make them look great and feel more intuitive. Most of the APIs in UIKit are not very difficult to use. For instance, you can create cool animations with a small amount of code.

In this chapter, you will learn about `UIViewPropertyAnimator`, a powerful object that can be used to replace the existing animations in the **Hello-Contacts** app. `UIViewPropertyAnimator` provides more control over your animations than the animations you implemented in previous chapters. You'll also learn about UIKit dynamics. UIKit dynamics can be used to make objects react to their surroundings by applying physics. Finally, you'll learn how to implement a custom transition when moving from one view controller to the next.

To sum everything up, this chapter covers the following topics:

- `UIViewPropertyAnimator`
- Vibrant animations using springs
- UIKit dynamics
- Customizing view controller transitions

Refactoring existing animations with UIViewPropertyAnimator

So far, you have seen animations that were implemented using the `UIView.animate` method. These animations are quite simple to implement and mostly follow the following format:

```
UIView.animate(withDuration: 1.5, animations:
{
    myView.backgroundColor = UIColor.red()
})
```

You have already seen this method implemented in slightly more complex ways, including one that used a closure that was executed upon completion of the animation. For instance, when a user taps on one of the contacts in the **Hello-Contacts** app, the following code is used to animate a bounce effect:

```
UIView.animate(withDuration: 0.1, delay: 0, options: [.curveEaseOut],
animations:
{
    cell.contactImage.transform = CGAffineTransform(scaleX: 0.9, y:
    0.9)
}, completion: { finished in
    UIView.animate(withDuration: 0.1, delay: 0, options:
    [.curveEaseIn], animations:
{
        cell.contactImage.transform = CGAffineTransform.identity
    }, completion: { [weak self] finished in
        self?.performSegue(withIdentifier: "detailViewSegue", sender:
self)
    })
})
```

It's not particularly pleasing to look at this code. The indentation is all over the place, and a lot is going on in a dense piece of code. If you dissect this code, you will find that the entire animation was implemented in a single method call. While this might be convenient for small animations, it's not very readable. This is especially true if the animation is more complex or if you want to chain several animations, which is the case for the preceding bounce animation.

One reason to favor `UIViewPropertyAnimator` over the implementation you just saw is readability. Let's see what the same bounce animation looks like when it's refactored to use `UIViewPropertyAnimator`:

```
// 1
let downAnimator = UIViewPropertyAnimator(duration: 0.1, curve:
.easeOut) {
    cell.contactImage.transform = CGAffineTransform(scaleX: 0.9, y:
0.9)
}

let upAnimator = UIViewPropertyAnimator(duration: 0.1, curve: .easeIn)
{
    cell.contactImage.transform = CGAffineTransform.identity
}

// 2
downAnimator.addCompletion { _ in
    upAnimator.startAnimation()
}

upAnimator.addCompletion { [weak self] _ in
    self?.performSegue(withIdentifier: "detailViewSegue", sender:
self)
}

// 3
downAnimator.startAnimation()
```

The first thing you should notice is how much longer this code is. The old implementation was only 9 lines, the new one is 17 lines if you count all the blank lines. The second thing you'll notice is how much more readable the code has become. Code readability is something you should never underestimate. You can write great code, but if you come back to it a week later and find yourself struggling with that great piece of code due to bad readability, your code suddenly isn't as great as it was when you first wrote it.

While it's great that `UIViewPropertyAnimator` makes your code more readable, it still doesn't teach you much about animations. Let's dissect the preceding implementation to see what's happening.

The first section of the preceding code is all about creating instances of `UIViewPropertyAnimator`. The `UIViewPropertyAnimator` class has several initializers. The simplest initializer takes no arguments, the duration property can be set later, and by calling the `addAnimation` method, new animations can be added to the animator. However, that version of the animator would be too basic for **Hello-Contacts**.

The example code uses a version of `UIViewPropertyAnimator` that accepts a timing function to make the final bounce animation more lively. If you look at the sample code, the first argument passed to the `UIViewPropertyAnimator` initializer is the duration of the animation in seconds. The second argument controls the timing function. A timing function describes how an animation should progress over time. For instance, the `easeIn` option describes how an animation starts off at a slow pace and speeds up over time. The following diagram describes some of the most commonly used timing functions:

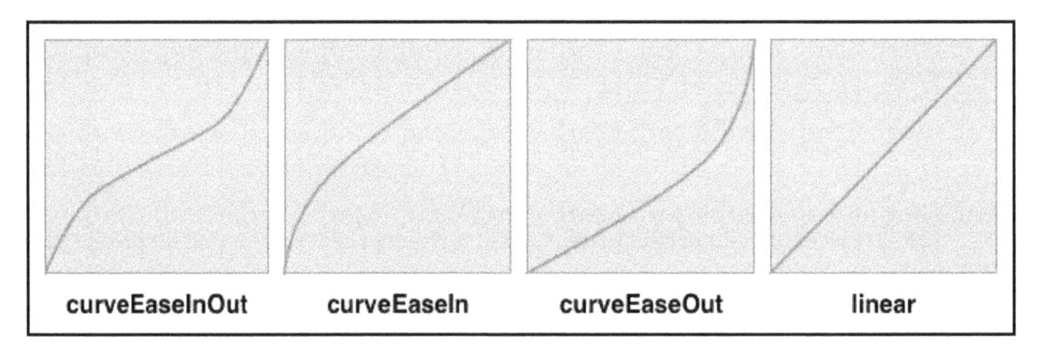

In these graphs, the *horizontal* axis represents the animation's progress. For each graph, the animation timeline is described from left to right on the x-axis. The animation's progress is visualized on the y-axis from bottom to top. At the bottom-left point, the animation hasn't started yet. At the right of the graph, the animation is completely done. The vertical axis represents time.

The final argument that is passed to the `UIViewPropertyAnimator` initializer is an optional argument for the animation that you wish to execute. This is quite similar to the `UIView.animate` way of doing things; the most significant difference is that you can add more animations after creating the animator, meaning that `nil` can be passed as the argument for the animations and you can add animations you wish to execute at a later time. This is quite powerful because you're even allowed to add new animations to `UIViewPropertyAnimator` while an animation is running!

The second section in the sample code you saw earlier adds completion closures to the animators. The completion closures both receive a single argument. The received argument describes at what point in the animation the completion closure was called. This property will usually have a value of `.end`, which indicates that the animation ended at the end position. However, this isn't always true because you can finish animations halfway through the animation if you desire. You could also reverse an animation, meaning that the completion position would be `.start`.

Once the completion closure is added, and the property animators are fully configured, the final step is to start the animation by calling `startAnimation()` on an animator object. Once the `startAnimation()` method is called, the animation begins executing immediately. If needed, you can make the animation start with a delay by calling `startAnimation(afterDelay:)`.

After replacing the animation in `collectionView(_:didSelectItemAt:)` with the `UIViewPropertyAnimator` version of the tap animation, you should be able to replace the remaining `UIView.animate` animations in the project. There are two animations in `ViewController.swift` that are used to animate the changing background color for the collection-view cell when you enter or exit edit mode.

There are also two animations in `ContactDetailViewController.swift` that you could replace. However, this animation is so short and simple that creating an instance of `UIViewPropertyAnimator` for it might be a bit much. However, as an exercise, it would be nice to try to find the simplest implementation possible to replace the `UIView.animate` calls in the `ContactDetailViewController` class.

Refer to this book's code bundle if you get stuck replacing the animations or want to see the solution to this challenge.

Understanding and controlling animation progress

One of the best features of `UIViewPropertyAnimator` is that you can use it to create animations that can be interrupted, reversed, or interacted with. Many of the animations you see in iOS are interactive animations. For instance, swiping on a page to go back to the previous page is an interactive transition. Swiping between pages on the home screen, opening the control center, or pulling down the notification center are all examples of animations that you manipulate by interacting with them.

While the concept of interactive animations might sound complicated, `UIViewPropertyAnimator` makes it quite simple to implement them. As an example, you'll see how to implement a drawer on the contact detail page in the **Hello-Contact**s app. First, you'll prepare the view, so the drawer is partially visible in the app. Once the view is all set up, you will write the code to perform an interactive show-and-hide animation for the drawer.

Open `Main.storyboard` in the **Hello-Contacts** project and add a plain view to the contact-detail view controller's view. Make sure that you do not add the drawer view to the scroll view. It should be added on top of the scroll view. Set up Auto Layout constraints to make sure the drawer view's width is equal to the main view's width. Also, align the view with the horizontal center of its container. Next, make the drawer `350` points in height. The last constraint that you must add is a bottom space to the Safe Area constraint. Set the constant for this constraint to `-305` so most of the drawer view is hidden from sight. Make sure this constraint is relative to the safe area margins so it doesn't overlap with the iPhone Xs' home screen indicator.

Next, add a button to the drawer. Align it horizontally and space it 8 points from the top of the drawer. Set the button's label to **Toggle Drawer**. It's a good idea to give the drawer view a background color so you can easily see it sliding over the contact detail page. The following screenshot shows the desired result:

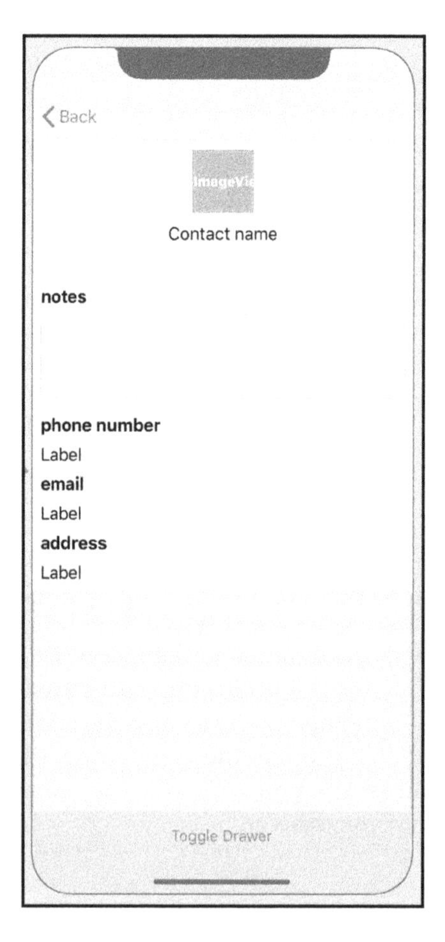

The layout is now prepared. The implementation of the drawer functionality should implement the following features:

- Toggle the drawer by tapping on the toggle button.
- Toggle the drawer interactively when swiping on the drawer.
- Allow the user to tap on the toggle button and then swipe the drawer to manipulate or reverse the animation.

Behavior such as this is not straightforward; without `UIViewPropertyAnimator`, you would have to write a lot of complex code, and you'd still be pretty far from your desired results. Let's see what `UIViewPropertyAnimator` does to make implementing this effect manageable.

To prepare for the implementation of the drawer, add the following two properties to `ContactDetailViewController`:

```
@IBOutlet var drawer: UIView!
var isDrawerOpen = false
var drawerPanStart: CGFloat = 0
var animator: UIViewPropertyAnimator!
```

Also, add an extension to `ContactDetailViewController` that holds an `@IBAction` for the tap action. `@IBAction` is similar to `@IBOutlet`, but it used to call a particular method in response to a specific user action. An extension is used so it's easy to group the animation code nicely:

```
extension ContactDetailViewController {
    @IBAction func toggleDrawerTapped() {
    }
}
```

Connect the outlet you've created to the drawer view in Interface Builder. Also, connect `@IBAction` to the toggle button's **Touch Up Inside** action. This is done by dragging from `toggleDrawerTapped` under the **Received Actions** header in the **Connections Inspector**. When you drag from the action to the button, a menu appears from which you can select the action for which `@IBAction` should trigger. To respond to a button tap, choose **Touch Up Inside** from this menu.

Lastly, add the following lines to the end of `viewDidLoad()`:

```
let panRecognizer = UIPanGestureRecognizer(target: self, action:
#selector(didPanOnDrawer(recognizer:)))
drawer.addGestureRecognizer(panRecognizer)
```

The preceding code sets up and adds a pan gesture-recognizer to the drawer view so the app can detect when a user starts dragging their finger on the drawer. Gesture-recognizers are a great way to respond to user interactions such as tapping, double-tapping, pinching, swiping, and panning.

Also, add the following method to the extension you created earlier for `@IBAction`. This is the method that is called when the user performs a pan gesture on the drawer:

```
@objc func didPanOnDrawer(recognizer: UIPanGestureRecognizer) {

}
```

Now that all of the placeholders are implemented, let's create a simple first version of the open drawer animation. When the user taps on the toggle button, the drawer should open or close depending on the drawer's current state. The following snippet implements such an animation:

```
@IBAction func toggleDrawerTapped() {
    animator = UIViewPropertyAnimator(duration: 1, curve: .easeOut) {
[unowned self] in
        if self.isDrawerOpen {
            self.drawer.transform = CGAffineTransform.identity
        } else {
            self.drawer.transform = CGAffineTransform(translationX: 0,
y: -305)
        }
    }

    animator?.addCompletion { [unowned self] _ in
        self.animator = nil
        self.isDrawerOpen = !(self.drawer.transform ==
CGAffineTransform.identity)
    }

    animator?.startAnimation()
}
```

The animation that is passed to the property animator uses the value of `isDrawerOpen` to determine whether the animation should open or close the drawer. When the drawer is currently open, it should close and vice versa. Once the animation finishes, the `isDrawerOpen` variable is updated to reflect the new state of the drawer. To determine the current state, the application reads the drawer's current transformation. If the drawer is not transformed, its transformation will equal `CGAffineTransform.identity` and the drawer is considered closed. Otherwise, the drawer is considered opened. You can build and run the app now to see this animation in action.

To allow the user to interrupt or start the animation by dragging their finger on the screen, the code must check whether an existing property animator is performing an animation. If no animator exists or if the current animator is not running any animations, a new instance of the animator should be created. In all other circumstances, it's possible to make use of the existing animator. Let's refactor the animator creation code from `toggleDrawerTapped()` so it reuses the animator if possible and creates a new animator if needed.

Add the following method to the extension and update `toggleDrawerTapped()` as shown in the following code:

```
func setUpAnimation() {
  guard animator == nil || animator?.isRunning == false
    else { return }

  animator = UIViewPropertyAnimator(duration: 1, curve: .easeOut) {
[unowned self] in
    if self.isDrawerOpen {
      self.drawer.transform = CGAffineTransform.identity
    } else {
      self.drawer.transform = CGAffineTransform(translationX: 0,
                                                y: -305)
    }
  }

  animator?.addCompletion { [unowned self] _ in
    self.animator = nil
    self.isDrawerOpen = !(self.drawer.transform ==
CGAffineTransform.identity)
  }
}

@IBAction func toggleDrawerTapped() {
  setUpAnimation()
  animator?.startAnimation()
```

```
}
```

Now add the following implementation for `didPanOnDrawer(recognizer: UIPanGestureRecognizer)`:

```
@objc func didPanOnDrawer(recognizer: UIPanGestureRecognizer) {
  switch recognizer.state {
  case .began:
    setUpAnimation()
    animator?.pauseAnimation()
    drawerPanStart = animator?.fractionComplete ?? 0
  case .changed:
    if self.isDrawerOpen {
      animator?.fractionComplete = (recognizer.translation(in:
drawer).y / 305) + drawerPanStart
    } else {
      animator?.fractionComplete = (recognizer.translation(in:
drawer).y / -305) + drawerPanStart
    }
  default:
    drawerPanStart = 0
    let timing = UICubicTimingParameters(animationCurve: .easeOut)
    animator?.continueAnimation(withTimingParameters: timing,
durationFactor: 0)

    let isSwipingDown = recognizer.velocity(in: drawer).y > 0
    if isSwipingDown == !isDrawerOpen {
      animator?.isReversed = true
    }
  }
}
```

This method is called for any change that occurs in the pan gesture-recognizer. When the pan gesture first starts, the animation is configured, and then `pauseAnimation()` is called on the animator object. This allows us to change the animation progress based on the user's pan behavior. Because the user might begin panning in the middle of the animation, for instance after tapping the toggle button first, the current `fractionComplete` value is stored in the `drawerPanStart` variable.

The value of `fractionComplete` is a value between `0` and `1` and it's decoupled from the time that your animation takes to run. So imagine that you are using an ease-in and ease-out timing parameter to animate a square from an `x` value of `0` to an `x` value of `100`, the `x` value of `10` is not at `10%` of the time the animation takes to complete. However, `fractionComplete` will be `0.1` which corresponds to the animation being `10%` complete. This is because `UIViewPropertyAnimator` converts the timescale for your animation to linear once you pause it. Usually, this is the best behavior for an interactive animation. However, you can change this behavior by setting the `scrubsLinearly` property on your animator to false. If you do this, `fractionComplete` will take any timing parameters you've applied into account. You can try playing with this to see what it feels like for the drawer animation.

Once the initial animation is configured and paused, the user can move their finger around. When this happens, the `fractionComplete` property is calculated and set on the animator by taking the distance traveled by the user's finger and dividing it by the total distance required. Next, the progress made by the animation before being interrupted is added to this new value.

Finally, if the gesture ends, is canceled, or anything else, the start position is reset. Also, a timing parameter to use for the rest of the animation is configured and the animation is set up to continue. By passing a `durationFactor` of `0`, the animator knows to use whatever time is left for the animation while taking into account its new timing function. Also, if the user tapped the toggle button to close the drawer, yet they catch it mid-animation and swipe upward, the animation should finish in the upward direction. The last couple of lines take care of this logic.

It's strongly recommended you experiment and play around with the code that you just saw. You can do powerful things with interruptible and interactive animations. Let's see how you can add some extra vibrancy to your animations with springs!

Adding vibrancy to animations

A lot of animations on iOS look bouncy and feel natural. For instance, when an object starts moving in the real world, it rarely does so smoothly. Often, something moves because something else applied an initial force to it, causing it to have a certain momentum. Spring animations help you to apply this sort of real-world momentum to your animations.

Spring animations are usually configured with an initial speed. This speed is the momentum an object should have when it begins moving. All spring animations require a damping to be set on them. The value of this property specifies how much an object can overflow its target value. A smaller damping will make your animation feel more bouncy because it will float around its end value more drastically.

The easiest way to explore spring animations is by slightly refactoring the animation you just created for the drawer. Instead of using an `easeOut` animation when a user taps the **Toggle Drawer** button, you can use a spring animation instead. The following code shows the changes you need to make to `setUpAnimation()`:

```
func setUpAnimation() {
  guard animator == nil || animator?.isRunning == false
    else { return }

  let spring: UISpringTimingParameters
  if self.isDrawerOpen {
    spring = UISpringTimingParameters(dampingRatio: 0.8,
initialVelocity: CGVector(dx: 0, dy: 10))
  } else {
    spring = UISpringTimingParameters(dampingRatio: 0.8,
initialVelocity: CGVector(dx: 0, dy: -10))
  }

  animator = UIViewPropertyAnimator(duration: 1, timingParameters:
spring)

  animator?.addAnimations { [unowned self] in
    if self.isDrawerOpen {
      self.drawer.transform = CGAffineTransform.identity
    } else {
      self.drawer.transform = CGAffineTransform(translationX: 0, y:
-305)
    }
  }

  // ...
}
```

When you implement a spring animation, you use a special initializer for `UIViewPropertyAnimator`. Since you can't pass animations to this initializer, you must add them by calling `addAnimations(_:)`. Adding spring animations did not require a considerable code change, but try running the app and tapping on the toggle button. The drawer will now feel more realistic because its animation curve is not as static as it was before.

Play around with the values for the spring damping and the velocity, if you use some extreme values you'll get interesting results. Keep in mind that the damping should be a value between `0` and `1` and that a value closer to `1` will make your animation bounce less.

The animation that is executed by the pan recognizer doesn't feel great at this point. It's very static and doesn't take into account how fast a user is panning on the drawer. When the user ends their pan gesture, you can set the sprint timing's `initialVelocity` based on the actual pan velocity. This will make the animation feel even more realistic because it will now use the actual pan speed as the initial speed for animation:

```
@objc func didPanOnDrawer(recognizer: UIPanGestureRecognizer) {
  // ...
  default:
    drawerPanStart = 0
    let currentVelocity = recognizer.velocity(in: drawer)
    let spring = UISpringTimingParameters(dampingRatio: 0.8,
initialVelocity: CGVector(dx: 0, dy: currentVelocity.y))

    animator?.continueAnimation(withTimingParameters: spring,
durationFactor: 0)
    let isSwipingDown = currentVelocity.y > 0
    if isSwipingDown == !isDrawerOpen {
      animator?.isReversed = true
    }
  }
}
```

As you've just seen, the use of spring animations can benefit your animations and they are not very hard to add to your apps. While they might not always be the best solution, their ease of implementation makes spring animations a worthy candidate to experiment with to determine whether your animation needs a spring.

While the animation you have just implemented is pretty lifelike and realistic, your animations might need even more realism. The next section covers UIKit Dynamics, which is a special method of animating objects that uses a physics engine and can detect collisions between objects.

Adding dynamism with UIKit Dynamics

Most apps implement simple animations, like the ones you saw in the last couple of sections. However, some animations might need even more realism. This is what UIKit Dynamics is for. With UIKit Dynamics, you can place one or more views in a scene that uses a physics engine to apply certain forces to the views it contains. For instance, you can apply gravity to a particular object, causing it to fall off the screen. You can even have objects bumping into each other, and if you assign a mass to your views, this mass is taken into account when two objects crash into each other. When you apply a certain force to an object with very little mass, it will be displaced more than an object with a lot of mass, just like you would expect in the real world.

Let's take a little break from building **Hello-Contacts** and use a separate app to implement a nice, little physics experiment! Create a new project and name it **Cradle**. Make sure to configure the project so it only runs in landscape orientation. In `Main.storyboard`, make sure to set the preview to landscape and add three square views to the view controller's view. Make each view about 100 x 100 points and give them a background color. Normally, you would set up constraints to position these views. However, since you're just experimenting right now and the resulting code will be a lot simpler if you don't set up constraints, you can skip setting up constraints for now.

Your layout should look similar to the following screenshot:

Create instances of `@IBOutlet` in `ViewController.swift` for the views you just added and connect them to the storyboard in the same way you did before. You can name the outlets `square1`, `square2`, and `square3` if you want to follow along with the code samples.

The simplest thing you can implement at this point is to set up a scene that contains the three squares and apply some gravity to them. This will cause the squares to fall off the screen because they'll start falling once gravity is applied and there is no floor to stop the squares from dropping off the screen.

To set up a scene like the one described here, add the following property to `ViewController.swift`:

```
var animator: UIDynamicAnimator!
```

Then, in `viewDidLoad()`, add the following:

```
override func viewDidLoad() {
  super.viewDidLoad()

  let squares: [UIDynamicItem] = [square1, square2, square3]
  animator = UIDynamicAnimator(referenceView: view)
  let gravity = UIGravityBehavior(items: squares)
  animator.addBehavior(gravity)
}
```

If you test your app now, you'll notice that your views start falling immediately. Setting up a simple scene such as this is easy with UIKit Dynamics. The downside of this simple example is that it's not particularly interesting to look at. Before you add features to make this sample more interesting, let's see what the preceding four lines of code do.

The views in a dynamic scene must be of the `UIDynamicItem` type. `UIView` can be used as `UIDynamicItem`, so by adding them to a list that has `[UIDynamicItem]` works automatically. Then, you create an instance of `UIDynamicAnimator` and you tell it the view to which it will apply its physics engine. The last step is to configure and apply a behavior. This sample uses `UIGravityBehavior` but there are several other behaviors you can use in your scenes.

For instance, you can create `UIAttachmentBehavior` to attach an item to another item or to some point on the screen. The following code implements an attachment behavior for every square on the screen and attaches it to the top of the screen. This will cause the squares to fall for a moment and then they will bounce and swing a little until they eventually come to a standstill. You can add the following code to `viewDidLoad()` to implement this:

```
var nextAnchorX = 250

for square in squares {
  let anchorPoint = CGPoint(x: nextAnchorX, y: 0)
  nextAnchorX -= 30
  let attachment = UIAttachmentBehavior(item: square,
attachedToAnchor: anchorPoint)
  attachment.damping = 0.7
  animator.addBehavior(attachment)
}
```

Every square is set up with a slightly different attachment point in this example. Note that the attachment behavior has a `damping` property. This damping is similar to the damping that is used in spring animations. Try experimenting with the value for `attachment.damping` to see what it does.

If you run the app now, you'll note that every square is attached to an invisible point on the screen that keeps it from falling. Some things are still missing though. The squares can now simply cross over each other. It would be a lot cooler if they bumped into each other instead.

All you need to do is add the following two lines to `viewDidLoad()`:

```
let collisions = UICollisionBehavior(items: squares)
animator.addBehavior(collisions)
```

Are you convinced that UIKit Dynamics are cool yet? I thought so; it's amazing how much you can do with just a little bit of code. Let's add some mass to the squares and make them more elastic to see whether this has any effect on how the squares collide.

Update the `for` loop from before by adding the following code:

```
for square in squares {
  //...

  let dynamicBehavior = UIDynamicItemBehavior()
  dynamicBehavior.addItem(square)
  dynamicBehavior.density = CGFloat(arc4random_uniform(3) + 1)
  dynamicBehavior.elasticity = 0.8
  animator.addBehavior(dynamicBehavior)
}
```

The preceding code should augment what you already have in the loop; it shouldn't replace the existing logic. By setting a `density` on `UIDynamicItemBehavior`, the engine can derive the mass of an item. This will change how the physics engine treats the item when it collides with another item.

If you build and run now, you won't see a huge difference. However, you can tell that the animation has changed because the variables that go into the physics simulation have changed. Even though this example was very simple, you should be able to implement some interesting behaviors by creating an animator and playing around with the different behaviors you can add to it. Have a look at Apple's documentation for a full overview of all available behaviors and possibilities.

Even though the UIKit Dynamics physics engine is powerful and performant, you should not use it to build games. If you want to build a game, have a look at SpriteKit. It has a similar, powerful physics engine except the framework is a lot better optimized for building games.

The last stop on your journey through animation land is view-controller transitions! Let's dive right in.

Customizing view-controller transitions

Implementing a custom view-controller transition is one of those things that can take a little while to get used to. Implementing custom transitions involves implementing several objects and it's not always easy to make sense of how this works. This section aims to explain exactly how custom view-controller transitions work so you can add one more powerful tool to your developer toolbox.

A nicely-implemented custom view controller transition will entertain and amaze your users. Making your transitions interactive could even ensure that your users spend some extra time playing around with your app, which is exactly what you want. You will implement a custom transition for the **Hello-Contacts** app.

First, you'll learn how you can implement a custom modal transition. Once you've implemented that, you will learn about custom transitions for `UINavigationController` so you can show and hide the contact details page with a custom transition. The dismissal of both the modal view controller and the contact detail page will be interactive, so users can swipe to go back to where they came from.

In this section, you will work through the following steps:

1. Implement a custom modal presentation transition.
2. Make the transition interactive.
3. Implement a custom `UINavigationController` transition.

Implementing a custom modal presentation transition

A lot of applications implement modally-presented view controllers. A modally-presented view controller is typically a view controller that is presented on top of the current screen as an overlay. By default, modally-presented view controllers animate upward from the bottom of the screen and are often used to present forms or other temporary content to the user. In this section, you'll take a look at the default modal presentation transition and how to customize it to suit your own needs.

The first thing you should do is create a view controller that will be presented modally. Start by creating a new Cocoa Touch Class and name it `CustomPresentedViewController`. Make sure that it subclasses `UIViewController`. After creating the file, open `Main.storyboard` and drag out a new `UIViewController` from the Object Library and set its class to `CustomPresentedViewController` in the **Identity Inspector** panel. Next, drag out a bar button item to the left side of the navigation bar on the contacts overview page. Set the bar button's label text to **Show Modal**.

Then press *Ctrl*, and drag from the bar button item to the new view controller to set up a segue from the contacts overview page to the view controller you just added. Select the **Present Modally** segue:

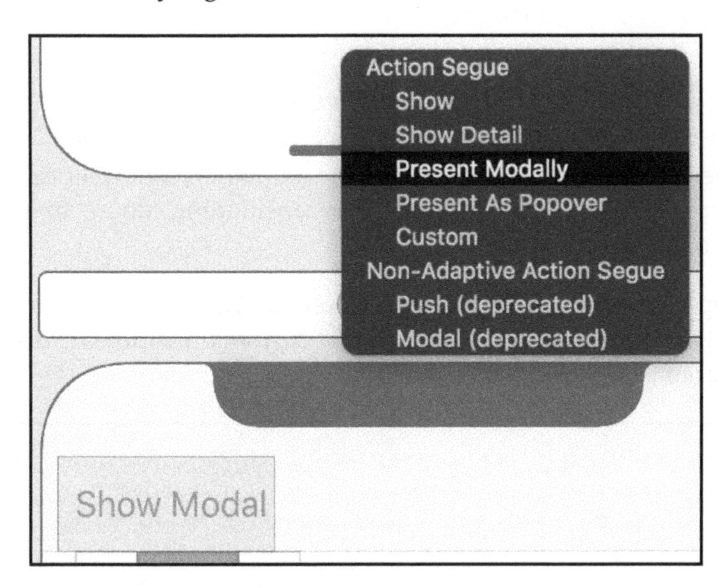

Finally, give the new view controller's view a bright blue background color, so it will be easier to see the transition later. If you run your app now, you can click on the **Show Modal** button and you'll see an empty view controller pop up from the bottom. You can't dismiss this view controller right now. That's okay; you will get to that later. Let's work on a custom transition to display this view controller first.

Custom view-controller transitions use several objects to facilitate the animation. The first object you will look at is `transitioningDelegate` for `UIViewController`. The `transitioningDelegate` property is responsible for providing an animation controller that provides the custom transition.

The animation controller uses a transitioning context object that provides information about the view controllers that are involved in the transition. Typically, these view controllers will be the current view controller and the view controller that is about to be presented.

The transitioning flow can be described in the following steps:

1. A transition begins; the target view controller is asked for its `transitioningDelegate`.
2. The `transitioningDelegate` is asked for an animation controller.
3. The animation controller is asked for the animation duration.
4. The animation controller is told to perform the animation.
5. When the animation is complete, the animation controller calls `completeTransition(_:)` on the transitioning context to mark the animation as completed.

If step 1 or step 2 returns `nil`, or isn't implemented at all, the default animation for the transition is used. The objects involved in a custom transition are displayed in the following diagram:

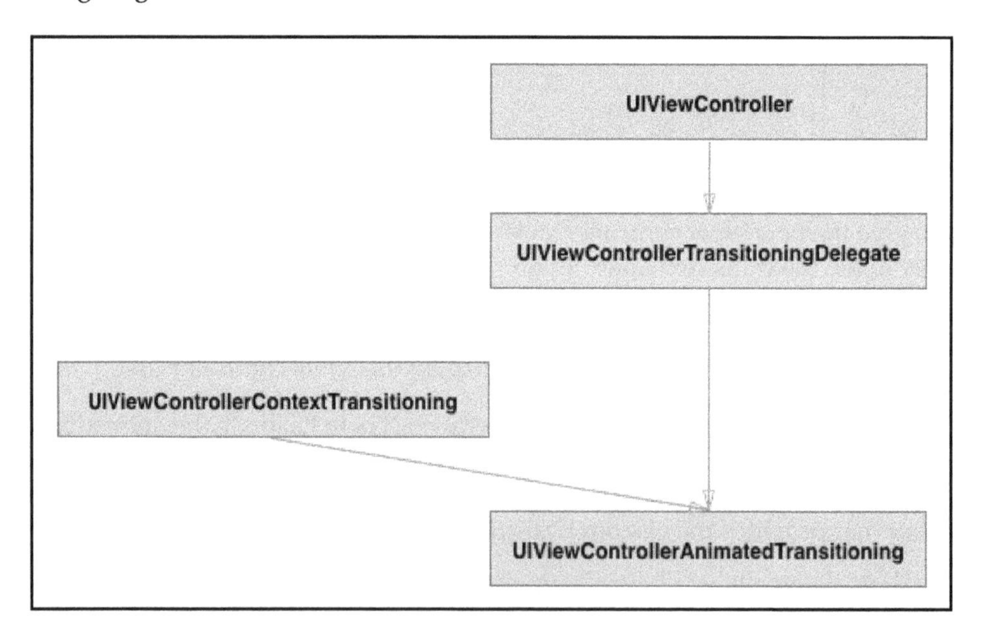

Creating a separate object to control the animation is often a good idea because it allows you to reuse a transition and it keeps your code nice and clean. The animation controller should be an object that conforms to `UIViewControllerAnimatedTransitioning`. This object will take care of animating the presented view onto the screen. Let's create the animation controller object next.

Create a new **Cocoa Touch** class and name it `CustomModalShowAnimator`. Pick `NSObject` as its superclass. This class will act as the animation controller. After creating the new file, open it and add the following extension to make `CustomModalShowAnimator` conform to `UIViewControllerAnimatedTransitioning`:

```
extension CustomModalShowAnimator:
UIViewControllerAnimatedTransitioning {

}
```

This makes the new class conform to the protocol that's required to be an animation controller. Xcode will show a build error because you haven't implemented all the methods to conform to `UIViewControllerAnimatedTransitioning` yet. Let's go over the methods one by one so you end up with a full implementation for the animation controller.

The first method that must be implemented for the animation controller is `transitionDuration(using:)`. The implementation of this method is shown here:

```
func transitionDuration(using transitionContext:
UIViewControllerContextTransitioning?) -> TimeInterval {
    return 0.6
}
```

This method is used to determine the total transition duration in seconds. In this case, the implementation is simple – the animation should last 0.6 seconds.

The second method that needs to be implemented is `animateTransition(using:)`. Its purpose is to take care of the actual animation for the custom transition. The implementation will take the target view controller and its view will be animated from the top of the screen downward to its final position. It will also do a little bit of scaling, and the opacity of the view will be animated; to do this, `UIViewPropertyAnimator` will be used. Add the following implementation to the animator:

```
func animateTransition(using transitionContext:
UIViewControllerContextTransitioning) {
  // 1
  guard let toViewController =
transitionContext.viewController(forKey:
UITransitionContextViewControllerKey.to)
    else { return }

  // 2
```

```
    let transitionContainer = transitionContext.containerView

    // 3
    var transform = CGAffineTransform.identity
    transform = transform.concatenating(CGAffineTransform(scaleX: 0.6,
                                                          y: 0.6))
    transform = transform.concatenating(CGAffineTransform(translationX:
      0, y: -200))

    toViewController.view.transform = transform
    toViewController.view.alpha = 0

    // 4
    transitionContainer.addSubview(toViewController.view)

    // 5
    let animationTiming = UISpringTimingParameters(
      dampingRatio: 0.8,
      initialVelocity: CGVector(dx: 1, dy: 0))

    let animator = UIViewPropertyAnimator(
      duration: transitionDuration(using: transitionContext),
      timingParameters: animationTiming)

    animator.addAnimations {
      toViewController.view.transform = CGAffineTransform.identity
      toViewController.view.alpha = 1
    }


    // 6
    animator.addCompletion { finished in
  transitionContext.completeTransition(!transitionContext.transitionWasC
  ancelled)
    }

    // 7
    animator.startAnimation()
  }
```

A lot is going on in the preceding code snippet. Let's go through the code step by step to see what's happening:

1. The target view controller is extracted from the transition context. This allows you to use the view controller's view in the animation that you're about to perform.

2. Obtain the animation's container view. The container view is a regular `UIView` and its intention is to contain all the animated views.

3. Prepare the target-view controller's view for the animation. The view is transformed so it's off the screen and the transparency is set to make the view completely transparent.
4. Once the view is prepared, it is added to the container view.
5. The animations are set up and added to a property animator.
6. The completion-handler for the property animator is configured so `completeTransition(_:)` is called on the context. The `transitionWasCancelled` variable is used to determine whether the animation completed normally.
7. Start the property animator so the animations begin.

Now that the animation controller is complete, the `UIViewControllerTransitioningDelegate` protocol should be implemented on `CustomPresentedViewController` so it can act as its own `transitioningDelegate`. Open the file and add the following implementation code:

```
class CustomPresentedViewController: UIViewController {

  override func viewDidLoad() {
    super.viewDidLoad()
    transitioningDelegate = self
  }
}

extension CustomPresentedViewController:
UIViewControllerTransitioningDelegate {

  func animationController(forPresented presented: UIViewController,
presenting: UIViewController, source: UIViewController) ->
UIViewControllerAnimatedTransitioning? {
    return CustomModalShowAnimator()
  }

  func animationController(forDismissed dismissed: UIViewController)
-> UIViewControllerAnimatedTransitioning? {
    return nil
  }
}
```

This code adds conformance to the `UIViewControllerTransitioningDelegate` protocol and assigns the view controller as its own transitioning delegate. The `animationController(forPresented:presenting:source:)` method returns the animation controller you created before. The `animationController(forDismissed:)` method returns `nil` for now. Go ahead and test your custom transition! This is all the code required to create a custom display transition. Now that the app can display the modal view controller with a custom transition, let's add an interactive dismissal transition.

Making an interactive dismissal transition

Implementing an interactive transition requires a bit more work than the non-interactive presentation animation, and the way it works is also somewhat harder to grasp. For the non-interactive transition, `transitioningDelegate` simply returns an animation controller from `animationController(forPresented:presenting:source:)`.

To implement an interactive dismiss transition, two methods should be implemented. These two methods work together to make the interactive animation happen. The first method is `animationController(forDismissed:)`. This method will return an object that performs and controls the interactive animation, similar to the animation controller you implemented before.

However, to make the animation interactive, you must also implement the `interactionControllerForDismissal(using:)` method. This method should return an object that works together with the object that is returned from `animationController(forDismissed:)`. The way this ties together can roughly be summed up as follows:

1. A `UIViewControllerAnimatedTransitioning` object is requested by calling `animationController(forDismissed:)`.
2. A `UIViewControllerInteractiveTransitioning` object is requested by calling `interactionControllerForDismissal(using:)`. The `UIViewControllerAnimatedTransitioning` object that was retrieved earlier is passed to this method. If this method returns `nil`, the transition will be executed without being interactive.
3. If both methods return a valid object, the transition is interactive.

Let's take a look at how this compares to the previous animation flow we looked at earlier:

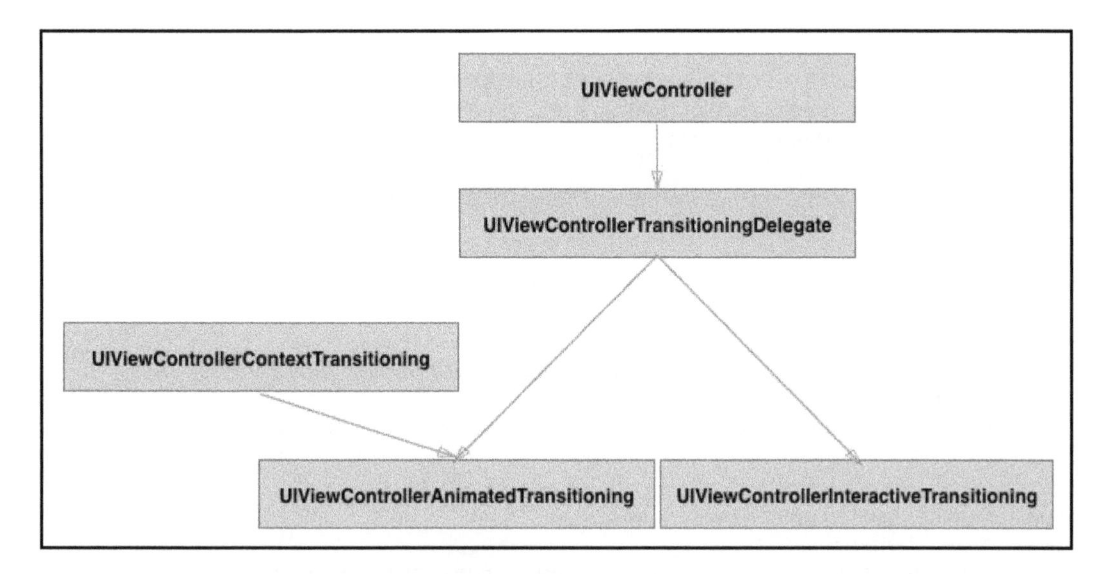

For convenience, you will implement both `UIViewControllerAnimatedTransitioning` and `UIViewControllerInteractiveTransitioning` in a single class. This will make it a little bit easier to see how everything ties together.

Create a new `Cocoa Touch` class and name it `CustomModalHideAnimator`. Choose `UIPercentDrivenInteractiveTransition` as its superclass. This class implements convenience methods to easily update the interactive transition. It also conforms to `UIViewControllerInteractiveTransitioning`, so you don't have to add conformance yourself. However, you should make sure to add conformance to `UIViewControllerAnimatedTransitioning` to `CustomModalHideAnimator` declaration yourself.

Let's start off by implementing a custom initializer that will connect the `CustomPresentedViewController` instance to `CustomModalHideAnimator`. This allows you to add a gesture-recognizer to the modal view and update the animation based on the status of the gesture-recognizer. Add the following code to the implementation for `CustomModalHideAnimator`:

```
let viewController: UIViewController

// 1
init(withViewController viewController: UIViewController) {
```

```
    self.viewController = viewController

    super.init()

    // 2
    let panGesture = UIScreenEdgePanGestureRecognizer(target: self,
  action: #selector(handleEdgePan(gestureRecognizer:)))
    panGesture.edges = .left

    viewController.view.addGestureRecognizer(panGesture)
  }

  @objc func handleEdgePan(gestureRecognizer:
  UIScreenEdgePanGestureRecognizer) {
    // 3
    let panTranslation = gestureRecognizer.translation(in:
  viewController.view)
    let animationProgress = min(max(panTranslation.x / 200, 0.0), 1.0)

    // 4
    switch gestureRecognizer.state {
    case .began:
      viewController.dismiss(animated: true, completion: nil)
    case .changed:
      update(animationProgress)
      break
    case .ended:
      if animationProgress < 0.5 {
        cancel()
      } else {
        finish()
      }
      break
    default:
      cancel()
      break
    }
  }
```

This snippet starts off with a custom initializer that immediately ties a `UIViewController` instance to itself. It completes the initialization by calling the superclass's initializer, and then the pan gesture is added to the view. `UIScreenEdgePanGestureRecognizer` is used so the pan gesture is bound to swiping from the left edge of the screen. This mimics the standard gesture that's usually used to go back one page in a navigation-view controller.

In `handleEdgePan(_:)`, the distance that the user has swiped is determined and this distance is translated into a completion percentage for the animation. Finally, the state of the gesture recognizer is used to determine the appropriate action. If the user just started the gesture, the view controller is told to begin performing the dismissal.

If the gesture has changed, the animation's progress is updated by calling the `update(_:)` method of `UIPercentDrivenInteractiveTransition`. If the gesture has ended, the current progress is determined. If there is enough progress, the transition is finished automatically. For all other states, it is assumed that the gesture got canceled so the transition is canceled as well. If you noticed some similarities between this implementation and the interactive drawer animation example you saw before, that means you're paying close attention! The implementations for tracking a gesture and updating an animation's progress are pretty similar.

So far, you have implemented the interactive part of the animation, but you don't have a complete animation controller just yet. To have a complete animation controller, you need to implement the actual dismissal animation and make `CustomModalHideAnimator` conform to the `UIViewControllerAnimatedTransitioning` protocol. Before you do this, add the following variable to the `CustomModalHideAnimator` class:

```
var currentAnimator: UIViewPropertyAnimator?
```

You didn't have to add a variable like this to the animator you wrote for the custom show animation. The reason you have to implement this property for the interactive animation is that we need to keep a reference to it so it can be returned from the `interruptibleAnimator(using:)` method. This method is part of the `UIViewControllerAnimatedTransitioning` protocol and is used specifically for interactive animations that use `UIViewPropertyAnimator`. To finish your custom dismiss transition, add the following extension to `CustomModalHideAnimator.swift`:

```
extension CustomModalHideAnimator:
UIViewControllerAnimatedTransitioning {
  func transitionDuration(using transitionContext:
UIViewControllerContextTransitioning?) -> TimeInterval {
    return 0.6
  }

  // 1
  func animateTransition(using transitionContext:
UIViewControllerContextTransitioning) {
    let animator = interruptibleAnimator(using: transitionContext)
    animator.startAnimation()
```

```
  }

  // 2
  func interruptibleAnimator(using transitionContext:
UIViewControllerContextTransitioning) -> UIViewImplicitlyAnimating {

    // 3
    if let animator = currentAnimator {
      return animator
    }

    guard let fromViewController = transitionContext.viewController(
      forKey: UITransitionContextViewControllerKey.from),
      let toViewController = transitionContext.viewController(
        forKey: UITransitionContextViewControllerKey.to) else {
          return UIViewPropertyAnimator()
    }

    let transitionContainer = transitionContext.containerView

    transitionContainer.addSubview(toViewController.view)
    transitionContainer.addSubview(fromViewController.view)

    let animationTiming = UISpringTimingParameters(
      dampingRatio: 0.8,
      initialVelocity: CGVector(dx: 1, dy: 0))

    let animator = UIViewPropertyAnimator(
      duration: transitionDuration(using: transitionContext),
      timingParameters: animationTiming)

    // 4
    self.currentAnimator = animator

    animator.addAnimations {
      var transform = CGAffineTransform.identity
      transform = transform.concatenating(CGAffineTransform(scaleX:
        0.6, y: 0.6))
      transform =
transform.concatenating(CGAffineTransform(translationX: 0, y: -200))

      fromViewController.view.transform = transform
      fromViewController.view.alpha = 0
    }

    animator.addCompletion { [weak self] finished in
      // 5
      self?.currentAnimator = nil
```

```
transitionContext.completeTransition(!transitionContext.transitionWasC
ancelled)
    }

    return animator
  }
}
```

The preceding code is very similar to the code you wrote for the custom show animation. However, instead of configuring and starting the animation in `animateTransition(using:)` (step 1), you obtain the animator object from `interruptibleAnimator(using:)` and then start the animations. This takes care of implementing a non-interactive version of the dismiss animation.

The second step highlights `interruptibleAnimator(using:)`; this method is expected to return an object that conforms to `UIViewImplicitlyAnimating`. `UIViewPropertyAnimator` conforms to this protocol automatically. The documentation for `interruptibleAnimator(using:)` states that you must return the same animator instance for the duration of a certain transition. This is why you had to store the current animator on the `CustomModalHideAnimator` class. When this method is called and an animator exists, the current animator is returned, as shown in step 3. Step 4 highlights storing the animator on `CustomModalHideAnimator`. The last step is to set `currentAnimator` to `nil` once the transition is completed.

The final step is to add and use the custom animator you just created in `CustomPresentedViewController`. Create a property for the hide animator and update `viewDidLoad()` as follows:

```
var hideAnimator: CustomModalHideAnimator?

override func viewDidLoad() {
  super.viewDidLoad()
  transitioningDelegate = self

  hideAnimator = CustomModalHideAnimator(withViewController: self)
}
```

The preceding code creates an instance of `CustomModalHideAnimator` and binds the view controller to it by passing it to the initializer. Next, update the code in `animationController(forDismissed:)` so it returns `hideAnimator` instead of `nil`:

```
func animationController(forDismissed dismissed: UIViewController) ->
UIViewControllerAnimatedTransitioning? {
```

```
        return hideAnimator
    }
```

Finally, add the `interactionControllerForDismissal(using:)` method to the `UIViewControllerTransitioningDelegate` extension on `CustomPresentedViewController` so the transition becomes interactive:

```
    func interactionControllerForDismissal(using animator:
    UIViewControllerAnimatedTransitioning) ->
    UIViewControllerInteractiveTransitioning? {
        return hideAnimator
    }
```

Try to run your app now and swipe from the left edge of the screen once you've presented your custom modal view. You can now interactively make the view go away by performing a gesture. Clever implementations of custom transitions can make users feel in control of the application and the way it responds to how they interact with it.

Implementing a custom transition isn't an easy task. There are a lot of moving parts involved and the amount of delegation and protocols used can be daunting. Take the time to go over the code you've written a few more times to figure out what exactly is going on if you need to. Again, this isn't an easy topic to grasp right away.

Implementing a custom UINavigationController transition

The view-controller transition technique that you just explored is very nice when you want to create a custom modal presentation. However, if you want to customize transitions in `UINavigationController` or `UITabBarController` that persist throughout your app, you need to implement the transitions in a slightly different way.

Let's take a look at how the setup for animating push transitions for `UINavigationController` differs from the setup that is used for the transition you saw earlier:

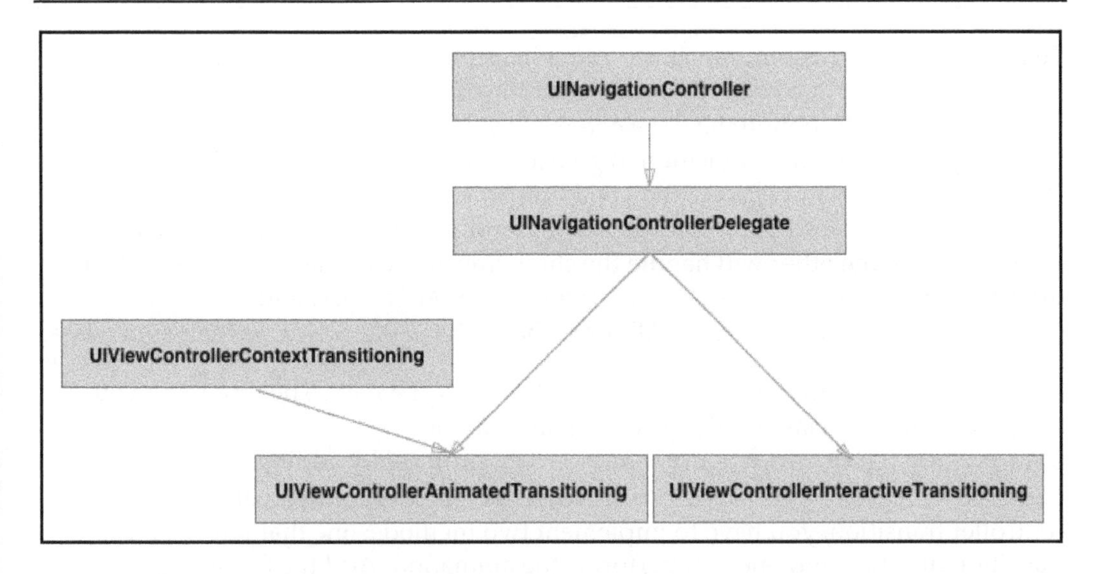

The depicted flow is one for an interactive transition. It's very similar to the way a normal view-controller transition works, except that for this type of transition, `UINavigationControllerDelegate` is the object that provides the `UIViewControllerAnimatedTransitioning` and `UIViewControllerInteractiveTransitioning` objects that are used to perform transitions between views as they are pushed onto the navigation stack and when they are popped off.

Because the delegate that is responsible for the transitions is set on the navigation controller instead of on a displayed view controller, every push and pop that is performed by the navigation controller that has a custom delegate will use the same custom transition. This can come in handy when you want to have consistent behavior throughout your app without manually assigning transitioning delegates all of the time.

To see how a custom navigation controller transition can be implemented, you will create a custom transition that zooms in on a contact from the contact overview page. When a user taps a contact, the contact's detail page will expand and grow from the contact's picture until the detail page covers the entire window, like it's supposed to. Pressing the Back button will shrink the view back down onto the tapped contact's image. Swiping from the left edge of the screen will interactively shrink the view, using the same animation that is triggered by tapping the back button.

To implement this custom transition, you will implement three classes. A `NavigationDelegate` class will implement the `UINavigationController` delegate and it will contain the `UIPercentDrivenInteractiveTransition` object to manage the interactive transition to go back to the overview page. The other two classes are the animator classes; they both implement the `UIViewControllerAnimatedTransitioning` protocol. One is responsible for the hide transition; the other will handle the show transition. Create three files and name them `NavigationDelegate`, `ContactDetailShowAnimator`, and `ContactDetailHideAnimator`. All three should have `NSObject` as their superclass.

Let's begin by implementing `ContactDetailShowAnimator`. The first thing you should do with this class is added conformance to the `UIViewControllerAnimatedTransitioning` protocol by adding an extension in `ContactDetailShowAnimator.swift`. Just like you did for the regular view-controller transition, you have to implement two methods: one that returns the transition duration and one that performs the animation. Add the following implementation to your extension:

```
extension ContactDetailShowAnimator:
UIViewControllerAnimatedTransitioning {
  func transitionDuration(using transitionContext:
UIViewControllerContextTransitioning?) -> TimeInterval {
    return 0.3
  }

  func animateTransition(using transitionContext:
UIViewControllerContextTransitioning) {
    // 1
    guard let toViewController =
transitionContext.viewController(forKey: .to),
      let fromViewController =
transitionContext.viewController(forKey: .from),
      let overviewViewController = fromViewController as?
ViewController,
      let tappedIndex =
overviewViewController.collectionView.indexPathsForSelectedItems?.firs
t,
      let tappedCell =
overviewViewController.collectionView.cellForItem(at: tappedIndex) as?
ContactCollectionViewCell
      else { return }

    // 2
    let contactImageFrame = tappedCell.contactImage.frame
    let startFrame =
```

```
overviewViewController.view.convert(contactImageFrame, from:
tappedCell)

    toViewController.view.frame = startFrame
    toViewController.view.layer.cornerRadius = startFrame.height / 2

    transitionContext.containerView.addSubview(toViewController.view)

    let animationTiming = UICubicTimingParameters(animationCurve:
.easeInOut)

    let animator = UIViewPropertyAnimator(duration:
transitionDuration(using: transitionContext), timingParameters:
animationTiming)

    animator.addAnimations {
      toViewController.view.frame = transitionContext.finalFrame(for:
toViewController)
      toViewController.view.layer.cornerRadius = 0
    }

    animator.addCompletion { finished in
transitionContext.completeTransition(!transitionContext.transitionWasC
ancelled)
    }

    animator.startAnimation()
  }
}
```

The first step in the preceding snippet shows how you can extract information about the tapped cell by casting `fromViewController` to an instance of `ViewController`, which is the page that contains an overview of all the contacts. This view controller's collection view is asked for for the selected index path, which is then used to determine the cell that the user has tapped on. All of the work that's done in this first part of the code deals with `Optionals`, which means that the values might not be present according to the compiler. Even though under normal conditions none of these operations should return `nil`, they are safely unwrapped and accessed using a `guard` statement.

Then, the detail view controller's initial frame is set up. To determine this frame, the frame for `contactImage` in `sourceCell` is extracted. Then, this frame is converted to the coordinates of `overviewViewController`. If you don't do this, the position of the frame will typically be off by about 64 points. That's because the collection view has a content inset of `64` so it can extend beneath the navigation bar. Converting to the proper coordinate space ensures that this won't be a problem for you.

After converting the image's frame, it's used as the starting frame for the target view. The target also gets rounded corners to aid the zooming-in effect. The animation is set up to remove the rounded corners and to adjust the frame to the planned end frame so the detail page covers the screen.

The next step is to implement the back transition. This transition is nearly identical to the "show" transition. Open the `ContactDetailHideAnimator.swift` file and add an extension to make `ContactDetailHideAnimator` conform to `UIViewControllerAnimatedTransitioning`. After adding the delegate, you should be able to implement `transitionDuration(using:)` on your own. Make sure it returns a duration of `0.3` seconds.

The following snippet contains the code you need to implement the back animation. Try implementing this on your own first; you can deduct all the code you need from the show animation and the custom modal hide transition you built:

```
func interruptibleAnimator(using transitionContext:
UIViewControllerContextTransitioning) -> UIViewImplicitlyAnimating {
  if let currentAnimator = self.animator {
    return currentAnimator
  }

  guard let toViewController =
transitionContext.viewController(forKey: .to),
    let fromViewController = transitionContext.viewController(forKey:
.from),
    let overviewViewController = toViewController as? ViewController,
    let tappedIndex =
overviewViewController.collectionView.indexPathsForSelectedItems?.firs
t,
    let tappedCell =
overviewViewController.collectionView.cellForItem(at: tappedIndex) as?
ContactCollectionViewCell
    else { return UIViewPropertyAnimator() }
  transitionContext.containerView.addSubview(toViewController.view)
  transitionContext.containerView.addSubview(fromViewController.view)

  let animationTiming = UICubicTimingParameters(animationCurve:
```

```
.easeInOut)

  let animator = UIViewPropertyAnimator(duration:
transitionDuration(using: transitionContext), timingParameters:
animationTiming)

  animator.addAnimations {
    let imageFrame = tappedCell.contactImage.frame
    let targetFrame = overviewViewController.view.convert(imageFrame,
from: tappedCell)
    fromViewController.view.frame = targetFrame
    fromViewController.view.layer.cornerRadius =
tappedCell.contactImage.frame.height / 2
  }

  animator.addCompletion { finished in
transitionContext.completeTransition(!transitionContext.transitionWasC
ancelled)
  }

  self.animator = animator

  return animator
}

func animateTransition(using transitionContext:
UIViewControllerContextTransitioning) {
  let animator = interruptibleAnimator(using: transitionContext)
  animator.startAnimation()
}
```

The animations are now fully implemented. The last object you need to implement is `UINavigationControllerDelegate`. As discussed before, this delegate is responsible for providing animations and managing interactive back gestures. First, you will implement the basics for your navigation delegate. Add the following code to the `NavigationDelegate` class:

```
let navigationController: UINavigationController
var interactionController: UIPercentDrivenInteractiveTransition?

init(withNavigationController navigationController:
UINavigationController) {
    self.navigationController = navigationController

    super.init()

    let panRecognizer = UIPanGestureRecognizer(target: self,
```

```
                                  action:
#selector(handlePan(gestureRecognizer:)))
    navigationController.view.addGestureRecognizer(panRecognizer)
}
```

The initializer for `NavigationDelegate` takes a navigation controller as an argument. This immediately associates a navigation controller with the navigation delegate instance. `UIPanGestureRecognizer` is added to the view of the navigation controller directly. This gesture recognizer will drive the interactive transition. The next step is to implement the handler for the pan gesture-recognizer:

```
@objc func handlePan(gestureRecognizer: UIPanGestureRecognizer) {
  guard let view = self.navigationController.view
    else { return }

  switch gestureRecognizer.state {
  case .began:
    let location = gestureRecognizer.location(in: view)
    if location.x < view.bounds.midX &&
      navigationController.viewControllers.count > 1 {

      interactionController = UIPercentDrivenInteractiveTransition()
      navigationController.popViewController(animated: true)
    }
    break
  case .changed:
    let panTranslation = gestureRecognizer.translation(in: view)
    let animationProgress = fabs(panTranslation.x / view.bounds.width)
    interactionController?.update(animationProgress)
    break
  default:
    if gestureRecognizer.velocity(in: view).x > 0 {
      interactionController?.finish()
    } else {
      interactionController?.cancel()
    }

    interactionController = nil
  }
}
```

This method is very similar to the one you saw before for the regular view-controller transition. The major difference here is that you create `UIPercentDrivenInteractiveTransition` when the gesture begins. The percentage-driven interactive transition is then destroyed when the gesture ends.

Add the following extension to `NavigationDelegate` to make it conform to `UINavigationControllerDelegate`:

```
extension NavigationDelegate: UINavigationControllerDelegate {
  func navigationController(_ navigationController:
UINavigationController, animationControllerFor operation:
UINavigationController.Operation, from fromVC: UIViewController, to
toVC: UIViewController) -> UIViewControllerAnimatedTransitioning? {
    if operation == .pop {
      return ContactDetailHideAnimator()
    } else {
      return ContactDetailShowAnimator()
    }
  }

  func navigationController(_ navigationController:
UINavigationController, interactionControllerFor animationController:
UIViewControllerAnimatedTransitioning) ->
UIViewControllerInteractiveTransitioning? {
    return interactionController
  }
}
```

These two methods are responsible for providing the required objects for the animations. Previously, we had one method that got called whenever a view controller was shown, and one when it was dismissed. `UINavigationControllerDelegate` has only one method for this. You can check whether the navigation controller is pushing or popping a view controller and, based on that, you can return a different animator.

The final step is to connect the animators to `ViewController`. Declare the following variable in `ViewController.swift`:

```
var navigationDelegate: NavigationDelegate?
```

Next, add the following lines of code to `viewDidLoad()`:

```
if let navigationController = self.navigationController {
    navigationDelegate = NavigationDelegate(withNavigationController:
navigationController)
    navigationController.delegate = navigationDelegate
}
```

If you run the app now, you can see your custom transition in action. However, there's still one problem. You'll notice that the drawer you added to the contact detail view controller earlier is visible outside the boundaries of the detail view controller. Also, the rounded-corner effect you added to the show and hide animations doesn't seem to be working yet. Add the following line of code to the `viewDidLoad()` method in `ContactDetailViewController`:

```
view.clipsToBounds = true
```

That's it! You have successfully implemented an interactive transition for a navigation controller. Build and run your app, tap on a cell, and see the freshly-created zoom-in and -out effect in action. Also, try swiping from the left edge of the screen to go back to the overview page. Pretty awesome, right?

Summary

In this chapter, you learned a lot about animation. You now know that you can easily animate a view's property using the powerful `UIViewPropertyAnimator` object. You learned what timing functions are and how they affect animations. Also, more importantly, you saw how to make use of springs and UIKit Dynamics to make your animations look more lifelike. After learning some animation basics, you also learned how to add custom animations to view-controller transitions. The complexity of this does not necessarily lie in the animations themselves. The web of collaborating objects makes it difficult to understand custom view-controller transitions at first. Once you have implemented this a few times, it will get easier to make sense of all the moving parts, and you'll see how all the building blocks fall into place.

Despite the amount of information already present in this chapter, it does not cover every single aspect of animation. Just the most important parts of animations are covered, so you can utilize them to build better, more engaging apps. If you want to learn more about animation, you should look for resources on Core Animation. This is the framework iOS uses under the hood to drive all of the animations we created.

With these animations in place, the **Hello-Contacts** app is complete. Also, with that, the first section of this book is complete. You've learned everything you need to know about creating layouts with Auto Layout, displaying data in lists, and pushing new views to the screen. The next chapter will focus on Swift and how you can make use of some of Swift's most powerful features, such as protocols and value types.

Questions

1. Name two ways to implement simple animations in iOS.

   a) `CoreGraphics` and UIKit Dynamics.
   b) `UIViewPropertyAnimator` and `UIView.animate`.
   c) `UIViewControllerTransitioning` and `UIPercentageDrivenAnimation`.

2. What does the damping property do in a spring animation?

   a) It is used for the bounciness of the animation.
   b) It sets up the initial velocity of the animation.
   c) It determines the initial acceleration of the animation.

3. Name an advantage of `UIViewPropertyAnimator`.

   a) Better performance.
   b) More animation capabilities.
   c) Easier to start, pause, and interact with.

4. What is a difference between implementing `UIViewControllerTransiting` for interactive and non-interactive transitions?

   a) The pan gesture-recognizer.
   b) You must use `UIViewPropertyAnimator`.
   c) You must implement `interruptibleAnimator(using:)`.

5. How do you prevent a view from showing out-of-bounds contents?

   a) By setting `clipToBounds` to true on the view.
   b) By setting `clipToBounds` to true on the view controller.
   c) By setting `overflowHidden` to true on the view.

6. Is it possible to add animations to a running property animator?

   a) No.
   b) Yes.

Further reading

- Core Animation: `https://developer.apple.com/documentation/quartzcore`
- UIKit Dynamics: https://developer.apple.com/documentation/uikit/animation*and*haptics/uikit_dynamics

5
Understanding the Swift Type System

The previous section left you with a solid foundation you can use to build great, adaptive apps on. At this point, it is a good idea to take a step back and look at the code you have written to gain a deeper understanding of Swift and how it works under. This section focuses on teaching you more about Swift as a language, regardless of what you intend to build.

In this chapter, you will learn about Swift's fantastic type system. Swift's type system is one of its most powerful features because it allows developers to express complex and flexible principles safely and predictably. This chapter takes you through the following steps:

- You will learn about the types that are available to you
- You will learn about the differences between the types, including their pros and cons
- You will learn how to make informed decisions about the types you use in your code to make sure your code is of great quality

Knowing what types are available

To write great code, you need to learn what tools are available in your toolbox. This applies to building apps and understanding the features `UIKit` has to offer, but it also applies to the language you use to write software in. Different languages come with various features, best practices, pros, and cons. The deeper your understanding of the language you work with, the better the decisions you can make about the code you write. As mentioned before, Swift's type system is one of the features that makes Swift such an excellent language for both experts and beginners to develop in.

Before you dive into the details of Swift's types and how they compare to each other, it's essential that you know what types Swift has to offer you. On a high level, you can argue that Swift has two types:

- Reference types
- Value types

Let's have a closer look at each type to see what they mean, how they work, and how you can use them.

Reference types

The types you have seen so far in this book were mostly, if not all, reference types. Two types of objects are classified as reference types:

- Classes
- Closures

You have seen both of these object types in this book already. For instance, all `UIViewController` subclasses you have created are reference types. All closures that you used as callbacks or to perform animations in are also reference types.

So what does it mean if something is a reference type, and why should it matter to you? Well, reference types come with behavior that can be both convenient and very frustrating depending on what you are trying to achieve in your code.

One feature that is unique to reference types and classes is the ability to subclass. The only type that can inherit functionality from another object is a class. This will be covered in more depth when you learn about the differences between types, but it's good to be aware of this information already. Let's examine reference types up close by writing some code in a Playground.

Create a new Playground project in Xcode and give it a name. Then add the following code to it:

```
class Pet {
  var name: String

  init(name: String) {
    self.name = name
  }
}
```

```
func printName(for pet: Pet) {
  print(pet.name)
}

let cat = Pet(name: "Misty")
printName(for: cat)
```

It's likely that you're not too excited about this little snippet of code. All it does is define a new `Pet` class, make an instance of it, and then it passes that instance into `printName(for:)`. However, this code is extremely well-suited to illustrating what a reference type is.

When you call `printName(for: cat)`, you pass a *reference* to your cat instance to `printName(for:)`. This means that it is possible for anybody who gets ahold of this reference to make changes to the object that is referenced. If this sounds confusing, that's okay.

Add the following code to the Playground you have created and then run it:

```
func printName2(for pet: Pet) {
  print(pet.name)
  pet.name = "Jeff"
}

let dog = Pet(name: "Bingo")
printName2(for: dog) // Bingo
print(dog.name)
```

What do you notice in the console after running this? If you noticed that the dog's name changes from `Bingo` to `Jeff`, you have just observed what it means to pass a reference to something around. Since `printName2(for:)` received a reference to your `Pet` instance, it was able to change its name. If you have programmed in other languages, this might be obvious to you. If not, this might be very surprising.

One more thing you should note is that `dog` was declared as a constant. Regardless, you were allowed to change the name of your instance from *Bingo* to *Jeff*. If you think this is obvious, add the following code to your Playground and run it:

```
import UIKit
let point = CGPoint(x: 10, y: 10)
point.x = 10
```

This code is very similar to what you did with the `Pet` instance. You make a constant instance of a thing, and then you change one of its properties. This time, however, when you try to run your Playground, you should see an error along the lines of:

```
Cannot assign to property:'point' is a 'let' constant.
```

Even though the code you implemented so far is pretty small, it does a great job of demonstrating reference types. You have currently seen two properties of a reference type in action:

- Anybody that receives an instance of a reference type can mutate it
- You can change properties on a reference type, even if the property that holds onto the reference type is declared as a constant

These two characteristics are typical of reference types. The reason reference types work like this is that a variable or constant that is assigned a reference type *does not contain or own* the object. The constant or variable only points to an address in memory where the instance is stored.

Any time you create an instance of a reference type, it is written to RAM where it will exist at a particular address. RAM is a special type of memory that is used by computers, such as an iPhone, to temporarily store data in that is used by a certain program. When you assign an instance of a reference type to a property, the property will have a **pointer** to the memory address for this instance. Have another look at the following line of code:

```
let dog = Pet(name: "Bingo")
```

The `dog` constant now points to a particular address in memory where the `Pet` instance is stored. You are allowed to change properties on the `Pet` instance as long as the underlying memory address isn't changed. In fact, you could theoretically put something entirely different at that memory address, and `let dog` won't care because it still points to the same address.

For this same reason, it is possible for `printName2(for:)` to change a pet's name. Instead of passing it an instance of `Pet`, you pass it the memory address at which the instance is expected to exist. It's okay for `printName2(for:)` to make changes to the `Pet` instance because it doesn't change the underlying address in memory.

If you tried to assign a new instance to `dog` by typing the following, you would get an error:

```
dog = Pet(name: "Nala")
```

The reason this would error is that you can't change the memory address `dog` points to since it's a constant. The following image should clarify this a little bit more by visualizing what it means to pass around a reference type:

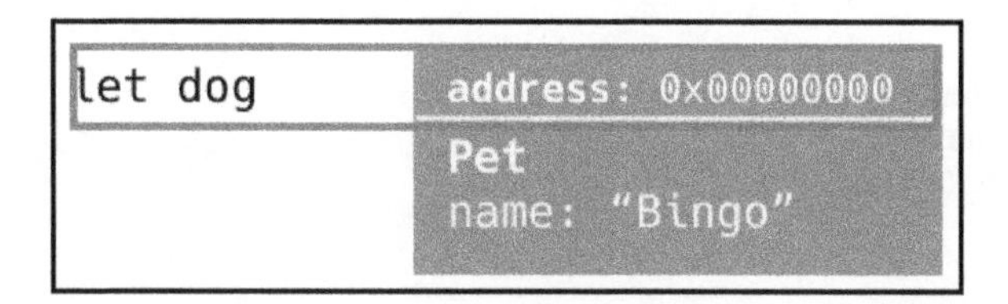

Now that you know what a reference type is and how it works, you might have already concluded that the `CGPoint` you saw in the preceding example must be a value type. Let's see what value types are all about next.

Value types

In the examples for references types, you saw the following snippet of code:

```
import UIKit
let point = CGPoint(x: 10, y: 10)
point.x = 10
```

At first sight, you might expect a value type to be a special kind of class because it looks like this snippet creates an instance of a class call `CGPoint`. You would be right in your observation, but your conclusion is wrong. `CGPoint` is not a class at all. Classes are inherently reference types, and they can't ever be something else. So what are value types, then?

There are two types of objects that are considered to be value types:

- Structs
- Enums

These two types are both very different, so let's make sure that you understand the basics of value types first, and then you'll learn what each of these two types is.

Let's have a look at the `Pet` example again, but use a struct instead of a class. Create a new Playground page in Xcode by opening the **Project Navigator** in your playground project and clicking the plus symbol in the bottom-left corner. This will allow you to add a new page to your Playground.

Add the following code to your new page:

```
struct Pet {
  var name: String
}

func printName(for pet: Pet) {
  print(pet.name)
  pet.name = "Jeff"
}

let dog = Pet(name: "Bingo")
printName(for: dog)
print(dog.name)
```

You will immediately notice that Xcode complains. The error you should see in the console tells you that `pet` is a `let` constant and you are not allowed to change its name. You can turn `pet` into a variable by updating `printName` as follows:

```
func printName(for pet: Pet) {
  var pet = pet
  print(pet.name)
  pet.name = "Jeff"
}
```

If you run your Playground now, make sure to look at the console closely. You'll notice that the pet's name remains unchanged in the second print. This demonstrates one of the key features of a value type. Instead of passing a reference to an address in memory around, a copy of the object is passed around. This explains why you aren't allowed to change properties on a value type that is assigned to a constant. Changing that property would change the value type's value, and it would, therefore, change the value of the constant. This also means that when you pass `dog` to `printName`, you pass a copy of the `Pet` instance to `printName`, meaning that any changes made to the instance are local to the `printName` function and won't be applied to `dog` in this case.

This behavior makes using value types extremely safe because it's tough for other objects or functions to make unwanted changes to a value type. And if you define something as a constant, it truly is a constant. Another characteristic of value types is that they're typically very fast and lightweight because they can exist on the `Stack` while reference types exist on the `Heap`. You'll learn more about this later when we compare reference types and value types.

Now that you have a basic understanding of value types, let's have a look at the specific value types: structs and enums.

Understanding structs

Structs are similar to classes in the way you define them. If you look at the Pet class you defined earlier, it might be easy to miss the fact that it's a struct. If you pay close attention, you will notice one big difference though. You didn't have to write an initializer for the struct! Swift can automatically generate initializers for structs. This is extremely convenient and can save you a lot of typing for larger structs.

Structs also can't inherit functionality from other objects. This means that structs always have a very flat and transparent set of properties and methods. This allows the compiler to make optimizations to your code that make structs extremely lightweight and fast.

A struct can, however, conform to protocols. The Swift standard library is full of protocols that define features for many of the built-in types, such as Array, Dictionary, and Collection. Most of these built-in types are implemented as structs that adopt one or more protocols.

One last thing you need to understand about structs is that they are very strict about whether they can be modified. Consider a struct that looks as follows:

```
struct Car {
  var gasRemaining: Double

  func fillGasTank() {
    gasRemaining = 1
  }
}
```

This struct will cause the compiler to throw an error. A struct itself is immutable by default, which means you cannot change any of its values. It's up to you to make it explicit to the compiler when a method can mutate, or change, a struct. You do this by adding the mutating keyword to a function, as follows:

```
mutating func fillGasTank() {
  gasRemaining = 1
}
```

When you create a constant instance of Car and call fillGasTank() on it, the compiler will error again. If you call a mutating function on a let instance, you mutate the instance, meaning the value of the property would change. Because of this, you can only call mutating functions on variable properties.

Understanding enums

An enum is a type that holds a finite set of predefined values. Enums are often used to represent a particular state or result of an operation. The best way to learn what this means is to look at an example of an enum that represents the state of a traffic light:

```
struct TrafficLight {
  var state: TrafficLightState
  // ...
}

enum TrafficLightState {
  case green, yellow, red
}
```

This sample shows a `TrafficLight` struct that has a `state` property. The type of this property is `TrafficLightState`, which is an enum. The `TrafficLightState` defines three possible states for a traffic light. This is very convenient because an enum such as this eliminates the possibility of a bad state because the compiler can now enforce that you never end up with an invalid value.

Enums can also contain properties and methods, just like structs can. However, an enum can also have an *associated value*. This means that each possible case can have a representation in a different type, such as a `String`. If you modify `TrafficLightState` as shown here, it will have a `String` as its `rawValue`:

```
enum TrafficLightState: String {
  case green, yellow, red
}
```

If Swift can infer the raw value, you don't have to do anything more than add the type of the raw value to the enum's type declaration. In this sample, the raw value for the `green` enum case will be the "green" string. This can be convenient if you need to map your enum to a different type, for instance, to set it as a label's text.

Just like structs, enums cannot inherit functionality from other objects, but they can conform to protocols. You make an enum conform to a protocol with an extension, just like you would do for classes and structs.

This wraps up the exploration of value types. Now that you know what value types and reference types are, let's explore some of their differences!

Understanding differences in types

Being aware of the available types in Swift, knowing their similarities, and, more importantly, their differences will help you make better decisions about the way you write your code. The preceding segments have listed several properties of value types and references types. More specifically, you learned a lot about classes, structs, and enums. Closures are also a reference type because they get passed around by their location in memory rather than their value, but there isn't much else to say about them in this context.

The most obvious comparison you can make is probably between structs and classes. They look very similar, but they have very different characteristics, as you have already seen. Enums are a special type altogether, they represent a value from a fixed number of possible values but are very similar to structs otherwise.

The most import difference you need to understand is the general difference between value types and reference types, and the difference between structs and classes specifically. Let's have a look at value types and reference types first, so you have the general picture. Then you'll learn about the specific differences between structs and classes.

Comparing value types to reference types

When comparing value types to reference types, it is essential to distinguish between the differences you can see as a developer and the differences that are internal to Swift and the way your app will end up working. Knowing these details will ensure that you can make a well-informed decision that considers all the implications instead of only focusing on memory usage or developer convenience.

Let's examine the more obvious and visible differences first. Afterward, you will learn about the memory implications for each type.

Differences in usage

Add a new page to your Playground and give it a name again. Something such as **Values vs References** would be a good name for your new page. Add the following code to your Playground:

```
protocol PetProtocol {
  var name: String { get }
  var ownerName: String { get set }
```

```
}

class Animal {
  let name: String

  init(name: String) {
    self.name = name
  }
}

class Pet: Animal, PetProtocol {
  var ownerName: String

  init(name: String, ownerName: String) {
    self.ownerName = ownerName

    super.init(name: name)
  }
}
```

This code defines a `PetProtocol` that requires two properties to exist on all objects that conform to this protocol. The `name` property is defined as a constant since it only needs it to be gettable, and `ownerName` is a variable since it requires both `get` and `set`. The code also defines an `Animal` and `Pet` class. `Pet` is a subclass of `Animal`, and it conforms to `PetProtocol` because `Animal` satisfies the `name` constant requirement and `Pet` itself satisfies the `ownerName` variable.

Try changing the `class` declarations to `struct`. Your Playground will not compile now because structs cannot inherit from other objects like classes can. This is a limitation that is sometimes frustrating because you can end up with a lot of code duplication. Imagine that, in addition to `Pet`, you would like to create more types of animals, such as a `WildAnimal`, `SeaCreature`, this would be easy to achieve with classes because you can inherit from `Animal`. This is not possible with structs, so you would implement all these types as structs they would need to duplicate their `Animal` logic.

Another difference between value types and reference types is how they act when they are passed around. Add the following code to your Playground:

```
class ImageInformation {
  var name: String
  var width: Int
  var height: Int

  init(name: String, width: Int, height: Int) {
```

```
    self.name = name
    self.width = width
    self.height = height
  }
}

struct ImageLocation {
  let location: String
  let isRemote: Bool
  var isLoaded: Bool
}

let info = ImageInformation(name: "ImageName", width: 100, height:
100)
let location = ImageLocation(location: "ImageLocation", isRemote:
false, isLoaded: false)
```

The declarations for `info` and `location` look very similar, but their underlying types are entirely different. Try writing a function that takes both `ImageLocation` and `ImageInformation` as an argument. And then try updating the `isLoaded` property of `location` and changing the `name` of `info`. The compiler will complain when you try to set `isLoaded` because the argument for `ImageLocation` is a `let constant`. The reason for this was described earlier in the discussion on value types.

Value types are passed around by value, meaning that changing a property of the argument will change the value altogether. Arguments for a function are always constants. This might not be obvious when you use a Reference type though, because it is perfectly fine to change the `name` property on `ImageInformation` inside of a function. This is because you don't pass the entire value around when you pass a reference type to a function. You pass the reference to the memory address around. This means that instead of the value being a constant, the underlying memory address is a constant. This, in turn, means that you can change whatever is in memory as much as you like, you just can't change the address that a constant points to.

Imagine that you must drive to somebody's house and they send you the address where they live. This is what it's like to pass around a reference type. Rather than sending you their entire house, they send you the address for their house. While you are driving to their house, the house can change in many ways. The owner could paint it, replace windows, doors, anything. In the end, you will still find the house because you received the address for this house and, as long as the owner doesn't move to a different address, you will find the correct house.

If you change this analogy to use value types, the person whose house you're looking for will simply send you a full copy of their house. So rather than you driving toward their house based on the address, they won't give you an address, they will just send you their whole house. If the owner makes changes to their copy of the house, you won't be able to see them reflected on your copy of the house unless they send you a new copy. This is also true for any modifications you make to your copy of the house.

You can imagine that in some cases it can be very efficient to send somebody a copy of something rather than the address. The sample of a house might be a bit extreme, but I'm pretty sure that if you order a parcel, you would much rather receive the parcel itself than receiving an address to fetch the parcel. This sort of efficiency is what you will learn about next by comparing how value types and reference types behave in terms of memory allocation.

Differences in memory allocation

To understand how using certain data types can affect your application's performance, you need to understand how and where they are placed into your application's memory and how they are accessed. Every application uses a certain amount of the host device's RAM (Random Access Memory). The memory that an application uses can be divided into two segments: the stack and the heap.

The heap is a section of reserved memory for your application that can grow and shrink as needed. The objects that are stored on the heap are stored in no particular order and the objects themselves can grow and shrink the memory they use on the heap as needed too. This means that, sometimes, a lot of memory will have to be reshuffled to accommodate the insertion of a new object in the heap. The following image visualizes this:

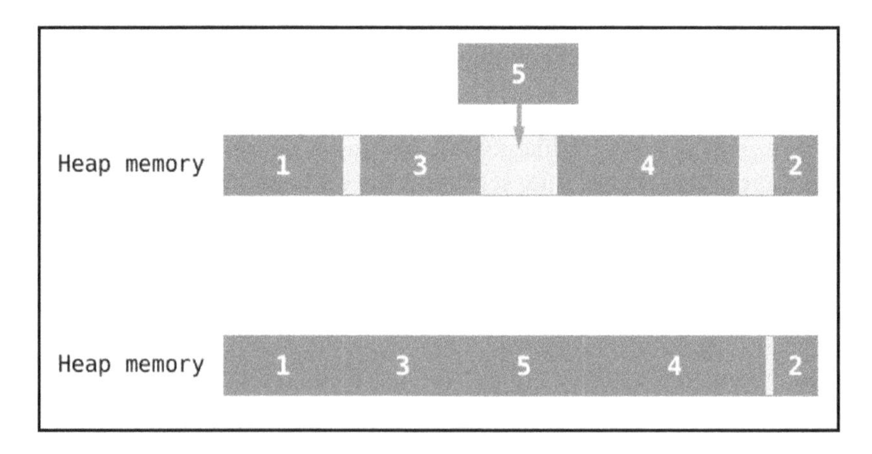

As you can see, the different objects in memory get moved around inside of the heap to make sure the new object fits. This dynamic nature of the heap makes inserting objects on the heap quite slow because you never know whether the object you are about to insert will fit immediately or whether something has to be reshuffled. The ability to reshuffle memory to fit new objects or to accommodate an existing one does make it a very good fit for objects that have a variable or growing size. The type of objects that are allocated on the heap are always reference types. This means that all class instances you create will be allocated on the heap. The same is true for closures since they too are reference types. Value types, on the other hand, are always allocated on the stack.

The stack is a section of RAM with a fixed size that is available to your application. Objects in the stack are always added and removed in a fixed manner, the last object to be added to the stack will be the first object that is removed from the stack. Because the stack behaves in such a predictable way, it is a very fast and efficient way to store and retrieve objects. The stack is always in the correct order and objects are never shuffled around to facilitate new objects. The fact that value types are allocated on the stack makes them very lightweight and performant when compared to reference types.

You might think that the performance of value types makes choosing between values types and references types a no-brainer. Unfortunately, choosing the correct type isn't *that* easy. The next section of this chapter will explain why, and provide you with enough information to make informed decisions about the types you use in your own apps.

Deciding which type you should use

Choosing the wrong type of object to use in your application can have bad implications for your app on several levels. For instance, your app could suffer from unwanted side-effects when a reference type is modified in some unexpected place. Or you could end up with a lot of duplicated logic if you use a struct instead of a class in certain places. Your app could even suffer in terms of performance when you choose a slow reference type where a value type would have been a better choice. You should always evaluate what type of object is best-suited for your current use case to make sure your code strikes a balanced trade-off between maintainability and performance.

When to use a reference type?

A great time to use a reference type is when you are subclassing a built-in class, such as `UIViewController`. In these cases, there is no point in fighting the system because that would definitely do more harm than good. Another time to use a reference type is when you are creating your own delegate protocols.

Delegates are best implemented as weak reference types. This means that the object that acts as a delegate is referenced weakly by an object to avoid memory leaks. Because value types are passed around by making copies, it does not make sense to have a weak reference to them. In this case, your only choice is to use a reference type.

You can read more about memory leaks and reference cycles in `Chapter 24`, *Discovering Bottlenecks with Instruments*.

You also need a reference type if it doesn't make sense to pass around copies or something. If you think back to the example of driving to somebody's house, it makes a lot more sense to pass around the address of a house than to give everybody full copies of the house. You might consider the house as having an identity. This means that each house is *unique*, there is only one house with that exact address and making copies of it makes no sense. If you are working with an object where copying it makes no sense, you likely want to implement it as a reference type, so everybody that receives an instance of that type is looking at the same instance.

One last reason to choose a reference type is if it can save you a lot of typing by subclassing. A lot of people consider subclassing bad, and you can often avoid it, but sometimes it just makes a lot more sense to work with a class hierarchy. The downside is that a lot of subclassing can lead to muddy classes that contain functionality to save typing on a couple of subclasses even though the functionality is not relevant to all subclasses. But just like many tools, subclassing can be quite convenient when used correctly; it's not inherently bad to use it.

When to use a value type?

It is often said that you should always start with a struct and change to a different type when needed. This is great advice for a lot of cases because structs are often fine for most cases. However, structs aren't the only value type, and it's always good to not default to using certain things blindly. If you need an object that represents a finite set of possible states or results, such as a network-connection state, a traffic-light state, or a limited set of valid configuration options for your app, you will likely need an enum. Regardless of the value semantics that make value types great, an enum is a great way to avoid typos and represent a state. It's often pretty clear when you should use an enum due to its nature.

Structs are used for objects that do not have an identity. In other words, it makes sense to pass copies of it around. A good example of this is when you create a struct that can communicate with the network or a database. This struct would have no identity because it's mostly a collection of properties and methods that aren't associated with a single version of the struct.

A good example of a struct is the `CGPoint` struct that you read about at the beginning of this structure. A `CGPoint` represents a location in a two-dimensional grid. It has no identity, and passing copies of it around makes sense. It only contains two properties, so it doesn't require any inheritance. These features make it a great candidate to be implemented as a value type.

If you follow the advice of always starting with a struct, try to figure out reasons for your new object to *not* be a struct. If you find a good reason to not use a struct, then make it a class. Often you won't be able to find a good reason to use a class instead of a struct. If this is the case, make your new object a struct; you can always switch to using a class later. It's usually harder to switch from a class to a struct due to the stricter rules regarding mutability and the lack of subclassing for structs.

Summary

You learned a lot about value types and reference types in this chapter. You learned what each type is and how you can use them. You learned that you can use classes, closures, structs, and enums in Swift and that each of these object types has its own pros and cons.

After learning about all types, you saw how value types and reference types compare to each other, which has shed some light on the sometimes subtle and sometimes obvious use cases for each type. You learned that structs can't be subclasses, while classes can. You also learned that passing around value types passes around copies of each instance, and passing around reference types does not copy each instance but rather passes around a pointer to the addresses in memory. Then you learned how each type is held in memory and what this means for the performance of the objects you created.

Lastly, you read about how to choose between value types and reference types by using several rules of thumb that should make choosing between structs, classes, and enums fairly straightforward without blindly picking one. The next chapter will take your Swift knowledge one step further by showing you how to write ultra-flexible code with Swift's generics.

Questions

1. Which statement about reference types is false?

   a) Reference types are stored on the heap.
   b) Reference types can always be subclassed.
   c) Reference types can conform to multiple protocols.

2. Which statement about value types is true?

   a) Value types are stored on the heap.
   b) Value types can always be subclassed.
   c) Values types can conform to multiple protocols.

3. What does heap allocation mean?

    a) An object allocated on the heap is stored at a fixed place in memory.
    b) An object allocated on the heap is stored with a variable size in RAM memory.
    c) It means that an object is stored in an optimized location.

4. You should always use a struct for your data models. True or false?

    a) True, because you don't need inheritance.
    b) False, it depends on whether your models have an identity.
    c) False, because data models should always be subclassed.

5. What keyword should you add to a function that mutates a property on a struct?

    a) `mutates`.
    b) `mutating`.
    c) `updates`.

6. Which type of object is not allocated on the heap?

    a) A closure.
    b) A class.
    c) An enum.

7. Which statement about the stack is true?

    a) The stack has a fixed size.
    b) The first object added to the stack will be the first object to be removed.
    c) It is slow to write objects to the stack.

8. In which of the following cases would a reference type be a good choice?

    a) When you want to create a new button in your app.
    b) When writing a networking object.
    c) When you write an object that represents a location on a map.

9. In which of the following cases would a value type be a good choice?

    a) When you want to create a new button in your app.
    b) When writing a networking object.
    c) When you're creating an object that represents a person.

Further reading

- `Chapter 24`, *Discovering Bottlenecks with Instruments*
- Structures and Classes: `https://docs.swift.org/swift-book/LanguageGuide/ClassesAndStructures.html`

6
Writing Flexible Code with Protocols and Generics

If you spend enough time around a Swift developer, you'll hear them mention protocol-oriented programming. Apple introduced this programming paradigm at *WWDC 2015* in a talk that generated a lot of buzz among developers. Suddenly, we learned that thinking in classes and hierarchies leads to code that's hard to maintain, modify, and expand. The talk introduced a way of programming that is focused on what an object can do instead of explicitly caring about what an object is.

This chapter is packed with complex and interesting information that is essential if you want to write beautiful Swift code. You will learn how you can make use of the powers of POP, and it will show you why it's an essential feature of Swift. You'll start off with some simple use cases, and then you'll take a deep dive into its associated types and generic protocols.

Understanding patterns and recognizing situations in which a protocol, protocol extension, or a generic protocol can help you improve your code will lead to code that is not only easier to maintain but also a joy to work with. The structure for this chapter is as follows:

- Defining your own protocols
- Checking for traits, not types
- Extending your protocols with default behavior
- Improving your protocols with associated types

By the end of this chapter, you might feel a bit overwhelmed. Don't worry, shifting your mindset from classical OOP to a protocol-oriented approach isn't easy. It's an entirely different way of thinking about structuring your code, which will take some getting used to. It's time to dive right in by defining some of your own protocols.

Defining your own protocols

Swift and `UIKit` have protocols at the core of their design. You might have noticed this when you were implementing custom `UIViewController` transitions, or when you worked on a table view or collection view. When you implement these features, you create objects that function as delegates for the transitions, table views, and collection views, and conform them to specific protocols. When you worked on view controller transitions in `Chapter 4`, *Immersing Your Users with Animation*, you also implemented an `NSObject` subclass that conformed to `UIViewControllerAnimatedTransitioning`.

It's possible for you to define and use your own protocols. Protocols are not confined to delegate behavior only. Defining a protocol is very similar to defining a class, struct, or enum. The main difference is that a protocol does not implement or store any values on its own. It acts as a contract between whoever calls an object that conforms to a protocol and the object that claims to conform to the protocol.

Create a new **Playground** (**File** | **New...** | **Playground**) if you want to follow along, or check out the Playground in this book's code bundle.

Let's implement a simple protocol that defines the expectations for any object that claims to be a pet. The protocol will be called the `PetType` protocol. Many protocols defined in `UIKit` and the Swift standard library use either `Type`, `Ing`, or `Able` as a suffix to indicate that the protocol defines a behavior rather than a concrete type. You should try to follow this convention as much as possible because it makes your code easier to understand for other developers:

```
protocol PetType {
  var name: String { get }
  var age: Int { get set }

  func sleep()

  static var latinName: String { get }
}
```

The definition for `PetType` states that any object that claims to be `PetType` must have a get-only variable (a constant) called `name`, an `age` that can be changed because it specifies both `get` and `set`, a method that makes the pet `sleep()`, and finally, a static variable that describes the Latin name of `PetType`.

Whenever you define that a protocol requires a certain variable to exist, you must also specify whether the variable should be gettable, settable, or both. If you specify that a certain method must be implemented, you write the method just as you usually would, but you stop at the first curly bracket. You only write down the method signature.

A protocol can also require that the implementer has a static variable or method. This is convenient in the case of `PetType` because the Latin name of a pet does not necessarily belong to a specific pet, but to the entire species that the pet belongs to, so implementing this as a property of the object rather than the instance makes a lot of sense.

To demonstrate how powerful a small protocol such as `PetType` can be, you will implement two pets: a cat and a dog. You'll also write a function that takes any pet and then makes them take a nap by calling the `sleep()` method.

To do this in OOP, you would create a `class` called `Pet`, and then you'd create two subclasses, `Cat` and `Dog`. The `nap` method would take an instance of `Pet`, and it would look a bit like this:

```
func nap(pet: Pet) {
    pet.sleep()
}
```

The object-oriented approach is not a bad one. Also, on such a small scale, no real problems will occur. However, when the inheritance hierarchy grows, you typically end up with base classes that contain methods that are only relevant to a couple of subclasses. Alternatively, you will find yourself unable to add certain functionalities to a certain class because the inheritance hierarchy gets in the way after a while.

Let's see what it looks like when you use the `PetType` protocol to solve this challenge without using inheritance at all:

```
protocol PetType {
  var name: String { get }
  var age: Int { get set }

  func sleep()
```

```
    static var latinName: String { get }
}

struct Cat: PetType {
    let name: String
    var age: Int

    static let latinName: String = "Felis catus"

    func sleep() {
        print("Cat: ZzzZZ")
    }
}

struct Dog: PetType {
    let name: String
    var age: Int

    static let latinName: String = "Canis familiaris"

    func sleep() {
        print("Dog: ZzzZZ")
    }
}

func nap(pet: PetType) {
    pet.sleep()
}
```

We just managed to implement a single method that can take both the `Cat` and `Dog` objects and makes them take a nap. Instead of checking for a class, the code checks that the pet that is passed in conforms to the `PetType` protocol, and if it does, its `sleep()` method can be called because the protocol dictates that any `PetType` must implement a `sleep()` method. This brings us to the next topic of this chapter: checking for traits instead of types.

Checking for traits instead of types

In classic OOP, you often create superclasses and subclasses to group objects with similar capabilities. If you roughly model a group of felines in the animal kingdom with classes, you end up with a diagram that looks like this:

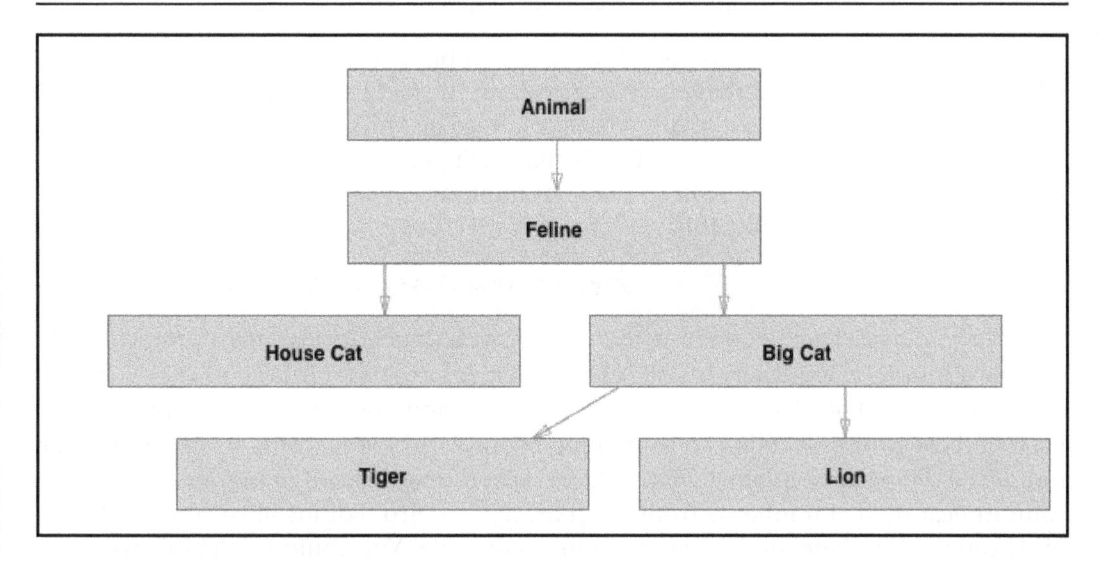

If you try to model more animals, you will find that it's a complex task because some animals share a whole bunch of traits, although they are quite far apart from each other in the class diagram.

One example would be that both cats and dogs are typically kept as pets. This means that they should optionally have an owner and maybe a home. But cats and dogs aren't the only animals kept as pets because fish, guinea pigs, rabbits, and even snakes are kept as pets. It would be tough to figure out a sensible way to restructure your class hierarchy in such a way that you don't have to redundantly add owners and homes to every pet in the hierarchy because it would be impossible to add these properties to the right classes selectively.

This problem gets even worse when you write a function or method that prints a pet's home. You would either have to make that function accept any animal or write a separate implementation of the same function for each type that has the properties you're looking for. Both don't make sense because you don't want to write the same function over and over again with just a different class for the parameter. Even if you choose to do this and you end up with a method that prints an animals home address that accepts a `Fish` instance, passing an instance of `GreatWhiteShark` to a function called `printHomeAddress()` doesn't make a lot of sense either, because sharks typically don't have home addresses. Of course, the solution to this problem is to use protocols.

In the situation described in the previous section, objects were mostly defined by what they *are*, not by what they *do*. We care about the fact that an animal is part of a particular family or type, not about whether it lives on land. You can't differentiate between animals that can fly and animals that can't because not all birds can fly. Inheritance isn't compatible with this way of thinking. Imagine a definition for a `Pigeon` struct that looks like this:

```
struct Pigeon: Bird, FlyingType, OmnivoreType, Domesticatable
```

Since `Pigeon` is a struct, you know that `Bird` isn't a struct or class; it's a protocol that defines a couple of requirements about what it means to be a bird. The `Pigeon` struct also conforms to the `FlyingType`, `OmnivoreType`, and `Domesticatable` protocols. Each of these protocols tells you something about `Pigeon` regarding its capabilities or traits. The definition explains what a pigeon is and does instead of merely communicating that it inherits from a certain type of bird. For instance, almost all birds can fly, but there are some exceptions to the rule. You could model this with classes, but this approach is tedious and might be inflexible, depending on your needs and how your code evolves over time. Setting the `Pigeon` struct up with protocols is powerful; you can now write a `printHomeAddress()` function and set it up so that it accepts any object that conforms to `Domesticatable`:

```
protocol Domesticatable {
  var homeAddress: String? { get }
}

func printHomeAddress(animal: Domesticatable) {
  if let address = animal.homeAddress {
    print(address)
  }
}
```

The `Domesticatable` protocol requires an optional `homeAddress` property. Not every animal that can be domesticated actually is. For example, think about the pigeon; some pigeons are kept as pets, but most aren't. This also applies to cats and dogs, because not every cat or dog has a home.

This approach is powerful, but shifting your mind from an object-oriented mindset, where you think of an inheritance hierarchy, to a protocol-oriented mindset, where you focus on traits instead of inheritance, isn't easy.

Let's expand the example code a bit more by defining `OmnivoreType`, `HerbivoreType`, and `CarnivoreType`. These types will represent the three main types of eaters in the animal kingdom. You can make use of inheritance inside of these protocols because `OmnivoreType` is both `HerbivoreType` and `CarnivoreType`, so you can make `OmnivoreType` inherit from both of these protocols:

```
protocol HerbivoreType {
  var favoritePlant: String { get }
}

protocol CarnivoreType {
  var favoriteMeat: String { get }
}

protocol OmnivoreType: HerbivoreType, CarnivoreType {}
```

Composing two protocols into one like you did in the preceding example is really powerful, but be careful when you do this. You don't want to create a crazy inheritance graph like you would when you do OOP; you just learned that inheritance could be wildly complex and inflexible. Imagine writing two new functions, one to print a carnivore's favorite meat and one to print a herbivore's favorite plant. Those functions would look like this:

```
func printFavoriteMeat(forAnimal animal: CarnivoreType) {
  print(animal.favoriteMeat)
}

func printFavoritePlant(forAnimal animal: HerbivoreType) {
  print(animal.favoritePlant)
}
```

The preceding code might be exactly what you would write yourself. However, neither of these methods accepts `OmnivoreType`. This is perfectly fine because `OmnivoreType` inherits from `HerbivoreType` and `CarnivoreType`. This works in the same way that you're used to in classical object-oriented programming, with the main exception being that `OmnivoreType` inherits from multiple protocols instead of just one. This means that the `printFavoritePlant` function accepts a `Pigeon` instance as its argument because `Pigeon` confirms to `OmnivoreType`, which inherits from `HerbivoreType`.

Using protocols to compose your objects like this can drastically simplify your code. Instead of thinking about complex inheritance structures, you can compose your objects with protocols that define certain traits. The beauty of this is that it makes defining new objects relatively easy.

Imagine that a new type of animal is discovered. One that can fly, swim, and lives on land. This weird new species would be really hard to add to an inheritance-based architecture since it doesn't fit in with other animals. When using protocols, you could add conformance to the `FlyingType`, `LandType`, and `SwimmingType` protocols and you'd be all set. Any methods or functions that take a `LandType` animal as an argument will happily accept your new animal since it conforms to the `LandType` protocol.

Getting the hang of this way of thinking isn't simple, and it will require some practice. But any time you're getting ready to create a superclass or subclass, ask yourself why. If you're trying to encapsulate a certain trait in that class, try using a protocol. This will train you to think differently about your objects, and before you know it, your code is cleaner, more readable, and more flexible, using protocols and checking for traits instead of taking actions based on what an object is.

As you've seen, a protocol doesn't need to have a lot of requirements; sometimes one or two are enough to convey the right meaning. Don't hesitate to create protocols with just a single property or method; as your projects grow over time and your requirements change, you will thank yourself for doing so.

Extending your protocols with default behavior

The previous examples have mainly used variables as the requirements for protocols. One slight downside of protocols is that they can result in a bit of code duplication. For example, every object that is `HerbivoreType` has a `favoriteMeat` variable. This means that you have to duplicate this variable in every object that conforms to `HerbivoreType`. Usually, you want as little code repetition as possible, and repeating a variable over and over again might seem like a step backward.

Even though it's nice if you don't have to declare the same property over and over again, there's a certain danger in not doing this. If your app grows to a large size, you won't remember every class, subclass, and superclass all of the time. This means that changing or removing a specific property can have undesired side-effects in other classes.

Declaring the same properties on every object that conforms to a certain protocol isn't that big a deal; it usually takes just a few lines of code to do this. However, protocols can also require certain methods to be present on objects that conform to them. Declaring those over and over again can be cumbersome, especially if the implementation is the same for most objects. Luckily, you can make use of **protocol extensions** to implement a certain degree of default functionality.

To explore protocol extensions, let's move the `printHomeAddress()` function into the `Domesticatable` protocol so all `Domesticatable` objects can print their own home addresses. The first approach you can take is to immediately define the method on a protocol extension without adding it to the protocol's requirements:

```
extension Domesticatable {
  func printHomeAddress() {
    if let address = homeAddress {
      print(address)
    }
  }
}
```

By defining the `printHomeAddress()` method in the protocol extension, every object that conforms to `Domesticatable` has the following method available without having to implement it with the object itself:

```
let myPidgeon = Pigeon(favoriteMeat: "Insects", favoritePlant:
"Seeds", homeAddress: "Leidse plein 12, Amsterdam")
myPidgeon.printHomeAddress() // "Leidse plein 12, Amsterdam"
```

This technique is very convenient if you want to implement default behavior that's associated with a protocol. You didn't even have to add the `printHomeAddress()` method as a requirement to the protocol. However, this approach will give you some strange results if you're not careful. The following snippet shows an example of such odd results by adding a custom implementation of `printHomeAddress()` to the `Pigeon` struct:

```
struct Pigeon: Bird, FlyingType, OmnivoreType, Domesticatable {
  let favoriteMeat: String
  let favoritePlant: String
  let homeAddress: String?

  func printHomeAddress() {
    if let address = homeAddress {
      print("address: \(address.uppercased())")
    }
  }
```

```
}

let myPigeon = Pigeon(favoriteMeat: "Insects", favoritePlant: "Seeds",
homeAddress: "Leidse plein 12, Amsterdam")

myPigeon.printHomeAddress() //address: LEIDSE PLEIN 12, AMSTERDAM

func printAddress(animal: Domesticatable) {
  animal.printHomeAddress()
}
printAddress(animal: myPigeon) // Leidse plein 12, Amsterdam
```

When you call `myPigeon.printHomeAddress()`, the custom implementation is used to print the address. However, if you define a function, such as `printAddress(animal:)`, that takes a `Domesticatable` object as its parameter, the default implementation provided by the protocol is used.

This happens because `printHomeAddress()` isn't a requirement of the protocol. Therefore, if you call `printHomeAddress()` on a `Domesticatable` object, the implementation from the protocol extension is used. If you use the same snippet as in the preceding section, but change the `Domesticatable` protocol as shown in the following code, both calls to `printHomeAddress()` print the same thing, that is, the custom implementation in the `Pigeon` struct:

```
protocol Domesticatable {
  var homeAddress: String? { get }

  func printHomeAddress()
}
```

This behavior is likely to be unexpected in most cases, so it's usually a good idea to define all methods you use in the protocol requirements unless you're absolutely sure you want the behavior you just saw.

Protocol extensions can't hold stored properties. This means that you can't add your variables to the protocol to provide a default implementation for them. Even though extensions can't hold stored properties, there are situations where you can still add a computed property to a protocol extension to avoid duplicating the same variable in multiple places. Let's take a look at an example:

```
protocol Domesticatable {
  var homeAddress: String? { get }
  var hasHomeAddress: Bool { get }

  func printHomeAddress()
}
```

```
extension Domesticatable {
  var hasHomeAddress: Bool {
    return homeAddress != nil
  }

  func printHomeAddress() {
    if let address = homeAddress {
      print(address)
    }
  }
}
```

If you want to be able to check whether a `Domesticatable` has a home address, you can add a requirement for a `Bool` value, `hasHomeAddress`. If the `homeAddress` property is set, `hasHomeAddress` should be `true`. Otherwise, it should be `false`. This property is computed in the protocol extension, so you don't have to add this property to all `Domesticatable` objects. In this case, it makes a lot of sense to use a computed property because the way its value is computed should most likely be the same across all `Domesticatable` objects.

Implementing default behaviors in protocol extensions makes the protocol-oriented approach we've seen before even more powerful; you can essentially mimic a feature called **multiple inheritance** without all the downsides of subclassing. Simply adding conformance to a protocol can add all kinds of functionality to your objects, and if the protocol extensions allow it, you won't need to add anything else to your code. Let's see how you can make protocols and extensions even more powerful with associated types.

Improving your protocols with associated types

One more awesome aspect of protocol-oriented programming is the use of associated types. An associated type is a generic, non-existing type that can be used in your protocol like any type that does exist. The real type of this generic is determined by the compiler based on the context it's used in. This description is abstract, and you might not immediately understand why or how an associated type can benefit your protocols. After all, aren't protocols themselves a very flexible way to make several unrelated objects fit certain criteria based on the protocols they conform to?

To illustrate and discover the use of associated types, you will expand your animal kingdom a bit. What you should do is give the herbivores an `eat` method and an array to keep track of the plants they've eaten, as follows:

```
protocol HerbivoreType {
  var plantsEaten: [PlantType] { get set }

  mutating func eat(plant: PlantType)
}

extension HerbivoreType {
  mutating func eat(plant: PlantType) {
    plantsEaten.append(plant)
  }
}
```

This code looks fine at first sight. A herbivore eats plants, and this is established by this protocol. The `PlantType` protocol is defined as follows:

```
protocol PlantType {
  var latinName: String { get }
}
```

Let's define two different plant types and an animal that will be used to demonstrate the problem with the preceding code:

```
struct Grass: PlantType{
  var latinName = "Poaceae"
}

struct Pine: PlantType{
  var latinName = "Pinus"
}

struct Cow: HerbivoreType {
  var plantsEaten = [PlantType]()
}
```

There shouldn't be a big surprise here. Let's continue with creating a `Cow` instance and feed it `Pine`:

```
var cow = Cow()
let pine = Pine()
cow.eat(plant: pine)
```

This doesn't really make sense. Cows don't eat pines; they eat grass! We need some way to limit this cow's food intake because this approach isn't going to work. Currently, you can feed `HerbivoreType` animals anything that's considered a plant. You need some way to limit the types of food your cows are given. In this case, you should restrict the `FoodType` to `Grass` only, without having to define the `eat(plant:)` method for every plant type you might want to feed a `HerbivoreType`.

The problem you're facing now is that all `HerbivoreType` animals mainly eat one plant type, and not all plant types are a good fit for all herbivores. This is where associated types are a great solution. An associated type for the `HerbivoreType` protocol can constrain the `PlantType` that a certain herbivore can eat to a single type that is defined by the `HerbivoreType` itself. Let's see what this looks like:

```
protocol HerbivoreType {
  // 1
  associatedtype Plant: PlantType

  var plantsEaten: [Plant] { get set }

  mutating func eat(plant: Plant)
}

extension HerbivoreType {
  mutating func eat(plant: Plant) {
    // 2
    print("eating a (plant.latinName)")
    plantsEaten.append(plant)
  }
}
```

The first highlighted line associates the generic `Plant` type, which doesn't exist as a real type, with the protocol. A constraint has been added to `Plant` to ensure that it's a `PlantType`.

The second highlighted line demonstrates how the `Plant` associated type is used as a `PlantType`. The plant type itself is merely an alias for any type that conforms to `PlantType` and is used as the type of object we use for `plantsEaten` and the `eat` methods. Let's redefine the `Cow` struct to see this associated type in action:

```
struct Cow: HerbivoreType {
  var plantsEaten = [Grass]()
}
```

Instead of making `plantsEaten` a `PlantType` array, it's now defined as an array of `Grass`. In the protocol and the definition, the type of plant is now `Grass`. The compiler understands this because the `plantsEaten` array is defined as `[Grass]`. Let's define a second `HerbivoreType` that eats a different type of `PlantType`:

```
struct Carrot: PlantType {
  let latinName = "Daucus carota"
}

struct Rabbit: HerbivoreType {
  var plantsEaten = [Carrot]()
}
```

If you try to feed a `Cow` some carrots, or if you attempt to feed the `Rabbit` a `Pine`, the compiler will throw errors. The reason for this is that the associated type constraint allows you to define the type of `Plant` in each struct separately.

One side note about associated types is that it's not always possible for the compiler to correctly infer the real type for an associated type. In our current example, this would happen if we didn't have the `plantsEaten` array in the protocol. The solution would be to define a `typealias` on objects that conform to `HerbivoreType` so that the compiler understands which type `Plant` represents:

```
protocol HerbivoreType {
  associatedtype Plant: PlantType

  mutating func eat(plant: Plant)
}

struct Cow: HerbivoreType {
  typealias Plant = Grass
}
```

Associated types can be really powerful when used correctly, but sometimes using them can also cause you a lot of headaches because of the amount of inferring the compiler has to do. If you forget a few tiny steps, the compiler can quickly lose track of what you're trying to do, and the error messages aren't always the most unambiguous messages. Keep this in mind when you're using associated types, and try to make sure that you're as explicit as possible about the type you're looking to be associated. Sometimes, adding a `typealias` to give the compiler a helping hand is better than trying to get the compiler to infer everything on its own correctly.

This type of flexibility is not limited to protocols. You can also add generic properties to functions, classes, structs, and enums. Let's see how this works and how it can make your code extremely flexible.

Adding flexibility with generics

Programming with generics is not always easy, but it does make your code extremely flexible. When you use something such as generics, you are always making a trade-off between the simplicity of your program and the flexibility of your code. Sometimes it's worth it to introduce a little bit of complexity to allow your code to be written in ways that were otherwise impossible.

For instance, consider the `Cow` struct you saw before. To specify the generic associated type on the `HerbivoreType` protocol, a type alias was added to the `Cow` struct. Now imagine that not all cows like to eat grass. Maybe some cows prefer flowers, corn, or something else. You would not be able to express this using the type alias.

To represent a case where you might want to use a different type of `PlantType` for every cow instance, you can add a generic to the `Cow` itself. The following snippet shows how you can do this:

```
struct Cow<Plant: PlantType>: HerbivoreType {
  var plantsEaten = [Plant]()
}
```

Between < and >, the generic type name is specified as `Plant`. This generic is constrained to the `PlantType` type. This means that any type that will act as `Plant` has to conform to `PlantType`. The protocol will see that `Cow` has a generic `Plant` type now, so there is no need to add a type alias. When you create an instance of `Cow`, you can now pass every instance its own `PlantType`:

```
let grassCow = Cow<Grass>()
let flowerCow = Cow<Flower>()
```

Applying generics to object instances like this is more common than you might think. An `Array` instance uses generics to determine what kind of elements it contains. The following two lines of code are identical in functionality:

```
let strings = [String]()
let strings = Array<String>()
```

The first line uses a convenient syntax to create an array of strings. The second line uses the `Array` initializer and explicitly specifies the type of element it will contain.

Sometimes, you might find yourself writing a function or method that can benefit from a generic argument or return type. An excellent example of a generic function is `map`. With `map`, you can transform an array of items into an array of different items. You can define your own simple version of `map` as follows:

```
func simpleMap<T, U>(_ input: [T], transform: (T) -> U) -> [U] {
  var output = [U]()

  for item in input {
    output.append(transform(item))
  }

  return output
}
```

`simpleMap(_:transform:)` has two generic types, `T` and `U`. These names are common placeholders for generics, so they make it clear to anybody reading this code that they are about to deal with generics. In this sample, the function expects an input of `[T]`, which you can read as an array of *something*. It also expects a closure that takes an argument, `T`, and returns `U`. You can interpret this as the closure taking an element out of that array of something, and it transforms it into something else. The function finally returns an array of `[U]`, or in other words, an array of *something else*.

You would use `simpleMap(_:transform:)` as follows:

```
let result = simpleMap([1, 2, 3]) { item in
  return item * 2
}

print(result) // [2, 4, 6]
```

Generics are not always easy to understand, and it's okay if they take you a little while to get used to. They are a powerful and complex topic that we could write many more pages about. The best way to get into them is to use them, practice them, and read as much as you can about them. For now, you should have more than enough to think about and play with.

Note that generics are not limited to structs and functions. You can also add generics to your enums and classes in the same way you add them to a struct.

Summary

In this chapter, you saw how you can leverage the power of protocols to work with an object's traits or capabilities, rather than just using its class as the only way of measuring its capabilities. Then, you saw how protocols can be extended to implement a default functionality. This enables you to compose powerful types by merely adding protocol conformance, instead of creating a subclass. You also saw how protocol extensions behave depending on your protocol requirements, and that it's wise to have anything that's in the protocol extension defined as a protocol requirement. This makes the protocol behavior more predictable. Finally, you learned how associated types work and how they can take your protocols to the next level by adding generic types to your protocols that can be tweaked for every type that conforms to your protocol. You even saw how you can apply generics to other objects, such as functions and structs.

The concepts shown in this chapter are pretty advanced, sophisticated, and powerful. To truly master their use, you'll need to train yourself to think regarding traits instead of an inheritance hierarchy. Once you've mastered this, you can experiment with protocol extensions and generic types. It's okay if you don't fully understand these topics right off the bat; they're completely different and new ways of thinking for most programmers with OOP experience. Now that we've explored some of the theory behind protocols and value types, in the next chapter, you will learn how you can put this new knowledge to use by shortly revisiting the **Hello-Contacts** app from previous chapters to improve the code you wrote there.

Questions

1. Which of the following is relevant in Protocol-Oriented Programming?

    a) The type an object has.
    b) The object's superclass.
    c) What an object can do.

2. Which of the following is not a downside of subclassing?

    a) Shared functionality.
    b) A fixed subclassing hierarchy.
    c) A complex inheritance graph.

3. Protocols can inherit from several other protocols. True or false?

    a) True.
    b) False.

4. An associated type has to be a struct or class. True or false?

    a) True.
    b) False.

5. In which two ways can a struct specify what type an associated type for a protocol is?

    a) By using a typealias.
    b) By using generics on the struct.
    c) By conforming to a protocol.

6. What does T mean in `genericFunction<T>()`?

    a) It refers to a type called T.
    b) It refers to a type that conforms to a T protocol.
    c) T is a placeholder for the generic type.

Further reading

Visit following links for more information:

- Jon Hofman – Swift 4 Protocol-Oriented Programming: `https://www.packtpub.com/web-development/swift-protocol-oriented-programming-third-edition`
- Swift documentation – Generics: `https://docs.swift.org/swift-book/LanguageGuide/Generics.html`

7

Improving the Application Structure

When you first built **Hello-Contacts** in `Chapters 1`, *UITableView Touch-up* to `Chapter 3`, *Creating a Detail Page*, you used classes and Object-Oriented Programming techniques. Now that you have seen how value types and protocols can improve your code, it's a good idea to revisit the **Hello-Contacts** application to see how you can improve it with this newfound knowledge. Even though the app is relatively small, there are a few places where you can make significant improvements to make it more flexible and future-proof.

This chapter is all about making the **Hello-Contacts** application *Swiftier* than it is now. You'll do this by implementing certain parts of the app with protocols and value types instead of classes and OOP. This chapter will cover the following topics:

- Properly separating concerns
- Adding protocols for clarity

Properly separating concerns

Before you can improve your project structure with value types and protocols, it's a good idea to improve upon your structure in general. We haven't really thought about the reuse possibilities of certain aspects of the **Hello-Contacts** app, which results in code that's harder to maintain in the long run. This is especially true for larger projects that are built over an extended period of time.

If you take a look at the source code for this project in the book's code bundle, you'll find that the project was slightly modified. First, all the project files were put together in sensible groups. Doing so makes it easier for you to navigate your project's files, and it creates a natural place for certain files, as shown in the following screenshot:

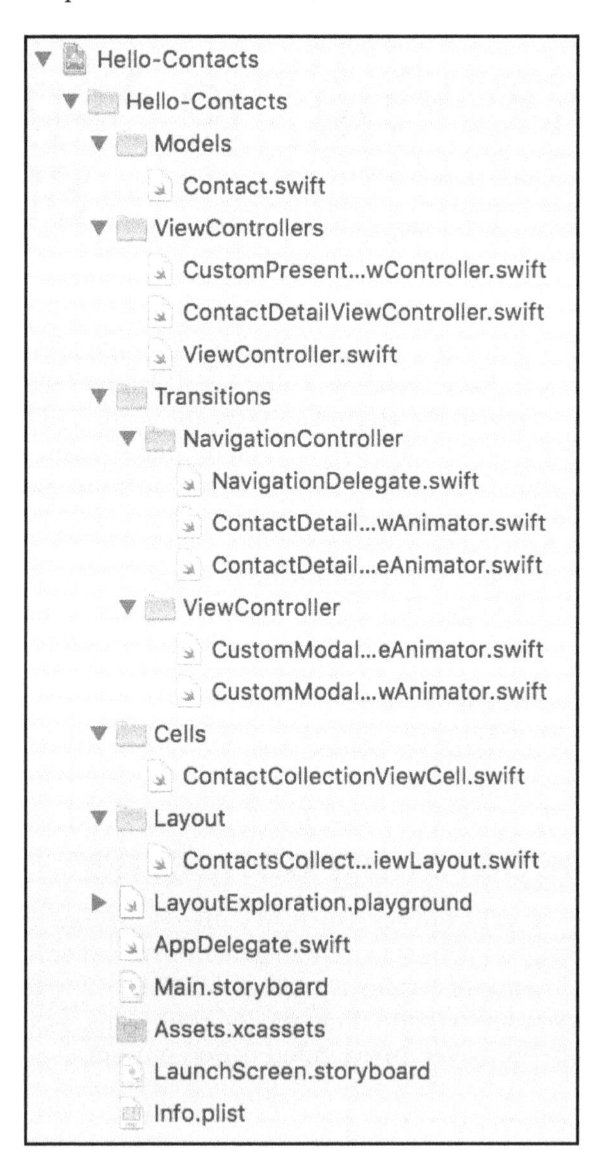

The structure applied in this project is merely a suggestion; if you feel that a different structure would suit you better, go ahead and make the change. The most important part is that you've thought about your project structure and set it up in a way that makes sense to you and helps you navigate your project.

With this improved folder structure, you may notice that there's a natural separation between certain files. Some files help with transitions, a model file, several view controllers and more. By separating files into groups, you've given yourself an overview of the file types in your project, and each file belongs to a certain group that describes its place in your app. This makes it easier for other developers (and yourself) to browse the code in your project. Now that your project's file structure is solid, let's refactor some code.

The file we will refactor first is `ViewController.swift`. This file contains the code for the contacts overview screen. Currently, this view controller fetches contacts, acts as a delegate and data source for the collection view, and takes care of the animations that play when a user taps on a cell.

You may consider this to be okay. Ideally, any given class shouldn't be responsible for that many things at once. What if you need to create a second kind of overview page; how can you reuse the code that fetches contacts? What if you want to add the bouncing cell image animation to another image? These are two scenarios that are pretty likely to happen at some point in the future. Let's extract the contact-fetching code and the animation code and place them in their own structs.

Extracting the contact-fetching code

Start off by creating a new Swift file called `ContactFetchHelper.swift`. After creating the file, add it to a new folder called `Helpers`. First, you will extract all the contact-fetching code to the `ContactFetchHelper` struct. Then, you'll refactor `ViewController.swift` by replacing all contact-fetching code, so it uses the new `ContactFetchHelper`. The following code shows the implementation for `ContactFetchHelper`:

```
import Contacts

struct ContactFetchHelper {
  // 1
  typealias ContactFetchCallback = ([Contact]) -> Void

  let store = CNContactStore()
```

```
    // 2
    func fetch(withCallback callback: @escaping
      ContactFetchCallback) {
      if CNContactStore.authorizationStatus(for: .contacts) ==
  .notDetermined {
        store.requestAccess(for: .contacts, completionHandler:
          {authorized, error in
            if authorized {
              self.retrieve(withCallback: callback)
            }
        })
      } else if CNContactStore.authorizationStatus(for: .contacts) ==
  .authorized {
        retrieve(withCallback: callback)
      }
    }

    // 3
    private func retrieve(withCallback callback: ContactFetchCallback) {
      let containerId = store.defaultContainerIdentifier()

      let keysToFetch =
        [CNContactGivenNameKey as CNKeyDescriptor,
         CNContactFamilyNameKey as CNKeyDescriptor,
         CNContactImageDataKey as CNKeyDescriptor,
         CNContactImageDataAvailableKey as CNKeyDescriptor,
         CNContactEmailAddressesKey as CNKeyDescriptor,
         CNContactPhoneNumbersKey as CNKeyDescriptor,
         CNContactPostalAddressesKey as CNKeyDescriptor]

      let predicate =
  CNContact.predicateForContactsInContainer(withIdentifier: containerId)

      guard let retrievedContacts = try? store.unifiedContacts(matching:
        predicate, keysToFetch: keysToFetch) else {
          // call back with an empty array if we fail to retrieve
  contacts
          callback([])
          return
      }

      let contacts: [Contact] = retrievedContacts.map { contact in
        return Contact(contact: contact)
      }

      callback(contacts)
    }
  }
```

This simple struct now contains all the required logic to fetch contacts. Let's go through some of the most interesting parts of code in the `ContactsFetchHelper` struct.

The first highlight points to an alias, named `ContactFetchCallback`, for a closure that receives an array of `Contact` instances and returns nothing. This is the closure that is passed to the `fetch(withCallback:)` method, and it's called after the fetch is performed.

The `fetch(withCallback :)` method is the method that should be called whenever you want to fetch contacts. The only argument it takes is a closure that will be called when the contacts are fetched. The fetch method performs the same authorization check that was in the view controller's `viewDidLoad` method before.

The third point of interest is a private method, called `retrieve(withCallback:)`, that retrieves the contacts. The `fetch(withCallback:)` method calls this method and passes it the `callback` that it received. Once `retrieve(withCallback:)` has retrieved the contacts, it calls the `callback` with the array of fetched contacts.

The `ContactFetchCallback` that `fetch(withCallback:)` receives is marked as `@escaping`. This keyword indicates that the supplied closure will escape the scope it was passed to. In this example, `callback` is passed to the scope of `fetch(withCallback:)`, which in turn passes this closure to the scope of `retrieve(withCallback:)`.

In `ViewController.swift`, all you need to do is use the following code to retrieve contacts:

```
let contactFetcher = ContactFetchHelper()
contactFetcher.fetch { [weak self] contacts in
    self?.contacts = contacts
    self?.collectionView.reloadData()
}
```

You can delete the `retrieveContacts` method entirely, and the preceding snippet replaces the code that checked for permissions in `viewDidLoad`. Also, because the `Contacts` framework is not used directly anymore, you can remove its `import` at the top of the file. You have now successfully extracted the contact-fetching code into a struct, and you're using a `typealias` to make your code more readable. This is already a big win for maintainability and reusability. Now, let's extract the bounce animation code as well.

Extracting the bounce animation

The process of extracting the bounce animation is a little more complicated than the process of extracting the contact-fetching code. The purpose of extracting this bounce animation is to make it so that it becomes possible to make other objects in other sections of the app bounce, just like the contact cell's image does.

To figure out what the bounce animation helper should do and how it should work, it's a great idea to think about how you want to use this helper at the call site. The **call site** is defined as the place where you plan to use your helper. So, in this case, the call site is considered the `ViewController`. Let's write some pseudocode to try to determine what you will program later:

```
let onBounceComplete = { [weak self] finished in
    self?.performSegue(withIdentifier: "contactDetailSegue", sender:
self)
}

let bounce = BounceAnimationHelper(targetView: cell.contactImage,
onComplete: onBounceComplete)
```

This looks pretty good already, and in reality, it's very close to the actual code you could end up with later. All you really want to do at the call site is configure a bounce animation by passing it a view to perform the bounce on and to have some control over what should happen once the animation is completed. The following two points should be considered before writing the `BounceAnimationHelper` implementation:

- It's not possible to set the bounce duration – is this acceptable?
- You have no control over the start time of the animation, so manually starting it would be nice.

To address the first point, you could implement two initializers: one that uses a default duration and another where the users of the helper can specify their own duration. Doing this makes use of Swift's powerful method overloading, which enables developers to write multiple initializers for the same object. This also allows developers to write methods with the same name but a different signature due to different parameter names.

The second point of concern is valid, and the helper should be written in a way that requires manually starting the animation, so it feels more like `UIViewPropertyAnimator`. Theoretically speaking, you could add a `Bool` value to the initializer that enables users of the helper to choose whether the animation should start automatically. This would make it less similar to `UIViewPropertyAnimator` so in this case manually calling a method to start the animation is preferred. The calling code you should end up with is shown in the following code snippet. You can go ahead and add it to the `ViewController.swift` file in place of the current bounce animation. You'll implement the helper shortly:

```
let onBounceComplete: BounceAnimationHelper.BounceAnimationComplete =
{ [unowned self] position in
    self.performSegue(withIdentifier: "detailViewSegue", sender: self)
}

let bounce = BounceAnimationHelper(targetView: cell.contactImage,
onComplete: onBounceComplete)
bounce.startAnimation()
```

Note that the preceding snippet explicitly defines the type of the `onBounceComplete` callback. This type will be a `typealias` on the helper you're about to implement.

Now that the call site is all finished, let's take a look at the implementation of `BounceAnimationHelper`. Create a new Swift file called **BounceAnimationHelper** and add it to the `Helpers` folder. Start by defining a struct named `BounceAnimationHelper` in the corresponding Swift file. Next, define a `typealias` for the completion handler and specify the properties we need in the struct, as follows:

```
import UIKit

struct BounceAnimationHelper {
  typealias BounceAnimationComplete = (UIViewAnimatingPosition) ->
Void

  let animator: UIViewPropertyAnimator
}
```

The initial implementation for the struct is pretty bare. A `typealias` is defined that passes a `UIViewAnimatingPosition` into a closure that has no return value. The struct also holds onto a `UIViewPropertyAnimator`, so its `startAnimation()` can be called whenever the helper's `startAnimation()` is called.

Next, let's add the two initializers that were described earlier. One with a default duration for the bounce effect and one with a custom duration. The second initializer is empty for now; you will implement it in a minute:

```
init(targetView: UIView, onComplete: @escaping
BounceAnimationComplete) {
  self.init(targetView: targetView, onComplete: onComplete, duration:
0.4)
}

init(targetView: UIView, onComplete: @escaping
BounceAnimationComplete, duration: TimeInterval) {

}
```

These two initializers provide the APIs you were looking for. The first initializer calls out to the second with a default `duration` value of `0.4`. Doing this allows you to write the animation in a single initializer. It is common to have multiple initializers on an object that all trickle down to a single initializer; the designated initializer. The designated initializer is responsible for setting up and configuring all properties of an object. The following code shows the implementation for the designated initializer. It replaces the empty initializer you saw in the previous snippet:

```
init(targetView: UIView, onComplete: @escaping
BounceAnimationComplete, duration: TimeInterval) {
  let downAnimationTiming = UISpringTimingParameters(dampingRatio:
0.9, initialVelocity: CGVector(dx: 20, dy: 0))

  // 1
  self.animator = UIViewPropertyAnimator(duration: duration/2,
timingParameters: downAnimationTiming)

  self.animator.addAnimations {
    // 2
    targetView.transform = CGAffineTransform(scaleX: 0.9, y: 0.9)
  }

  self.animator.addCompletion { position in
    let upAnimationTiming = UISpringTimingParameters(dampingRatio:
0.3, initialVelocity:CGVector(dx: 20, dy: 0))

    let upAnimator = UIViewPropertyAnimator(duration: duration/2,
timingParameters: upAnimationTiming)

    upAnimator.addAnimations {
      targetView.transform = CGAffineTransform.identity
    }
```

```
        // 2
        upAnimator.addCompletion(onComplete)

        upAnimator.startAnimation()
    }
}
```

This snippet is very similar to the old animation; the main differences are highlighted with comments. Instead of hardcoding a duration, half of the total duration of the bounce animation is used for the downward motion, and the other half is for the upward motion. Also, instead of using the cell's image directly, the specified `targetView` is used as the animation target.

Finally, instead of passing an inline `callback` to the `upAnimator` closure's completion, the `onComplete` closure that was passed to the initializer is passed to `upAnimator`. Note that the down animation isn't started in the initializer; this should be implemented in a separate method:

```
func startAnimation() {
  animator.startAnimation()
}
```

Add the preceding method to `BounceAnimationHelper` and run your app. The contact images should bounce just like they did before, except the animation is reusable now and the code in `ViewController.swift` looks a lot cleaner.

With the cleaned-up `ViewController` in place, let's see where **Hello-Contacts** could benefit from protocols.

Adding protocols for clarity

You already know that protocols can be used to improve code by removing complex inheritance hierarchies. You also know how powerful the Protocol-Oriented paradigm is when it is when it comes to checking for protocol conformance instead of checking whether a particular object is of a certain type. Let's see how you can improve and future-proof the **Hello-Contacts** application by adding some protocols.

You will define two protocols for now: one that specifies the requirements for any object that claims to be able to add a special animation to a view, and one that defines what it means to be able to be displayed as a contact.

Defining the ViewEffectAnimatorType protocol

The first protocol you will define is called `ViewEffectAnimatorType`. This protocol should be applied to any object that implements the required behaviors to animate a view. This protocol does not necessarily give you a direct advantage, but there are a few considerations that make this a very useful protocol.

A protocol is not only used to check whether an object can do something. It can also formalize a certain API that you came up with. In this case, `BounceAnimationHelper` needed certain initializers. It also needs to hold on to an animator, and it has a `startAnimation` method. These traits are not specific to the bounce animation and might be relevant for several other animation effects.

Adding a protocol to this helper makes sure that any other helpers that conform to the same protocol have the same interface. This helps you, the developer, make sense of what you should minimally implement for your new animation helper. It also makes adding new effects or swapping one effect for another effect very easy and straightforward.

Another advantage is that the `startAnimation` method can be moved to a protocol extension. Its implementation is simple and straightforward, and you typically won't need to customize it, so it's a great candidate to provide a default implementation for. Create a new Swift file named `ViewEffectAnimatorType`, and add it to a new folder called `Protocols`. Now add the following implementation for the protocol:

```
import UIKit

typealias ViewEffectAnimatorComplete = (UIViewAnimatingPosition) ->
Void

protocol ViewEffectAnimatorType {

  var animator: UIViewPropertyAnimator { get }

  init(targetView: UIView, onComplete: @escaping
ViewEffectAnimatorComplete)
  init(targetView: UIView, onComplete: @escaping
ViewEffectAnimatorComplete, duration: TimeInterval)

  func startAnimation()
}

extension ViewEffectAnimatorType {
  func startAnimation() {
    animator.startAnimation()
```

```
    }
}
```

This protocol defines all of the requirements for an animation helper. Note that a globally-available `typealias` named `ViewEffectAnimatorComplete` has been defined. This means that you can replace the type declaration for `onBounceComplete` in `ViewController`, so it is called `ViewEffectAnimatorComplete` instead of `BounceAnimationHelper.BounceAnimationComplete`. This enables you to use the same completion-handler type across the app, which enhances code consistency. To use this protocol, update the initializers for `BounceAnimationHelper` to use the new `typealias` and remove the old one. Also, remove the `startAnimation` method, and finally, add `ViewEffectAnimatorType` to the `BounceAnimationHelper` definition, as shown in the following code:

```
struct BounceAnimationHelper: ViewEffectAnimatorType
```

By conforming `BounceAnimationHelper` to `ViewEffectAnimatorType`, it uses the protocol extension's default implementation for `startAnimation`, and you have a predictable, formalized interface for `BounceAnimationHelper` and any future effects that you may wish to add to the app. Let's add a protocol to our contact object as well.

Defining a contact-displayable protocol

Many apps display lists of contents that are almost the same, but not quite. Imagine displaying a list of contacts: a placeholder for a contact that can be tapped to add a new contact and other cells that could suggest people you may know. Each of these three cells in the collection view could look the same, yet the underlying models can be very different.

A certain sense of unity among these three models can be achieved with a simple protocol that defines what it means to be displayed in a certain way. It's a perfect example of a situation where you're more interested in an object's capabilities than its concrete type. To determine what it means to be displayed in the contact overview, you should look inside `ViewController.swift`. The following code is used to configure a cell in the contact-overview page:

```
let contact = contacts[indexPath.row]

cell.nameLabel.text = "\(contact.givenName) \(contact.familyName)"
contact.fetchImageIfNeeded { image in
  cell.contactImage.image = image
}
```

From this code, you can extract four things a contact-displayable item should contain:

- A `givenName` property
- A `familyName` property
- A `fetchImageIfNeeded` method
- A `contactImage` property

Since `givenName` and `familyName` are specific to a real person, it's wise to combine the two in a new property: `displayName`. This provides some more flexibility regarding what kinds of objects can conform to this protocol without having to resort to crazy tricks. Create a new Swift file named `ContactDisplayable` and add it to the `Protocols` folder. Add the following implementation:

```
import UIKit

protocol ContactDisplayable {
  var displayName: String { get }
  var image: UIImage? { get set }

  func fetchImageIfNeeded()
  func fetchImageIfNeeded(completion: @escaping ((UIImage?) -> Void))
}
```

Now add the following computed property to `Contact` and make sure that you add conformance to `ContactDisplayable` in the `Contact` class's definition:

```
var displayName: String {
   return "\(givenName) \(familyName)"
}
```

You may have noticed that the protocol contains two `fetchImageIfNeeded()` declarations, one with the completion closure and one without. Unfortunately, you can't provide default parameters for function arguments in protocols, so to keep code changes to a minimum, you must specify both versions of `fetchImageIfNeeded()`. Update `Contact` by adding `fetchImageIfNeeded()` without arguments, as follows:

```
func fetchImageIfNeeded() {
  fetchImageIfNeeded(completion: { _ in })
}
```

Also, update the signature for `fetchImageIfNeeded(completion:)` by removing its default completion closure:

```
func fetchImageIfNeeded(completion: @escaping ((UIImage?) -> Void)) {
  // existing implementation
}
```

Next, update the declaration for the contacts array in `ViewController.swift` to look as follows (this will allow you to add any object that can be displayed as a contact to the array):

```
var contacts = [ContactDisplayable]()
```

The next change you need to make in `ViewController` is in `prepare(for:sender:)`. Because the contacts are now `ContactDisplayable` instead of `Contact`, you can't assign them to the detail view controller right away. Update the implementation as follows to typecast the `ContactDisplayable` item to `Contact` so it can be set on the detail view controller:

```
override func prepare(for segue: UIStoryboardSegue, sender: Any?) {
  if let contactDetailVC = segue.destination as?
ContactDetailViewController,
    segue.identifier == "detailViewSegue",
    let selectedIndex =
collectionView.indexPathsForSelectedItems?.first,
    let contact = contacts[selectedIndex.row] as? Contact {
      contactDetailVC.contact = contact
  }
}
```

You're almost done. Just a few more changes to make sure that the project compiles again. The issues you see right now are all related to the change from a class to a struct and to the addition of the `ContactDisplayable` protocol. In `ViewController.swift`, update the `collectionView(_:cellForItemAt:)` method to look as follows:

```
func collectionView(_ collectionView: UICollectionView,
                    cellForItemAt indexPath: IndexPath) ->
UICollectionViewCell {
  let cell = collectionView.dequeueReusableCell(withReuseIdentifier:
    "ContactCollectionViewCell", for: indexPath) as!
ContactCollectionViewCell

  let contact = contacts[indexPath.row]
  // 1
  cell.nameLabel.text = "\(contact.displayName)"
  contact.fetchImageIfNeeded { image in
    cell.contactImage.image = image
  }
```

```
        return cell
    }
```

Lastly, make sure to update the following line in `previewingContext(_:viewControllerForLocation:)`:

```
    viewController.contact = contact as? Contact
```

Summary

This chapter wraps up your exploration of protocol-oriented programming, value types, and reference types. In the previous two chapters, you saw some hypothetical situations that explain the power of these features in Swift. This chapter tied it all together by applying your newfound knowledge to the **Hello-Contacts** app you worked on before. You now know how you can bump up the quality of your code and future-proof an existing application by implementing protocols. To implement protocols, you have to improve your application structure by making sure that `ViewController` didn't contain too much functionality. This in itself was a vast improvement that you were able to take to the next level with a protocol.

Now that you have explored some of the best practices of Swift and applied them to an existing app, it's time to deep dive into Core Data. The next chapters outline how Core Data works and how you can take advantage of it in your applications. If all this value type and protocol-oriented programming talk has made your head reel, don't worry. Take some time to review this chapter and to experiment. The best way to learn these principles is by practicing.

Questions

Good news! No questions for this chapter, you will need your energy once you start working with data in the next chapter.

Further reading

Refer to the previous two chapters for further reading; this chapter was all about applying knowledge from those chapters to an existing app.

8
Adding Core Data to Your App

Core Data is Apple's data persistence framework. You can utilize this framework whenever your application needs to store data. Simple data can often be stored in `UserDefaults`, but when you're handling data that's more complex, has relationships, or needs some form of efficient searching, Core Data is much better suited to your needs.

You don't need to build a very complex app or have vast amounts of data to make Core Data worth your while. Regardless of your app's size, even if it's tiny with only a couple of records, or if you're holding onto thousands of records, Core Data has your back.

In this chapter, you'll learn how to add Core Data to an existing app. The app you will build keeps track of a list of favorite movies for all members of a family. The main interface is a table view that shows a list of family members. If you tap on a family member, you'll see their favorite movies. Adding family members can be done through the overview screen, and adding movies can be done through the detail screen.

You won't build the screens in this app from scratch. The code bundle for this chapter includes a starter project called **MustC**. The starter project contains all of the screens, so you don't have set up the user interface before you get around to implementing Core Data.

In this chapter, the following topics are covered:

- Understanding the Core Data Stack
- Adding Core Data to an application
- Modeling data in the model editor

- Storing and querying data
- Using multiple managed object contexts

Understanding the Core Data Stack

Before you dive right into the project and add Core Data to it, let's take a look at how Core Data actually works, what it is, and what it isn't. In order to make efficient use of Core Data, it's essential that you know what you're working with.

When you work with Core Data, you're actually utilizing a stack of layers that starts with managed objects and ends with a data store. This is often a SQLite database, but there are different storage options you can use with Core Data, depending on your application needs. Let's take a quick look at the layers involved with Core Data and discuss their roles in an application briefly:

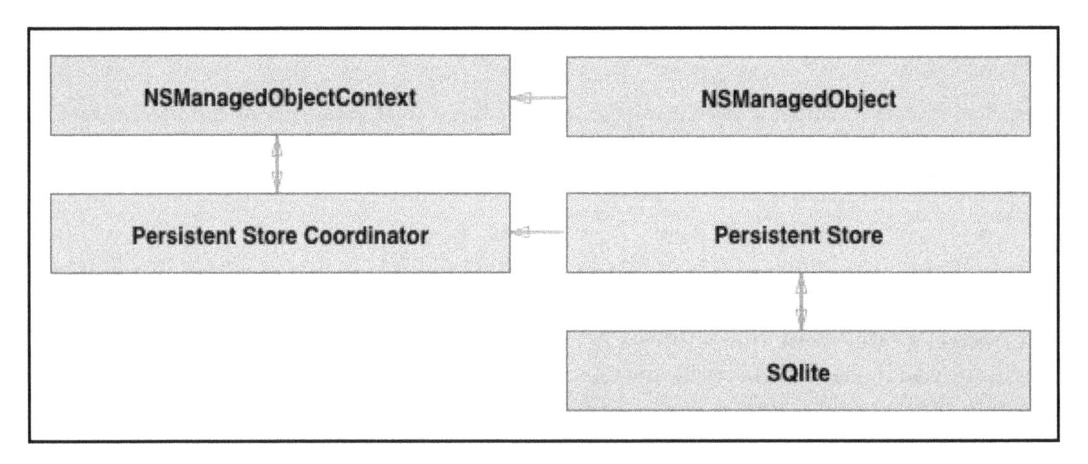

At the top-right of this diagram is the `NSManagedObject` class. When you use Core Data, this is the object you'll interact with most often since it's the base class for all Core Data models your app contains. For instance, in the app you will build in this chapter, the family member and movie models are subclasses of `NSManagedObject`.

Each managed object belongs to an `NSManagedObjectContext`. The managed object context is responsible for communicating with the persistent store coordinator. Often, you'll only need a single managed object context and a single Persistent Store Coordinator. However, it is possible to use multiple persistent store coordinators and multiple managed object contexts. It's even possible to have multiple managed object contexts for the same persistent store coordinator.

A setup with multiple managed object contexts can be particularly useful if you're performing costly operations on your managed objects; for example, if you're importing or synchronizing large amounts of data. Usually, you will stick to using a single managed object context and a single persistent store coordinator because most apps don't need more than one.

The persistent store coordinator is responsible for communicating with the persistent store. In most scenarios, the persistent store uses SQLite as its underlying storage database. However, you can also use other types of storage, such as an in-memory database. An in-memory database is especially useful if you're writing unit tests or if your app has no need for long-term storage. More information on testing can be found in `Chapter 23`, *Ensuring App Quality with Tests*.

If you've worked with MySQL, SQLite, or any other relational database, it is tempting to think of Core Data as a layer on top of a relational database. Although this isn't entirely false since Core Data can use SQLite as its underlying storage, Core Data does not work the same as using SQLite directly; it's an abstraction on top of this.

One example of a difference between SQLite and Core Data is the concept of primary keys. Core Data doesn't allow you to specify your own primary keys. Also, when you define relationships, you don't use foreign keys. Instead, you simply define the relationship and Core Data will figure out how to store this relationship in the underlying database. You will learn more about this later. It's important to know that you should not directly translate your SQL experiences to Core Data. If you do, you will run into issues, simply because Core Data is not SQL. It just so happens that SQLite is one of the ways that data can be stored but the similarities really do end right there.

To recap, all core data apps have a **persistent store**. This store is backed by an in-memory database or a SQLite database. A **persistent store coordinator** is responsible for communicating with the **persistent store**. The object communicating with the **persistent store coordinator** is the **managed object context**. An application can have multiple **managed object context** instances talking to the same **persistent store coordinator**. The objects that a **managed object context** retrieves from the **persistent store coordinator** are **managed objects**.

Now that you have an overview of the Core Data stack and where all the parts involved with its usage belong, let's add the Core Data stack to the **MustC** application.

Adding Core Data to an existing application

When you create a new project in Xcode, Xcode asks whether you want to add Core Data to your application. If you check this checkbox, Xcode will automatically generate some boilerplate code that sets up the Core Data stack. For practicing purposes, **MustC** was set up without Core Data so you'll have to add this to the project yourself. Start by opening `AppDelegate.swift` and add the following `import` statement:

```
import CoreData
```

Next, add the following `lazy` variable to the implementation of `AppDelegate`:

```
private lazy var persistentContainer: NSPersistentContainer = {
      let container = NSPersistentContainer(name: "MustC")
      container.loadPersistentStores(completionHandler: {
(storeDescription, error) in
        if let error = error {
          fatalError("Unresolved error (error), (error.userInfo)")
        }
      })
      return container
  }()
```

If you declare a variable as `lazy`, it won't be initialized until it is accessed. This is particularly useful for variables that are expensive to initialize, rely on other objects, or are not always accessed. The fact that the variable is initialized just in time comes with a performance penalty since the variable needs to be set up the first time you access it. In certain cases this is fine, but in other cases, it might negatively impact the user experience. When used correctly, `lazy` variables can offer great benefits.

The preceding code snippet creates an instance of `NSPersistentContainer`. The persistent container is a container for the persistent store coordinator, persistent store, and managed object context. This single object manages different parts of the Core Data stack, and it ensures that everything is set up and managed correctly.

If you let Xcode generate the Core Data code for your app, it adds a similar property to create an `NSPersistentContainer`. Xcode also adds a method called `saveContext()` to `AppDelegate`. This method is used in `applicationWillTerminate(_:)` to perform a last-minute save of changes and updates when the application is about to terminate. Since you're setting up Core Data manually, this behavior isn't added by Xcode so it must be added by you manually.

Instead of placing the `saveContext()` method in `AppDelegate`, you will add this method as an extension to `NSPersistentContainer`. This makes it easier for other parts of your code to use this method, without having to rely on `AppDelegate`.

Create a new folder in the **Project Navigator** and name it `Extensions`. Also, create a new Swift file and name it `NSPersistentContainer.swift`. Add the following implementation to this file:

```
import CoreData

extension NSPersistentContainer {
  func saveContextIfNeeded() {
    if viewContext.hasChanges {
      do {
        try viewContext.save()
      } catch {
        let nserror = error as NSError
        fatalError("Unresolved error \(nserror), \(nserror.userInfo)")
      }
    }
  }
}
```

This code adds a new method to `NSPersistentContainer` instances by extending it. This is really convenient because it decouples the save method from `AppDelegate` entirely. This is much nicer than the default save mechanism provided for Core Data apps by Xcode. Add the following implementation of `applicationWillTerminate(_:)` to `AppDelegate` to save the context right before the app terminates:

```
func applicationWillTerminate(_ application: UIApplication) {
    persistentContainer.saveContextIfNeeded()
}
```

Now, whenever the application terminates, the persistent store will check whether there are any changes to the managed object context that the `viewContext` property points to. If there are any changes, an attempt to save them is made. If this attempt fails, the app will crash with a `fatalError`. In your own app, you might want to handle this scenario a bit more gracefully. It could very well be that failing to save data before the app terminates is no reason to crash in your app. You can modify the error-handling implementation of `saveContextIfNeeded()` if you think a different behavior is more appropriate for your app.

Now that you have the Core Data stack set up, you need a way to provide this stack to the view controllers in the app. A common technique to achieve this is called *dependency injection*. In this case, dependency injection means that `AppDelegate` will pass the persistent container to `FamilyMemberViewController`, which is the first view controller in the app. It then becomes the job of `FamilyMemberViewController` to pass the persistent container to the next view controller that depends on it, and so forth.

In order to inject the persistent container, you need to add a property to `FamilyMembersViewController` that holds the persistent container. Don't forget to import Core Data at the top of the file and add the following code:

```
var persistentContainer: NSPersistentContainer!
```

Now, in `AppDelegate`, modify the `application(_:didFinishLaunchingWithOptions:)` method as follows:

```
func application(_ application: UIApplication,
didFinishLaunchingWithOptions launchOptions:
[UIApplication.LaunchOptionsKey: Any]?) -> Bool {

  if let navVC = window?.rootViewController as?
UINavigationController,
    let initialVC = navVC.viewControllers[0] as?
FamilyMembersViewController {

    initialVC.persistentContainer = persistentContainer
  }

  return true
}
```

Even though this code does exactly what it should, you can make one major improvement. You know that there might be more view controllers that depend on a persistent container. You also learned that checking whether something is a certain type is something you should generally avoid. As an exercise, attempt to improve the code by adding a protocol called `PersistenContainerRequiring`. This protocol should add a requirement for an implicitly-unwrapped `persistentContainer` property. Make sure that `FamilyMembersViewController` conforms to this protocol and fix the implementation of `application(_:didFinishLaunchingWithOptions:)` as well so it uses your new protocol.

You have just put down the foundation that is required to use Core Data in your app. Before you can use Core Data and store data in it, you must define what data you would like to save by creating your data model. Let's go over how to do this next.

Creating a Core Data model

So far, you have worked on the persistence layer of your app. The next step is to create your models so you can actually store and retrieve data from your Core Data database. All models in an application that uses Core Data are represented by `NSManagedObject` subclasses. When you retrieve data from your database, `NSManagedObjectContext` is responsible for creating instances of your managed objects and populating them with the relevant fetched data.

The **MustC** application requires two models: a family-member model and a movie model. When you define models, you can also define relationships. For the models in **MustC**, you should define a relationship that links multiple movies to a single family member.

Creating the models

In order for Core Data to understand which models your application uses, you must define them in Xcode's model editor. Let's create a new model file so you can add your own models to the **MustC** application. Create a new file, and from the file template selection screen, pick **Data Model**. Name your model file `MustC`. First, you will set up the basic models, and then see how you can define a relationship between family members and their favorite movies:

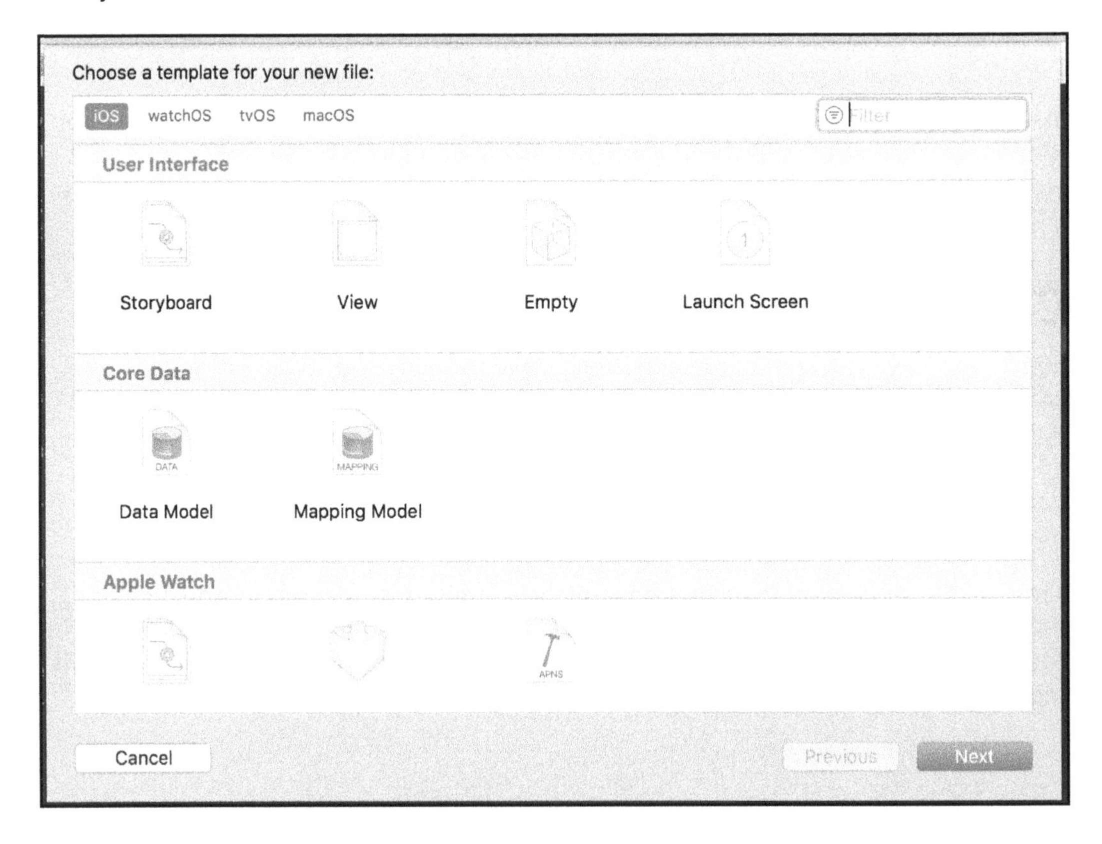

Your project now contains a file called `MustC.xcdatamodeld`. Open this file to go to the model editor. In the bottom-left corner of the editor, you'll find a button labeled **Add Entity**. Click this button to add a new Entity and name it `FamilyMember`.

When you select an Entity by clicking it, you can see all of its attributes, relationships, and fetched properties. Let's add a **name** property to the family member. Click on the plus (+) icon at the bottom of the empty attributes list and add a new attribute called **name**. Make sure that you select `String` as the type for this attribute:

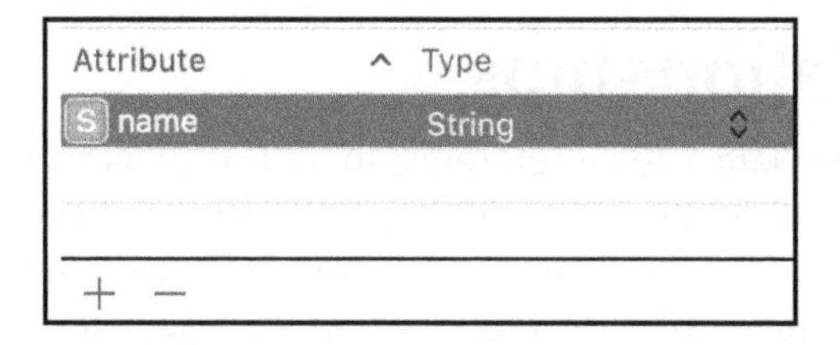

Click on this new property to select it. In the sidebar on the right, select the third tab to open the **Data Model inspector**. This is where you can see more detailed information on this attribute. For instance, you can configure a property to be indexed for faster lookups. You can also choose whether you want the attribute to be optional. For now, you shouldn't care too much about indexing since you're not performing lookups by family members' names, and, even if you were, a family doesn't tend to have hundreds or thousands of members. By default, the **Optional** checkbox is checked. Make sure that you uncheck this box because you don't want to store family members without a name.

Some other options you have for attributes are adding validation, adding a default value, and enabling indexing in Spotlight. For now, leave all those options in their default setting:

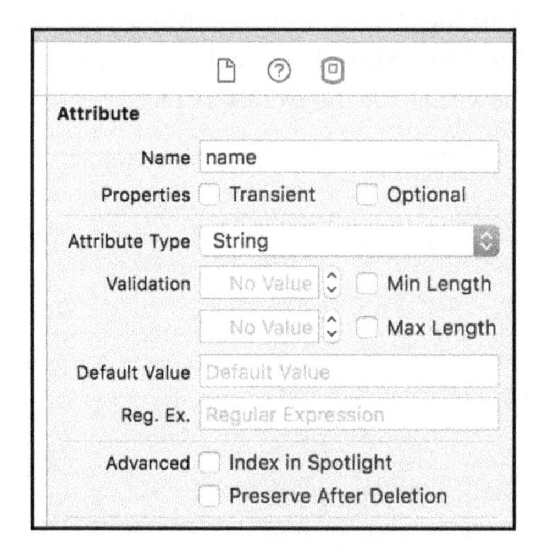

In addition to a `FamilyMember` Entity, **MustC** also needs a `Movie` Entity. Create this entity using the same steps as before and give it a single property: `title`. This property should be a string and it shouldn't be optional. Once you've added this property, you can set up a relationship between family members and their favorite movies.

Defining relationships

A relationship in Core Data adds a reference to an Entity as a property on an Entity. In this case, you want to define a relationship between `FamilyMember` and `Movie`. The best way to describe this relationship is a one-to-many relationship. This means that every movie will have only one family member associated with it and every family member can have multiple favorite movies.

Configuring your data model with a one-to-many relationship from `Movie` to `FamilyMember` is not the most efficient way to define this relationship. A many-to-many relationship is likely a better fit because that would allow multiple family members to add the same movie instance as their favorite. A one-to-many relationship is used in this example to keep the setup simple and make it easy to follow along with the example.

Select the `FamilyMember` entity and click on the plus icon at the bottom of the **Relationships** list. Name the relationship `movies` and select `Movie` as the destination. Don't select an **Inverse** relationship yet because the other end of this relationship is not defined yet. The **Inverse** relationship will tell the model that `Movie` has a property that points back to the `FamilyMember`. Make sure that you select **to many** as the relationship type in the **Data Model Inspector** panel for the `movies` property. Also, select **Cascade** as the value for the **delete rule**:

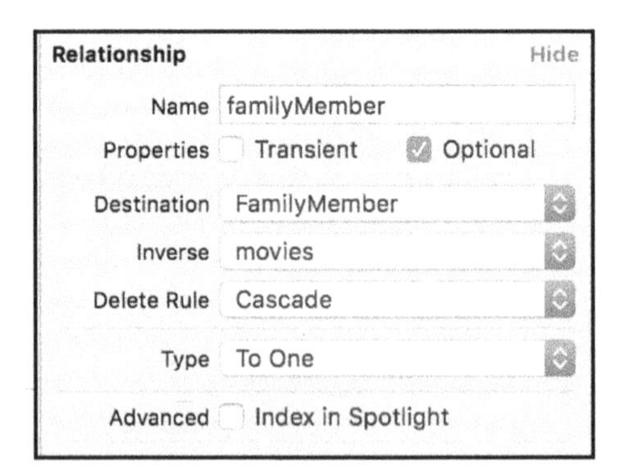

The **delete rule** is a very important property to be set correctly. Not paying attention to this property could result in a lot of orphaned, and even corrupted, data in your database. For instance, setting this property to nullify simply sets the **Inverse** of the relationship to nil. This is the correct behavior when deleting a movie because deleting a movie shouldn't delete the entire family member who added this movie as their favorite. It should simply be removed from the list of favorites.

However, if a `FamilyMember` gets deleted and the relationship is nullified, you would end up with a bunch of movies that don't have a family member associated with them. In this application, these movies are worthless; they won't be used anymore because every movie only belongs to a single `FamilyMember`. For this app, it's desirable that when a `FamilyMember` gets deleted, Core Data also deletes their favorite movies. This is precisely what the cascade option does; it cascades the deletion over to the relationship's **Inverse**.

After setting the delete rule to cascade, select the `Movie` entity and define a relationship called `familyMember`. The destination should be `FamilyMember` and the **Inverse** for this relationship is `favoriteMovies`. After adding this relationship, the **Inverse** will be automatically set on the `FamilyMember` entity:

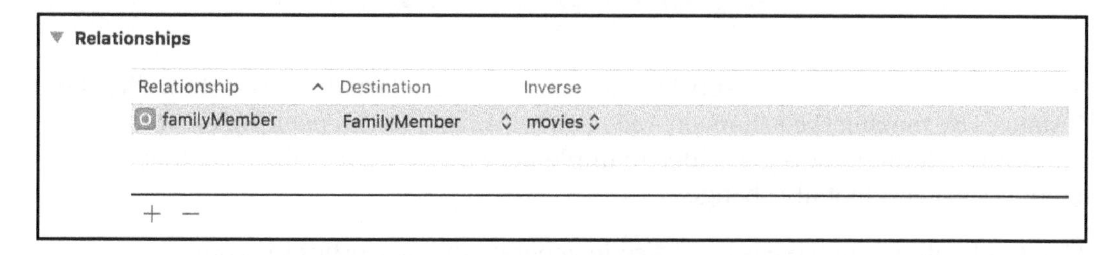

Using your entities

As mentioned before, every model or Entity in your Core Data database is represented by an `NSManagedObject`. There are a couple of ways to create or generate `NSManagedObject` subclasses. In the simplest of setups, an `NSManagedObject` subclass contains just the properties for a certain managed object and nothing else. If this is the case, you can let Xcode generate your model classes for you.

This is actually what Xcode does by default. If you build your project now and add the following code to `viewDidLoad()` in `FamilyMembersViewController`, your project should compile just fine:

```
let fam = FamilyMember(entity: FamilyMember.entity(), insertInto:
persistentContainer.viewContext)
```

This works automatically; you don't have to write any code for your models yourself. Don't worry about what the preceding code does just yet, we'll get into that very soon. The point is that you see that a `FamilyMember` class exists in your project even though you didn't have to create one yourself.

If the default behavior doesn't suit the approach you want in your app – for instance, if you want to prevent your code from modifying your models by defining your variables as `private(set)` – you may want to create a custom subclass instead of making Xcode generate the classes for you. A custom `NSManagedObject` subclass for `FamilyMember` could look like this:

```
class FamilyMember: NSManagedObject {
    @NSManaged private(set) var name: String
    @NSManaged private(set) varfavoriteMovies: [Movie]?
}
```

This custom `FamilyMember` subclass makes sure that external code can't modify the instances by making the setters on `FamilyMember` private. Depending on your application, it might be a good idea to implement this since it will ensure that your models can't accidentally change.

One final option you have is to let Xcode generate the properties for your `NSManagedObject` as an extension on a class you define. This is particularly useful if you have some custom stored properties that you'd like to define on your model or if you have a customized `NSManagedObject` subclass that you can use as the base for all of your models.

All code that Xcode generates for your Core Data models is added to the `Build` folder in Xcode's Derived Data. You shouldn't modify it, or access it directly. These files will be automatically regenerated by Xcode whenever you perform a build, so any functionality you add inside the generated files will be overwritten.

For the **MustC** app, it's okay if Xcode generates the model definition classes since there are no custom properties that you need to add. In the model editor, select each entity and make sure that the **Codegen** field is set to **Class Definition**; you can find this field in the Data Model inspector panel:

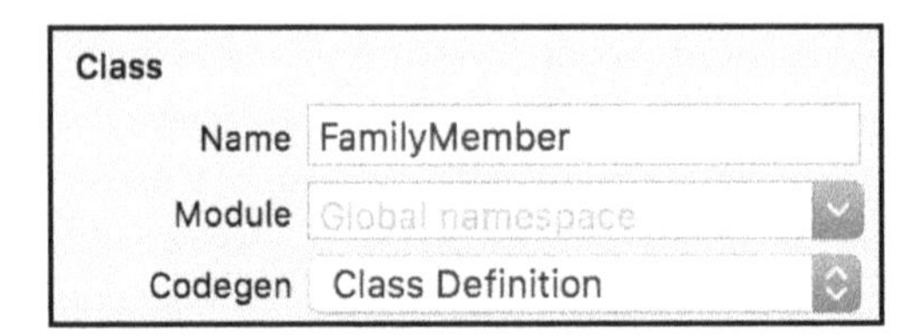

At this point, you are all set up to store your first data in the Core Data database.

Reading and writing data in a Core Data database

The first step to implement data persistence for your app is to make sure that you can store data in the database. You have defined the models that you want to store in your database so the next step is to actually store your models. Once you have implemented a rough version of your data persistence, you will refine the code to make it more reusable. The final step will be to read data from Core Data and dynamically respond to changes in the database.

Understanding data persistence

Whenever you want to persist a model with Core Data, you must insert a new `NSManagedObject` into an `NSManagedObjectContext`. Doing this does not immediately persist the model. It merely stages the object for persistence in the current `NSManagedObjectContext`. If you don't properly manage your managed objects and contexts, this is a potential source of bugs. For example, not persisting your managed objects results in the loss of your data once you refresh the context. Even though this might sound obvious, it could lead to several hours of frustration if you aren't aware of this and have bugs in managing your managed object context.

If you want to save managed objects correctly, you must tell the managed-object context to persist its changes to the persistent store coordinator. The persistent store coordinator will take care of persisting the data in the underlying SQLite database.

Extra care is required when you use multiple managed object contexts. If you insert an object in one managed object context and persist it, you will manually need to synchronize the changes into the other managed object contexts. Also, managed objects are not thread-safe. This means that you must make sure that you create, access, and store a managed object on a single thread at all times. The managed-object context has a helper method called `perform(_:)` to help you with this.

Inserting new objects, updating them, or adding relationships between objects should always be done using the `perform(_:)` method. The reason is that the helper method makes sure that all code in the closure you want to perform are executed on the same thread that the managed object context is on.

Now that you're aware of how data persistence works in Core Data, it's time to start implementing the code to store family members and their favorite movies. You will implement the family member persistence first. Then you'll expand the app so you can safely add movies to family members.

Persisting your models

The first model you will persist is the family member model. The app is already set up with a form that asks for a family member name and a delegate protocol that informs `FamilyMembersViewController` whenever the user wants to store a new family member.

Note that none of the input data is validated; usually, you'd want to add some checks that make sure that the user is not trying to insert an empty family member name, for instance. For now, we'll skip that because this type of validation isn't Core-Data-specific.

The code to persist new family members should be added to the `saveFamilyMember(withName:)` method. Add the following implementation to `FamilyMembersViewController`; we'll go over it line by line after adding the code:

```
func saveFamilyMember(withName name: String) {
  // 1
  let moc = persistentContainer.viewContext

  // 2
  moc.perform {
    // 3
    let familyMember = FamilyMember(context: moc)
    familyMember.name = name

    // 4
    do {
      try moc.save()
    } catch {
      moc.rollback()
    }
  }
}
```

The first comment in this code marks where the managed object context is extracted from `persistentContainer`. All `NSPersistentContainer` objects have a `viewContext` property. This property is used to obtain a managed object context that exists on the main thread.

The second comment marks the call to `perform(_:)`. This ensures that the new `FamilyMember` instance is created and stored on the correct thread. When you create an instance of a managed object context, you must pass the managed object context to the instance of `NSManagedObject`.

Lastly, saving the managed object context can fail, so you must wrap the call to `save()` in a `do {} catch {}` block, so it correctly handles potential errors. If the managed object context can't be saved, all unsaved changes are rolled back.

This code is all you need to persist family members. Before you implement the required code to read existing family members and respond to the insertion of new family members, let's set up `MoviesViewController` so it can store movies for a family member.

The code to store movies for a family member is very similar to the code you wrote earlier. Before you implement the following snippets, make sure that you conform `MoviesViewController` to `PersistenContainerRequiring` and import `CoreData`. Also, add a `persistentContainer` property to `MoviesViewController`.

In order to connect a new movie to a family member, you also need a variable to hold the family member in `MoviesViewController`. Add the following declaration to `MoviesViewController`:

```
var familyMember: FamilyMember?
```

After doing this, add the following implementation for `saveMovie(withName:)`:

```
func saveMovie(withName name: String) {
  guard let familyMember = self.familyMember
    else { return }

  let moc = persistentContainer.viewContext

  moc.perform {
    let movie = Movie(context: moc)
    movie.title = name

    // 1
    let newFavorites: Set<AnyHashable> =
```

```
familyMember.movies?.adding(movie) ?? [movie]

      // 2
      familyMember.movies = NSSet(set: newFavorites)

      do {
        try moc.save()
      } catch {
        moc.rollback()
      }
    }
  }
```

The most important differences between adding the movie and the family member are highlighted with comments. Note that the `movies` property on a family member is `NSSet`. This is an immutable object so you need to create a copy and add the movie to that copy. If no copy could be made because there is no set created yet, you can create a new set with the new movie in it. Next, this new updated `Set` is converted back to an `NSSet` so it can be the new value for `movies`.

As you can see, both save methods share about half of the implementation. You can make some clever use of extensions and generics in Swift to avoid writing this duplicated code. Let's refactor the app a bit.

Refactoring the persistence code

Many iOS developers dislike the amount of boilerplate code that is involved with using Core Data. Simply persisting an object requires you to repeat several lines of code, which can become quite a pain to write and maintain over time. The approach to refactoring the persistence code presented in the following examples is heavily inspired by the approach taken in the *Core Data* book written by *Florian Kugler* and *Daniel Eggert*.

If you're interested in learning more about Core Data outside of what this book covers, and if you'd like to see more clever ways to reduce the amount of boilerplate code, you should pick up this book.

One pattern you can find in both save methods is the following:

```
moc.perform {
  // create managed object

  do {
```

```
    try moc.save()
  } catch {
    moc.rollback()
  }
}
```

It would be great if you could write the following code to persist data instead:

```
moc.persist {
    // create managed object
}
```

This can be achieved by writing an extension for `NSManagedObjectContext`. Add a file called `NSManagedObjectContext` to the extensions folder, and add the following implementation:

```
import CoreData

extension NSManagedObjectContext {
  func persist(block: @escaping () -> Void) {
    perform {
      block()

      do {
        try self.save()
      } catch {
        self.rollback()
      }
    }
  }
}
```

The preceding code allows you to reduce the amount of boilerplate code, which is something that you should always try to achieve. Reducing boilerplate code greatly improves your code's readability and maintainability. Update both the family overview and the movie list view controllers to make use of this new persistence method.

Reading data with a simple fetch request

The simplest way to fetch data from your database is to use a fetch request. The managed object context forwards fetch requests to the persistent store coordinator. The persistent store coordinator will then forward the request to the persistent store, which will then convert the request to a SQLite query. Once the results are fetched, they are passed back up this chain and converted to `NSManagedObject` instances.

By default, these objects are called faults. When an object is a fault, it means that the actual properties and values for the object are not fetched yet, but they will be fetched once you access them. This is an example of a good implementation of lazy variables because fetching the values is a pretty fast operation, and fetching everything up front would greatly increase your app's memory footprint because all values would have to be loaded into memory right away.

Let's take a look at an example of a simple fetch request that retrieves all `FamilyMember` instances that were saved to the database:

```
let request: NSFetchRequest<FamilyMember> =
FamilyMember.fetchRequest()

let moc = persistentContainer.viewContext

guard  let results = try? moc.fetch(request)
  else { return }
```

As you can see, it's not particularly hard to fetch all of your family members. Every `NSManagedObject` has a class method that configures a basic fetch request that can be used to retrieve data. If you have large amounts of data, you probably don't want to fetch all of the persisted objects at once. You can configure your fetch request to fetch data in batches by setting the `fetchBatchSize` property. It's recommended that you use this property whenever you want to use fetched data in a table view or collection view. You should set the `fetchBatchSize` property to a value that is just a bit higher than the number of cells you expect to display at a time. This makes sure that Core Data fetches plenty of items to display while avoiding loading everything at once.

Now that you know how to fetch data, let's display some data in the family member's table view. Add a new variable called `familyMembers` to `FamilyMembersViewController`. Give this property an initial value of `[FamilyMember]()`, so you start off with an empty array of family members. Also, add the example fetch request you saw earlier to `viewDidLoad()`. Next, assign the result of the fetch request to `familyMembers` as follows:

```
familyMembers = results
```

Finally, update the table view delegate methods so `tableView(_:numberOfRowsInSection:)` returns the number of items in the `familyMembers` array. Also, update the `tableView(_:cellForRowAtIndexPath:)` method by adding the following two lines before returning the cell:

```
let familyMember = familyMembers[indexPath.row]
cell.textLabel?.text = familyMember.name
```

If you build and run your app now, you should see the family members you already saved. New family members won't show up right away. However, when you quit the app and run it again, new members will show up.

You could manually reload the table view right after you insert a new family member so it's always up to date but this isn't the best approach. You will see a better way to react to the insertion of new data soon. Let's finish the family member detail view first so it shows a family member's favorite movies. Add the following code to the end of the `prepare(for:sender:)` method in the overview view controller:

```
if let moviesVC = segue.destination as? MoviesViewController {
  moviesVC.persistentContainer = persistentContainer
  moviesVC.familyMember = familyMembers[selectedIndex.row]
}
```

The preceding lines of code pass the selected family member and the persistent container to `MoviesViewController` so it can display and store the current family member's favorite movies.

All you need to do to show the correct movies for a family member is use the family member's favorite movies in the `MovieViewController` class's table-view data-source methods, as follows:

```
func tableView(_ tableView: UITableView, numberOfRowsInSection
section: Int) ->Int {
  return familyMember?.movies?.count ?? 0
}

func tableView(_ tableView: UITableView, cellForRowAt indexPath:
IndexPath) -> UITableViewCell {
  guard let cell = tableView.dequeueReusableCell(withIdentifier:
"MovieCell"),
    let movies = familyMember?.movies
      else { fatalError("Wrong cell identifier requested or missing
family member") }

  let moviesArray = Array(movies as! Set<Movie>)
  let movie = moviesArray[indexPath.row]
  cell.textLabel?.text = movie.title

  return cell
}
```

You don't need to use a fetch request here because you can simply traverse the `movies` relationship on the family member to get their favorite movies. This isn't just convenient for you as a developer, it's also good for your app's performance. Every time you use a fetch request, you force a query to the database. If you traverse a relationship, Core Data will attempt to fetch the object from memory instead of asking the database.

Again, adding new data won't immediately trigger the table view to update its contents. We'll get to that after we take a look at how to filter data. If you want to check whether your code works, build and rerun the app so all the latest data is fetched from the database.

Filtering data with predicates

A typical operation you'll want to perform on your database is filtering. In Core Data, you use predicates to do this. A **predicate** describes a set of rules that any object that gets fetched has to match.

When you model your data in the model editor, it's wise to think about the types of filtering you need to do. For instance, you may be building a birthday calendar where you'll often sort or filter by date. If this is the case, you should make sure that you have a Core Data index for this property. You can enable indexing with the checkbox you saw earlier in the model editor. If you ask Core Data to index a property, it will significantly improve performance when filtering and selecting data in large datasets.

Writing predicates can be confusing, especially if you try to think of them as the `where` clause from SQL. Predicates are very similar, but they're not quite the same. A simple predicate looks as follows:

```
NSPredicate(format: "name CONTAINS[n] %@", "Gu")
```

A predicate has a format; this format always starts with a key. This key represents the property you want to match on. In this example, it would be the name of a family member. Then, you specify the condition, for instance ==, >, <, or `CONTAINS[n]`. There are more conditions available, but the ones listed are some examples of conditions you'll commonly use. Finally, you will specify a placeholder that is substituted with the true value. This placeholder is `%@` in the preceding sample. If you have written any Objective-C before you picked up this book, the `%@` placeholder might look familiar to you because it's used as a placeholder in format strings there.

The example predicate is very simple and bare; it could be the template for a search feature you're building. Usually, a simple search doesn't have to be much more complicated than this as long as there's an index added to the properties you search for.

If you have multiple predicates you want to match on, you can combine them using `NSCompoundPredicate`. This class combines different predicates using either an `and`, `or`, or `not` clause. A typical use case for this approach is when you build a complex filter in your app where the predicate is hard to express in a single statement.

To use a predicate in a fetch request, you assign it to the `predicate` property of a fetch request. Every fetch request has a predicate property that you can set. It can handle both a single predicate and a compound predicate. If you set this property before executing the fetch request, the predicate is applied to the request, and you will receive a filtered dataset instead of the full dataset.

Predicates are powerful, and they have many options available.

If you're interested in an in-depth overview of predicates and all of the ways in which you can make use of format strings, I recommend that you read *Apple's Predicate Programming Guide at* `http://apple.co/2fF3qHc`. It provides a well-documented overview of predicates and their applications.

Next up, you will learn how to respond to changes in the managed object context. For instance, when you add new family members and movies.

Reacting to database changes

In its current state, the **MustC** app doesn't update its list when a new managed object is persisted. One possible solution for this is to manually reload the table right after a new family member is inserted. Although this might work well for some time, it's not the best solution to this problem. If the app grows, you might add a functionality that imports new family members from the network. Manually refreshing the table view would be problematic because the networking logic should not be aware of the table view. Luckily, there is a better solution to react to changes in your data.

One way to respond to database changes is using `NSFetchedResultsController`. This object is perfect for listening to the insertion of new family members. A second way to respond to notifications is through notifications. You will implement this approach in `MoviesViewController`.

Implementing NSFetchedResultsController

`NSFetchedResultsController` is a helper object that specializes in fetching data and managing this data. It listens to changes in its managed object context and notifies a delegate whenever the data it has fetched changes. This is incredibly helpful because it allows you to respond to specific changes in the dataset rather than reloading the table view entirely.

Being a delegate for the fetched results controller involves the following four important methods:

- `controllerWillChangeContent(_:)`
- `controllerDidChangeContent(_:)`
- `controller(_:didChange:at:for:newIndexPath:)`
- `controller(_:didChange:atSectionIndex:for:)`

The first method, `controllerWillChangeContent(_:)`, is called right before the controller passes updates to the delegate. If you're using a table view with a fetched-results controller, this is the perfect method to begin updating the table view.

Next, `controller(_:didChange:at:for:newIndexPath:)` and `controller(_:didChange:atSectionIndex:for:)` are called to inform the delegate about updates to the fetched items and sections, respectively. This is where you should handle updates in the fetched data. For instance, you could insert new rows in a table view if new items were inserted in the dataset.

Finally, `controllerDidChangeContent(_:)` is called. This is the point where you should let the table view know you've finished processing the updates so all the updates can be applied to the table view's interface.

For **MustC**, it doesn't make sense to implement all four methods because the table view that shows family members only has a single section. This means `controller(_:didChange:atSectionIndex:for:)` does not have to be implemented.

To use a fetched-results controller to fetch the stored family members, you need to create an instance of `NSFetchedResultsController` and assign `FamilyMembersViewController` as its delegate so it can respond to changes in the underlying data. You can then implement the delegate methods so you can respond to changes in the fetched-results dataset. Remove the `familyMembers` array from the variable declarations in `FamilyMembersViewController` and add the following `fetchedResultsController` property:

```
var fetchedResultsController:
NSFetchedResultsController<FamilyMember>?
```

The `viewDidLoad` method should be adjusted as follows:

```
override func viewDidLoad() {
  super.viewDidLoad()

  let moc = persistentContainer.viewContext
  let request = NSFetchRequest<FamilyMember>(entityName:
"FamilyMember")

  request.sortDescriptors = [NSSortDescriptor(key: "name", ascending:
true)]
  fetchedResultsController = NSFetchedResultsController(fetchRequest:
request, managedObjectContext: moc, sectionNameKeyPath: nil,
cacheName: nil)

  fetchedResultsController?.delegate = self

  do {
    try fetchedResultsController?.performFetch()
  } catch {
    print("fetch request failed")
  }
}
```

This implementation initializes `NSFetchedResultsController`, assigns its delegate, and tells it to execute the fetch request. Note that the `sortDescriptors` property of the fetch request is set to an array that contains `NSSortDescriptor`. A fetched-request controller requires this property to be set, and for the list of family members, it makes sense to order family members by name.

Now that you have a fetched-results controller, you should implement the delegate methods on `FamilyMembersViewController` and make it conform to `NSFetchedResultsControllerDelegate`. Add the following extension to `FamilyMembersViewController.swift`:

```
extension FamilyMembersViewController:
NSFetchedResultsControllerDelegate {
  func controllerWillChangeContent(_ controller:
NSFetchedResultsController<NSFetchRequestResult>) {
    tableView.beginUpdates()
  }

  func controllerDidChangeContent(_ controller:
NSFetchedResultsController<NSFetchRequestResult>) {
```

```
        tableView.endUpdates()
    }
}
```

The implementation for this extension is fairly straightforward. The table view gets notified when the fetched-result controller is about to process changes to its data and when the fetched-results controller is done processing changes. The bulk of the work needs to be done in `controller(_:didChange:at:for:newIndexPath)`. This method is called when an update has been processed by the fetched-result controller. In **MustC**, the goal is to update a table view, but you could also update a collection view or store all of the updates in a list and do something else with them.

Let's take a look at how you can process changes to fetched data in the following method:

```
func controller(_ controller:
NSFetchedResultsController<NSFetchRequestResult>, didChange anObject:
Any, at indexPath: IndexPath?, for type: NSFetchedResultsChangeType,
newIndexPath: IndexPath?) {

  switch type {
  case .insert:
    guard let insertIndex = newIndexPath
      else { return }
    tableView.insertRows(at: [insertIndex], with: .automatic)
  case .delete:
    guard let deleteIndex = indexPath
      else { return }
    tableView.deleteRows(at: [deleteIndex], with: .automatic)
  case .move:
    guard let fromIndex = indexPath,
      let toIndex = newIndexPath
      else { return }
    tableView.moveRow(at: fromIndex, to: toIndex)
  case .update:
    guard let updateIndex = indexPath
      else { return }
    tableView.reloadRows(at: [updateIndex], with: .automatic)
  }
}
```

This method contains quite a lot of code, but it's actually not that complex. The preceding method receives a `type` parameter. This parameter is an `NSFetchedResultsChangeType` that contains information about the kind of update that was received. The following are the four types of updates that can occur:

- `insert`
- `delete`
- `move`
- `update`

Each of these change types corresponds to a database action. If an object was inserted, you will receive an `insert` change type. The proper way to handle these updates for **MustC** is to simply pass them on to the table view. Once all updates are received, the table view will apply all of these updates at once.

If you had implemented `controller(_:didChange:atSectionIndex:for:)` as well, it would also have received a change type; however, the sections only deal with the following two types of changes:

- `Insert`
- `Delete`

Sections don't update or move, so if you implement this method, you don't have to account for all cases because you won't encounter any, other than the two listed types of changes.

If you take a close look at the implementation for `controller(_:didChange:at:for:newIndexPath)`, you'll notice that it receives two index paths. One is named `indexPath`, and the other is named `newIndexPath`. They're both optional, so you will need to make sure that you safely unwrap them if you use them. For new objects, only the `newIndexPath` property will be present. For delete and update, the `indexPath` property will be set. When an object is moved from one place in the dataset to another, both `newIndexPath` and `indexPath` will have a value.

The last thing you need to do is update the code in `FamilyMembersViewController` so it uses the fetched results controller instead of the `familyMembers` array that it used earlier. First, update the `prepare(for:sender:)` method as follows:

```
if let moviesVC = segue.destination as? MoviesViewController,
  let familyMember = fetchedResultsController?.object(at:
selectedIndex) {
```

```
    moviesVC.persistentContainer = persistentContainer
    moviesVC.familyMember = familyMember
  }
```

This makes sure that a valid family member is passed to the movies view controller. Update the table view data source methods as shown in the following code. A fetched-results controller can retrieve objects based on an index path. This makes it a great fit to use in combination with table views and collection views:

```
func tableView(_ tableView: UITableView, numberOfRowsInSection
section: Int) -> Int {
  return fetchedResultsController?.fetchedObjects?.count ?? 0
}

func tableView(_ tableView: UITableView, cellForRowAt indexPath:
IndexPath) -> UITableViewCell {
  guard let cell = tableView.dequeueReusableCell(withIdentifier:
"FamilyMemberCell"),
    let familyMember = fetchedResultsController?.object(at: indexPath)
    else { fatalError("Wrong cell identifier requested") }


  cell.textLabel?.text = familyMember.name

  return cell
}
```

If you run your app now, the interface updates automatically when you add a new family member to the database. However, the list of favorite movies doesn't update yet. That page does not use a fetched-results controller so it must listen to changes to the dataset directly.

The reason `MoviesViewController` doesn't use a fetched-results controller for the movie list is that fetched-result controllers will always need to drop down all the way to your persistent store (SQLite in this app). As mentioned before, querying the database has a significant memory overhead compared to traversing the relationship between family members and their movies; it's much faster to read the `movies` property than fetching them from the database.

Whenever a managed object context changes, a notification is posted to the default `NotificationCenter`. `NotificationCenter` is used to send events inside of an app so other parts of the code can react to those events.

It can be very tempting to use notifications instead of delegates, especially if you're coming from a background that makes heavy use of events such as JavaScript. Don't do this; delegation is better-suited to most cases, and it will make your code much more maintainable. Only use notifications if you don't care who's listening to your notifications or if setting up a delegate relationship between objects would mean you'd create very complex relationships between unrelated objects just to set up the delegation.

Let's subscribe `MoviesViewController` to changes in the managed object context so it can respond to data changes if needed. Before you implement this, add the following method, which should be called when changes in the managed object context occur:

```
extension MoviesViewController {
  @objc func managedObjectContextDidChange(notification:
NSNotification) {
    guard let userInfo = notification.userInfo,
      let updatedObjects = userInfo[NSUpdatedObjectsKey] as?
Set<FamilyMember>,
      let familyMember = self.familyMember
      else { return }

    if updatedObjects.contains(familyMember) {
      tableView.reloadData()
    }
  }
}
```

This method reads the notification's `userInfo` dictionary to access the information that's relevant to the current list. You're interested in changes to the current `familyMember` because when this object changes, you can be pretty sure that a new movie was just inserted. The `userInfo` dictionary contains keys for the inserted, deleted, and updated objects. In this case, you should look for the updated objects because users can't delete or insert new family members in this view. If the family member was updated, the table view is reloaded so it shows the new data.

The following code subscribes `MoviesViewController` to changes in the persistent container's managed object context:

```
override func viewDidLoad() {
  super.viewDidLoad()

  NotificationCenter.default.addObserver(self, selector:
#selector(self.managedObjectContextDidChange(notification:)), name:
```

```
.NSManagedObjectContextObjectsDidChange, object: nil)
}
```

When the view loads, the current `MoviesViewController` instance is added as an observer to the `.NSManagedObjectContextObjectsDidChange` notification. Go ahead and build your app; you should now see the user interface update whenever you add new data to your database.

Understanding the use of multiple instances of NSManagedObjectContext

It has been mentioned several times in this chapter that you can use multiple managed object contexts. In many cases, you will only need a single managed object context. Using a single managed object context means that all of the code related to the managed object context is executed on the main thread. If you're performing small operations, that's fine. However, imagine importing large amounts of data. An operation such as that could take a while. Executing code that runs for a while on the main thread will cause the user interface to become unresponsive. This is not good, because the user will think your app has crashed. So how do you work around this? The answer is using multiple managed object contexts.

In the past, using several managed object contexts was not easy to manage; you had to create instances of `NSManagedObjectContext` using the correct queues yourself. Luckily, `NSPersistentContainer` helps to make complex setups a lot more manageable. If you want to import data on a background task, you can obtain a managed object context by calling `newBackgroundContext()` on the persistent container. Alternatively, you can call `performBackgroundTask` on the persistent container and pass it a closure with the processing you want to do in the background.

One important thing to understand about Core Data, background tasks, and multithreading is that you must always use a managed object context on the same thread it was created on. Consider the following example:

```
let backgroundQueue = DispatchQueue(label: "backgroundQueue")
let backgroundContext = persistentContainer.newBackgroundContext()
backgroundQueue.async {
  let results = try? backgroundContext.fetch(someRequest)

  for result in results {
    // use result
```

```
        }
    }
```

The behavior of this code can cause you a couple of headaches. The background context was created on a different queue than the one it's used it on. It's always best to make sure to use a managed object context on the same queue it was created on by using the `perform(_:)` method of `NSManagedObject`. More importantly, you must also make sure to use the managed objects you retrieve on the same queue that the managed object context belongs to.

Often, you'll find that it's best to fetch data on the main queue using the `viewContext` persistent containers. Storing data can be delegated to background contexts if needed. If you do this, you must make sure that the background context is a child context of the main context. When this relationship is defined between the two contexts, your main context will automatically receive updates when the background context is persisted. This is quite convenient because it removes a lot of manual maintenance, which keeps your contexts in sync. Luckily, the persistent container takes care of this for you.

When you find that your app requires a setup with multiple managed object contexts, it's essential to keep the rules mentioned in this section in mind. Bugs related to using managed objects or managed object contexts in the wrong places are often tedious to debug and hard to discover. When implemented carefully, complex setups with multiple managed object contexts can increase your application's performance and flexibility.

Summary

This chapter has shown you how to implement a relatively simple Core Data database that stores family members and their favorite movies. You used the Core Data model editor in Xcode to configure the models you wanted to store and define the relationship between these models. Once the models were set up, you implemented code that created instances of your models so they could be stored in the database and retrieved later.

Next, you retrieved the data from the database and saw that your table view's don't automatically update when the underlying data changes. You used an `NSFetchedResult` controller to fetch family members and listen to changes on the list of family members. You saw that this setup is very powerful because you can respond to changes in your data quite easily.

In the next chapter, you will learn how to enrich the data your users add to the database by fetching and storing data from the web.

Questions

1. What does a persistent store coordinator do?

    a) It talks to the SQLite database.
    b) It talks to the persistent store.
    c) It manages the data model file.

2. What is the function of the managed object context?

    a) It shows instances of managed objects on the screen.
    b) It mediates between managed objects and the persistent store coordinator.
    c) It responds to changes in the persistent store.

3. How many managed object contexts can you use in an app?

    a) Three.
    b) It depends on the number of CPU cores in the device.
    c) An unlimited amount.

4. What is the correct way to call `save()` on a managed object context?

    a) `try! save()`.
    b) `do save()`.
    c) `try save()`.

5. When should you use a fetched-results controller?

    a) Any time you want to show data on the screen and react to changes.
    b) Only when you need to fetch data from the database and want to react to changes.
    c) Any time you want to fetch data.

6. Does a fetched-results controller always go to the database to fetch data?

    a) No, Core Data implements clever caching.
    b) No, only when there is new data.
    c) Yes, all fetch requests must go to the database.

Further reading

- The **Core Data** book, by Florian Kugler and Daniel Eggert
- Apple's Predicate Programming Guide: `http://apple.co/2fF3qHc`

9
Fetching and Displaying Data from the Network

Most modern applications communicate with a web service. Some apps rely on them heavily, acting as a layer that merely reads data from the web and displays it in app form. Other apps use the web to retrieve and sync data to make it locally available, and others only use the web as backup storage. Of course, there are a lot more reasons to use data from the internet than the ones mentioned.

In this chapter, you will expand the **MustC** application, so it uses a web service to retrieve popularity ratings for the movies that family members have added as their favorites. These popularity ratings will be stored in the Core Data database and displayed together with the names of the movies.

In this chapter, you'll learn about the following topics:

- `URLSession`
- Working with JSON in Swift
- Updating Core Data objects with fetched data

Fetching data from the web

Retrieving data from the web is something that you will often do as an iOS professional. You won't just fetch data from a web service; you'll also send data back to it. For example, you might have to make an HTTP `POST` request as part of a login flow or to update a user's profile information. Over time, iOS has evolved quite a bit in the web requests department, making it easier to use web services in apps.

HTTP (or HTTPS) is a protocol that almost all web traffic uses for communication between a client, such as an app, and a server. The HTTP protocol supports several methods that signal the request's intent. `GET` is used to retrieve information from a server. A `POST` request indicates the intention to push new content to a server, for instance, submitting a form.

When you want to perform a web request in iOS, you will typically use the `URLSession` class. The `URLSession` class makes asynchronous web requests on your behalf. This means that iOS loads data from the web on a background thread, ensuring that the user interface remains responsive throughout the entire request. If a web request is performed synchronously, the user interface is unresponsive for the duration of the network request because a thread can only do one thing at a time, so if it's waiting for a response from the network, it can't respond to touches or other user input.

If your user has a slow internet connection, a request could take several seconds. You don't want your interface to freeze for several seconds. Even a couple of milliseconds will create a noticeable drop in its responsiveness and frame rate. This can be easily avoided with `URLSession` to perform asynchronous network requests.

First, you will experiment with basic network requests in a playground. You can create a new playground or use the one provided in this book's code bundle. After you've seen the basics of `URLSession`, you'll implement a way to fetch movies from an open source movie database and put this implementation to use in the **MustC** app.

Understanding URLSession basics

The following code snippet shows a sample of a network request that loads the `https://apple.com` homepage:

```
import Foundation

let url = URL(string: "https://apple.com")!
let task = URLSession.shared.dataTask(with: url)
{
  data, response, error in
  print(data)
  print(response)
  print(error)
}

task.resume()
```

This is an elementary example: A URL is created, and then the shared `URLSession` instance is used to create a new `dataTask`. This `dataTask` is an instance of `URLSessionDataTask` and allows you to load data from a remote server. Alternatively, you could use a download task if you're downloading a file, or an upload task if you're uploading files to a web server. After creating the task, you must call resume on the task, because new tasks are always created in a suspended state.

If you run this sample in an empty playground, you'll find that the example doesn't work. Because the network request is made asynchronously, the playground finishes its execution before the network request is complete. To fix this, you should make sure that the playground runs indefinitely. Doing so will allow the network request to finish. Add the following lines to the top of the playground source file to enable this behavior:

```
import PlaygroundSupport
PlaygroundPage.current.needsIndefiniteExecution = true
```

Now that the playground runs indefinitely, you'll find that there isn't a lot of useful data printed to the console. In this case, you're not interested in the raw data, HTTP headers, or the fact that the error is `nil`. When you load data from a URL, you're often most interested in the response's body. The body of a response usually contains the string representation of the data you requested. In the case of the preceding example, the body is the HTML that makes up Apple's homepage. Let's see how you can extract this HTML from the response. Replace the data task's completion callback with the following:

```
{ data, response, error in
  guard let data = data, error == nil
    else { return }

  let responseString = String(data: data, encoding: .utf8)
  print(responseString)
}
```

The preceding callback closure makes sure that there are no errors returned by the web service and that there is data present. Then, the raw data is converted to a string, and that string is printed to the console. If you use this callback instead of the old one, you'll see the HTML for the Apple homepage printed. Simple requests to a web server like the one you just saw are relatively simple to implement with `URLSession`.

You can take more control over the request that's executed by creating a `URLRequest` instance. The example you saw is one where you let `URLSession` create the `URLRequest` on your behalf. This is fine if you want to perform a simple HTTP `GET` request with no custom headers, but if you're going to post data or include specific headers, you will need to have more control over the request that's used. Let's take a look at what a `GET` request with some parameters and a custom header looks like.

The following code uses an API key from `https://www.themoviedb.org/`. If you want to try this code example, create an account on their website and request an API key on your account page. Setting this up should only take a couple of minutes, and if you want to follow along with this chapter, you will need to have your own API key:

```
let api_key = "YOUR_API_KEY_HERE"
var urlString = "https://api.themoviedb.org/3/search/movie/"
urlString = urlString.appending("?api_key=\(api_key)")
urlString = urlString.appending("&query=Swift")

let movieURL = URL(string: urlString)!

var urlRequest = URLRequest(url: movieURL)
urlRequest.httpMethod = "GET"
urlRequest.setValue("application/json", forHTTPHeaderField: "Accept")

let movieTask = URLSession.shared.dataTask(with: urlRequest) { data,
response, error in
  print(response)
}

movieTask.resume()
```

The preceding code is a bit more complicated than the example you saw before. In this example, a more complex URL is configured that includes some HTTP `GET` parameters. The `httpMethod` for `URLRequest` is specified, and a custom header is provided to inform the receiver of this request about the type of response it would like to receive.

The flow for executing this URL request is the same as the one you saw earlier. However, the URL that is loaded responds with a JSON string instead of an HTML document. JSON is used by many APIs as the preferred format to pass data around on the web. In order to use this response, the raw data must be converted to a useful data structure. In this case, a dictionary will do. If you haven't seen or worked with JSON before, it's a good idea to take a step back and read up on the JSON data format because this chapter will continue under the assumption that you are at least somewhat familiar with JSON.

Working with JSON in Swift

The following snippet shows how you can convert raw data to a JSON dictionary. Working with JSON in Swift can be a little tedious at times, but overall, it's an alright experience. Let's look at an example:

```
guard let data = data,
  let json = try? JSONSerialization.jsonObject(with: data, options:
[])
  else { return }

print(json)
```

The preceding snippet converts the raw data that is returned by a URL request to a JSON object. The print statement prints a readable version of the response data, but it's not quite ready to be used. Let's see how you gain access to the first available movie in the response.

If you look at the type of object returned by the `jsonObject(with:options:)` method, you'll see that it returns `Any`. This means that you must typecast the returned object to something you can work with, such as an array or a dictionary. When you inspect the JSON response that the API returned, for instance by using print to make it appear in the console like you did with Apple's homepage HTML, you'll notice that there's a dictionary that has a key called `results`. The `results` object is an array of movies. In other words, it's an array of `[String: Any]`, because every movie is a dictionary, where strings are the keys and the value can be a couple of different things, such as `Strings`, `Int`, or `Booleans`. With this information, you can access the first movie's title in the JSON response, as shown in the following code:

```
guard let data = data,
  let json = try? JSONSerialization.jsonObject(with: data, options:
[]),
  let jsonDict = json as? [String: AnyObject],
  let resultsArray = jsonDict["results"] as? [[String: Any]]
  else { return }

let firstMovie = resultsArray[0]
let movieTitle = firstMovie["title"] as! String
print(movieTitle)
```

Working with dictionaries to handle JSON isn't the best experience. Since the JSON object is of the `AnyObject` type and you need to typecast every element in the dictionary you want to access, there's a lot of boilerplate code you need to add. Luckily, Swift has better ways to create instances of objects from the JSON data. The following example shows how you can quickly create an instance of a `Movie` struct without having to cast all the keys in the JSON dictionary to the correct types for the `Movie` struct.

First, let's define two structs, one for the `Movie` itself, and one for the response that contains the array of `Movie` instances:

```
struct MoviesResponse: Codable {
  let results: [Movie]
}

struct Movie: Codable {
  let id: Int
  let title: String
  let popularity: Float
}
```

Next, you can use the following snippet to quickly convert the raw data from a URL request to an instance of `MoviesResponse`, where all movies are converted to instances of the `Movie` struct:

```
let decoder = JSONDecoder()
guard let data = data,
    let movies = try? decoder.decode(MoviesResponse.self, from: data)
    else { return }

print(movies.results[0].title)
```

You might notice that both `MoviesResponse` and `Movie` conform to the `Codable` protocol. The `Codable` protocol was introduced in Swift 4, and it allows you to easily encode and decode data objects. The only requirement is that all properties of a `Codable` object conform to the `Codable` protocol. A lot of built-in types, such as `Array`, `String`, `Int`, `Float`, and `Dictionary` conform to `Codable`. Because of this, you can easily convert an encoded JSON object into a `MoviesResponse` instance that holds `Movie` instances.

By default, each property name should correspond to the key of the JSON response it is mapped to. However, sometimes you might want to customize this mapping. For instance, the `poster_path` property in the response we've been working with so far would be best mapped to a `posterPath` property on the `Movie` struct. The following example shows how you would tackle these circumstances:

```
struct Movie: Codable {

    enum CodingKeys: String, CodingKey {
        case id, title, popularity
        case posterPath = "poster_path"
    }

    let id: Int
    let title: String
    let popularity: Float
    let posterPath: String?
}
```

By specifying a `CodingKeys` enum, you can override how the keys in the JSON response should be mapped to your `Codable` object. You must cover all keys that are mapped, including the ones you don't want to change. As you've seen, the `Codable` protocol provides powerful tools for working with data from the network. Custom key mapping makes this protocol even more powerful because it allows you to shape your objects exactly how you want them instead of having the URL responses dictate the structure to you.

If the only conversion you need to apply in the coding keys is converting from snake case (`poster_path`) to camel case (`posterPath`), you don't have to specify the coding keys yourself. The `JSONEncoder` object can automatically apply this type of conversion when decoding data if you set its `keyDecodingStrategy` to `.convertFromSnakeCase`, as shown in the following code:

```
let decoder = JSONDecoder()
decoder.keyDecodingStrategy = .convertFromSnakeCase
```

Try implementing this in your playground and remove `CodingKeys` from the `Movie` object to ensure your JSON decoding still works.

Now let's move on to storing fetched data in the Core Data database.

Updating Core Data objects with fetched data

So far, the only thing you have stored in Core Data is movie names. You will expand this functionality by performing a lookup for a certain movie name through the movie database API. The fetched information will be used to display and store a popularity rating for the movies in the Core Data database.

A task such as this seems straightforward at first; you could come up with a flow such as the one shown in the following steps:

1. A user indicates their favorite movie.
2. The movie's popularity rating is fetched.
3. The movie and its rating are stored in the database.
4. The interface updates with the new movie.

At first glance, this is a fine strategy; insert the data when you have it. However, it's important to consider that API calls are typically done asynchronously so the user interface stays responsive. More importantly, API calls can be really slow if your user doesn't have a good internet connection. This means that you would be updating the interface with very noticeable lag if the preceding steps are executed one by one.

The following would be a much better approach to implement the feature at hand:

1. The users indicates their favorite movie.
2. The users store the movie.
3. Update the interface with the new movie.
4. Begin popularity fetching.
5. Update the movie in the database.
6. Update the interface with the popularity.

This approach is somewhat more complex, but it will give the user a responsive experience. The interface will respond to new movies immediately by showing them, and then automatically updates as soon as new data is retrieved. Before you can fetch the data and update the models, the Core Data model must be updated in order to store the movie's popularity rating.

Open the Core Data model editor and select the `Movie` entity. All you have to do is add a new property and name it **popularity**. Select the `Double` type for this property because the **popularity** is stored as a decimal value. You have to make sure that this property is optional since you won't be able to provide a value for it straight away:

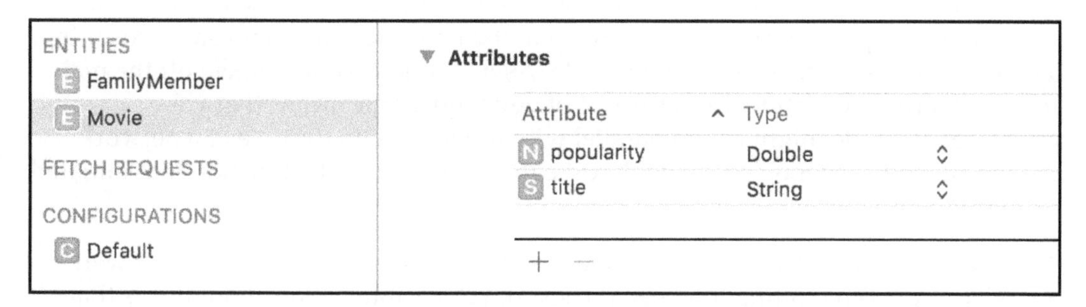

If you've worked with Core Data prior to when iOS 10 was released, this is the part where you expect to read about migrations and how you can orchestrate them. However, for simple changes such as this, we don't need to manage migrations. All you need to do is simply build and run your application to regenerate your model definitions, and for a simple change, such as the one we performed just now, Core Data will automatically manage the migration on its own.

If you want to support iOS versions earlier than 10, make sure you read up on Core Data migrations. Whenever you update your models, you have to make sure that your database can properly migrate from one model version to another. During development, this isn't extremely important: you just reinstall the app whenever your models change. However, app updates will crash on launch if the Core Data model isn't compatible with the previous model.

Now that the model is updated, let's figure out how to implement the flow that was described earlier.

Implementing the fetch logic

The asynchronous nature of network requests makes certain tasks, such as the one you're about to implement, quite complex. Usually, when you write code, its execution is very predictable. Your app typically runs line by line, so any line that comes after the previous one can assume that the line before it has finished executing. This isn't the case with asynchronous code. Asynchronous code is taken off the main thread and runs separately from the rest of your code. This means that your asynchronous code might run in parallel with other code. In the case of a network request, the asynchronous code might execute seconds after the function that initiated the request.

This means that you need to figure out a way to update and save movies that were added as soon as the rating has been retrieved. What's interesting about this is that once you see the code that implements this feature, it will feel natural to you that this is how it works. However, it's important that you're aware of the fact that it's not as straightforward as it may seem at first.

It's also important that you're aware of the fact that the code you're about to look at is executed on multiple threads. This means that even though all pieces of the code are defined in the same place, they are not executed on the same thread. The callback for the network request is executed on a different thread than the code that initiated the network request. You have already learned that Core Data is not thread-safe. This means that you can't safely access a Core Data object on a different thread than the thread it was created on.

If this confuses you, that's okay. You're supposed to be a bit confused right now. Asynchronous programming is not easy, and fooling you into thinking it is will cause frustration once you run into concurrency-related troubles (and you will). Whenever you work with callbacks, closures, and multiple threads, you should be aware that you're doing complex work that isn't straightforward.

Now that you understand that asynchronous code is hard, let's take a closer look at the feature you're about to implement. It's time to start implementing the network request that fetches popularity ratings for movies. You will abstract the fetching logic into a helper named `MovieDBHelper`. Go ahead and create a new helper folder in Xcode and add a new Swift file called `MovieDBHelper.swift` to it.

Abstracting this logic into a helper has multiple advantages. One of them is simplicity; it will keep our view controller code nice and clean. Another advantage is flexibility. Let's say that you want to combine multiple rating websites, or a different API, or compute popularity based on the number of family members who added this same title to their list; it will be easier to implement since all the logic for ratings is in a single place.

Add the following skeleton implementation to the `MovieDBHelper` file:

```
struct MovieDBHelper {
  typealias MovieDBCallback = (Double?) -> Void
  let apiKey = "YOUR_API_KEY_HERE"

  func fetchRating(forMovie movie: String, callback: @escaping
MovieDBCallback) {

  }

  private func url(forMovie movie: String) -> URL? {
    guard let query =
movie.addingPercentEncoding(withAllowedCharacters: .urlHostAllowed)
      else { return nil }

    var urlString = "https://api.themoviedb.org/3/search/movie/"
    urlString = urlString.appending("?api_key=\(apiKey)")
    urlString = urlString.appending("&query=\(query)")

    return URL(string: urlString)
  }
}
```

The preceding code starts off with an interesting line:

```
typealias MovieDBCallback = (Double?) -> Void
```

This line specifies the type that's used for the callback closure that's called when the rating is fetched. This callback will receive an optional `Double` as its argument. If the network request fails for any reason, the `Double` will be `nil`. Otherwise, it contains the rating for the movie that the request was created for.

The snippet also contains a dummy method that performs the fetch; you will implement this method soon. Finally, there's a method that builds a URL. This method is private because it's only supposed to be used inside of the helper struct. Note that the movie is converted to a percent-encoded string. This is required because if your user adds a movie with spaces in it, you would end up with an invalid URL if the spaces aren't properly encoded.

Before you implement `fetchRating(forMovie:callback)`, add a new file named `MovieDBResponse.swift` to the helper folder. This file will be used to define a struct that represents the response we expect to receive from the `Moviedb` API. Add the following implementation to this file:

```
struct MovieDBLookupResponse: Codable {

  struct MovieDBMovie: Codable {
    let popularity: Double?
  }

  let results: [MovieDBMovie]
}
```

The preceding code uses a nested struct to represent the movie objects that are part of the response. This is similar to what you saw in the playground example. Structuring the response this way makes the intent of this helper very obvious, which usually makes code easier to reason about. With this struct in place, let's see what the implementation of `fetchRating(forMovie:callback)` looks like in the following code:

```
func fetchRating(forMovie movie: String, callback: @escaping
MovieDBCallback) {
  guard let searchUrl = url(forMovie: movie) else {
    callback(nil)
    return
  }

  let task = URLSession.shared.dataTask(with: searchUrl) { data,
response, error in
    var rating: Double? = nil

    defer {
      callback(rating)
    }

    let decoder = JSONDecoder()

    guard error == nil, let data = data,
      let lookupResponse = try?
decoder.decode(MovieDBLookupResponse.self, from: data),
      let popularity = lookupResponse.results.first?.popularity
      else { return }

    rating = popularity
  }
```

```
    task.resume()
}
```

This implementation looks very similar to what you experimented with earlier in the playground. The URL-building method is used to create a valid URL. If this fails, it makes no sense to attempt requesting the movie's rating, so the callback is called with a `nil` argument. This will inform the caller of this method that the execution is done and no result was retrieved.

Next, a new data task is created and `resume` is called on this task to kick it off. There is an interesting aspect to how the callback for this data task is called, though. Let's take a look at the following lines of code:

```
var rating: Double? = nil

defer {
  callback(rating)
}
```

A `rating` double is created here, and it is given an initial value of `nil`. Then there's a `defer` block. The code inside of the `defer` block is called right before exiting the scope. In other words, it's executed right before the code returns from a function or closure.

Since this defer block is defined inside the callback for the data task, the callback for the `fetchRating(forMovie:callback:)` method is always called just before the data task callback is exited. This is convenient because all you must do is set the value for the rating to a double, and you don't have to manually invoke the callback for each possible way the scope can be exited. This also applies when you return because of unmet requirements. For instance, if there is an error while calling the API, you don't need to invoke the callback. You can simply return from the closure, and the callback is called automatically. This strategy can also be applied if you instantiate or configure objects temporarily and you want to perform some clean-up when the method, function, or closure is done.

The rest of the code should be fairly straightforward since most of it is nearly identical to the code used in the playground. Now that you have the networking logic down, let's take a look at how to actually update the movie object with a popularity rating.

Updating a movie with a popularity rating

To update the movie object, you will implement the final step of the approach that was outlined earlier. You need to asynchronously fetch a rating from the movie database and then use that rating to update the movie. The following code should be added to `MoviesViewController.swift`, right after adding a new movie to a family member, inside of the `persist` block, approximately at line 35 (depending on how you have formatted your code):

```
let helper = MovieDBHelper()
helper.fetchRating(forMovie: name) { rating in
  guard let rating = rating
    else { return }

  moc.persist {
    movie.popularity = rating
  }
}
```

You can see that the helper abstraction provides a nice interface for the view controller. You can simply use the helper and provide it a movie to fetch the rating for with a callback and you're all set. Abstracting code like this can make maintaining your code a lot more fun in the long run.

The most surprising thing in the preceding snippet is that `moc.persist` is called again inside of the helper callback. This must be done because this callback is actually executed long after the initial persist has finished. Actually, this callback isn't even executed on the same thread as the code it's surrounded by.

To see how your code fails if you don't properly persist your model, try replacing the `moc.persist` block in the rating retrieval callback with the following code:

```
movie.popularity = rating
do {
  try moc.save()
} catch {
  moc.rollback()
}
```

If you add a new movie now, the rating will still be fetched. However, you will suddenly run into issues when reloading your table view. This is because the managed object context was saved on a background thread. This means that the notification that informs the table view about updates is also sent on a background thread. You could resolve the issue by pushing the `reloadData()` call onto the main thread as you've done before, but in this case, doing so would only make the problem worse. Your app might work fine for a while, but once your app grows in complexity, using the same managed object context in multiple threads will most certainly cause crashes. Therefore, it's important to always make sure that you access managed objects and their contexts on the correct thread by using a construct, such as the `persist` method we implemented for this app.

Now that you have looked at all the code involved, let's see what all this threading talk means in a more visual way.

Visualizing multiple threads

The following diagram will help you understand multiple threads:

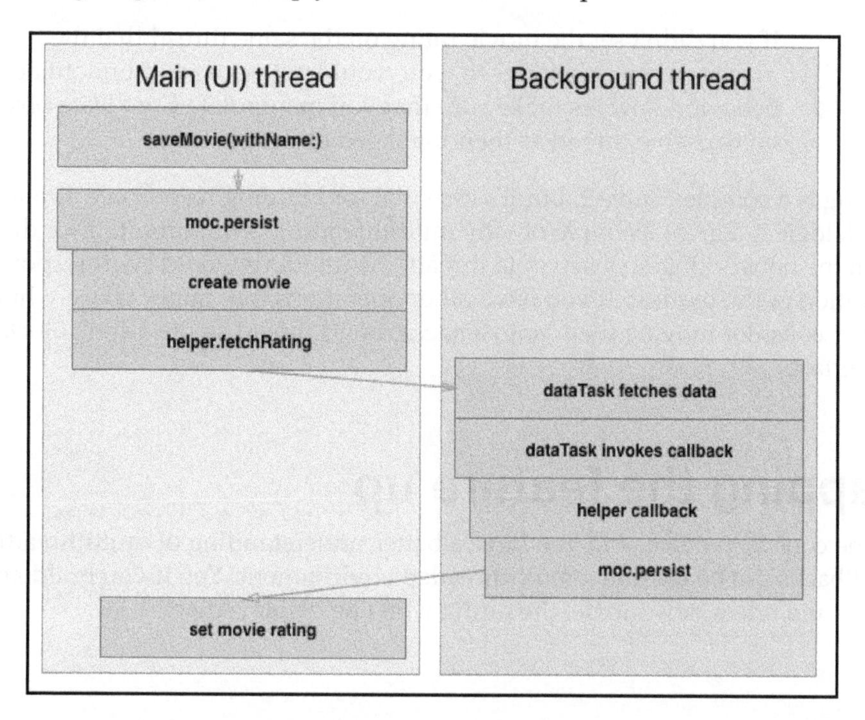

When `saveMovie(withName:)` is called, the execution is still on the main thread. The persistence block is opened, the movie is created, its name is set, a helper is created, and then `fetchRating(forMovie:callback:)` is called on the helper. This call itself is still on the main thread. However, the fetching of data is pushed to a background thread. This was discussed earlier when you experimented with fetching data in a playground.

The callback that's invoked by `dataTask` is called on the same background thread that the task itself is on. The code will do its thing with the JSON and finally, the callback that was passed to `fetchRating(forMovie:callback:)` is called. The code inside of this callback is executed on the background thread as well.

You can see that the set movie-rating step in the update flow is somehow pushed back to the main thread. This is because of the `persist` method that you added as an extension to the managed object context. The context uses the `perform` method internally to ensure that any code we execute inside of the `persist` block is executed on the thread the managed object context is on. Also, since the managed object context was created on the main thread, the movie rating will be set on the main thread.

If you didn't set the movie rating on the same thread that the managed object belongs to, you would get errors and undefined behavior. Always make sure that you manipulate Core Data objects on the same thread as their managed object context.

Threading is a complex subject, but it's essential for building responsive applications. Network logic is a great example of why multithreading is important. If we didn't perform the networking on a separate thread, the interface would be unresponsive for the duration of the request. If you have other operations that might take a while in your app, consider moving them onto a background thread so they don't block the user interface.

Wrapping the feature up

All of the code is in place, and you have a better understanding of multithreading and how callbacks can be used in a multithreaded environment. Yet, if you build and run your app and add a new movie, the rating won't be displayed yet.

The following are the three reasons why this is happening:

- The table view cell that shows the movie isn't updated yet.
- The network request doesn't succeed because of App Transport Security.
- Updates to movie objects aren't observed yet.

Let's solve these issues in order, starting with the table view cell.

Adding the rating to the movie cell

Currently, the movie table view displays cells that have a title. `UITableViewCell` has a built-in option to display a title and a subtitle for a cell.
Open `Main.storyboard` and select the prototype cell for the movies. In the `Attributes Inspector` field, change the cell's style from basic to subtitle. This will allow you to use `detailTextLabel` on the table view cell. This is where we'll display the movie rating.

In `MoviesViewController`, add the following line to `tableView(_:cellForRow:atIndexPath:)`, right after you set the cell's title:

```
cell.detailTextLabel?.text = "Rating: \(movie.popularity)"
```

This line will put the movie's popularity rating in a string and assign it as a text for the detail text label.

If you build and run your app now, all movies should have a popularity of 0.0. Let's fix this by resolving the networking issue.

Understanding App Transport Security

With iOS 9, Apple introduced **App Transport Security** (**ATS**). ATS makes applications safer and more secure by prohibiting the use of non-HTTPS resources. This is a great security feature, as it protects your users from a wide range of attacks that can be executed on regular HTTP connections.

If you paid close attention to the URL that's used to fetch movies, you may have noticed that the URL should be an HTTPS resource, so it should be fine to load this URL. However, the network requests are still blocked by ATS. Why is this?

Well, Apple has strict requirements. At the time of writing this book, the movie database uses the SHA-1 signing of certificates, whereas Apple requires SHA-2. Because of this, you will need to circumvent ATS for now. Your users should be safe regardless, since the movie database supports HTTPS, just not the version Apple considers to be secure enough.

To do this, open the `Info.plist` file and add a new dictionary key named **App Transport Security Settings**. In this dictionary, you will need an **Exception Domains** dictionary. Add a new dictionary key named **themoviedb.orgr** to this dictionary and add two Booleans to this dictionary. Both should have **YES** as their values, and they should be named **NSIncludesSubdomains** and **NSTemporaryExceptionAllowsInsecureHTTPLoads**. Refer to the following screenshot to make sure that you've set this up correctly:

Key	Type	Value
▼ App Transport Security Settings	Dictionary	(1 item)
▼ Exception Domains	Dictionary	(1 item)
▼ themoviedb.org	Dictionary	(2 items)
NSIncludesSubdomains	Boolean	YES
NSTemporaryExceptionAllow...	Boolean	YES

If you add a new movie to a family member now, nothing updates yet. However, if you go back to the family overview and then back to the family member, you'll see that the rating for the most recent movie is updated. Great! Now, all you need to do is make sure that we observe the managed object context for updates to the movies so they are reloaded if their rating changes.

Observing changes to movie ratings

You're already observing the managed object context for changes, but they are only processed if the family member that is shown on the current page has updated. This logic should be replaced so that it will reload the table view if either the family member or their favorite movies change. Update the `managedObjectContextDidChange(_:)` method in `MoviesViewController.swift` as follows:

```
@objc func managedObjectContextDidChange(notification: NSNotification)
{
  guard let userInfo = notification.userInfo
    else { return }

  if let updatedObjects = userInfo[NSUpdatedObjectsKey] as?
Set<FamilyMember>,
```

```
      let familyMember = self.familyMember,
      updatedObjects.contains(familyMember) {

      tableView.reloadData()
    }

    if let updatedObjects = userInfo[NSUpdatedObjectsKey] as? Set<Movie>
  {
      for object in updatedObjects {
        if object.familyMember == familyMember {
          tableView.reloadData()
          break
        }
      }
    }
  }
```

The logic for observing the family member hasn't changed; its conditions simply moved from the guard statement to an `if` statement. An extra `if` statement was added for the movies. If the updated object set is a list of movies, we loop through the movies and check whether one of the movies has the current family member as its family member. If so, the table is refreshed immediately and the loop is exited.

It's important that the loop in the second `if` statement is set up like this because you might have just added a movie for family member A and then switched to family member B while the new movie for family member A is still loading its rating. Also, breaking out of the loop early ensures that you don't loop over any more objects than needed. All you want to do is refresh the table view if one of the current family members' favorite movies is updated.

Okay, build and run your app to take it for a spin! You'll notice that everything works as you'd want it to right now. Adding new movies triggers a network request; as soon as it finishes, the UI is updated with the new rating. Sometimes, this update will be done in an instant, but it could take a short while if you have a slow internet connection. Great! That's it for this feature.

Summary

This chapter was all about adding a small, simple feature to an existing app. We added the ability to load real data from an API. You saw that networking is made pretty straightforward by Apple with `URLSession` and data tasks. You also learned that this class abstracts away some very complex behavior regarding multithreading, so your apps remain responsive while data is loaded from the network. Next, you implemented a helper struct for networking and updated the Core Data model to store ratings for movies. Once all this was done, you could finally see how multithreading worked in the context of this app. This wasn't everything we needed to do, though. You learned about ATS and how it keeps your users secure. You also learned that you sometimes need to circumvent ATS, and we covered how you can achieve this.

Even though the feature itself wasn't very complex, the concepts and theory involved can be quite overwhelming. You suddenly had to deal with code that would be executed asynchronously in the future. And not just that. The code even used multiple threads to make sure that its performance was optimal. The concepts of multithreading and asynchronous programming are arguably two of the more complex aspects of programming. Practice them a lot and try to remember that any time you're passing around a closure, you could be writing some asynchronous code that gets executed on a different thread.

Now that the list of movies is updated with data from the web, let's take it one step further in the next chapter. You can use background-fetching to make sure that you always fetch the most up-to-date information for movies and update them without the user even noticing.

Questions

1. What object should you use to make web requests?

    a) `URLSession`.
    b) `URLRequest`.
    c) `WebTask`.

2. How are web requests executed?

    a) Asynchronously.
    b) Synchronously.

3. What is a typealias?

    a) A closure callback.
    b) A tuple.
    c) An alias for any other type, for instance a callback.

4. What happens if you use managed objects on the wrong thread?

    a) Nothing.
    b) They won't save.
    c) Your app might crash.

5. Which object do you need if you make an HTTP POST request?

    a) `POSTURLRequest`.
    b) `URLRequest`.
    c) `POSTRequest`.

6. What is ATS?

    a) An encryption protocol.
    b) Apple's way to enforce secure web requests.
    c) Apple's name for HTTPS.

Further reading

- `Chapter 25`, Offloading Tasks with Operations and GCD

10
Being Proactive with Background Fetch

So far, the **MustC** application is shaping up quite nicely. You can add family members and associate movies with them, and these movies are automatically enriched with the first rating available from the movie database API. You may still have some movies in your app that were added earlier and don't have a rating, although the ratings for the existing movies might be correct for now, they might not be correct in a couple of days, weeks, or months from now. You could update the ratings whenever the user accesses them, but it's not very efficient to do this. You might reload the same movie a couple of times in a single session, which isn't very useful.

Ideally, the movies would be updated in the background, when the user isn't using the app. Functionality such as this can be implemented using background fetch. This chapter is all about fetching data on behalf of the user while the app isn't active in the foreground. Implementing this feature can significantly benefit your users because your app will have fresh content every time the user opens up the app. No pull to refresh is needed, and users love these little touches of magic in the apps they use.

This chapter covers the following topics:

- Understanding how background fetching works
- Implementing the prerequisites for background fetch
- Updating movies in the background

Besides implementing the background fetch feature, you will do a lot of refactoring. It's important to understand how and when to refactor your code, because it's highly unlikely that you'll be able to exactly nail a flexible setup for your app that can adapt to any requirement that is thrown at it. Being able to refactor your code whenever you see duplication or readability issues is an essential skill that you will utilize almost every day as your understanding of your code base, the requirements, and the iOS platform grows.

Understanding how background fetch works

Any application that provides users with some form of data that updates over time can implement background fetch. An application that implements background fetch is woken up by iOS periodically, and it's given a small window of time to fetch and process new data that has become available. The OS expects applications to call a callback when they're done with fetching and processing the data. The application uses this callback to inform the OS about whether or not it retrieved new data. First, you will learn about background fetch at a high level. Then, you'll learn about some of the details involved with background fetch, and lastly, you will implement background fetch in the **MustC** app.

Looking at background fetch from a distance

Background fetch allows apps to download new content in the background without draining the battery because iOS manages the wake-up intervals as efficiently as possible. Since iOS will not allow your application to stay active for a very long time, you must keep in mind that your app might not be able to perform all the tasks you would like it to perform. If this is the case, you're probably implementing background fetch for the wrong reasons. Sometimes, it's a better idea to split work between your background fetch and application launch.

One example of this would be to download a large amount of data and perform some expensive processing with the downloaded data. If your app takes too long to process the data, it won't be able to finish. It may be a good idea to continue processing once the app launches. An even better approach would be to download only the data and perform all of the expensive processing only when the app launches. Background fetch is intended to pull in quick, small updates and not to process large amounts of data.

The time intervals that iOS uses to wake your app up aren't entirely transparent. You can tell iOS how often you would like your app to be woken up to do some background work, but ultimately the OS itself decides whether or not your app is woken up and how much time it has to perform its tasks.

A neat feature of background fetch is that iOS learns how often your app has updated data because you have to report this to the completion callback. In addition to understanding the moments where your app is most likely to retrieve updates, iOS also learns when your users are most likely to use your app. If a user tends to open your app every morning around 10:00 A.M., chances are that your app will be woken up in the background sometime before 10:00 A.M. This feature makes sure that your users will see recent content for your application. Similarly, if iOS notices that your app rarely has new content within a specified time interval, it's likely that your app isn't woken up as often as you'd like.

Leaving this interval in the hands of the OS might seem a bit scary at first; you often want as much control over these things as possible. What if you know that once a month, at a particular time, you will publish a bunch of new content and you want your app to fetch this in the background? How will you know that all users will receive this new content in time? The answer to this is simple: you don't know. Background fetch is a service that attempts to help your users to receive content in the background. There are no guarantees.

You should always be aware that your app might have to update content in the foreground since users can turn off background fetching entirely. One more important aspect of background fetch is that it will only work if your app isn't completely closed. If a user opens the multitasking window and swipes your app upward to close it, it won't be woken up for a background refresh, and your app will have to refresh when the user launches it the next time.

An excellent way to facilitate this in your code is to implement your background fetch logic in a separate method or a helper that you can call. You could invoke a method that's responsible for fetching new content any time your application is launched, or when it's awakened to perform a background fetch. This strategy ensures that your users always receive the latest content when they open your app.

This is all you need to know about background fetching on the surface. It's a feature that periodically wakes up your application so it can perform a task for a short time. You don't have tight control over when and how your app is awakened, which is okay because iOS learns how often your app should be woken up, and more importantly, iOS learns the best time to wake your app up. All you have to do is make sure that iOS knows that your app wants to be woken up.

Looking at background fetch in more depth

Now that you have a broad understanding of background fetch; let's investigate what your app needs to do to be compatible with background fetch. First of all, you should let iOS know that your app wants to be woken up periodically. To do this, you will need to turn on the **Background Modes** capability for your app. Within this capability, you can opt in to background fetch. Enabling this mode will automatically add a key for this service to your app's `Info.plist`.

When this capability is enabled, iOS will allow your app to be woken up periodically. However, it won't magically work right away. In the `application(_:didFinishLaunchingWithOptions:)` method of `AppDelegate`, you must inform the system that you want the app to become active every once in a while and also specify the minimum time between fetches.

Once you've done this, iOS will periodically wake your app up. When it does, the `application(_:performFetchWithCompletionHandler:)` method is called. This method should be implemented in `AppDelegate`, and it's expected to call the provided completion handler once the work is completed. If you fail to call this method in time, iOS will make a note of this, and missing the window to call this completion-handler could result in your app not being woken up as often as you'd like, so make sure to always call it when the background work completes.

After calling the completion-handler, your app will be put back to sleep until the user opens the app or until the next time your app is allowed to perform a background fetch. Calling the completion-handler shouldn't be done until all of the work that you intended to do is done because once the completion-handler is called, there are no guarantees about how and when your app will make a transition to the background. The only thing you know for sure is that it will happen soon.

Implementing the prerequisites for background fetch

There are three steps involved in supporting background fetch in your app:

1. Add the background fetch capability to your app.
2. Ask iOS to wake your app up.
3. Implement `application(_:performFetchWithCompletionHandler:)` in `AppDelegate`.

You can implement step 1 and 2 right now; step 3 will be implemented separately because this step will involve writing the code to fetch and update the movies using a helper struct.

Adding the background fetch capability

Every application has a list of capabilities they can opt in to. Some examples of these capabilities are using Maps, HomeKit, and Background Modes. You can find the **Capabilities** tab in your project settings. If you select your project in the file navigator, you can see the **Capabilities** tab right next to your app's General settings.

If you select this tab, you will see a list of all the capabilities your app can implement. If you expand a capability, you'll find some information about what the capability does and what happens automatically if you enable it. If you expand the **Background Modes** capability, you'll see the following information:

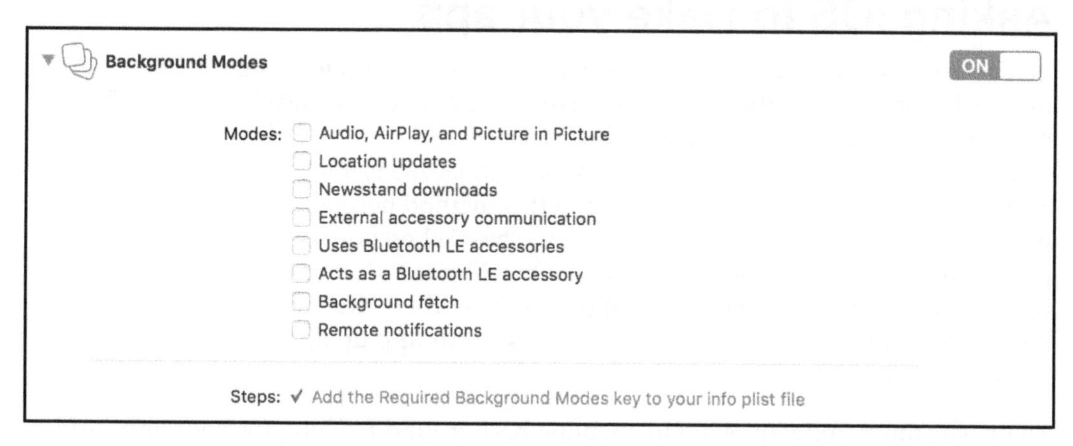

There are several options that you can enable, but there's just one that you need to focus on for now. The capability you need to enable is **Background Fetch**. Click the on/off checkbox to enable the capability and then check the **Background Fetch** option.

Enabling **Background Fetch** isn't something that magically changes some invisible settings for your application. You can enable background fetch manually if you'd like to.

Open the `Info.plist` file and search for the **Required background modes** key. You can see that it contains a single item with a value of **App downloads content from the network**. This entry in the `Info.plist` file enables your app to ask iOS to wake it up periodically:

Required background modes	Array	(1 item)
Item 0	String	App downloads content from the network

However, just because you can do this manually does not mean you should. The capabilities tab is a very convenient place to manage capabilities from and manually adding the right key to the `Info.plist` file is tedious and error-prone. It's interesting to know that there's no magic involved in this process, and it makes the entire feature feel more transparent.

Asking iOS to wake your app

You're halfway there in the process of enabling background fetching for **MustC**. All that's left is inform iOS that it should wake the app up every once in a while. This behavior should be implemented in the `application(_:didFinishLaunchingWithOptions:)` method in `AppDelegate`. You must ask iOS to wake your app up in this method because you should ask iOS to wake your app up every time your app launches. When your app transitions from the foreground to the background normally, iOS will know that your app should wake up at certain times. However, if the app is killed entirely, iOS won't be able to wake the app up again, so you must ask iOS to wake your app up again as soon as the app launches again.

There isn't much code involved in enabling background fetching for your app. Add the following line to `application(_:didFinishLaunchingWithOptions:)`:

```
application.setMinimumBackgroundFetchInterval(UIApplication.background
FetchIntervalMinimum)
```

This line will ask iOS to wake the app up at a minimum interval. This means that the app is woken up as often as iOS will allow it to be woken up. It's impossible to predict how often this will be since iOS throttles this interval as it sees fit. If your app often has new data available or is used many times throughout the day, the odds are that your app will be woken up more often than it would be if you rarely have new data available or if the user opens your app once every couple of days.

Alternatively, you could set the fetch interval to `UIApplication.backgroundFetchIntervalNever` if you want to prevent your app from being woken up at all, or you can use a custom value. If you provide a custom value, iOS will attempt to honor your custom interval, but again, there are no guarantees with background fetch because the system remains in charge concerning the intervals that are used.

The only thing iOS does guarantee is that your app is not woken up more often than the value you specified. So, let's say you define a custom interval for your app to be woken up once every three hours, iOS won't wake your app more often than that, but it's possible that your app is woken up only once every eight hours. This depends on several factors that are not transparent to you as a developer.

Your app is now ready to fetch data in the background. Let's go ahead and implement the fetching of data.

Updating movies in the background

The final step in enabling background fetch for **MustC** is to add the `application(_:performFetchWithCompletionHandler:)` method to `AppDelegate`. As explained before, this method is called by iOS whenever the app is woken up in the background, and it allows you to perform an arbitrary amount of work. Once the app is done performing its task, it must call the completion-handler that was passed to this method.

Upon calling the completion-handler, you should inform iOS about the results of the operations you performed. It's important to correctly report this status because background fetch is intended to improve the user experience. If you falsely report to iOS that you have new data all the time, so your app is woken up more often, you're degrading the user experience. You should trust the system to judge when your app is woken up. It's in the best interest of your users, their battery life, and ultimately your app to not abuse background fetch.

To efficiently implement background fetch, you will work through the following steps:

1. Updating the data model so you can query the movie database more efficiently.
2. Refactoring the existing code to use the improved data model.
3. Implementing background fetch with the existing helper struct.

The first two steps are not directly tied to implementing background fetch, but they do illustrate that an efficient background fetch strategy may involve refactoring some of your app's existing code. Remember, there is nothing wrong with refactoring old code to implement a new feature. Both the new feature and the old code will usually benefit from refactoring.

Updating the data model

The data model you currently have associates movies with a single family member. This means that the app potentially stores the same movie over and over again. When we were only storing data, this wasn't that big of a deal. This wasn't a big deal when you first implemented your database. However, now that the app will query the movie database in a background fetch task, it would be a waste of resources to ask the movie database for the same ratings multiple times. Also, you most certainly don't want to use the movie database search API in the same way you did before; you should refer to the movie you want as precisely as possible.

To facilitate this, the relationship between movies and family members must be changed to a many-to-many relationship. You'll also add a new field to the movie entity: `remoteId`. `remoteId` will hold the identifier the movie database uses for a particular movie so it can be used directly in later API calls.

Open the model editor in Xcode and add the new `remoteId` property to **Movie**. Make sure that it's a 64-bit integer and that it's optional. Also, select the **familyMember** relationship and change it to a **To Many** relationship in the sidebar. It's also a good idea to rename the relationship to **familyMembers**, since it's now related to more than one family member. Your model should resemble the model in the following screenshot:

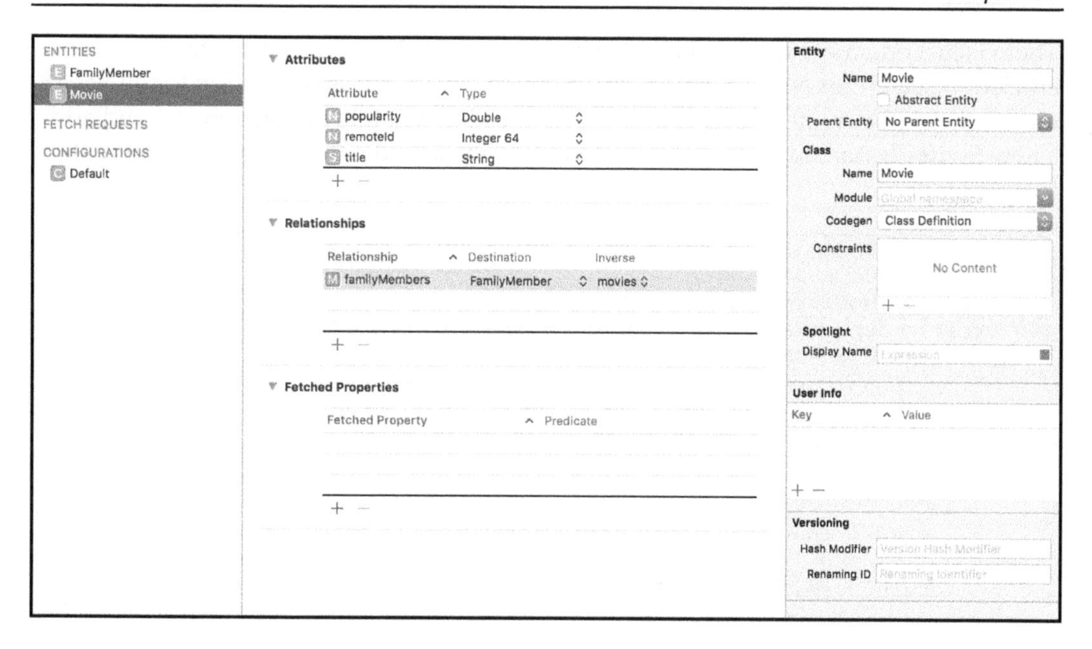

Great, the model has been updated. There's still a lot of work to be done though. Because the name and nature of the family member relationship were changed, the project won't compile. Make the following modifications to the `managedObjectContextDidChange(_:)` method in `MoviesViewController.swift`:

```
if let updatedObjects = userInfo[NSUpdatedObjectsKey] as? Set<Movie> {
  for object in updatedObjects {
    if let familyMember = self.familyMember,
      let familyMembers = object.familyMembers,
      familyMembers.contains(familyMember) {

      tableView.reloadData()
      break
    }
  }
}
```

There is just one more model-related change that you should incorporate. To efficiently search for an existing movie or create a new one, an extension to the `Movie` model should be created. Create a new group called **Models** in your project and add a new Swift file named `Movie.swift` to it. Finally, add the following implementation to the file:

```
import CoreData

extension Movie {
  static func find(byName name: String, orCreateIn moc:
NSManagedObjectContext) -> Movie {
    let predicate = NSPredicate(format: "title ==[dc] %@", name)
    let request: NSFetchRequest<Movie> = Movie.fetchRequest()
    request.predicate = predicate

    guard let result = try? moc.fetch(request)
      else { return Movie(context: moc) }

    return result.first ?? Movie(context: moc)
  }
}
```

The preceding code queries Core Data for an existing movie with the provided name. The movies are matched case-insensitive by passing `[dc]` to the == operator in `NSPredicate`. It's important that this lookup is case-insensitive because people might write the same movie name with different capitalizations. If no movies could be found, or if the results come back empty, a new movie is created. Otherwise, the first, and presumably the only, result that Core Data has is returned. This wraps up the changes that need to be made to the app's data layer.

Refactoring the existing code

The existing code compiles, but it's not optimal yet. `MovieDBHelper` doesn't pass the movie's remote ID to its callback, and the movie-insertion code doesn't use this remote ID yet. When the user wants to save a new movie, the app still defaults to creating a new movie instead of using the handy new helper you just wrote to avoid data duplication. You should update the code, so the callback is called with the fetched remote ID.

Let's update `MovieDBHelper` first. Replace the following lines in the `fetchRating(forMovie:callback:)` method:

```
typealias MovieDBCallback = (Int?, Double?) -> Void
let apiKey = "d9103bb7a17c9edde4471a317d298d7e"

func fetchRating(forMovie movie: String, callback: @escaping
MovieDBCallback) {
  guard let searchUrl = url(forMovie: movie) else {
    callback(nil, nil)
    return
  }

  let task = URLSession.shared.dataTask(with: searchUrl) { data,
response, error in
    var rating: Double? = nil
    var remoteId: Int? = nil

    defer {
      callback(remoteId, rating)
    }

    let decoder = JSONDecoder()

    guard error == nil, let data = data,
      let lookupResponse = try?
decoder.decode(MovieDBLookupResponse.self, from: data),
      let movie = lookupResponse.results.first
      else { return }

    rating = movie.popularity
    remoteId = movie.id
  }

  task.resume()
}
```

These updates change the callback-handler, so it takes both the remote ID and the rating as parameters. A variable to hold the remote ID is added and incorporated into the callback. With this code, `MovieDBHelper` is entirely up to date.

You should also update the response struct, so the `MovieDBMovie` struct includes the ID from the API response:

```
struct MovieDBMovie: Codable {
  let popularity: Double?
  let id: Int?
}

moc.persist {
  // 1
  let movie = Movie.find(byName: name, orCreateIn: moc)

  // 2
  if movie.title == nil || movie.title?.isEmpty == true {
    movie.title = name
  }

  let newFavorites: Set <AnyHashable> =
familyMember.movies?.adding(movie) ?? [movie]
  familyMember.movies = NSSet(set: newFavorites)

  let helper = MovieDBHelper()
  helper.fetchRating(forMovie: name) { remoteId, rating in
    guard let rating = rating,
      let remoteId = remoteId
      else { return }

    moc.persist {
      movie.popularity = rating
      movie.remoteId = Int64(remoteId)
    }
  }
}
```

First, the preceding code either fetches an existing movie or creates a new one with the `find(byName:orCreateIn:)` method you just created. Next, it checks whether the returned movie already has a name. If it doesn't have a name yet, the name is set. Also, if it does have a name, you can safely assume you were handed an existing movie object, so there is no need to set the name. Next, the rating and ID are fetched, and the corresponding properties on the movie object are set to their correct values in the callback.

This is all the code needed to prepare **MustC** for background fetch. Let's implement this feature now.

Updating movies in the background

Almost all of the building blocks involved with implementing background fetch are in place. All you need now is a way to fetch movies from the movie database using their remote ID instead of using the movie database search API.

To enable this way of querying movies, another fetch method is required on `MovieDBHelper`. The simplest way to do this would be to copy and paste both the fetch and URL-building methods and adjust them to enable fetching movies by ID. While it might be the easiest way to go about this, it isn't the best idea. If you add another fetch method or require more flexibility later, you will be in trouble. It's much better to refactor the movie db helper so it's better suited to multiple types of requests.

Preparing the helper struct

To maintain a clear overview of the available API endpoints, you can add a nested enum to `MovieDBHelper`. Doing this will make other parts of the code more readable, help to avoid errors, and abstract away code duplication. An associated value will be used on the enum to hold onto the ID of a movie. This is quite convenient because the movie ID is part of the API endpoint.

Add the following code inside of the `MovieDBHelper` struct:

```
static let apiKey = "d9103bb7a17c9edde4471a317d298d7e"

enum Endpoint {
  case search
  case movieById(Int64)

  var urlString: String {
    let baseUrl = "https://api.themoviedb.org/3/"

    switch self {
    case .search:
      var urlString = "\(baseUrl)search/movie/"
      urlString =
urlString.appending("?api_key=\(MovieDBHelper.apiKey)")
      return urlString
    case let .movieById(movieId):
      var urlString = "\(baseUrl)movie/\(movieId)"
      urlString =
urlString.appending("?api_key=\(MovieDBHelper.apiKey)")
      return urlString
```

```
        }
    }
}
```

Note that the `apiKey` constant has been changed from an instance property to a static property. Making it a static property makes it available to be used inside of the nested `Endpoint` enum. Note that the value associated with the `movieById` case is `Int64` instead of `Int`. This is required because the movie ID is a 64-bit integer type in Core Data.

With this new `Endpoint` enum in place, you can refactor the way you build the URLs as follows:

```
private func url(forMovie movie: String) -> URL? {
  guard let query = movie.addingPercentEncoding(withAllowedCharacters:
.urlHostAllowed)
    else { return nil }

  var urlString = Endpoint.search.urlString
  urlString = urlString.appending("&query=\(query)")

  return URL(string: urlString)
}

private func url(forMovieId id: Int64) -> URL? {
  let urlString = Endpoint.movieById(id).urlString
  return URL(string: urlString)
}
```

The `url(forMovie:)` method was updated to make use of the `Endpoint` enum. The `url(forMovieId:)` method is new and uses the `Endpoint` enum to obtain a movie-specific URL easily.

When you fetch data, there are some pieces of code that you will have to write regardless of the URL that will ultimately be used to fetch data. When you look at the `fetchRating(forMovie:)` method, there are a couple of things that apply to both endpoints you will eventually use to retrieve movie information. The following list is an overview of these things:

- Checking whether we're working with a valid URL
- Creating the data task
- Extracting the JSON
- Calling the callback

If you think about it, the only real difference is the API response object that is used. When you examine the JSON responses returned by the search API and fetching a movie by ID, the difference is that the search API returns an array of movies where you're interested in the first result. The movie-by-id API returns the correct movie as part of the root object.

With this in mind, the refactored code should be able to retrieve the desired data using just a URL, a data-extraction strategy, and a callback. Based on this, you can write the following code:

```
// 1
typealias IdAndRating = (id: Int?, rating: Double?)
typealias DataExtractionStrategy = (Data) -> IdAndRating

// 2
private func fetchRating(fromUrl url: URL?, withExtractionStrategy
extractionStrategy: @escaping DataExtractionStrategy, callback:
@escaping MovieDBCallback) {
  guard let url = url else {
    callback(nil, nil)
    return
  }

  let task = URLSession.shared.dataTask(with: url) { data, response,
error in
    var rating: Double? = nil
    var remoteId: Int? = nil

    defer {
      callback(remoteId, rating)
    }

    guard error == nil
      else { return }

    guard let data = data
      else { return }

    // 3
    let extractedData = extractionStrategy(data)
    rating = extractedData.rating
    remoteId = extractedData.id
  }

  task.resume()
}
```

There is quite a lot going on in the preceding snippet. Most of the code will look familiar, but some of the details might be new. Let's go over the comments in this code one by one:

1. This part defines a couple of type aliases that will make the code a little bit more readable. The first alias is for a named tuple that contains the movie's ID and rating. The second alias defines the signature for the data extraction closure.
2. The second highlight is for `fetchRating(fromUrl:withExtractionStrategy:callback:)`. This method will be used to obtain movie data. It's marked private because it's not very useful to call this method directly. Instead, it will be called by two methods that will be implemented soon.
3. The data for the movie is extracted from the raw data by passing the raw data to the data extraction closure.

Let's use this method to implement both the old way of fetching a movie through the search API and the new way that uses the movie ID to request the resource directly, as follows:

```
func fetchRating(forMovie movie: String, callback: @escaping
MovieDBCallback) {
  let searchUrl = url(forMovie: movie)
  let extractData: DataExtractionStrategy = { data in
    let decoder = JSONDecoder()

    guard let response = try?
decoder.decode(MovieDBLookupResponse.self, from: data),
      let movie = response.results.first
      else { return (nil, nil) }

    return (movie.id, movie.popularity)
  }

  fetchRating(fromUrl: searchUrl, withExtractionStrategy: extractData,
callback: callback)
}

func fetchRating(forMovieId id: Int64, callback: @escaping
MovieDBCallback) {
  let movieUrl = url(forMovieId: id)
  let extractData: DataExtractionStrategy = { data in
    let decoder = JSONDecoder()

    guard let movie = try?
```

```
decoder.decode(MovieDBLookupResponse.MovieDBMovie.self, from: data)
      else { return (nil, nil) }

    return (movie.id, movie.popularity)
  }

  fetchRating(fromUrl: movieUrl, withExtractionStrategy: extractData,
callback: callback)
}
```

The code duplication is minimal in these methods, which means that this attempt at refactoring the code was a huge success. If you add new ways to fetch movies, all you will need to do is obtain a URL, define how to retrieve the data you're looking for from the `data` object, and finally, kick off the fetching.

You're now finally able to fetch movies using their ID without duplicating a lot of code. The final step in implementing the background update feature is to implement the code that updates movies. Let's go!

Updating the movies

The process of updating movies is a strange one. As you saw earlier, network requests are performed asynchronously, which means that you can't rely on the network request being finished by the time a function is finished executing. Because of this, a callback is used. The callback is then called when the network request is done.

But what happens if you need to wait for multiple requests? How do you know that all requests to update movies have been completed? Since the movie database doesn't allow developers to fetch multiple movies at once, a bunch of requests must be made. When all of these requests are complete, the background fetch `completionHandler` should be called with the result of the operation.

To achieve this, you can make use of the grand central dispatch. More specifically, you can use a dispatch group. A dispatch group keeps track of an arbitrary number of tasks, and it won't consider itself as completed until all of the tasks that are added to the group have finished executing.

This behavior is precisely what's needed to wait for all movies to be updated. Whenever we fetch a movie from the network, you can add a new task to the dispatch group that will be marked as completed once the underlying movie is updated. Finally, when all of the movies are updated, `completionHandler` can be called to inform iOS about the result of the background fetch. Let's take a step-by-step look at how to achieve this behavior. Start by adding the following code to `AppDelegate`:

```
func application(_ application: UIApplication,
performFetchWithCompletionHandler completionHandler: @escaping
(UIBackgroundFetchResult) -> Void) {

  let fetchRequest: NSFetchRequest<Movie> = Movie.fetchRequest()
  let managedObjectContext = persistentContainer.viewContext
  guard let allMovies = try? managedObjectContext.fetch(fetchRequest)
else {
    completionHandler(.failed)
    return
  }
}
```

This first part of the background fetch is relatively straightforward. All it does is retrieve all movie objects from the database. If this fails, `completionHandler` is called with a `.failed` status.

All the following code snippets should be added to the `application(_:performFetchWithCompletionHandler:)` method inside of `AppDelegate` in the same order as they are presented. A full overview of the implementation will be provided at the end:

```
let queue = DispatchQueue(label: "movieDBQueue")
let group = DispatchGroup()
let helper = MovieDBHelper()
var dataChanged = false
```

These lines create a dispatch queue and a dispatch group. The dispatch queue represents the background thread on which the fetch operations will be executed. Next, add the following snippet:

```
for movie in allMovies {
  queue.async(group: group) {
    group.enter()
    helper.fetchRating(forMovieId: movie.remoteId) { id, popularity in
      guard let popularity = popularity,
        popularity != movie.popularity else {
          group.leave()
          return
```

```
        }

        dataChanged = true

        managedObjectContext.persist {
          movie.popularity = popularity
          group.leave()
        }
      }
    }
  }
```

This part of the implementation loops through the fetched movies. For every movie, a task is added to the dispatch queue and `group.enter()` is called to tell the dispatch group that a new task has just been added. The next step is to fetch the rating. If this fails, `group.leave()` is called to tell the dispatch group that this task has finished. If the data was retrieved successfully, the movie is updated with the fetched rating. Once the managed object has persisted the changes, `group.leave()` is called, and the operation has finished.

The next and final snippet that must be added executes when all the tasks in the queue are performed; at this point, the code should check whether new data has been fetched by reading the `dataChanged` property, and based on this property, call `callbackHandler`:

```
group.notify(queue: DispatchQueue.main) {
  if dataChanged {
    completionHandler(.newData)
  } else {
    completionHandler(.noData)
  }
}
```

The `group.notify` method takes a queue and a block of code that it should execute. The queue is set to the main queue, which means that the code inside of the block is performed on the main queue. Then, the `dataChanged` variable is read, and `completionHandler` is called accordingly.

As promised, the full implementation for `application(_:performFetchWithCompletionHandler:)` is as follows:

```
func application(_ application: UIApplication,
performFetchWithCompletionHandler completionHandler: @escaping
(UIBackgroundFetchResult) -> Void) {

  let fetchRequest: NSFetchRequest<Movie> = Movie.fetchRequest()
```

```
    let managedObjectContext = persistentContainer.viewContext
    guard let allMovies = try? managedObjectContext.fetch(fetchRequest)
  else {
      completionHandler(.failed)
      return
    }

    let queue = DispatchQueue(label: "movieDBQueue")
    let group = DispatchGroup()
    let helper = MovieDBHelper()
    var dataChanged = false

    for movie in allMovies {
      queue.async(group: group) {
        group.enter()
        helper.fetchRating(forMovieId: movie.remoteId) { id, popularity
  in
          guard let popularity = popularity,
            popularity != movie.popularity else {
              group.leave()
              return
          }

          dataChanged = true

          managedObjectContext.persist {
            movie.popularity = popularity
            group.leave()
          }
        }
      }
    }

    group.notify(queue: DispatchQueue.main) {
      if dataChanged {
        completionHandler(.newData)
      } else {
        completionHandler(.noData)
      }
    }
  }
```

To test whether background fetching is working as expected, you can build and run your app. Then, add a new movie, so you have a movie for which the ID is stored. Finally, you can use the debug menu item in Xcode that's at the top of your screen to simulate a background refresh. This will trigger a background refresh.

Summary

This chapter opened with an overview of background fetch, how it works, and how it benefits your users. You also learned about the prerequisites and best practices for this feature. After establishing this basic understanding, you learned how background fetch works in concert with your application. Then, we continued to implement the required permissions and asked iOS to wake up the **MustC** application periodically so we could update the movies' ratings.

Once you did this, you had to refactor a good portion of the application to accommodate the new feature. It's important to be able to recognize scenarios where refactoring is a good idea, especially if it enables the smooth implementation of a feature later on. It also demonstrates that you don't have to think about every possible scenario for your code every time you implement a feature. After refactoring the application, you were finally able to implement background-fetching behavior. To do this, you quickly glanced over dispatch groups and how they allow you to group an arbitrary amount of asynchronous work together and be notified when all of the tasks in a group are completed.

Now that you know how to fetch data from the network, store it locally, and update it in the background, wouldn't it be great if you could sync data across devices and automatically update it in the background as soon as new data is available? That's precisely what the next chapter is all about!

Test your knowledge

1. For how long can you perform work when your app is woken up for background refresh?

    a) About two minutes.
    b) There is no way to be sure.
    c) You have no time limit.

2. How do you enable background fetch for your app?

    a) By requesting it in `AppDelegate`.
    b) By enabling the Background Refresh capability.
    c) Both.

3. Which statement about the minimum refresh interval is true?

   a) Your app will be woken up every time the interval passes.
   b) iOS will determine when it's the best time wake up your app. It will be no more often that you specify.
   c) iOS will ask your app whether it has any work to do when the interval passes.

4. What did you use a dispatch group for in this chapter?

   a) To take work off the main thread.
   b) To be notified when all network calls were done.
   c) To make sure all network calls are fired at the same time.

5. When should you refactor your code?

   a) When you get frustrated with your code.
   b) When you find strange bugs.
   c) Any time you think you can improve something.

6. What do you call the ID that is passed to the movieById case on the `MovieDBHelper.Endpoint` enum?

   a) Associated value.
   b) Attached value.
   c) Enum value.

Further reading

- `Chapter 25`, Offloading Tasks with Operations and GCD
- Updating your App with Background App Refresh: `https://developer.apple.com/documentation/uikit/core_app/managing_your_app_s_life_cycle/preparing_your_app_to_run_in_the_background/updating_your_app_with_background_app_refresh`

11
Syncing Data with CloudKit

Storing data locally in Core Data is pretty sweet, but storing data locally on a device has its limitations. When a user gets a new device or uninstalls your app, their database is gone. This could be very inconvenient for the user because they might want to hold on to their data somehow.

Another inconvenience of a local database is that the database isn't shared across a user's devices. To work around this problem, you could implement your own backend server and store your user's data there. However, this comes with great responsibility because you would have to ensure that your user's data is protected and stored securely, not to mention the amount of effort and costs that will go into building and maintaining your own backend.

If you're looking for an excellent way to synchronize data across a user's device without having to build your own backend or invest vast amounts of time into storing data securely, **CloudKit** is precisely what you need. In this chapter, you will learn how you can set up a simple data store on Apple's CloudKit service to allow data-sharing between a user's devices and even how you can share data across multiple users!

In this chapter you will learn about the following topics:

- Getting familiar with CloudKit
- Storing and retrieving data with CloudKit
- Combining CloudKit and Core Data

By the end of this chapter, you will know everything you need to know about implementing CloudKit in an application and keeping a local Core Data database in sync with the remote data.

Getting familiar with CloudKit

CloudKit is a service that Apple created as a simple tool for application developers to store data on a remote server. The beauty of this service is that any user with an iCloud account can be identified automatically by CloudKit, meaning you can begin connecting data to your user immediately without having them first sign up for your service.

Signup screens are often considered to be dealbreakers for users because they don't want to share their email address with certain apps, or just because they think signing up for a particular app isn't worth the effort. With CloudKit, you can remove all this friction because users are discretely logged into your app using the iCloud account that they already use on their device.

When you add CloudKit to your app, you will find that it offers a lot of great functionalities to help you keep the number of web requests as low as possible, and you can even have CloudKit notify your app when something changes on the remote server even when your app isn't running. Apple uses features such as these extensively in their own applications to try to make sure that their apps provide a seamless experience between devices.

To add CloudKit to the **MustC** app, you don't need to do a lot of work. All you need to do is make sure your project is set up correctly and enable the required capabilities for your app.

Adding CloudKit to your project

First, make sure the Xcode project is set up correctly. Note that the project files in this book's code bundle have been modified in preparation of this chapter. If you want to follow along with the steps in this chapter, make sure to grab the **MustC** starter project for this chapter from the code bundle. You need to have a unique bundle identifier for the **MustC** app. The code bundle that goes with this book uses a bundle identifier of `com.donnywals.MustC`. Make sure that you change this to something unique. Also, make sure that you have selected a valid developer team under Xcode's signing options. Refer to the following screenshot to see an example of what your **General** settings should look like:

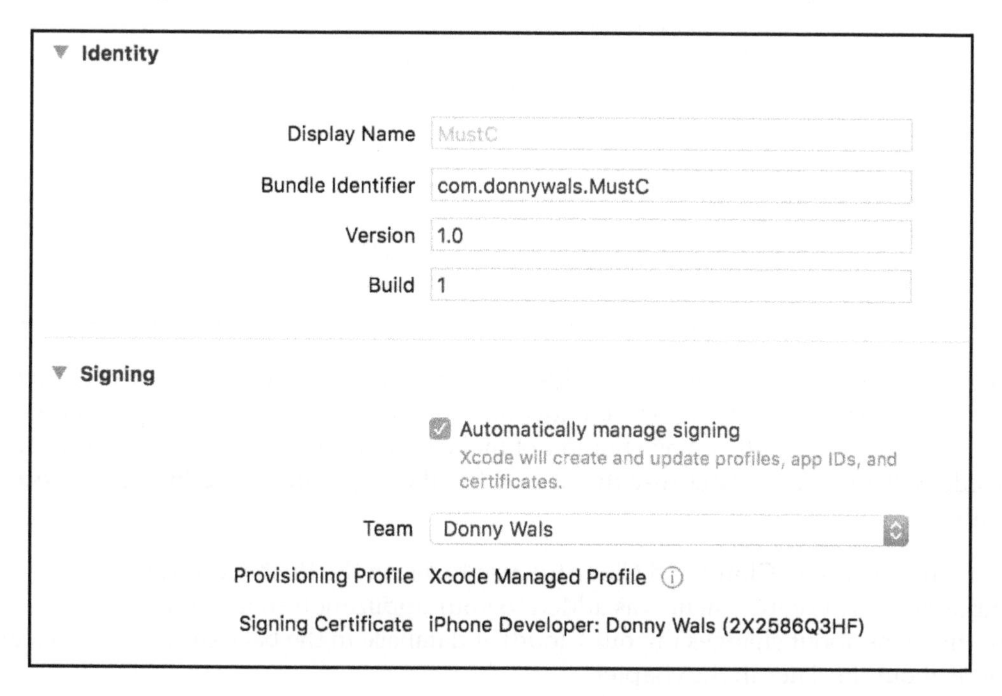

Next, open the **Capabilities** tab in the project settings. Scroll down until you see the **iCloud** capability. Make sure to toggle it so it's turned on and expand the iCloud section. Check the **CloudKit** checkbox to enable CloudKit for your application, as shown in the following screenshot:

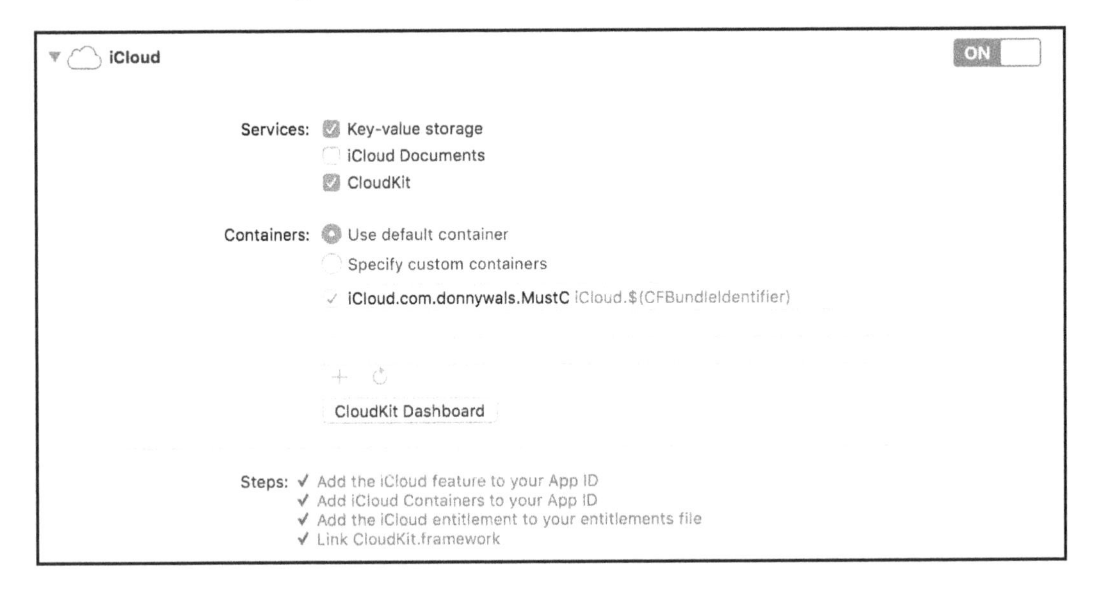

Toggling the iCloud capability has added an entitlements file to your project. This file is coupled to your Xcode signing identity and it makes sure that your app is allowed to use certain capabilities. You should not make manual changes to this file because Xcode will automatically ensure that it matches the **Capabilities** tab in your project settings.

Note that enabling iCloud and CloudKit also turns on the Push Notifications entitlement. This entitlement was added to your application so it can receive notifications about changes in your CloudKit database in the background. You'll learn more about this later in the chapter.

Technically this is all you need to do to add CloudKit to your app. Before you learn how to perform read and write operations on your CloudKit database, you should familiarize yourself with the heart of your CloudKit backend, the dashboard.

Exploring the CloudKit dashboard

If you look at the **Capabilities** tab in your project settings, there is a **CloudKit Dashboard** button underneath your iCloud capability configuration. Click this button or navigate to `https://icloud.developer.apple.com/dashboard/` in your web browser to go to the CloudKit dashboard. After logging into the dashboard, you'll see your developer teams and the containers that you have set up for each team.

Sometimes it takes a little while for a new container to show up in the CloudKit dashboard. If your project doesn't show up right away, don't worry. Just take a five-minute break and then check whether your container shows up.

A CloudKit container uses a unique identifier that is based on your app's bundle identifier. To explore your CloudKit container, click it. You should see a screen that resembles the following screenshot:

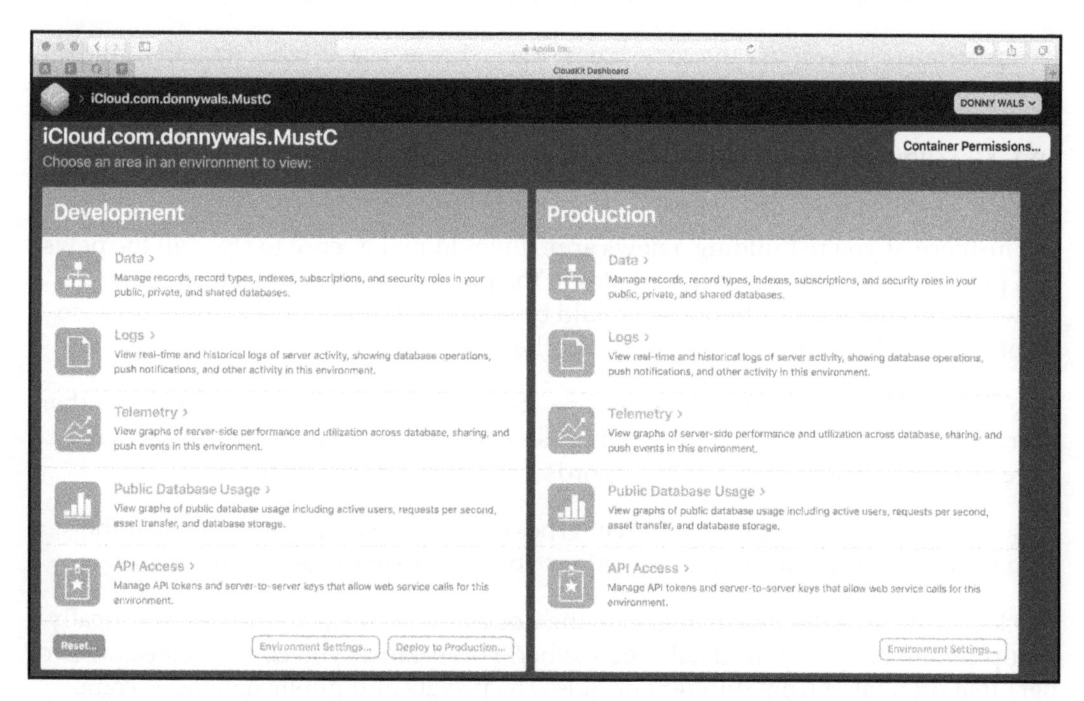

This screen is the starting point for your container. Every container has a development environment and a production environment. This is convenient because it allows you to experiment as much as you'd like in your development database without disrupting the production database where your user's data will live.

From this screen, you can monitor what your container is doing, how much it's used, and you can even access logs that help you with debugging issues if you encounter them. For now, go to the **Data** section so you can explore your CloudKit container's data store.

Understanding your CloudKit database

In order to efficiently use CloudKit to store data, you must understand how it stores data and what terminology is used for the different components that are used throughout your database. Every CloudKit container comes with a couple of different databases:

- A private database for each user
- A shared database for each user
- A public database that's shared for all users

It's important that you use the correct database whenever you write data to CloudKit, you wouldn't want to accidentally store some sensitive user information in the public database because that could potentially expose the information to all users of your app. You should only use the public database for data that is suited for *all* users of an application.

For instance, if you're building a news app, it would make sense to store all the news and possibly comments on news articles in the public database for your container. However, saving a user's to-do list should be stored in the private database because no other users have any business with another user's to-do list.

One exception to this is when a user shares their list or individual items with another user. This shared data can be stored in the shared database where only users that have explicitly gained access to the records are allowed to access them.

Now that you know about the different types of databases, let's have a look at the data page itself to see what kind of options you have to manage data there.

The leftmost tab in the data manager is the **Zones** tab. You can use zones to logically group certain parts of your database together. For instance, you could create zones for data that originated from different origins. The private and public databases come with a default zone. You cannot create custom zones for the public database so all public data is automatically stored in the default zone. For the private and shared databases, you can create custom zones.

Note that the zones page has the option to **Fetch zone changes since...** This is a performance optimization that makes sure apps can load a minimal amount of data whenever they want to synchronize with the data from CloudKit. If you want to add a custom zone, you can do so by clicking on the **Create new Zone...** button. While you're there, create a new zone for the private database and name it `moviesZone`. This is where you'll store the data from the **MustC** app later.

The second tab, **Records**, is more interesting. On this tab, you can perform queries on the data that is stored in your databases and zones. This allows you to explore your data and make sure that everything is saved as you would expect. For privacy reasons, you can only see the data for your own private and shared database and the public database. You can never access a user's private database.

The records tab provides a window into your CloudKit database. Note that the records page also has a **Changes** tab. You can use this tab to retrieve changes to all records in the selected zone. Note that you can't do this for the default zone, so if you want to use this functionality in your app, you will need to create a custom zone on the **Zones** tab.

The third tab is the **Record Types** tab. This is where you define the objects you want to store in your CloudKit database and what properties they have. Note that record types are not connected to a zone. This means that you can write a certain record to any zone you'd like. Keep in mind though that records from different zones cannot reference each other. So if you consider the **MustC** app, a family member and their favorite movies would have to exist in the same zone. If this isn't the case, the objects are not allowed to reference each other in CloudKit.

To prepare for storing data later, you should create your record types now. Click the **Create New Type** button and name it `FamilyMember`. Add a field to it by clicking the **Add field** button and call the field **name**. The type should be **String**. Add another field and call it **movies**, give it a type of **Reference (list)**. The following screenshot shows what your fields should look like:

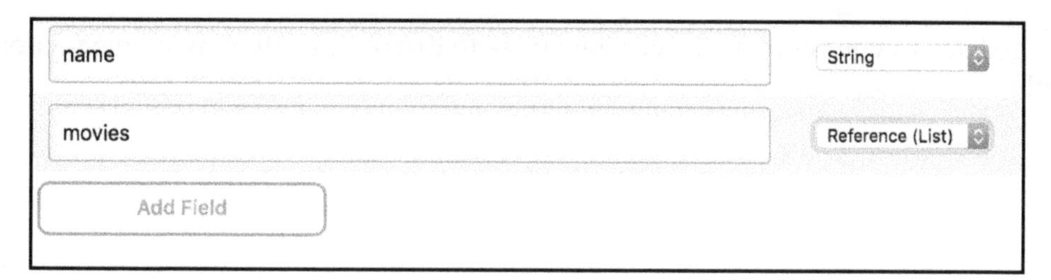

Click the **Create Record Type** button to store your record type and create a second record type for the movie records. Use **Movie** as the name for your record type and give it three properties, named **title**, which should be a String, **rating** (Double), and a **remoteId** (Int64). Make sure to save this record type as well.

After you save a record type, CloudKit automatically adds some extra metadata properties to it. These properties are used internally by CloudKit to figure out what objects changed and when. You can't remove or modify these properties because CloudKit needs them.

The fourth tab in the dashboard is the **Indexes** tab. This tab is used to specify which properties of a record will be optimized for querying. If you don't specify any indexes, you won't be able to easily fetch all records from your CloudKit database. Typically you will want to at least add an index for the **recordName** property. To do this, select your record type and click the **Add Index** button. Select the **recordName** property in the first dropdown and **QUERYABLE** from the second. Do this for both record types you created earlier.

The fifth tab is **Subscriptions**. This is where you can see the subscriptions that exist for a particular database. You can only look at the subscriptions for your own private and shared database, and the public database. You can use this tab to check whether you are correctly subscribing apps to database updates.

The next tab is **Subscription types**. There isn't much to do here other than explore the automatically-generated subscription types and check whether they match what you expect to find there.

The last tab is the **Security Roles** tab. This tab allows you to manage what permissions users have for record types in the public database. This allows you to limit the actions that a certain user can perform. For instance, you might want to allow all users to read news articles from the public database but restrict writing new articles to the database to your administrative users. You can manage all this in the security roles tab. The default values will suffice for the **MustC** app.

Let's go back to your container overview page to discuss the other available sections briefly.

Exploring the rest of your container

From your container overview page, you can gain insights into several interesting and helpful metrics. You can use the **Logs** page to see real-time logs about what's going in your application. You can use these logs to make sure specific data is sent and received as you would expect, or to troubleshoot problems your users are having. Note that you can't see the private data that your users are storing, but you can look at the operations they are trying to perform on the CloudKit database.

The **Telemetry** page gives you insight into what's going on inside of your application. You can see performance-related statistics and learn how your CloudKit database is used. You can use this page together with **Public Database Usage** to understand where your implementation could be improved and how users use your app.

The last section is **API Access**. This section is relevant when you want to use the CloudKit store outside of iOS, for instance from a web page or web server.

This information should cover the most interesting bits of the CloudKit dashboard. Let's see how you can make your app communicate with CloudKit.

Storing and retrieving data with CloudKit

CloudKit is meant to be used as a backend for your app that syncs across devices and provides a seamless experience for users. This section will go over the principles that you should keep in mind when implementing CloudKit in your app. By the end of this section, you will know precisely how storing, retrieving, and updating data in CloudKit works. You will then go on to implement CloudKit in the **MustC** app by combining it with the existing Core Data database.

Communicating with CloudKit for the first time

When your app launches for the very first time, it should immediately make itself known to CloudKit so it can subscribe to changes. Technically you don't have to do this because you can implement your own logic to retrieve data from CloudKit as you need it. However, an efficient CloudKit implementation will work flawlessly when there is no internet connection, which means that it's probably a good idea to store all data locally by writing it to a database such as Core Data.

The recommended way to subscribe to database changes is to keep track of whether your app is already subscribed locally. You can keep track of this by storing a variable in the local `UserDefaults` store. `UserDefaults` is perfect for storing variables such as this. Open `CloudStore.swift` in the **MustC** project and add the following computed property to it:

```
var isSubscribedToDatabase: Bool {
  get { return UserDefaults.standard.bool(forKey:
"CloudStore.isSubscribedToDatabase") }
  set { UserDefaults.standard.set(newValue, forKey:
"CloudStore.isSubscribedToDatabase") }
}
```

This property proxies the standard `UserDefaults` through its custom getters and setters. A construction such as this is very convenient since it makes using this property simple and straightforward even though there's a bit more logic going on behind the scenes.

With the `isSubscribedToDatabase` property in place, you should also implement a method that's responsible for subscribing the app to changes in the CloudKit database. Add the following implementation to `CloudStore.swift`:

```
extension CloudStore {
  func subscribeToChangesIfNeeded(_ completion: @escaping (Error?) ->
Void) {
    // 1
    guard isSubscribedToDatabase == false else {
      completion(nil)
      return
    }

    // 2
    let notificationInfo = CKSubscription.NotificationInfo()
    notificationInfo.shouldSendContentAvailable = true

    // 3
    let subscription = CKDatabaseSubscription(subscriptionID:
"private-changes")
    subscription.notificationInfo = notificationInfo

    // 4
    let operation =
CKModifySubscriptionsOperation(subscriptionsToSave: [subscription],
subscriptionIDsToDelete: [])

    // 5
```

```
    operation.modifySubscriptionsCompletionBlock = { [unowned self]
subscriptions, subscriptionIDs, error in
      if error == nil {
        self.isSubscribedToDatabase = true
      }

      completion(error)
    }

    // 6
    privateDatabase.add(operation)
  }
}
```

Let's go over the preceding code step by step, so you understand what happens in this snippet:

1. If the application has already subscribed to changes, there is no need to subscribe again. This means that you can call the completion closure without any errors.
2. When there are changes in the database, CloudKit will send a silent push notification to the app. Silent push notifications can be used to let the app know that new data is available. This mechanism is very similar to background fetch, which is covered in `Chapter 10`, *Being Proactive with Background Fetch.*
3. Every subscription in CloudKit is created as an instance of `CKDatabaseSubscription`. The notification settings that were configured in step two are attached to the database subscription.
4. This step creates an instance of `CKModifySubscriptionsOperation`. CloudKit uses a concept called `Operations` to manage the operations that you want the database to execute. Subscribing to changes is an example of such an operation, but fetching changes from the database or sending data to the database are also examples of operations. You can read more about `Operations` in `Chapter 25`, *Offloading Tasks with Operations and GCD.*
5. This step defines a closure that is called when the operation to subscribe to changes has finished. If this operation completes without error, the app is subscribed to changes and the status is persisted to `UserDefaults`. Every CloudKit operation has one or more of these closure callbacks that you can implement to respond to certain events that can occur during an operation.

6. The last step is to add the operation to the private database. Since all data will be stored in the private `moviesZone` that you created earlier, the appropriate database to use for the subscription is the private database.

Your app is now subscribed to changes from the CloudKit database. The next step is to make sure your app receives and handles updates from the CloudKit server.

Listening for changes in the database

When new data is available on the CloudKit server, it sends a notification to all subscribed apps. A great way to handle these notifications is to add a method to `CloudStore` that should be called when there are new changes in the CloudKit database. Add the following extension to `CloudStore.swift`:

```
extension CloudStore {
  func handleNotification(_ dict: [String: AnyObject],
completionHandler: @escaping (UIBackgroundFetchResult) -> Void) {
    let notification =
CKNotification(fromRemoteNotificationDictionary: dict)

    if notification.subscriptionID == "private-changes" {
      fetchDatabaseChanges { error in
        if error == nil {
          completionHandler(.newData)
        } else {
          completionHandler(.failed)
        }
      }
    }
  }
}
```

This method takes a dictionary and a completion-handler as its arguments. Note that the completion-handler has one argument, which is `UIBackgroundFetchResult`. As mentioned before, CloudKit sends a silent push notification when data has changed to trigger a process that's very similar to background fetch.

The previous method uses the notification dictionary to create a `CKNotification` instance and then checks whether the notification originated through the subscription for the private database. If this is the case, the changes should be fetched and ultimately get persisted to the local database.

When you subscribe to database changes, CloudKit will notify your app when any data in the database changes. When you fetch database changes, CloudKit will not tell you what records have changed straight away. Instead, CloudKit will send you the database zones that have changed. A flexible CloudKit setup should be able to handle incoming data from different zones, which means that you might add a new zone to your database and the app should be able to handle this.

If your app receives a notification to inform it of new changes, CloudKit won't provide any additional information. Instead, it is your job to ask CloudKit for the relevant changes. Remember that you were able to request only data that has changed since a specific moment in the CloudKit dashboard? The same mechanism is used when you fetch changes for your app.

Retrieving changes from CloudKit

To fetch changes that occurred after a certain point in time, you need to a change token along with your fetch operations. Add the following computed property to the `CloudStore` class so you can keep track of the change token that represents the last time the app fetched changes from `CloudKit`:

```
var privateDatabaseChangeToken: CKServerChangeToken? {
  get { return UserDefaults.standard.serverChangeToken(forKey:
"CloudStore.privateDatabaseChangeToken") }
  set { UserDefaults.standard.set(newValue, forKey:
"CloudStore.privateDatabaseChangeToken") }
}
```

This computed property uses a helper method on `UserDefaults` to store `CKServerChangeToken`.

If you're curious to see how the computed property is implemented, make sure to look at the `UserDefaults.swift` file in the project from the code bundle.

Add the following code to `CloudStore.swift` to implement the logic to fetch database changes:

```
extension CloudStore {
  func fetchDatabaseChanges(_ completionHandler: @escaping (Error?) ->
Void) {
    // 1
    let operation =
```

```
CKFetchDatabaseChangesOperation(previousServerChangeToken:
privateDatabaseChangeToken)

    // 2
    var zoneIds = [CKRecordZone.ID]()
    operation.recordZoneWithIDChangedBlock = { zoneId in
      zoneIds.append(zoneId)
    }

    // 3
    operation.changeTokenUpdatedBlock = { [weak self] changeToken in
      self?.privateDatabaseChangeToken = changeToken
    }

    // 4
    operation.fetchDatabaseChangesCompletionBlock = { [weak self]
changeToken, success, error in
      self?.privateDatabaseChangeToken = changeToken

      if zoneIds.count > 0 && error == nil {
        self?.fetchZoneChangesInZones(zoneIds, completionHandler)
      } else {
        completionHandler(error)
      }
    }

    privateDatabase.add(operation)
  }
}
```

The previous method retrieves only the record changes from the CloudKit database in the following steps:

1. `CKFetchDatabaseChangesOperation` is created. This operation will retrieve changes in the database zones. The latest change token is passed to this operation to make sure it only returns data that has changed since the last successful fetch.
2. The fetch operation will not return all changes zones at once. Instead, every changed zone is provided one by one, so they should be stored in an array for easy access at a later time.

3. At certain times in the refresh cycle, the operation will receive a change token from CloudKit. A single fetch operation might result in more than one change token. This could occur if there are many changes to be fetched from the server. In that case, the server will return the changed data in batches, allowing you to process the changes as they come in. It's essential that you store the latest token you have received from the server.
4. Provide a closure to the operation that should be executed when the fetch operation completes. This method also receives a change token, so you should store this token locally since that token represents the completed fetch operation. If no errors occurred and at least one database zone has changed, a new method is called. This method is responsible for fetching all changes to the database records, and you will implement it next.

Once the server has informed the app about all of the changed zones in the database, the app should then ask CloudKit for the relevant record changes. A single operation can retrieve data across multiple zones so your app won't have to issue many requests to CloudKit if there are several zones that have changes. Add the following extension to `CloudStore.swift`:

```
extension CloudStore {
  func fetchZoneChangesInZones(_ zones: [CKRecordZone.ID], _
completionHandler: @escaping (Error?) -> Void) {
    // 1
    var fetchConfigurations = [CKRecordZone.ID:
CKFetchRecordZoneChangesOperation.ZoneConfiguration]()

    for zone in zones {
      if let changeToken =
UserDefaults.standard.zoneChangeToken(forZone: zone) {
        let configuration =
CKFetchRecordZoneChangesOperation.ZoneConfiguration(previousServerChan
geToken: changeToken, resultsLimit: nil, desiredKeys: nil)
        fetchConfigurations[zone] = configuration
      }
    }

    let operation = CKFetchRecordZoneChangesOperation(recordZoneIDs:
zones, configurationsByRecordZoneID: fetchConfigurations)

    // 2
    var changedMovies = [CKRecord]()
    var changedFamilyMembers = [CKRecord]()
    operation.recordChangedBlock = { record in
      if record.recordType == "Movie" {
        changedMovies.append(record)
```

```
      } else if record.recordType == "FamilyMember" {
        changedFamilyMembers.append(record)
      }
    }

    operation.fetchRecordZoneChangesCompletionBlock = { [weak self]
error in
      for record in changedMovies {
        print(record["title"])
      }

      for record in changedFamilyMembers {
        print(record["name"])
      }

      completionHandler(error)
    }

    // 3
    operation.recordZoneFetchCompletionBlock = { recordZone,
changeToken, data, moreComing, error in
      UserDefaults.standard.set(changeToken, forZone: recordZone)
    }

    privateDatabase.add(operation)
  }
}
```

The preceding method is quite long but it follows a flow that is similar to the other two. Let's walk through this code step by step again:

1. When you fetch changes for record zones, each zone will have its own change token. This step uses another helper that was added to `UserDefaults` to conveniently store change tokens for each individual zone. The zones and their change tokens should be provided to `CKFetchRecordZoneChangesOperation` through a configuration dictionary. The operation itself is initialized with the dictionary and a list of zones that have changed.
2. The operation executes a callback for every changed record that it receives from the server. To make processing at a later time easier, the records for family members and movies are stored in their own lists. Note that both movies and family members are sent to the app as `CKRecord` instances. `CKRecord` has several specific CloudKit properties and methods, but otherwise, it behaves a lot like a dictionary. You will learn more about `CKRecord` instances later when you store your objects in Core Data.

3. `CKFetchRecordZoneChangesOperation` will call a closure for every zone that it processes. This callback receives a lot of arguments, such as the database zone that the closure was called for, the corresponding change token, and whether more data is coming in. The most interesting cases are the database zone and the change token. These two properties are used to store the latest change token for the corresponding zone in `UserDefaults`.

At this point, the `CloudStore` object contains all of the logic required to subscribe to database changes, retrieve them, and then print the objects that were fetched. To use `CloudStore`, you will need to update some code in `AppDelegate`.

Configuring your AppDelegate

As mentioned, silent push notifications work very similar to the way background fetch works. Just like background fetch, there is a special `AppDelegate` method that is called when a silent push notification arrives. You need to implement this method so you can handle the incoming notifications and fetch the change data from CloudKit. Add the following method to `AppDelegate`:

```
func application(_ application: UIApplication,
didReceiveRemoteNotification userInfo: [AnyHashable : Any],
fetchCompletionHandler completionHandler: @escaping
(UIBackgroundFetchResult) -> Void) {

  guard let dict = userInfo as? [String: NSObject] else {
    completionHandler(.failed)
    return
  }

  cloudStore.handleNotification(dict) { result in
    completionHandler(result)
  }
}
```

This method is called by iOS when your app receives a silent push notification from CloudKit. This message doesn't do much apart from telling `CloudStore` to process the notification and then it calls `completionHandler` to inform iOS about the result of the fetch. As you can see, this is very similar to background fetch apart from the fact that this method is called when the CloudKit server sends a push message to the app instead of when iOS decides to do so.

Lastly, add the following code at the end of `application(_:didFinishLaunchingWithOptions:)`:

```
cloudStore.subscribeToChangesIfNeeded { [weak self] error in
  DispatchQueue.main.async {
    application.registerForRemoteNotifications()
  }

  if error == nil {
    self?.cloudStore.fetchDatabaseChanges { fetchError in
      if let error = fetchError {
        print(error)
      }
    }
  }
}
```

The preceding code takes care of telling `CloudStore` to subscribe to changes. It also registers the app for remote notifications so CloudKit can send silent push messages to the app. Note that this type of push messaging does not require you to explicitly ask the user to send them push notifications. This type of permission is only required if you want to send the user visual push notifications. You can read more about push notifications in `Chapter 20`, *Implementing Rich Push Notifications*.

At this point, you can add a couple of records to your CloudKit database. Since you haven't implemented sending data to CloudKit yet, this is the best way for you to verify that your code works as expected. You should see the names of the movies and family members you have added in the CloudKit dashboard printed to the console. Once you have the app running, try adding a new family member while the app is open. You should see the name of the new family member appear in the console right after you add the record to the CloudKit dashboard. Pretty neat, right?

To use the private database in CloudKit, a user that is logged in to CloudKit on their device is required. Make sure to run your app on a physical device that has a logged-in iCloud user (preferably the same account you use as your Apple developer account) or log an iCloud user in on the simulator by navigating to the Settings app on the simulator.

Storing data in CloudKit

To store objects in CloudKit, you use `CKModifyRecordsOperation`. This operation takes a list of records that should be created, updated, or removed. Every record is connected to a certain zone, should have a unique identifier, and can have different properties associated with it. If you add a property to an object that doesn't exist in CloudKit yet, CloudKit will automatically update the data model on the server using the data you sent. You can even create entirely new models this way. You can add the following code to `CloudStore.swift` to store a very basic version of a family member's record:

```
extension CloudStore {
  func storeFamilyMember(_ familyMember: FamilyMember, _
completionHandler: @escaping (Error?) -> Void) {
    let defaultZoneId = CKRecordZone.ID(zoneName: "moviesZone",
ownerName: CKCurrentUserDefaultName)
    let recordId = CKRecord.ID(recordName: UUID().uuidString, zoneID:
defaultZoneId)
    let record = CKRecord(recordType: "FamilyMember", recordID:
recordId)
    record["name"] = familyMember.name!

    let operation = CKModifyRecordsOperation(recordsToSave: [record],
recordIDsToDelete: [])
    operation.modifyRecordsCompletionBlock = { records, recordIds,
error in
      guard let records = records, error == nil else {
        completionHandler(error)
        return
      }

      print(records)
    }

    privateDatabase.add(operation)
  }
}
```

The preceding code creates a zone ID and a record ID that will be used to identify the record in the database. Next, a family member record is created and the family member name is added to it. Lastly, `CKModifyRecordsOperation` is created and added to the database to send the record to CloudKit. To call this method when a new family member is added, update the `saveFamilyMember(withName:)` method in `FamilyMembersViewController` as follows:

```
func saveFamilyMember(withName name: String) {
  let moc = persistentContainer.viewContext

  moc.persist {
    let familyMember = FamilyMember(context: moc)
    familyMember.name = name
    familyMember.uuidString = UUID().uuidString

    self.cloudStore.storeFamilyMember(familyMember) { _ in
      // no action
    }
  }
}
```

This snippet passes the family member to the cloud store so it can convert it to `CKRecord` and store it in CloudKit. Try adding a new family member in the app, you should see your newly-inserted record in the console in Xcode; when you refresh the list of family members in the CloudKit dashboard, a new entry should have appeared.

This implementation of adding a family member to CloudKit is too basic because there is no way to associate one of the local family members to the family members stored in CloudKit. The next section will implement a proper local cache of the data in CloudKit by adapting the existing Core Data model a little bit so you can import data from the CloudKit server, send new family members and movies to CloudKit, and update existing family members when they add new movies to their favorites.

Since your app stores database change tokens in `UserDefaults`, you should delete **MustC** from your device to remove the existing change tokens. This allows you to start fresh and receive all data again.

Combining CloudKit and Core Data

When you implement CloudKit in your app, it is always highly recommended to have a local cache of all CloudKit data. The previous section showed you how to fetch changes from the CloudKit database and how to process them. You also saw a very basic sample of how to send data to CloudKit. In this section, you will implement some helper methods on the `Movie` and `FamilyMember` classes. You will also add some extra properties to the Core Data models to facilitate CloudKit, and finally, you will implement the importing and exporting of CloudKit data.

Preparing the Core Data models for CloudKit

You might have noticed the following line in the code for creating a new `CKRecord` that was shown in the previous section:

```
let recordId = CKRecord.ID(recordName: UUID().uuidString, zoneID:
defaultZoneId)
```

The `recordName` that is set on `CKRecord.ID` is the unique identifier that CloudKit uses to store records. When you want to import data from CloudKit, you can use `recordName` to check whether you have already saved a record in your local database. In addition to this unique identifier, CloudKit itself stores metadata about records. You saw this metadata when you added a new record type in the CloudKit dashboard and all these new properties that you didn't add yourself showed up.

Apart from the unique identifier, none of the metadata is very relevant to use in the **MustC** app, so adding all fields to the Core Data model would be a waste. Luckily, `CKRecord` has a convenient method that allows you to encode all of the automatically-added metadata into a `Data` object that you can easily store in Core Data. This means that you'll need to add two new properties to both the `FamilyMember` and `Movie` models:

- `recordName` (`String`)
- `cloudKitData` (`Data`)

Go ahead and add these two properties to your models in the Core Data model editor.

After adding the `recordName` and `cloudKitData` attributes, you should add a couple of helpers to your models. Create a new file in the Models folder and name it `FamilyMember.swift`. Add the following extension to this file:

```
extension FamilyMember {
  // 1
  func recordIDForZone(_ zone: CKRecordZone.ID) -> CKRecord.ID {
    return CKRecord.ID(recordName: self.recordName!, zoneID: zone)
  }

  // 2
  func recordForZone(_ zone: CKRecordZone.ID) -> CKRecord {
    let record: CKRecord

    // 3
    if let data = cloudKitData, let coder = try?
NSKeyedUnarchiver(forReadingFrom: data) {
      coder.requiresSecureCoding = true
      record = CKRecord(coder: coder)!
    } else {
      record = CKRecord(recordType: "FamilyMember", recordID:
recordIDForZone(zone))
    }

    record["name"] = name!

    // 4
    if let movies = self.movies as? Set<Movie> {
      let references: [CKRecord.Reference] = movies.map { movie in
        let movieRecord = movie.recordForZone(zone)
        return CKRecord.Reference(record: movieRecord, action: .none)
      }

      record["movies"] = references
    }

    return record
  }

  // 5
  static func find(byIdentifier recordName: String, in moc:
NSManagedObjectContext) -> FamilyMember? {
    let predicate = NSPredicate(format: "recordName == %@",
recordName)
    let request: NSFetchRequest<FamilyMember> =
FamilyMember.fetchRequest()
    request.predicate = predicate
```

```
        guard let result = try? moc.fetch(request)
          else { return nil }

        return result.first
      }
    }
```

A lot is going on in this snippet, so let's go over it step by step:

1. The first comment highlights a convenient helper method that creates a record ID. This helper method receives an existing zone ID and uses the new `recordName` property that you just added in the Core Data model editor.
2. This method is used to convert the `FamilyMember` model into a `CKRecord` instance. This method will be used when you send data to CloudKit.
3. `CKRecord` objects can be created in different ways. One way is through the `CKRecord(recordType:recordID:)` initializer. You can also use an instance of `NSCoder` to create a `CKRecord` with the `CKRecord(coder:)` initializer. An `NSCoder` can convert objects into data and vice versa. So, in this case, a special version of `NSCoder`, called `NSKeyedUnarchiver`, is used to convert the metadata that is stored as `Data` in Core Data back into a `CKRecord` instance. If the Core Data object has just been added, it won't have any CloudKit metadata yet, so a new `CKRecord` instance should be created.
4. To make sure all movies for the family member are sent to CloudKit, it is required to loop over each movie and create a `CKRecord.Reference` to the object. This list of references is then assigned to the movie record.
5. To import CloudKit objects, the code must be able to look up existing movie records using the record's `recordName`. This method is used to look up family members by their record name.

The `Movie` object should receive similar helper methods to the ones you just added to `FamilyMember`. Add the following methods to the extension in `Movie.swift`:

```
func recordIDForZone(_ zone: CKRecordZone.ID) -> CKRecord.ID {
  return CKRecord.ID(recordName: self.recordName!, zoneID: zone)
}

func recordForZone(_ zone: CKRecordZone.ID) -> CKRecord {
  let record: CKRecord

  if let data = cloudKitData, let coder = try?
NSKeyedUnarchiver(forReadingFrom: data) {
```

```
    coder.requiresSecureCoding = true
    record = CKRecord(coder: coder)!
  } else {
    record = CKRecord(recordType: "Movie", recordID:
recordIDForZone(zone))
  }

  record["title"] = title!
  record["rating"] = popularity
  record["remoteId"] = remoteId

  return record
}

static func find(byIdentifier recordName: String, in moc:
NSManagedObjectContext) -> Movie? {
  let predicate = NSPredicate(format: "recordName == %@", recordName)
  let request: NSFetchRequest<Movie> = Movie.fetchRequest()
  request.predicate = predicate

  guard let result = try? moc.fetch(request)
    else { return nil }

  return result.first
}

static func find(byIdentifiers recordNames: [String], in moc:
NSManagedObjectContext) -> [Movie] {
  let predicate = NSPredicate(format: "ANY recordName IN %@",
recordNames)
  let request: NSFetchRequest<Movie> = Movie.fetchRequest()
  request.predicate = predicate

  guard let result = try? moc.fetch(request)
    else { return [] }

  return result
}
```

The code in the preceding snippet should speak for itself. One interesting addition is the `find(byIdentifiers:)` method. Instead of taking just a single record name, this method takes a list of record names. When you import family members from CloudKit, a single family member could have multiple movies in their favorites. Instead of fetching each movie individually, this method allows you to retrieve all matching movies at once.

Your Core Data models are now fully compatible with CloudKit, and you're ready to write the code that will import the data from the CloudKit servers and add them to your local Core Data database.

Importing CloudKit data

You have already written most of the code to retrieve changed records from the CloudKit database. The next step is to implement the code that takes the `CKRecord` instances that you receive from CloudKit and convert them into the correct Core Data models. Before you implement the import methods, update `fetchRecordZoneChangesCompletionBlock` in `fetchZoneChangesInZones(_:_:)` on `CloudStore` as follows:

```
let backgroundContext = persistentContainer.newBackgroundContext()

operation.fetchRecordZoneChangesCompletionBlock = { [weak self] error
in
  for record in changedMovies {
    self?.importMovie(withRecord: record, withContext:
backgroundContext)
  }

  for record in changedFamilyMembers {
    self?.importFamilyMember(withRecord: record, withContext:
backgroundContext)
  }

  completionHandler(error)
}
```

The preceding snippet uses `persistentContainer` to obtain a background managed object context. This type of managed object context runs off the main thread, making it ideal for operations that could take a while. Since you can't ever be sure about the amount of data that CloudKit sends your way, it's a good idea to perform all import work on a background managed object context. When using background contexts, it's important that you pay extra attention to the Core Data objects you're working with. You should always make sure to create and use Core Data objects on a single thread. You can't create an object in one managed object context and then save it in the other. If you do accidentally mix up threads, you could end up with some very difficult-to-debug crashes.

Inside of the completion block, two new methods are called: one that imports movies and one that imports family members. The reason they are called in this order is to make sure all movies are imported first and the family members second. This ensures that all the family member's favorite movies exist in the database so they can be associated with the family member immediately.

Add the following extension to `CloudStore.swift` to implement the importers:

```
extension CloudStore {
  func importMovie(withRecord record: CKRecord, withContext moc:
NSManagedObjectContext) {
    moc.persist {
      let identifier = record.recordID.recordName
      let movie = Movie.find(byIdentifier: identifier, in: moc) ??
Movie(context: moc)
      movie.title = record["title"] ?? "unkown-title"
      movie.remoteId = record["remoteId"] ?? 0
      movie.popularity = record["rating"] ?? 0.0
      movie.cloudKitData = record.encodedSystemFields
      movie.recordName = identifier
    }
  }

  func importFamilyMember(withRecord record: CKRecord, withContext
moc: NSManagedObjectContext) {
    moc.persist {
      let identifier = record.recordID.recordName
      let familyMember = FamilyMember.find(byIdentifier: identifier,
in: moc) ?? FamilyMember(context: moc)
      familyMember.name = record["name"] ?? "unkown-name"
      familyMember.cloudKitData = record.encodedSystemFields
      familyMember.recordName = identifier

      if let movieReferences = record["movies"] as?
[CKRecord.Reference] {
        let movieIds = movieReferences.map { reference in
          return reference.recordID.recordName
        }

        familyMember.movies = NSSet(array: Movie.find(byIdentifiers:
movieIds, in: moc))
      }
    }
  }
}
```

The first method in this extension imports movie objects. If you look at this method carefully, you'll find that it's relatively straightforward. The record's unique identifier is used to look up a movie in Core Data. If no movie could be found, a new one is created. All the following steps extract data from `CKRecord` and apply it to the `Movie` instance. Since the all code is wrapped in a `moc.persist` closure, the new movie will be saved immediately.

The method to import family members is slightly more complex. The basics are the same as they are for the movie import. However, the family member importer has to extract the references to movie records and then map the references to record names. These record names are then used to look up all movies in the list of favorites for this family member, and then assigns this list to the family member.

Try deleting and reinstalling the app now. You should automatically see the list of family members appear.

Does the list of family members only appear for you if you start the app for the second time? Your `viewContext` might be misconfigured. Make sure that you set `container.viewContext.automaticallyMergesChangesFromParent` to `true` in the lazy variable that creates the persistent container in `AppDelegate`.

Just one more step until you have a complete CloudKit-enabled application that sends Core Data models to CloudKit.

Sending Core Data models to CloudKit

To send Core Data models to CloudKit, you must convert them to `CKRecord` instances. So far, you've learned how to create an instance of `CKRecord` and you even implemented a helper method on the `Movie` and `FamilyMember` objects. Take another look at `recordForZone(_:)` on both objects to make sure you understand how the models are converted to records. When you convert Core Data objects to CloudKit models, it's important that you properly use the CloudKit metadata. If you attempt to store an existing object with the correct record name, but you omit the metadata, CloudKit won't be able to save your object because it can't accurately compare the version that it currently has stored and the new version you are trying to send it.

Before you update the `storeFamilyMember(_:_)` method, update the `saveMovie(withName:)` method in `MoviesViewController` so family members get stored when you add a new family member or when you update one by assigning it a new movie. Update the method as follows:

```
func saveMovie(withName name: String) {
  // ...

    let helper = MovieDBHelper()
    helper.fetchRating(forMovie: name) { remoteId, rating in
      guard let rating = rating,
        let remoteId = remoteId
        else { return }

      moc.persist {
        movie.popularity = rating
        movie.remoteId = Int64(remoteId)

        self.cloudStore.storeFamilyMember(familyMember) { _ in
          // no action
        }
      }
  // ...
}
```

The only thing that has changed in this method is that the family member is now stored after the movie's popularity rating is fetched. The next and final step is to implement the new version of `storeFamilyMember(_:_)`, as follows:

```
func storeFamilyMember(_ familyMember: FamilyMember, _
completionHandler: @escaping (Error?) -> Void) {
  // 1
  guard let movies = familyMember.movies as? Set<Movie> else {
    completionHandler(nil)
    return
  }

  let defaultZoneId = CKRecordZone.ID(zoneName: "moviesZone",
ownerName: CKCurrentUserDefaultName)

  // 2
  var recordsToSave = movies.map { movie in
    movie.recordForZone(defaultZoneId)
  }
  recordsToSave.append(familyMember.recordForZone(defaultZoneId))

  let operation = CKModifyRecordsOperation(recordsToSave:
```

```
recordsToSave, recordIDsToDelete: nil)

  operation.modifyRecordsCompletionBlock = { records, recordIds, error
in

    guard let records = records, error == nil else {
      completionHandler(error)
      return
    }

    // 3
    for record in records {
      if record.recordType == "FamilyMember" {
        familyMember.managedObjectContext?.persist {
          familyMember.cloudKitData = record.encodedSystemFields
        }
      } else if record.recordType == "Movie",
        let movie = movies.first(where: { $0.recordName ==
record.recordID.recordName }) {

        familyMember.managedObjectContext?.persist {
          movie.cloudKitData = record.encodedSystemFields
        }
      }
    }

    completionHandler(error)
  }

  privateDatabase.add(operation)
}
```

Even though the preceding method is quite long, it should be relatively straightforward; there are some parts that are worth taking a closer look at:

1. Because the `movies` property on `FamilyMember` is an `NSSet`, it needs to be converted to a `Set<Movie>` so the movies can be looped over. This conversion should never fail, so if it does, something is wrong and it makes no sense to continue saving the family member to CloudKit.
2. Extract all movie records from the family member to obtain a list of records that should be saved. The family member itself is then also added to this list, so it's also saved to CloudKit.

3. Once the save operation is created, and all records are saved, the CloudKit metadata is added to their corresponding Core Data objects, and the objects are then saved. Note that the managed object context is obtained by calling `familyMember.managedObjectContext`. Doing this ensures that the family member and its movies are updated in the correct managed object context.

If you run your app now, you should be able to import records from CloudKit automatically. When you add records to CloudKit while the app is running, it will automatically update the user interface. And when you add new family members and movies to the Core Data database, they are automatically saved to CloudKit.

If you want to test your app on multiple devices at once, you can use different simulators that are signed in to the same iCloud account. When you add a family member on one simulator, it will automatically appear on all other simulators that have the same iCloud user.

Summary

In this chapter, you learned a lot about CloudKit. You learned what CloudKit is and what it's used for. You added the required capabilities to your app, and then you explored your new CloudKit container in the CloudKit dashboard. You used the dashboard to define some models that you will later use in your app.

After configuring the CloudKit dashboard, you got your hands dirty by implementing a CloudKit sync feature for the **MustC** app. You learned that CloudKit makes extensive use of `Operations` to manage database operations. You also had to set up `AppDelegate` so it would subscribe the app to changes in the user's database. Then you learned how you can fetch updates from CloudKit using a change token, and how to send new records to CloudKit.

To wrap the feature up, you adapted your Core Data store to make it act as a local cache for CloudKit. You saw how to convert a Core Data model object to a CloudKit record and vice versa. This chapter made some extensive use of extensions on objects, such as `UserDefaults` to make working with them more convenient. It's always a good idea to extract code you don't want to write too often into an extension if possible.

This chapter wraps up the **MustC** app for now. In the next chapters, you will dive into some of the excellent frameworks iOS has to offer, such as ARKit, Core Location, and SiriKit.

Questions

1. How does the CloudKit server notify apps about updates?

    a) Background Fetch.
    b) Silent push notifications.
    c) It doesn't. Apps should ask for changes on launch.

2. Which object do you send along with a fetch request to CloudKit so you only receive new changes?

    a) A change token.
    b) A timestamp.
    c) The last object you received updates for.

3. How does your code know whether a certain `CKRecord` is a movie, family member, or something else?

    a) Through `recordName`.
    b) By checking what properties are available.
    c) Through `recordType`.

4. What kind of data do you have to store alongside your objects in Core Data to be able to properly sync local data with CloudKit?

    a) Their change tokens.
    b) The encoded CloudKit metadata.
    c) The zone they were retrieved from.

5. What's a good place to store server change tokens?

    a) Core Data.
    b) In an array.
    c) `UserDefaults`.

6. Why is it smart to import CloudKit data on a background managed object context?

    a) Because you don't know how long the import will take and you don't want to block the main thread.
    b) Because background contexts are faster at writing data.
    c) To prevent simultaneous read and write operations.

Further reading

- Refer to `Chapter 10`, *Being Proactive with Background Fetch,* for more information about background fetch features
- Refer to `Chapter 20`, *Implementing Rich Push Notifications,* for more information about push notifications
- Refer to `Chapter 25`, *Offloading Tasks with Operations and GCD,* to learn more about GCD and threading

12
Using Augmented Reality

One of the major features that Apple shipped as part of iOS 11 was **ARKit**. ARKit enables developers to create amazing Augmented Reality experiences with only a minimal amount of code. Apple has continuously worked on improving ARKit, resulting in the release of ARKit 2 at WWDC 2018.

In this chapter, you will learn what ARKit is, how it works, what you can do with it, and how you can implement an Augmented Reality Art gallery that uses several of ARKit 2 new features, such as image tracking. This chapter covers the following ARKit topics:

- Understanding ARKit
- Using ARKit Quicklook
- Exploring SpriteKit
- Exploring SceneKit
- Implementing an Augmented Reality gallery

By the end of this chapter, you will be able to integrate ARKit in your apps and implement your own ARKit experiences.

Understanding ARKit

Augmented Reality (**AR**) is a topic that has captured the interest of app developers and designers for a long time now. Implementing an excellent AR experience had not been easy though, and many applications haven't lived up to the hype. Small details such as lighting, and detecting walls, floors, and other objects have always been extremely complicated to implement and getting these details wrong has a negative impact on the quality of an AR experience.

Augmented reality apps usually have at least some of the following features:

- They show a camera view.
- Content is shown as an overlay on the camera.
- Content responds appropriately to device movement.
- Content is attached to a specific location in the world.

Even though this list of features is simple, they aren't all trivial to implement. An AR experience relies heavily on reading the motion sensors from the device, as well as using image analysis to determine exactly how a user is moving and to learn what a 3D map of the world should look like.

ARKit is Apple's way of giving developers the power to create great AR experiences. ARKit takes care of all the motion and image analysis to make sure you can focus on designing and implementing great content rather than getting slowed down by the intricate details involved in building an AR app.

Unfortunately, ARKit comes with a hefty hardware requirement for the devices that can run ARKit apps. Only devices with Apple's A9 chip or newer can run ARKit. This means that any device older than the iPhone 6s or first iPad Pro cannot run ARKit apps.

Understanding how ARKit renders content

ARKit itself only takes care of the massive calculations related to keeping track of the physical world the user is in. To render content in an ARKit app, you must use one of the following three rendering methods:

- `SpriteKit`
- `SceneKit`
- `Metal`

Later in this chapter, you will have a quick look at SpriteKit and SceneKit, and you will ultimately implement your Augmented Reality gallery using SceneKit. If you already have experience with any of the available rendering techniques, you should feel right at home when using ARKit.

Implementing ARKit in your app is not limited to manually rendering the contents you want to show in AR. In iOS 12, Apple has added a feature called **ARKit Quicklook**. You can implement a special view controller in your app that takes care of placing a 3D model you supply in a scene. This is ideal if you're implementing a feature that allows users to preview products or other objects in the real world.

Understanding how ARKit tracks the physical environment

To understand how ARKit renders content, it's essential that you understand how ARKit makes sense of the physical environment a user is in. When you implement an AR experience, you use an ARKit session. An ARKit session is represented by an instance of `ARSession`. Every `ARSession` uses an instance of `ARSessionConfiguration` to describe the tracking that it should do on the environment. The following diagram depicts the relationship between all objects involved in an ARKit session:

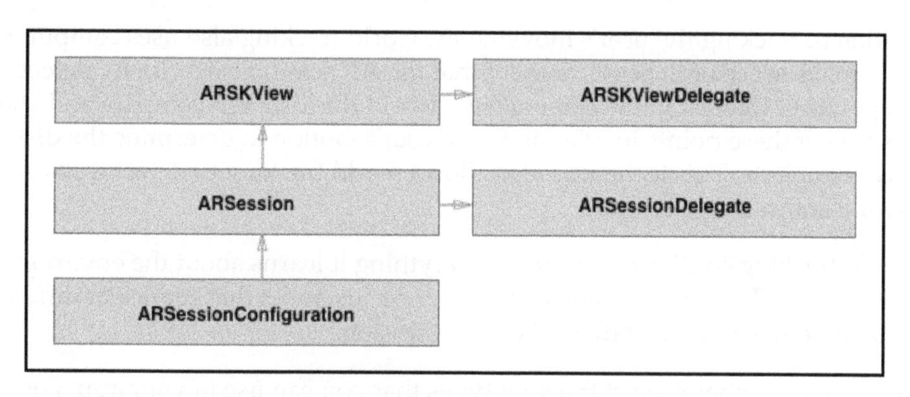

The preceding image shows how the session configuration is passed to the session. The session is then passed to a view that is responsible for rendering the scene. If you use SpriteKit to render the scene, the view is an instance of `ARSKView`. When you use `SceneKit`, this would be an instance of `ARSCNView`. Both the view and session have a delegate that will be informed about certain events that can occur during an ARKit session. You will learn more about these delegates later when you implement your AR gallery.

There are several different tracking options that you can configure on a session. One of the most basic tracking configurations is `AROrientationTrackingConfiguration`. This configuration only tracks the device's orientation, so not the user's movement in the environment. This kind of tracking monitors the device using three degrees of freedom. To be more specific, this tracking tracks the device's x, y, and z orientation. This kind of tracking is perfect if you're implementing something such as a 3D video where the user's movements can be ignored.

A more complex tracking configuration is `ARWorldTrackingConfiguration`, also known as **World tracking**. This type of configuration tracks the user's movements as well as the device's orientation. This means that a user can walk around an AR object to see it from all different sides. World tracking uses the device's motion sensors to determine the user's movements and the device orientation. This is very accurate for short and small movements, but not accurate enough to track movements over long periods of time and distances. To make sure the AR experience remains as precise as possible, world tracking also performs some advanced computer vision tasks to analyze the camera feed to determine the user's location in an environment.

In addition to tracking the user's movements, world tracking also uses computer vision to make sense of the environment that the AR session exists in. By detecting certain points of interest in the camera feed, world tracking can compare and analyze the position of these points in relation to the user's motion to determine the distances and sizes of objects. This technique also allows world tracking to detect walls and floors for instance.

The world tracking configuration stores everything it learns about the environment in an `ARWorldMap`. This map contains all `ARAnchor` instances that represent different objects and points of interest that exist in the session.

There are several other special tracking types that you can use in your app. For instance, you can use `ARFaceTrackingConfiguration` on devices with a TrueDepth camera to track a user's face. This kind of tracking is perfect if you want to recreate Apple's Animoji feature that was added to the iPhone X and newer in iOS 12.

You can also configure your session, so it automatically detects certain objects or images in a scene. To implement this, you can use `ARObjectScanningConfiguration` to scan for specific items or `ARImageTrackingConfiguration` to identify still images.

Before you get your hands dirty with implementing an ARKit session, let's explore the new ARKit Quicklook session to see how simple it is for you to allow users of your app to preview items in AR.

Using ARKit Quicklook

One of the great benefits that AR brings to end users is that it is now possible to preview certain objects in the real world. For instance, when you buy a new sofa, you might want to see what it looks like in the real world. Of course, it was possible to implement features such as this in iOS 11 using ARKit, and many developers have, but it wasn't as easy as it could be.

iOS users can preview content using a feature called Quicklook. **Quicklook** can be used to preview certain types of content without having to launch any specific applications. This is convenient for users because they can quickly determine whether a particular document is the document they are looking for by previewing it in Quicklook.

In iOS 12, Apple added the USDZ file format to the content types that can be previewed using Quicklook. Apple's USDZ format is a 3D file format based on Pixar's USD format that is used to represent 3D objects. Using Quicklook for 3D models is not just available in apps, ARKit Quicklook can also be integrated on the web. Developers can use a special HTML tag on their web pages to link to a USDZ and Safari will display the model in an ARKit Quicklook view controller.

Before you implement your AR gallery, it's a good idea to get a feeling for how AR works on iOS by implementing the ARKit Quicklook view controller to show one of the models that Apple provides on `https://developer.apple.com/arkit/gallery/`. To download a model you like, all you need to do is navigate to this page on your Mac and click on an image. The USDZ file should start downloading automatically.

Navigate to the ARKit gallery on a device that supports ARKit and tap on one of the models to see what ARKit Quicklook in Safari looks like.

Implementing the ARKit Quicklook view controller

After obtaining a USDZ file from Apple's gallery, also make sure to capture the image that belongs to this file. Taking a screenshot of the model should be fine for testing purposes. Make sure to prepare your image in the different required sizes by scaling your screenshot up to two and three times the size of your screenshot.

Create a new project in Xcode and pick a name for your project. The sample project in this book's code bundle is called **ARQuickLook**. Add your prepared image to the `Assets.xcassets` file. Also, drag your USDZ file into Xcode and make sure to add it to the app target by checking your app's checkbox when importing the file:

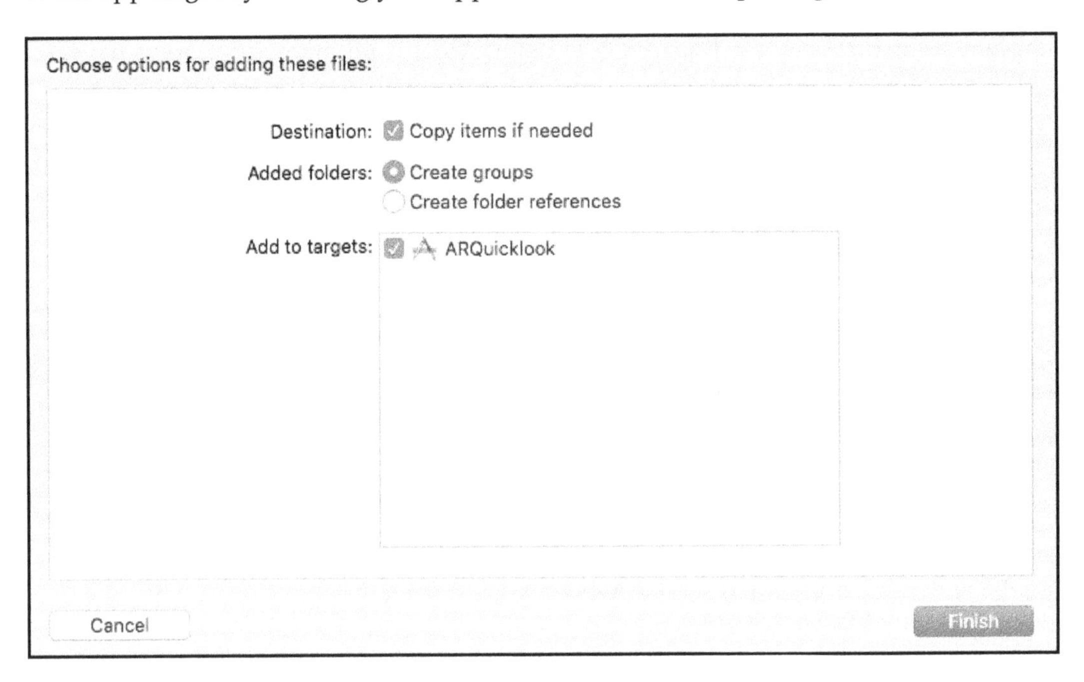

Next, open the storyboard file and drag an image view to the view controller. Add the proper constraints to the image, so it's centered in the view controller and give it a width and height of 200 points. Make sure to check the **User Interaction Enabled** checkbox in the **Attributes Inspector** and set your model image as the image for the image view.

After doing this, open `ViewController.swift`, add `@IBOutlet` for the image view, and connect the image in the storyboard to this outlet. If the details regarding outlets are a little bit fuzzy right now, refer to the sample project in the code bundle for a refresher. The image view in the sample project uses an outlet called `guitarImage`.

The next steps to implement Quicklook for the USDZ model are to add a tap gesture recognizer to the image view and then trigger the Quicklook view controller when a user taps on the image.

Quicklook uses delegation to object one or more items that it should preview from a data source. It also uses a delegate to obtain the source view from which the Quicklook preview should animate. This flow applies to all different kinds of files that you can preview using Quicklook.

To begin implementing Quicklook, you must import the `QuickLook` framework. Add the following import statement to the top of `ViewController.swift`:

```
import Quicklook
```

Next, set up the tap-gesture recognizer for the image by adding the following code to `viewDidLoad()`:

```
let tapGesture = UITapGestureRecognizer(target: self, action:
#selector(presentQuicklook))
guitarImage.addGestureRecognizer(tapGesture)
```

The next step is to implement `presentQuicklook()`. This method will create a Quicklook view controller, set the delegate and data source, and then present the Quicklook view controller to the user. Add the following implementation for this method to the `ViewController` class:

```
@objc func presentQuicklook() {
  let previewViewController = QLPreviewController()
  previewViewController.dataSource = self
  previewViewController.delegate = self

  present(previewViewController, animated: true, completion: nil)
}
```

This implementation should not contain any surprises for you. `QLPreviewController` is a `UIViewController` subclass that is responsible for displaying the content it receives from its data source. It is presented in the same way you would present any other view controller, by calling `present(_:animated:completion:)`.

The final step is to implement the data source and delegates. Add the following extensions to `ViewController.swift`:

```
extension ViewController: QLPreviewControllerDelegate {
  func previewController(_ controller: QLPreviewController,
transitionViewFor item: QLPreviewItem) -> UIView? {
    return guitarImage
  }
}

extension ViewController: QLPreviewControllerDataSource {
  func numberOfPreviewItems(in controller: QLPreviewController) -> Int
{
    return 1
  }

  func previewController(_ controller: QLPreviewController,
previewItemAt index: Int) -> QLPreviewItem {
    let fileUrl = Bundle.main.url(forResource: "stratocaster",
withExtension: "usdz")!
    return fileUrl as QLPreviewItem
  }
}
```

The first extension you added makes `ViewController` conform to `QLPreviewControllerDelegate`. When the preview controller is about to present the 3D model, it wants to know which view is the source for the transition that is about to happen. It's recommended to return the view that acts as a preview for the Quicklook action from this method. In this case, the preview is the image of the 3D model.

The second extension acts as the Quicklook data source. When you implement Quicklook for ARKit, you can only return a single item. So when the preview controller asks for the number of items in the preview, you should always return 1. The second method in the data source provides the item that should be previewed in the preview controller. All you need to do here is obtain the file URL for the item you wish to preview. In the sample app, the Stratocaster model from Apple's gallery is used. If your model has a different name, make sure to use the correct filename.

After obtaining the `URL` that points to the image in the app bundle, it should be returned to the preview controller as a `QLPreviewItem`. Luckily, `URL` instances can be converted to `QLPreviewItem` instances automatically.

If you run your app now, you can tap on your image of the 3D model to begin previewing it. You can preview the image on its own, or you can choose to preview it in AR. If you tap this option, the preview controller will tell you to move your device around:

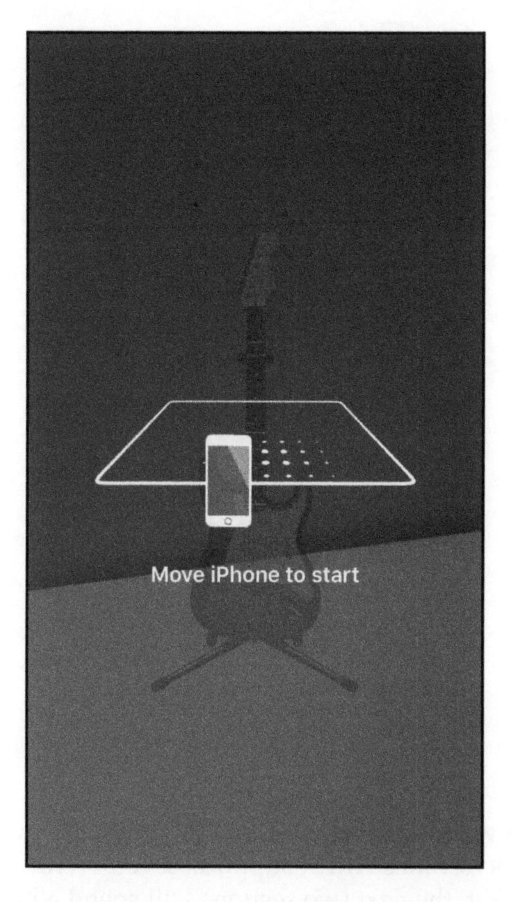

To make a mapping of the world around you, ARKit requires some samples of the environment. When you move your device around, make sure not just to tilt it, but physically move it. Doing this will help ARKit discover trackable features in your surroundings.

Once ARKit has enough data about your surroundings, you can place the 3D model in the environment, scale it by pinching, rotate it, and move it around in the space. Note that the model is placed on a flat surface such as a table or the floor automatically rather than awkwardly floating around.

Also note that ARKit applies very realistic lighting to your object. The visual data that ARKit gathers about the environment is used to create a lighting map that is applied to the 3D model to make it properly blend in with the context in which the object was placed:

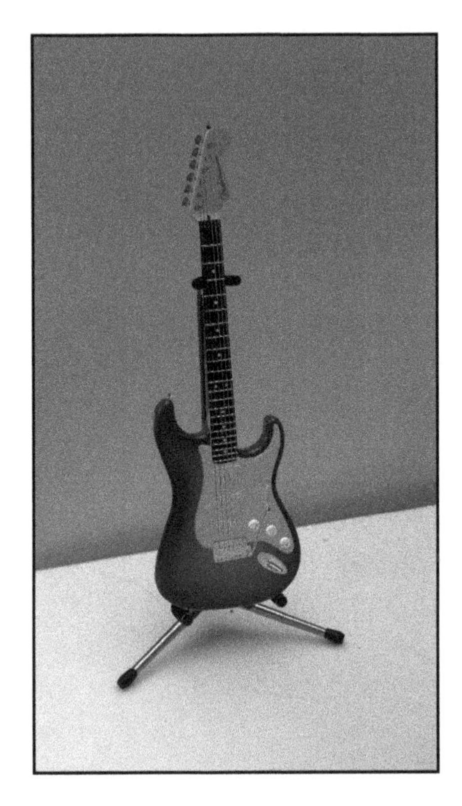

While playing around with ARKit like this is a lot of fun, it's even more fun to create your own AR experiences. Since ARKit supports several rendering techniques, such as SpriteKit and SceneKit, the next two sections will spend a little bit of time explaining the very basics of SpriteKit and SceneKit. You won't learn how to build complete games or worlds with these frameworks. Instead, you will learn just enough to get you started with implementing either rendering engine in an ARKit app.

Exploring SpriteKit

SpriteKit is mostly used by developers to build two-dimensional games. SpriteKit has been around for quite some time already, and it has helped developers to create many successful games over the years. SpriteKit contains a full-blown physics simulation engine, and it can render many sprites at a time. A **sprite** represents a graphic in a game. A sprite could be an image for the player, but also a coin, an enemy, or even the floor that a player walks on. When sprites are mentioned in the context of SpriteKit, it is meant to refer to one of the nodes that are visible on the screen.

Because SpriteKit has a built-in physics engine, it can detect collisions between objects, apply forces to them, and more. This is pretty similar to what UIKit Dynamics is capable of. If you're a bit rusty on what UIKit Dynamics are and how they work, be sure to have a look at `Chapter 4`, Immersing Your Users with Animation for a refresher.

To render content, SpriteKit uses scenes. These scenes can be considered levels or major building parts of a game. In the context of AR, you will find that you typically only need a single scene. A SpriteKit scene is responsible for updating the position and state of the scene. As a developer, you can hook into the rendering of frames through the `update(_:)` method of `SKScene`. This method is called every time SpriteKit is about to render a new frame for your game or ARKit scene. It is essential that this method's execution time is as short as possible, as a slow implementation of the `update(_:)` method will cause frames to drop, which is considered bad. You should always aim to maintain a steady 60 frames per second. This means that the `update(_:)` method should always perform its work in less than 1/60th of a second.

To begin exploring SpriteKit, create a new project in XCode and choose the **Game** template. Pick SpriteKit as the underlying game technology and give the project a name. For instance, **SpriteKitDefault**, as shown in the following screenshot:

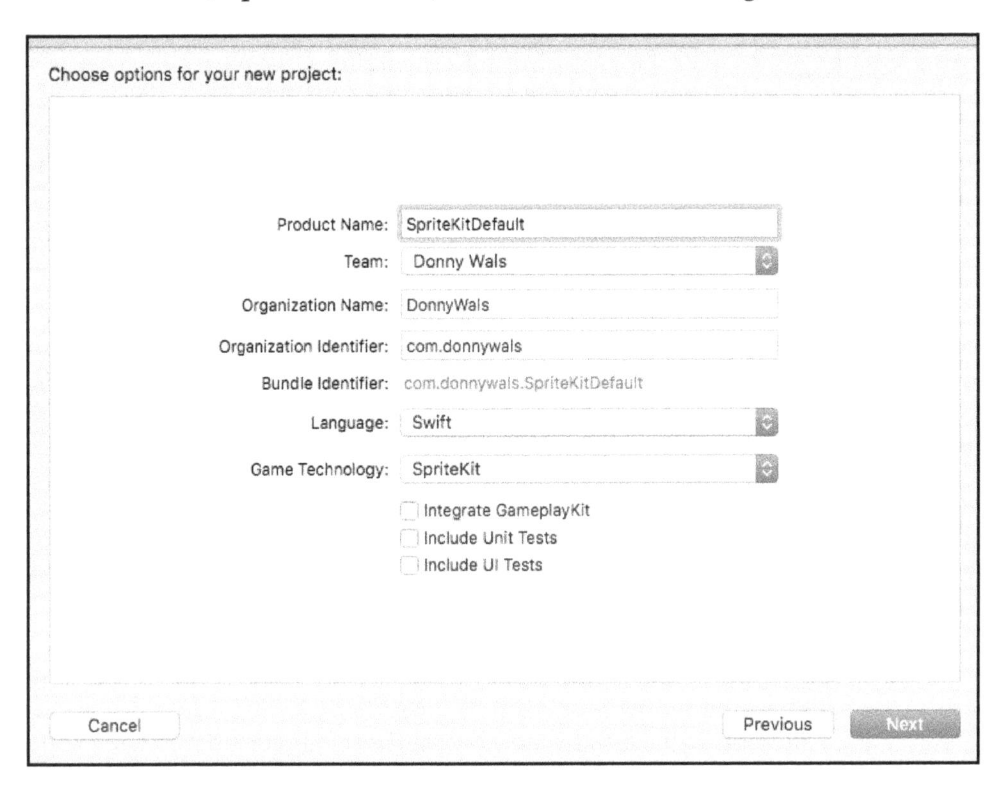

When Xcode generates this project for you, you should notice some new files that you haven't seen before:

- `GameScene.sks`
- `Actions.sks`

These two files are to SpriteKit games what storyboards are to regular apps. You can use these to set up all nodes for your game scene or to set up reusable actions that you can attach to your nodes. We will not get into these files for now as they are pretty specific to game development.

If you are interested in learning all the ins and outs of SpriteKit, go ahead pick up **Swift 3 Game Development - Second Edition, by Stephen Haney**. This book goes in-depth about game development on iOS with SpriteKit.

If you build and run the sample project that Xcode provides, you can tap the screen to make new sprite nodes appear on the screen. Each node performs a little animation before it disappears. This isn't very special in itself, but it does contain a lot of valuable information. For instance, it shows you how to add something to a scene and how to animate it. Let's see exactly how this project is set up so you can apply this knowledge in case you wish to build an AR experience with SpriteKit at some point.

Creating a SpriteKit scene

SpriteKit games use a special type of view to render its contents. This special view is always an instance or subclass of `SKView`. If you want to use SpriteKit with ARKit, you should use `ARSKView` instead because that view implements some special AR-related behavior, such as rendering the camera feed.

The view itself usually doesn't do much work regarding managing the games or its child views. Instead, the `SKScene` that contains the view is responsible for doing this work. This is similar to how you usually work with view controllers in other apps.

When you have created a scene, you can tell an `SKView` to present the scene. From this moment on, your game is running. In the sample code for the game project you created earlier, the following lines take care of loading and presenting the scene:

```
if let scene = SKScene(fileNamed: "GameScene") {
    scene.scaleMode = .aspectFill
    view.presentScene(scene)
}
```

When you create your scenes, you can choose whether you want to use `.sks` files or create scenes programmatically.

When you open the `GameScene.swift` file that Xcode created for you, most of the code should be pretty self-explanatory. When the scene is added to a view, a couple of `SKNode` instances are created and configured. The most interesting lines of code in this file are the following:

```
spinnyNode.run(SKAction.repeatForever(SKAction.rotate(byAngle:
CGFloat(Double.pi), duration: 1)))
spinnyNode.run(SKAction.sequence([SKAction.wait(forDuration: 0.5),
SKAction.fadeOut(withDuration: 0.5), SKAction.removeFromParent()]))
```

These lines set up an animation sequence for the spinning squares that get added when you tap the screen. In SpriteKit, actions are the preferred way to set up animations. You can group, chain, and combine actions to achieve pretty complicated effects. This is one of the many powerful tools that SpriteKit has to offer.

If you examine the code a little bit more, you'll find that copies of `spinnyNode` are created every time the user taps on the screen, moves their finger, or lifts their finger. Each interaction produces a slightly different copy of `spinnyNode` so you can determine why `spinnyNode` was added to the scene by looking at its appearance.

Study this code, play around with it, and try to make sure that you grasp what it does. You don't have to become a SpriteKit expert by any means. Let's have a look at how SceneKit works to prepare to implement your Augmented Reality gallery.

Exploring SceneKit

When you're looking for a game framework that has excellent support for 3D games, SceneKit is a great candidate. SceneKit is Apple's framework for creating 3D games, and it is structured very similarly to how SpriteKit is set up.

Of course, SceneKit is entirely different from SpriteKit because it's used for 3D games rather than 2D games. Because of this, SceneKit also has very different ways of creating views and positioning them on screen. For instance, when you want to create a simple object and place it on the screen, you will see terms such as geometry and materials. These terms should be familiar to game programmers, but if you're an AR enthusiast, you will probably have to get used to the terminology.

This section will walk you through setting up a straightforward SceneKit scene that closely resembles a part of the AR gallery you will implement later. This should provide you with enough information to begin experimenting with SceneKit.

If you're left longing for more SceneKit knowledge by the end of this chapter, make sure to have a look at **3D iOS Games by Tutorials written by Chris Language**. That book goes in-depth regarding 3D game development using SceneKit.

Creating a basic SceneKit scene

To practice your SceneKit knowledge, create a new project and instead of choosing the **Game** template, pick the **Single View Application** template. Of course, you are free to explore the default project Xcode creates for you when you choose the **Game** template with SceneKit, but it's not terribly useful for the AR gallery.

After creating your project, open the main storyboard and look for a **SceneKit view**. Drag this view into the view controller. You should notice that the view you just added to the view controller has replaced the default view entirely. Because of this, the `view` property on `ViewController` will not be a regular `UIView`, it will be an instance of `SCNView` instead. This is the view that will be used to render the SceneKit scene in.

Add the following code to `viewDidLoad()` in `ViewController.swift` to cast the `view` property from a `UIView` to an `SCNView`:

```
guard let sceneView = self.view as? SCNView
      else { return }
```

Similar to how SpriteKit works, SceneKit uses a scene to render its nodes in. Create an instance of `SCNScene` right after `guard` in `viewDidLoad()`, as shown:

```
let scene = SCNScene()
sceneView.scene = scene
sceneView.allowsCameraControl = true
sceneView.showsStatistics = true
sceneView.backgroundColor = UIColor.black
```

The preceding code creates a simple scene that will be used render all elements in. In addition to creating the scene, several debugging features are enabled to monitor the performance of the scene. Also, note that the `allowsCameraControl` property on the scene view is set to `true`. This will allow users to move a virtual camera around so they can explore the scene by swiping around in it.

Every SceneKit scene is viewed as if you're looking at it through a camera. You will need to add this camera to the scene yourself, and you must set it up appropriately for your purpose. The fact that SceneKit uses a camera is very convenient because the camera that you are going to set up in a second is replaced by the actual camera of a device when the scene is run with ARKit.

Add the following lines of code to `viewDidLoad()` to create and configure the camera:

```
let cameraNode = SCNNode()
cameraNode.camera = SCNCamera()
cameraNode.position = SCNVector3(x: 0, y: 0, z: 15)
scene.rootNode.addChildNode(cameraNode)
```

Setting up a basic camera isn't very complicated. All you need is an `SCNNode` to add the camera to and an `SCNCamera` that will be used to view your scene through. Note that the camera is positioned using an `SCNVector3` object. All nodes in a SceneKit scene use this object to express their positions in 3D space.

In addition to using a simulated camera, SceneKit also simulates real lighting conditions. When you run your scene with ARKit, the lighting conditions will be automatically managed by ARKit, making your objects look as if they truly are part of the environment. When you create a plain scene, however, you will need to add the lights yourself. Add the following lines of code to implement some ambient lighting:

```
let ambientLightNode = SCNNode()
ambientLightNode.light = SCNLight()
ambientLightNode.light!.type = .ambient
ambientLightNode.light!.color = UIColor.orange
scene.rootNode.addChildNode(ambientLightNode)
```

You can add different types of lights to a SceneKit scene. You can use ambient light as this sample does, but you can also add directional lights that focus on a particular direction, spotlight, or light points that light in all directions.

Now that you have lighting and a camera in place, you can add an object to the scene. You can use several pre-made shapes, also known as geometries, in your scene. Alternatively, you could import an entire 3D model in your scene. If you take a look at the default SceneKit app that Xcode generates if you create a new project with the **Game** template, you can see that it imports a 3D model of an airplane.

In the AR gallery you will build later, the artwork is augmented with digital information signs that are attached to the piece of art they belong to. To practice building such a sign, you will add a rectangular shape, or plane, to your SceneKit scene and place some text on top of it.

Add the following code to create a simple white plane, a node that renders the plane, and add it to the scene:

```
let plane = SCNPlane(width: 15, height: 10)
plane.firstMaterial?.diffuse.contents = UIColor.white
```

```
plane.firstMaterial?.isDoubleSided = true
plane.cornerRadius = 0.3

let planeNode = SCNNode(geometry: plane)
planeNode.position = SCNVector3(x: 0, y: 0, z: -15)
scene.rootNode.addChildNode(planeNode)
```

If you were to build and run your app now, you would see a white square that is positioned in front of the camera. By swiping on the scene, you can make the camera move around the plane to view it from all possible sides. Note that the plane appears to be quite large even though it was only set to be 15 wide and 10 height. You might have guessed that these numbers represent points on the screen, just like in other apps. In SceneKit, there is no concept of points. All values for size and distance must be specified in meters. This means that everything you do is done relative to other objects or their real-world sizes. Using real sizes is essential when you take your SceneKit knowledge to ARKit.

To add some text to the plane you just created, use the following code:

```
let text = SCNText(string: "Hello, world!", extrusionDepth: 0)
text.font = UIFont.systemFont(ofSize: 2.3)
text.isWrapped = true
text.containerFrame = CGRect(x: -6.5, y: -4, width: 13, height: 8)
text.firstMaterial?.diffuse.contents = UIColor.red

let textNode = SCNNode(geometry: text)
planeNode.addChildNode(textNode)
```

The preceding code creates a text geometry. Since all values in SceneKit are in meters, the text size will be a lot smaller than you would probably expect. To make sure the text is positioned properly in the plane, text wrapping is enabled, and a `containerFrame` is used to specify the bounds for the text. Since the origin for the text field will be in the center of the plane it is displayed on, the x and y positions are offset negatively from the center to make sure the text appears in the correct place. You can try to play with this frame to see what happens. After configuring the text, it is added to a node, and the node is added to the plane node.

If you run your app now, you can see the *Hello, World!* text rendered on the white plane you created before. This sample is an excellent taste of what you're going to create next. Let's dive straight into building your AR gallery!

Implementing an Augmented Reality gallery

Creating an excellent AR experience has been made a lot simpler with the great features that exist in ARKit. However, there still are several things to keep in mind if you want to build an AR experience that users will love.

Certain conditions, such as lighting, the environment, and even what the user is doing, can have an impact on the AR experience. In this section, you will implement an AR gallery, and you will discover firsthand how ARKit is both amazingly awesome and sometimes a little bit fragile.

First, you'll set up a session in ARKit so you can implement image tracking to discover certain predefined images in the world, and you'll show some text above the found picture. Then, you'll implement another feature that allows users to place art from a gallery in the app in their own room.

If you want to follow along with the steps to implement the ARKit gallery, make sure to grab the **ARGallery** start project from the book's code bundle. Before you move on to implementing the AR gallery, explore the starter project for a little bit. The user interface that is prepared contains an instance of `ARSCNView`; this is the view that will be used to render the AR experience. A collection view has been added in preparation of the user adding their own images to the gallery, and a view for error messages was added to inform the user about certain things that might be wrong.

You'll find that the project is quite basic so far. All the existing code does is set up the collection view, and some code was added to handle errors during the AR session. Let's implement image tracking, shall we?

Adding image tracking

When you add image tracking to your ARKit app, it will continuously scan the environment for images that match the ones you added to your app. This feature is great if you want users to look for specific images in their environment so you can provide more information about them or as part of a scavenger hunt. But more elaborate implementations might exist as part of a textbook or magazine where scanning a particular page would cause the whole page to come alive as part of a unique experience.

Before you can implement the image-tracking experience, you must prepare some images for your users to find in the app. Once the content is ready, you're ready to build the AR experience itself.

Preparing images for tracking

Adding images to your app that are eligible for image tracking is relatively straightforward. The most important part is that you pay close attention to the images you add to your app. It's up to you to make sure that the images you add are high-quality and well-saturated. ARKit will scan for special features in an image to try to match it, so it's important that your image has enough details, contrast, and colors. An image of a smooth gradient might look like a recognizable image to you, but it could be tough for ARKit to detect.

To add images to your project, go to the `Assets.xcassets` folder, click the + icon in the bottom-left corner, and select **New AR Resource Group**, as shown in the following screenshot:

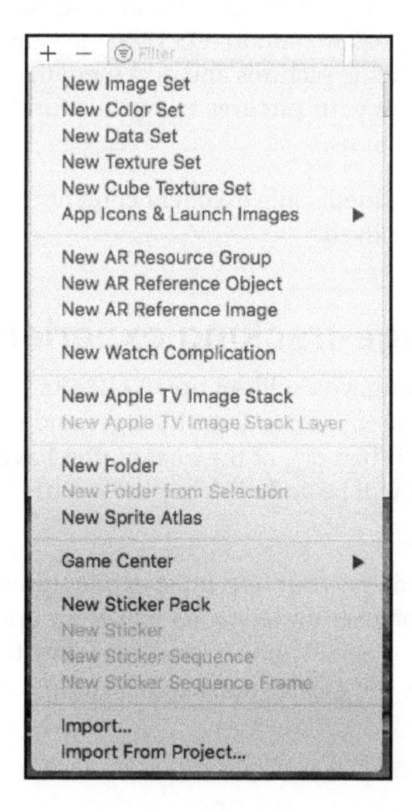

After adding a new resource group, you can drag images into the folder that was created. Each resource group will be loaded and monitored by ARKit all at once, so make sure you don't add too many images to a single resource group because that could negatively impact the performance of your app. Apple recommends you add up to about 25 images to a single resource group.

After you add an image to a resource group, Xcode will analyze the images and warn you if it thinks something is wrong with your image. Usually, Xcode will inform you as soon as you add a new image because ARKit requires the physical size of the image you want to detect to be known. So if you're going to detect a specific painting or a page in a magazine, you must add the dimensions for these resources in centimeters as they exist in the real world.

The start project from the code bundle comes with a couple of prepared images that you can explore to see some examples of the kinds of images that you could use in your own apps.

If you want to have some content of your own, take photos of artwork or pictures that you have around the house or office. You can use the Measure app in iOS 12 to measure the physical dimensions of the pictures and add them to your AR gallery project. Make sure that your pictures are well-saturated and free from any glare or reflections.

Once you have found and added some excellent content to use in your AR gallery, it's time to build the experience itself.

Building the image-tracking experience

To implement image tracking, you will set up an `ARSession` that uses `ARWorldTrackingConfiguration` to detect images and track a user's movement through the environment. When one of the images you have prepared is discovered in the scene, an `SCNPlane` will be added above the picture with a short description of the picture itself.

Because ARKit uses the camera, your app must explicitly provide a reason for accessing the camera, so the user understands why your app needs permission to use their camera. Add the `NSCameraUsageDescription` key to the `Info.plist` file and add a short text about why the gallery needs access to the camera.

If you open `ViewController.swift`, you will find a property called `artDescriptions`. Make sure to update this dictionary with the names of the images you added to the resource group, and add a short description for each image.

Next, update `viewDidLoad()` so `ViewController` is set as the delegate for both `ARSCNView` and `ARSession`. Add the following lines of code to do this:

```
arKitScene.delegate = self
arKitScene.session.delegate = self
```

The scene delegate and session delegate are very similar. The session delegate provides very fine-grained control of the content that is displayed in the scene, and you'll usually use this protocol extensively if you build your own rendering. Since the AR gallery is rendered using SceneKit, the only reason to adopt `ARSessionDelegate` is to respond to changes in the session's tracking state.

All of the interesting methods that you should adopt are part of `ARSCNViewDelegate`. This delegate is used to respond to specific events. For instance, when new features are discovered in the scene or when new content was added.

Currently, your AR gallery doesn't do much. You must configure the `ARSession` that is part of the scene to begin using ARKit. The best moment to set this all up is right before the view controller becomes visible. Therefore, you should do all of the remaining setup in `viewWillAppear(_:)`. Add the following implementation for this method to `ViewController`:

```
override func viewWillAppear(_ animated: Bool) {
  super.viewWillAppear(animated)

  // 1
  let imageSet = ARReferenceImage.referenceImages(inGroupNamed: "Art",
bundle: Bundle.main)!

  // 2
  let configuration = ARWorldTrackingConfiguration()
  configuration.planeDetection = [.vertical, .horizontal]
  configuration.detectionImages = imageSet

  // 3
  arKitScene.session.run(configuration, options: [])
}
```

The first step in this method is to read the reference image from the app bundle. These are the images you added to `Assets.xcassets`. Next, `ARWorldTrackingConfiguration` is created, and it's configured to track both horizontal and vertical planes, as well as the reference images. Lastly, the configuration is passed to the session's `run(_:options:)` method. If you run your app now, you should already be prompted for camera usage, and you should see the error-handling working. Try covering the camera with your hand, that should make an error message appear.

Keeping an AR session alive if a view isn't visible anymore is quite wasteful, so it's a good idea to pause the session if the app is closed or if the view controller that contains the AR scene becomes invisible. Add the following method to `ViewController` to achieve this:

```
override func viewWillDisappear(_ animated: Bool) {
  super.viewWillDisappear(animated)

  arKitScene.session.pause()
}
```

In the current setup, the AR session detects your images, but it does nothing to visualize this. When one of the images you added is identified, `ARSCNViewDelegate` is notified of this. To be specific, the `renderer(_:didAdd:for:)` method is called on the scene delegate when a new `SCNNode` is added to the view. For instance, when the AR session discovers a flat surface, it adds a node for `ARPlaneAnchor`, or when it detects one if the image you're tracking, a node for `ARImageAnchor` is added. Since this method can be called with different reasons, it's essential that you add logic to differentiate between the various reasons that could cause a new `SCNNode` to be added to the scene.

Because the AR gallery will implement several other features that could trigger the addition of a new node, you should separate the different actions you want to take for each different type of anchor into specialized methods. Add the following method to `ARSCNViewDelegate` to add the information plane next to a detected image:

```
func placeImageInfo(withNode node: SCNNode, for anchor: ARImageAnchor)
{
  let referenceImage = anchor.referenceImage

  // 1
  let infoPlane = SCNPlane(width: 15, height: 10)
  infoPlane.firstMaterial?.diffuse.contents = UIColor.white
  infoPlane.firstMaterial?.transparency = 0.5
  infoPlane.cornerRadius = 0.5
```

```
  // 2
  let infoNode = SCNNode(geometry: infoPlane)
  infoNode.localTranslate(by: SCNVector3(0, 10, -
referenceImage.physicalSize.height / 2 + 0.5))
  infoNode.eulerAngles.x = -.pi / 4

  // 3
  let textGeometry = SCNText(string:
artDescriptions[referenceImage.name ?? "flowers"], extrusionDepth:
0.2)
  textGeometry.firstMaterial?.diffuse.contents = UIColor.red
  textGeometry.font = UIFont.systemFont(ofSize: 1.3)
  textGeometry.isWrapped = true
  textGeometry.containerFrame = CGRect(x: -6.5, y: -4, width: 13,
height: 8)

  let textNode = SCNNode(geometry: textGeometry)

  // 4
  node.addChildNode(infoNode)
  infoNode.addChildNode(textNode)
}
```

The preceding code should look somewhat familiar to you. First, an instance of `SCNPlane` is created. Then, this plane is added to `SCNNode`. This node is translated slightly to position it above the detected image. This translation uses `SCNVector3` so it can be translated into three dimensions. The node is also rotated a little bit to create a nice-looking effect.

Next, add the following implementation for `renderer(_:didAdd:for:)`:

```
func renderer(_ renderer: SCNSceneRenderer, didAdd node: SCNNode, for
anchor: ARAnchor) {
  if let imageAnchor = anchor as? ARImageAnchor {
    placeImageInfo(withNode: node, for: imageAnchor)
  }
}
```

This method checks whether the anchor that was discovered is an image anchor; if it is, `placeImageInfo(withNode:for:)` is called to display the information sign.

Go ahead and run your app now! When you find one of the images that you added to your resource group, an information box should appear on top of it as shown in the following screenshot:

Pretty awesome, right? Let's take it one step further and allow users to position some of the pictures from the collection view wherever they want in the scene.

Placing your own content in 3D space

To spice up the AR gallery a little bit, it would be great to be able to add some new artwork to the environment. Using ARKit, doing this becomes relatively simple. There are a couple of gotchas to take into account when implementing a feature such as this, but overall Apple did a great job making ARKit an accessible platform to work with for developers.

When a user taps on one of the images in the collection view at the bottom of the screen, the image they tapped should be added to the environment. If possible, the image should be attached to one of the walls surrounding the user. If this isn't possible, the image should still be added, except it will float in the middle of the space.

To build this feature, you should implement `collectionView(_:didSelectItemAt:)` since this method is called when a user taps on one of the items in a collection view. When this method is called, the code should take the current position of the user in the environment and then insert a new `ARAnchor` that corresponds to the location where the new item should be added.

Also, to detect nearby vertical planes, such as walls, some hit testing should be done to see whether a vertical plane exists in front of the user. Add the following implementation of `collectionView(_:didSelectItemAt:)` to implement this logic:

```
func collectionView(_ collectionView: UICollectionView,
didSelectItemAt indexPath: IndexPath) {
  // 1
  guard let camera = arKitScene.session.currentFrame?.camera
    else { return }

  // 2
  let hitTestResult = arKitScene.hitTest(CGPoint(x: 0.5, y: 0.5),
types: [.existingPlane])
  let firstVerticalPlane = hitTestResult.first(where: { result in
    guard let planeAnchor = result.anchor as? ARPlaneAnchor
      else { return false }

    return planeAnchor.alignment == .vertical
  })

  // 3
  var translation = matrix_identity_float4x4
  translation.columns.3.z = -Float(firstVerticalPlane?.distance ?? -1)
  let cameraTransform = camera.transform
  let rotation = matrix_float4x4(cameraAdjustmentMatrix)
  let transform = matrix_multiply(cameraTransform,
matrix_multiply(translation, rotation))

  // 4
  let anchor = ARAnchor(transform: transform)
  imageNodes[anchor.identifier] = UIImage(named:
images[indexPath.row])!
```

```
        arKitScene.session.add(anchor: anchor)
    }
```

Even though there are only four steps in this snippet, a lot is going on. First, the camera is grabbed from the current frame in the AR session so it can be used later to determine the user's location in the scene.

Next, a hit test is performed to see whether any planes were already detected in the scene. Since this hit test will return both vertical and horizontal planes, the results are filtered to find the very first vertical plane that was found in the hit test.

Since the location of every `ARAnchor` is represented as a transformation from the world origin, the third step is to determine the transformation that should be applied to position the new artwork in the correct place. The world origin is the place where the AR session first became active.

After creating a default translation, the z value for the translation is adjusted, so the object is added either in front of the user or against the nearest vertical plane. Next, the current position of the user is retrieved through the camera. The rotation for the camera will have to be adjusted in the next steps because the camera does not follow the device's orientation. This means that the camera will always assume that the x-axis runs across the length of the device, starting at the top and moving downward towards the home indicator area. A computed property to determine how the rotation should be adjusted is already added to the AR gallery starter project.

After setting up the correct transformation properties for the anchor, an instance of `ARAnchor` is created. The unique identifier and image that the user tapped are then stored in the `imageNodes` dictionary so the image can be added to the scene after the new anchor is registered on the scene.

To add the image to the scene, you should implement a helper method that will be called from `rendered(_:didAdd:for:)`, similar to the helper method you added to show the information card for the image-tracking feature. Add the following code to `ViewController` to implement this helper:

```
func placeCustomImage(_ image: UIImage, withNode node: SCNNode) {
  let plane = SCNPlane(width: image.size.width / 1000, height:
image.size.height / 1000)
  plane.firstMaterial?.diffuse.contents = image

  node.addChildNode(SCNNode(geometry: plane))
}
```

To make it easier to see whether an appropriate vertical plane exists, you can implement a helper method that visualizes the planes that the AR session discovers. Add the following code to the `ViewController` class to implement this helper:

```
func vizualise(_ node: SCNNode, for planeAnchor: ARPlaneAnchor) {
  let infoPlane = SCNPlane(width: CGFloat(planeAnchor.extent.x),
height: CGFloat(planeAnchor.extent.z))
  infoPlane.firstMaterial?.diffuse.contents = UIColor.orange
  infoPlane.firstMaterial?.transparency = 0.5
  infoPlane.cornerRadius = 0.2

  let infoNode = SCNNode(geometry: infoPlane)
  infoNode.eulerAngles.x = -.pi / 2

  node.addChildNode(infoNode)
}
```

The previous method takes a node and anchor to create a new `SCNPlane`, which is added to the exact position where the new plane anchor was discovered.

The final step to implementing this feature is to call the helper methods when needed. Update the implementation for `renderer(_:didAdd:for:)` as follows:

```
func renderer(_ renderer: SCNSceneRenderer, didAdd node: SCNNode, for
anchor: ARAnchor) {
  if let imageAnchor = anchor as? ARImageAnchor {
    placeImageInfo(withNode: node, for: imageAnchor)
  } else if let customImage = imageNodes[anchor.identifier] {
    placeCustomImage(customImage, withNode: node)
  } else if let planeAnchor = anchor as? ARPlaneAnchor {
    vizualise(node, for: planeAnchor)
  }
}
```

If you run your app now, you should see orange squares appear in areas where ARKit detected a flat surface. Note that ARKit needs textures and visual markers to work well. If you try to detect a solid white wall, it's unlikely that ARKit will properly recognize the wall due to a lack of textures. However, a brick wall or a wall that has wallpaper with some graphics on it should work well for this purpose.

The following image shows an example where an image is attached to a wall, together with the plane indicator:

This wraps up the implementation of your own personal AR gallery. There still is much to learn about the things you can do with AR, so make sure to keep on experimenting and learning so you can create amazing experiences for your users.

Summary

In this chapter, you learned a lot. You gained some insight into what AR is, the basic workings of AR, and what you can do with it. Then you learned about the components that make up an excellent AR experience, and you implemented your first small AR experience by adopting Quicklook in an app to preview AR content in a real AR session.

Then you explored different ways to render content in an AR scene. You took a quick look at SpriteKit and SceneKit, and learned that SpriteKit is Apple's 2D game-development framework. You also learned that SceneKit is Apple's 3D game framework, which makes it extremely suited for usage in an AR app.

Then you implemented an AR gallery that uses image tracking, plane detection, and allows users to add their own contents to their gallery. In the process of doing this, you saw that it's not always easy to get ARKit to work well. Bad lighting and other factors can make AR experiences less than ideal.

In the next chapter, you will learn about location services and how you can use them to create better experiences for your users.

Questions

1. What is SceneKit typically used for?

   a) 2D games.
   b) Only for AR.
   c) 3D games.

2. What is a view in a SpriteKit game called?

   a) A view.
   b) A sprite.
   c) A node.

3. What unit of measurement do SceneKit and ARKit use?

   a) Points.
   b) Inches.
   c) Centimeters.

4. Which tracking mode tracks both device movement and rotation?

   a) Image tracking.
   b) Orientation tracking.
   c) World tracking.

5. What should you keep in mind when you prepare images for image tracking in ARKit?

   a) The images should be colorful.
   b) The images should be big.
   c) The images should have sufficient features.

6. What does ARKit need to accurately track objects?

   a) A low-light environment.
   b) Smooth surfaces.
   c) Surfaces with lots of textures.

Further reading

- Swift 3 Game Development – Stephen Haney: `https://www.packtpub.com/application-development/swift-3-game-development-second-edition`
- 3D iOS Games by Tutorials – Chris Language: `https://store.raywenderlich.com/products/3d-apple-games-by-tutorials`
- Apple's MultipeerConnectivity framework: `https://developer.apple.com/documentation/multipeerconnectivity`

13
Improving Apps With Location Services

All iOS devices come with a huge variety of chips and sensors that can be used to enhance the user experience. Augmented reality applications make heavy use of sensors such as the gyroscope, accelerometer, and camera. These sensors are great if you want to grab a picture or want to know how a device is moving. Other apps require different data, like, for instance, the user's GPS location at a given time. In this chapter, you will learn how to use the `CoreLocation` framework to do just that.

Core Location is a framework that allows developers to gain access to a user's current location, but it also allows developers to track whether a user has entered or exited a specific area, or even to monitor a user's location over time. A proper implementation of Core Location can be the core of many great features in your app, but a lousy implementation could drain a user's battery in no time.

In this chapter, you will learn about the following location-related topics:

- Requesting a user's location
- Subscribing to location changes
- Setting up geofences

If you want to follow along with the code samples and implementations that are provided throughout this chapter, make sure to grab the **LocationServices** start project from this book's code bundle. By the end of the chapter, you should be able to make educated decisions about how and when you implement Core Location in your apps.

Requesting a user's location

As you can imagine, giving an application access to your exact location is quite a big deal. In the wrong hands, this data could allow people with malicious intentions to know exactly where you are at any given time, and abuse this knowledge in many different ways. For this reason, it's essential that you only request a user's location if you absolutely have to. Simply using it for a small feature, or to make sure a user is in some arbitrary location before they sign up for a service, might not always be a good enough reason to ask for a user's location.

Asking for permission to access location data

When you do need access to a user's location, you have to ask permission first. Similar to how you have to add a reason for needing the camera or a user's contacts to the `Info.plist` file, you must also provide a reason for requesting location data. In the case of location data, there are two keys you can add to the `Info.plist`:

- Privacy-Location When In Use Usage Description (`NSLocationWhenInUseUsageDescription`)
- Privacy-Location Always And When In Use Usage Description (`NSLocationAlwaysAndWhenInUseUsageDescription`)

When your app asks for permission to use a user's location data, they have the option to only allow your app access to their location when the app is in use, or they can choose to allow your app to access their location all the time, even when the app is in the background. You can also configure the type of access that you want to ask yourself. If you only need the user's location when they are using the app, make sure to configure your permission request properly, so the user isn't asked to provide their location to your app when it's in the background.

After adding the required keys to the `Info.plist` file in the LocationServices app, you will need to write some code to ask the user for permission to use their location. Before doing this, let's quickly examine the sample project's structure and content, so you are aware of what information can be found where.

First, open the `Main.storyboard` file in the project. You will find a tab bar controller with two view controllers in it. Throughout this chapter, you will implement the features to populate these view controllers with appropriate data. Next, look at the `AppDelegate` implementation. The implementation here follows the Dependency Injection pattern that you have seen used in earlier projects, so there shouldn't be any surprises for you here. Now, go ahead and examine the view controllers. The `GeofenceViewController` is the one you will work on first, to make the user's current location appear on the screen.

You will notice that a lot of code has already been implemented in this view controller. Examine the existing code for a bit, and you'll find that all the code makes calls to empty methods in `LocationHelper.swift`. Most of your focus in this chapter will be on implementing the Core Location code required to work with a user's location data, so the UI work has already been set up. As you add code to `LocationHelper`, you'll find that the user interface for LocationServices comes to life bit by bit.

Now that you have a better understanding of how the LocationServices app was set up, let's see what steps are involved in asking the user for permission to use their location. Since this app will eventually track location changes in the background, you should ask the user access to their location even when the app is in the background. To do this, add the following `viewDidAppear(_:)`
code to `GeofenceViewController`:

```
locationHelper.askPermission { [weak self] status in
  if status == .authorizedAlways {
    self?.showCurrentLocation()
  } else {
    // handle the case where you don't always have access
  }
}
```

This is the first view controller the user will see, so asking the user for their location as soon is this view appears is a good idea. If it's not obvious that you will be prompting the user for their location, it's often a good idea to inform the user about why you are going to ask them for location permissions before actually showing the location access dialog. To actually make the permission dialog appear, you will need to add some code to `LocationHelper.swift`.

All location service-related requests are performed through an instance of `CLLocationManager`. The location manager is responsible for obtaining the user's GPS location, asking for permission to access the user's location, and more. When the location manager receives updates about the user's location, authorization status, or other events, it will notify its delegate. A location manager delegate should conform to the `CLLocationManagerDelegate` protocol. Note that the `LocationHelper` already conforms to `CLLocationManagerDelegate`, and that an instance of `CLLocationManager` is already created on this object. All that's left to do is assign the helper as the delegate for the location manager. Add the following line at the end of the `init()` method in `LocationHelper` to set it as the location manager delegate:

```
locationManager.delegate = self
```

Next, add the following implementation for the `askPermission(_:)` method:

```
func askPermission(_ completion: @escaping (CLAuthorizationStatus) ->
Void) {
  let authorizationStatus = CLLocationManager.authorizationStatus()
  if authorizationStatus != .notDetermined {
    completion(authorizationStatus)
  } else {
    askPermissionCallback = completion
    locationManager.requestAlwaysAuthorization()
  }
}
```

This implementation checks whether a current authorization status exists. If it does, the completion callback is called with the current status. If the current status has not been determined yet, the location manager is asked to request authorization to access the user location using the `requestAlwaysAuthorization()` method. This will prompt the user for their location permissions. The reason you need to have permanent access to a user's location in this app is to ensure you can implement geofencing later in this chapter. Add the following method to the `CLLocationManagerDelegate` to retrieve the user's response to the authorization prompt:

```
func locationManager(_ manager: CLLocationManager,
didChangeAuthorization status: CLAuthorizationStatus) {
  askPermissionCallback?(status)
  askPermissionCallback = nil
}
```

The preceding code immediately passes the user's response to the stored completion callback that was passed to `askPermission(_:)`. After calling the callback, it is set to `nil` to avoid accidentally calling it again. At this point, you have done all the work required to request access to a user's location. Let's see how you can retrieve a user's current location next.

Obtaining a user's location

Once your app has access to location data, you can use the location manager to begin observing a user's location, the direction in which a user is heading, and more. For now, you will focus on obtaining the user's current location. The `GeofenceViewController` already contains a method, called `showCurrentLocation()`, that is responsible for asking the location helper for a current location. If you examine this method closely, you'll find that it also asks the location helper for a location name by calling `getLocationName(for:_:)` and passing the obtained location to this method. The `showCurrentLocation()` method also uses the obtained location to focus a map view on the user's location by calling `setRegion(_:animated:)` on the map view.

Since the view controller is already fully prepared to handle location updates, all you need to do is add the proper implementations for `getLatestLocation(_:)` and `getLocationName(for:_:)`. Begin by adding the following implementation for `getLatestLocation(_:)`:

```
func getLatestLocation(_ completion: @escaping (CLLocation) -> Void) {
  if let location = trackedLocations.last {
    completion(location)
  } else if CLLocationManager.locationServicesEnabled() {
    latestLocationObtainedCallback = completion
    locationManager.startUpdatingLocation()
  }
}
```

The preceding method first checks whether a location has already been obtained. If it has, then the latest obtained location is returned. If there is no existing location, the code checks whether location services are enabled. It's always good practice to check whether the location service you are about to use is actually available. If location services are available, the completion callback is stored in the helper, and the location manager is told to start monitoring the user's location by calling `startUpdatingLocation()`.

Calling `startUpdateLocation()` will make the location observer continuously monitor the user's GPS location, and will send any relevant updates to its delegate by calling `locationManager(_:didUpdateLocations:)`. This method will always receive one or more new locations that the manager has obtained, where the latest location will be the last item in the list of obtained locations. Add the following implementation for this method to the `CLLocationManagerDelegate` extension of `LocationHelper`:

```
func locationManager(_ manager: CLLocationManager, didUpdateLocations
locations: [CLLocation]) {
  latestLocationObtainedCallback?(locations.last!)
  latestLocationObtainedCallback = nil
  locationManager.stopUpdatingLocation()

  trackedLocations += locations
}
```

The implementation for `locationManager(_:didUpdateLocations:)` is fairly straightforward: the latest location is passed to the callback, and the callback is removed to prevent subsequent location updates from triggering the callback unexpectedly. Also, the location manager is told to stop monitoring the user's location by calling `stopUpdatingLocation()`. Lastly, the obtained locations are stored for later use.

It's always good practice to make the location manager stop monitoring location updates if you won't be needing updates any time soon. Monitoring location updates has a pretty significant impact on battery life, so you shouldn't spend more time tracking a user's location than needed.

Now that you can retrieve the user's location, the last step is to also retrieve the location name, by implementing `getLocationName(for:_:_)` in the location helper. Add the following implementation for this method to the location helper:

```
func getLocationName(for location: CLLocation, _ completion: @escaping
(String) -> Void) {
  let geocoder = CLGeocoder()
  geocoder.reverseGeocodeLocation(location) { placemarks, error in
    guard error == nil else {
      completion("An error ocurred: \(error?.localizedDescription ??
"Unknown error")")
      return
    }

    completion(placemarks?.first?.name ?? "Unkown location")
```

```
        }
    }
```

The preceding code uses a `CLGeocoder` to find a placemark that corresponds with the user's current location. Note that this feature uses an internet connection, so the name lookup will only work if the user has an internet connection. Regular GPS-related features do not require internet access, so your app can monitor and track a user's location even if they don't have an active internet connection.

Try running your app now—you should be able to see the user's current location on the map, and the location name, latitude, and longitude should be displayed on the screen as well. Now that you know how to obtain a user's location, let's see how you can efficiently subscribe your app to follow changes in a user's location.

Subscribing to location changes

One way of subscribing to changes in a user's location has already been covered in the previous section of this chapter. When you call `startUpdatingLocation()` on a location manager, it will automatically subscribe to the user's location. This method of tracking a user's location is excellent if you need very detailed reporting on a user's location, but usually, you don't need this level of detail. More importantly, using this kind of location tracking for an extended period will drain the user's battery.

Luckily, there are better ways to listen to location changes. One way is to subscribe to locations that the user visits by calling `startMonitoringVisits()`. This method is used if you aren't interested in the user's detailed movement but only want to know whether the user spent an extended period in a particular area. This type of tracking of a user's location is perfect if you need a low-power way to track very rough location changes. This kind of tracking even works well if your app is running in the background, because your app will automatically be woken up or launched if a visit event occurs.

If your app is relaunched due to a location-related event, then `UIApplication.LaunchOptionsKey.location` will be present in the application's launch options dictionary. When it is, you are expected to create an instance of a location manager and assign it a delegate to receive the relevant location update.

If the visits monitoring is a bit too inaccurate for your purposes but you don't need continuous location tracking, you can use significant location change tracking. This type of tracking triggers when a user has moved a significant distance over time, providing your app with updates only when the user is truly moving. This is a lot more power efficient than making your app track the user's location even when their current location hasn't changed. Just like visits tracking, significant location changes will wake up your app or even relaunch it when they occur. When an event is delivered to the app like this, you should re-enable the significant location changes monitoring. Let's implement significant location changes in the LocationServices sample app to see exactly how they work .

If you look at `SignificantChangesViewController`, you'll note that the view controller is fully set up to begin monitoring significant location changes. The `monitorSignificantChanges(_:)` method, defined on the location helper, takes a callback that's called every time a significant location change occurs. Every time new location data is retrieved, the table view is reloaded to display the latest available data. Since significant location updates can wake the app up with a special key in the app's launch options, let's update `AppDelegate` so it can handle this scenario. Add the following code `application(_:didFinishLaunchingWithOptions:)` right before the return statement:

```
if launchOptions?[UIApplication.LaunchOptionsKey.location] != nil {
  locationHelper.monitorSignificantChanges { _ in
    // continue monitoring
  }
}
```

Since `AppDelegate` already has a reference to the location helper, all it needs to do is re-enable significant location changes monitoring. This small change to `AppDelegate` is quite powerful because it allows your app to respond to changes in the user's location even when the app is not running. Let's implement the appropriate code in the location helper next.

Add the following implementation for `monitorSignificantLocationChanges(_:)` to `LocationHelper`:

```
func monitorSignificantChanges(_ locationHandler: @escaping
(CLLocation) -> Void) {
  guard
CLLocationManager.significantLocationChangeMonitoringAvailable()
    else { return }

  significantChangeReceivedCallback = locationHandler
  locationManager.startMonitoringSignificantLocationChanges()
```

```
    isTrackingSignificantLocationChanges = true
}
```

This method is very similar to the location helper methods you have seen before. When a significant location change is detected, the location manager calls `locationManager(_:didUpdateLocations:)` on its delegate. Since this method is already implemented, you should update the implementation as follows:

```
func locationManager(_ manager: CLLocationManager, didUpdateLocations
locations: [CLLocation]) {
  latestLocationObtainedCallback?(locations.last!)
  latestLocationObtainedCallback = nil

  if isTrackingSignificantLocationChanges == false {
    locationManager.stopUpdatingLocation()
  }

  significantChangeReceivedCallback?(locations.last!)

  trackedLocations += locations
}
```

Note that the location manager is only told to stop updating the user's location when significant location tracking is not active. When you call `stopUpdatingLocation()`, the location manager will cease to deliver any location updates to this delegate method. Also, note that `significantChangeReceivedCallback` is not removed after calling it. The reason for this is that the caller of `monitorSignificantChanges(_:)` is interested in continuous location updates, so any time this method is called, the `SignificantChangesViewController` view controller that initiated significant location tracking should always be called.

One last thing you need to do, so your app receives significant location changes while it's not in the foreground, is to set the `allowsBackgroundLocationUpdates` property to `true`. Add the following line of code to the location helper's `init()`:

```
locationManager.allowsBackgroundLocationUpdates = true
```

In addition to subscribing to significant location changes or visits, you can also respond to the user entering or leaving a certain area with geofences.

Setting up geofences

Sometimes, your app doesn't really need to know the details of the user's whereabouts. Sometimes, you're only interested in tracking whether the user has exited or left a certain area, in order to show certain content in your app or to unlock some kind of special feature. Core Location has great support for monitoring geofences. A geofence is a certain area that is defined using a certain GPS coordinate and a circular radius around this point. In Core Location, geofences are set up using `CLRegion` subclasses. Core Location provides two different region types that you can use:

- `CLCircularRegion`
- `CLBeaconRegion`

A `CLCircularRegion` type is used to set up a geofence, as described before. A `CLBeaconRegion` type is used with physical BLE iBeacons, and essentially provides geofencing in a very small radius, for instance, just a couple of meters. In this section you will learn how to set up a `CLCircularRegion` type that is set up around a user's first detected location. Setting up geofencing, or region monitoring, with both types of regions is very similar so all principles for monitoring a circular region also applies to beacon regions.

If you look at the `GeofenceViewController`, you'll notice that it has a button labelled **Set Geofence**. The `@IBAction` for this button does quite a lot of the work already, but one key element is missing—it doesn't inform the location manager about the region that should be monitored. Add the following code to the end of `setGeofence()` in `GeofenceViewController`:

```
let region = CLCircularRegion(center: location.coordinate, radius: 30,
identifier: "current-location-geofence")
locationHelper.setGeofence(at: region, exitHandler, enterHandler)
```

The preceding code uses the location that was obtained from the user before, and uses it to create a circular region with a radius of 30 meters. The identifier that is passed to the region should be an identifier that uniquely defines the region. If you reuse an identifier, Core Location will stop monitoring the old region with that identifier, and will monitor the new region instead. For the LocationServices app, this is perfect, but if you want your app to observe multiple regions, you must make sure every region has its own unique identifier.

Next, add the following implementation for `setGeofence(at:_:_:)` to the `LocationHelper`:

```
func setGeofence(at region: CLRegion, _ exitHandler: @escaping () ->
Void, _ enterHandler: @escaping () -> Void) {
  guard CLLocationManager.isMonitoringAvailable(for:
CLCircularRegion.self)
    else { return }

  geofenceExitCallback = exitHandler
  geofenceEnterCallback = enterHandler
  locationManager.startMonitoring(for: region)
}
```

The preceding method is again very similar to the other location helper methods. Let's move right on to implementing the `CLocationManagerDelegate` methods that the location manager will call:

```
func locationManager(_ manager: CLLocationManager, didEnterRegion
region: CLRegion) {
  geofenceEnterCallback?()
}

func locationManager(_ manager: CLLocationManager, didExitRegion
region: CLRegion) {
  geofenceExitCallback?()
}
```

The preceding two methods are part of the `CLocationManagerDelegate` protocol, and are called when a user enters or exits a certain area. Since there's no extra work to be done by the helper, the corresponding callbacks are immediately called so the `GeofenceViewController` can update its interface accordingly.

Try opening the app and tapping the **Set Geofence** button. An orange circle should now appear on the map to visualize the geofence you have set up. If you exit or enter the region, the status label should update accordingly, to show whether you have just entered or left the geofence. Note that it might take up to five minutes for iOS to properly register, monitor, and report updates about your geofence. Note that your user should have an active internet connection for region monitoring to work optimally.

Summary

In this chapter, you have learned several techniques to obtain and respond to a user's location. You have implemented a `LocationHelper` class that provided a simple interface for view controllers to use the location manager that is contained in the helper. You learned about the best practices in the area of asking the user for access to their location data, and you have learned that asking for a user's location is a pretty privacy-sensitive question that shouldn't be asked without a good reason.

You learned that there are different ways, each with different levels of detail, that you can use to track a user's location. You saw that you can subscribe to continuous changes, which has a bad impact on battery life. You also learned about subscribing to visits and significant location changes. In addition to learning about tracking a user's location, you also learned about monitoring whether a user has entered or exited a certain area by implementing geofencing. When you implement Core Location in your own apps, always make sure to keep the user's privacy in mind. If you don't really need the location data, then don't request access to it. And if you do, make sure to handle your user's location data with great care.

In the next chapter, you will learn about the CoreML framework and how you can use it to enhance your apps.

Questions

1. What different types of location access exist?

    a) Always and When in use
    b) Always and When in background
    c) When in use and When in background

2. When should you ideally ask for access to location services?

    a) When your app launches
    b) As soon as you want access to a user's location
    c) Once the user understands why you need access to their location

3. How do you obtain a user's current location?

    a) By calling `getCurrentLocation()` on the location manager
    b) By calling `startUpdatingLocation()` on the location manager
    c) By calling `startMonitoring(for:)` on the location manager

4. Which of the following location tracking techniques is the most battery friendly?

    a) Significant location change tracking
    b) Continuous location tracking
    c) Visit tracking

5. What different types of regions can you monitor?

    a) Geofences and iBeacons
    b) Geofences and WiFi networks
    c) WiFi networks and iBeacons

6. Up to how long could it take for region monitoring to start?

    a) 5 minutes
    b) 10 minutes
    c) 2 minutes

7. How can you check whether your app launched due to a significant location change?

    a) By checking whether a new location is available
    b) By inspecting the app's launch options
    c) By implementing a special AppDelegate method

Further reading

- Apple's Core Location documentation can be found at `https://developer.apple.com/documentation/corelocation`

14
Making Smarter Apps with CoreML

Over the past few years, machine learning has gained in popularity. However, it has never been easy to implement in mobile applications—that is, until Apple released the **CoreML** framework as part of iOS 11. CoreML is Apple's solution to all of the problems they have run into themselves while implementing machine learning for iOS. As a result, CoreML should have the fastest, most efficient implementations for working with complex machine learning models, through an interface that is as simple and flexible as possible.

In this chapter, you will learn what machine learning is, how it works, and how you can use trained machine learning models in your apps. You will also learn how you can use Apple's *Vision framework* to analyze images, and you'll see how it integrates with CoreML for powerful image detection. Lastly, you'll learn how to use the new **CreateML** tool to train your own models. You will learn about these topics in the following sections:

- Understanding machine learning and CoreML
- Combining CoreML and computer vision
- Training your own models with CreateML

By the end of this chapter, you will be able to train and use your own CoreML models to make the apps you build more intelligent and compelling.

Understanding machine learning and CoreML

Machine learning and CoreML go hand in hand, but they're not quite the same. Machine learning is all about teaching a machine how it can recognize, analyze, or apply certain things. The result of all this teaching is a trained model that can be used by CoreML to analyze specific inputs and produce an output based on the rules that were established during the training phase.

Before you learn about CoreML, it's good to obtain some knowledge about machine learning to make sure you're familiar with some of the terms that are used, and so you know what machine learning is.

Understanding what machine learning is

A lot of developers will hear about machine learning, deep learning, or neural networks at some point in their career. You may have already heard about these topics. If you have, you know that machine learning is a complex field that requires particular domain knowledge. However, machine learning is becoming more prominent and popular by the day, and it is used to improve many different types of applications.

For instance, machine learning can be used to predict what type of content a particular user might like to see in a music app, based on music that they already have in their library, or to automatically tag faces in photos to connect them to people in the user's contact list. It can even be used to predict costs for specific products or services based on past data. While this might sound like magic, the flow for creating machine learning experiences like these can be split roughly into two phases:

1. Training a model
2. Using inference to obtain a result from the model

Large amounts of high-quality data must be collected to perform the first step. If you're going to train a model that should recognize cats, you will need large amounts of pictures of cats. You must also collect images that do not contain cats. Each image must then be appropriately tagged to indicate whether the image includes a cat or not.

If your dataset only contains images of cats that face towards the camera, the chances are that your model will not be able to recognize cats from a sideways point of view. If your dataset does contain cats from many different sides, but you only collected images for a single breed or with a solid white background, your model might still have a tough time recognizing all cats. Obtaining quality training data is not easy, yet it's essential.

During the training phase of a model, it is imperative that you provide a set of inputs that are of the highest quality possible. The smallest mistake could render your entire dataset worthless. It's in part due to the process of collecting data that training a model is a tedious task. One more reason is that training a model typically takes a lot of time. Certain complex models could take a couple of hours to crunch all the data and train themselves.

A trained model comes in several types. Each type of model is suitable for a different kind of task. For instance, if you are working on a model that can classify certain email messages as spam, your model might be a so-called **support vector machine**. If you're training a model that recognizes cats in pictures, you are likely training a **neural network**.

Each model comes with its own pros and cons, and each model is created and used differently. Understanding all these different models, their implications, and how to train them is extremely hard, and you could likely write a book on each kind of model.

In part, this is why CoreML is so great. CoreML enables you to make use of pre-trained models in your own apps. On top of this, CoreML standardizes the interface that you use in your own code. This means that you can use complex models without even realizing it. Let's learn more about CoreML, shall we?

Understanding CoreML

Due to the complex nature of machine learning and using trained models, Apple has built CoreML to make incorporating a trained model as simple as possible. On top of this, another goal was to ensure that whenever you implement machine learning using CoreML, your implementation is as fast and energy efficient as possible. Since Apple has been enhancing iOS with machine learning for a couple of years now, they have loads of experience of implementing complex models in apps.

If you have ever researched machine learning, you might have come across cloud-based solutions. Typically, you send a bunch of data to such a cloud-based solution, and the result is passed back as a response to your request. CoreML is very different, since the trained model lives on the device, instead of in the cloud. This means that your user's data never has to leave the device, which is very good for your user's privacy. Also, having your trained model on the device means that no internet connection is required to use CoreML, which saves both time and precious data. And since there is no potential bottleneck regarding response latency, CoreML is capable of calculating results in real time.

In the previous section, you learned that there are several types of trained models. Each type of model is used slightly differently, so if you were to implement machine learning in your app manually, you would have to write different wrappers around each of the different models your app uses. CoreML makes sure that you can use each type of model without even being aware of this in your app; they all share the same programming interface. A CoreML model is domain agnostic.

To be domain agnostic, all trained models that you use with CoreML must be in a particular format. Since machine learning already has a vibrant community with several popular formats, Apple has made sure that the most popular models can be easily converted to Apple's own `.mlmodel` format. Let's see how to obtain `.mlmodel` files for you to use in your own apps.

Obtaining CoreML models

The are two ways to obtain a model for you to use in your apps. The simplest way is to find an existing `.mlmodel` file. You can find several ready-to-use `.mlmodel` files on Apple's machine learning website, at `https://developer.apple.com/machine-learning/`. This website contains several of the most popular models. At the time of writing, most of these models are focused on recognizing dominant objects in an image, and chances are that you have different needs for your app.

If you're looking for something that isn't already converted by Apple, you can try to look in several places online for a pre-converted `.mlmodel` file, or you can convert an existing model you have found online. Apple has created converters for several popular machine learning formats, such as *caffe*. The conversion tools for converting an existing model to a `.mlmodel` file are written in Python, and they ship as part of Xcode. If your needs do not fit the converters that Apple provides, you can extend the **toolchain**, since the conversion tools are open source. This means that everybody can add their own converters, or tweak existing converters.

Converting CoreML models using Apple's tools can usually be done with a couple of lines of Python. Writing a good conversion script does typically involve a little bit of domain knowledge in the area of machine learning, because you'll need to make sure that the converted model works just as well as the original model.

Once you have obtained a CoreML model for your app, either by converting one or finding an existing one, you're ready to add it to your project and begin using it. Let's see how to do this next.

Using a CoreML model

Applications can utilize CoreML for many different purposes. One of these purposes is text analysis. You can use a trained model to detect whether a particular piece of text has a positive or negative sentiment. To implement a feature like this, you can use a trained and converted CoreML model.

The code bundle for this chapter includes a project named `TextAnalyzer`. If you open the start version of this project, you'll find a project that has an implementation of a simple layout along with a button that is hooked up to an `@IBAction`, named `analyze()`. The project folder also contains a file called `SentimentPolarity.mlmodel`. This file is a trained CoreML model that analyzes the sentiment associated with a certain text. Drag this file into Xcode to add the CoreML model to your project.

After adding the model to your project, you can click it in the **Project Navigator** to see more information about the model, as illustrated in the following screenshot:

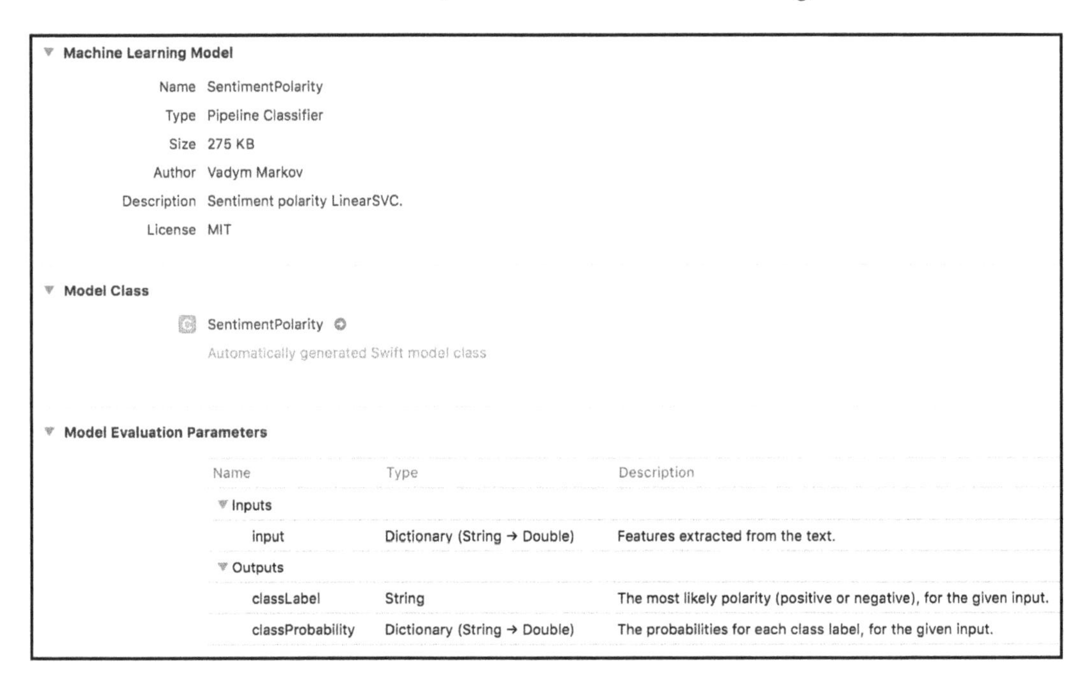

You can see that this model is provided by *Vadym Markov* under the **MIT** license. In the bottom section, you can find out which **inputs** and **outputs** you can expect this model to work with. In this case, the **input** is a dictionary of the `[String: Double]` type. This means that we should feed this model a dictionary of word counts. If you add this model to Xcode, the center section that lists the **Model Class** might notify you that the model isn't part of any targets yet. If this is the case, fix it as you have done previously, by adding this model to your app target in the **Utilities** sidebar on the right side of the window.

Now that your model is implemented, it's time to take it for a spin. First, implement a method that extracts the word count from any given string. You can implement this using the `NLTokenizer` object from the new `NaturalLanguage` framework. `NLTokenizer` is a text analysis class that is used to split a string into words, sentences, paragraphs, or even whole documents. In this example, the tokenizer is set up to detect individual words. Implement the word count method as follows:

```
func getWordCounts(from string: String) -> [String: Int] {
  let tokenizer = NLTokenizer(unit: .word)
  tokenizer.string = string
```

```
  var wordCount = [String: Int]()

  tokenizer.enumerateTokens(in: string.startIndex..<string.endIndex) {
range, attributes in
    let word = String(string[range])
    wordCount[word] = (wordCount[word] ?? 0) + 1

    return true
  }

  return wordCount
}
```

The previous code iterates over all the words that the tokenizer has recognized, and stores it in a dictionary of the `[String: Double]` type. You might wonder why a `Double` type is used for the word count, rather than an `Int` type, since the word counts won't have to deal with decimals. This is true, but the `SentimentPolarity` model requires its input to be a dictionary of the `[String: Double]` type, so you must prepare the data accordingly.

Now that you have the code to prepare the input data for the `SentimentPolarity` model, let's see how you can use this model to analyze the user's input. Add the following implementation for the `analyze()` method:

```
@IBAction func analyze() {
  let wordCount = getWordCounts(from: textView.text)
  let model = SentimentPolarity()

  guard let prediction = try? model.prediction(input: wordCount)
    else { return }

  let alert = UIAlertController(title: nil, message: "Your text is
rated: \(prediction.classLabel)", preferredStyle: .alert)
  let okayAction = UIAlertAction(title: "Okay", style: .default,
handler: nil)
  alert.addAction(okayAction)
  present(alert, animated: true, completion: nil)
}
```

You might be surprised that this method is so short, but that's how simple CoreML is! First, we retrieve the `wordCount` using the method we implemented earlier. Then, an instance of the CoreML model is created. When you added the `SentimentPolarity` model to the app target, Xcode generated a class interface that abstracted away all complexities involving the model. Because the model is now a simple class, you can obtain a prediction for the sentiment of the text by calling `prediction(input:)` on the model instance.

The `prediction` method returns an object that contains the processed prediction in the `classLabel` property, as well as an overview of all available predictions and how certain the model is about each option in the `classProbability` property. You can use this property if you want to be a bit more transparent to the user about the different options that the model suggested and how certain it was about these options.

In the last section of this chapter, you will learn how you can use **CreateML** to train your own natural language model to analyze texts that use domain-specific language relevant to your own app.

Using CoreML to perform text analysis was quite simple. Now let's see how you can use computer vision together with CoreML to determine the type of object that exists in a particular picture.

Combining CoreML and computer vision

When you're developing an app that works with photos or live camera footage, there are several things you might like to do using computer vision. For instance, it could be desirable to detect faces in an image. Or, maybe you would want to identify certain rectangular areas of photographs, such as traffic signs. You could also be looking for something more sophisticated, like detecting the dominant object in a picture.

To work with computer vision in your apps, Apple has created the **Vision** framework. You can combine Vision and CoreML to perform some pretty sophisticated image recognition. Before you implement a sample app that uses dominant object recognition, let's take a quick look at the Vision framework, so you have an idea of what it's capable of and when you might like to use it.

Understanding the Vision framework

The Vision framework is capable of many different tasks that revolve around computer vision. It is built upon several powerful deep learning techniques to enable state-of-the-art facial recognition, text recognition, barcode detection, and more. When you use Vision for facial recognition, you get much more information than just the location of a face in an image. The framework can recognize several facial landmarks, such as eyes, noses, or mouths. All this is possible due to the extensive use of deep learning behind the scenes at Apple.

For most tasks, using Vision consists of three stages:

1. You create a request that specifies what you want. For instance, a `VNDetectFaceLandmarksRequest` request to detect facial features.
2. You set up a handler that can analyze the images.
3. The resulting observation contains the information you need.

The following code sample illustrates how you might find facial landmarks in an image:

```
let handler = VNImageRequestHandler(cgImage: image, options: [:])
let request = VNDetectFaceLandmarksRequest(completionHandler: {
request, error in
  guard let results = request.results as? [VNFaceObservation]
    else { return }

  for result in results where result.landmarks != nil {
    let landmarks = result.landmarks!

    if let faceContour = landmarks.faceContour {
      print(faceContour.normalizedPoints)
    }

    if let leftEye = landmarks.leftEye {
      print(leftEye.normalizedPoints)
    }

    // etc
  }
})

try? handler.perform([request])
```

For something as complex as detecting the contour of a face or the exact location of an eye, the code is quite simple. You set up a `handler` and a `request`. Next, the handler is asked to perform one or more requests. This means that you can run several requests on a single image.

In addition to enabling computer vision tasks like this, the Vision framework also tightly integrates with CoreML. Let's see just how tight this integration is, by adding an image classifier to the augmented reality gallery app you have been working on!

Implementing an image classifier

The code bundle for this chapter contains an app called **ImageAnalyzer**. This app uses an image picker to allow a user to select an image from their photo library to use it as an input for the image classifier you will implement. Open the project and explore it for a little bit to see what it does and how it works. Use the starter project if you want to follow along with the rest of this section.

To add an image classifier, you need to have a CoreML model that can classify images. On Apple's machine learning website (`https://developer.apple.com/machine-learning/build-run-models/`) there are several models available that can do image classification. An excellent lightweight model you can use is the **MobileNet** model; go ahead and download it from the machine learning page. Once you have downloaded the model, drag the model into Xcode to add it to the **ImageAnalyzer** project. Make sure to add it to your app target so that Xcode can generate the class interface for the model.

After adding the model to Xcode, you can open it to examine the **Model Evaluation Parameters**. The parameters tell you the different types of inputs and outputs the model will expect and provide. In the case of `MobileNet`, the input should be an image that is 224 points wide and 224 points high, as shown in the following screenshot:

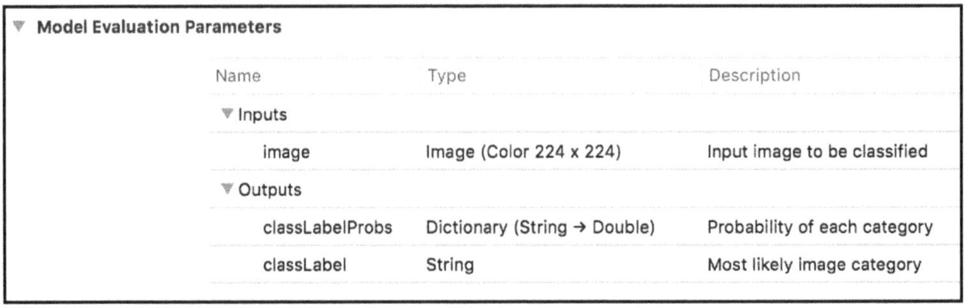

Model Evaluation Parameters

Name	Type	Description
Inputs		
image	Image (Color 224 x 224)	Input image to be classified
Outputs		
classLabelProbs	Dictionary (String → Double)	Probability of each category
classLabel	String	Most likely image category

After generating the model, the code to use the model is very similar to the code used to detect facial features with Vision earlier. The most significant difference is that the type of request that is used is a special `VNCoreMLRequest`. This type of request takes the CoreML model you want to use, in addition to a completion handler.

When combining CoreML and Vision, Vision will take care of image scaling and converting the image to a type that is compatible with the CoreML model. You should make sure that the input image has the correct orientation. If your image is rotated in an unexpected orientation, CoreML might not be able to analyze it correctly.

Add the following implementation for `analyzeImage(_:)` to the `ViewController` class in the **ImageAnalyzer** project:

```
func analyzeImage(_ image: UIImage) {
  guard let cgImage = image.cgImage,
    let classifier = try? VNCoreMLModel(for: MobileNet().model)
    else { return }

  let request = VNCoreMLRequest(model: classifier, completionHandler:
{ [weak self] request, error in
    guard let classifications = request.results as?
[VNClassificationObservation],
      let prediction = classifications.first
      else { return }

    DispatchQueue.main.async {
      self?.objectDescription.text = "\(prediction.identifier)
(\(round(prediction.confidence * 100))% confidence"
    }
  })

  let handler = VNImageRequestHandler(cgImage: cgImage, options: [:])

  try? handler.perform([request])
}
```

The previous method takes a `UIImage` and converts it to a `CGImage`. Also, a `VNCoreMLModel` is created, based on the `MobileNet` model. This particular model class wraps the CoreML model, so it works seamlessly with Vision. The request is very similar to the request you have seen before. In the `completionHandler`, the results array and first prediction of the image classifications are extracted and shown to the user. Every prediction made by the classifier will have a label that's stored in the `identifier` and a confidence rating with a value between 0 and 1 stored in the `confidence` property. Note that the value of the description label is set on the main thread to avoid crashes.

You have already implemented two different types of CoreML models that were trained for general purposes. Sometimes, these models won't be specific enough for your purposes. For instance, take a look at the following screenshot, where a machine learning model labels a certain type of car as a sports car with only 30% confidence:

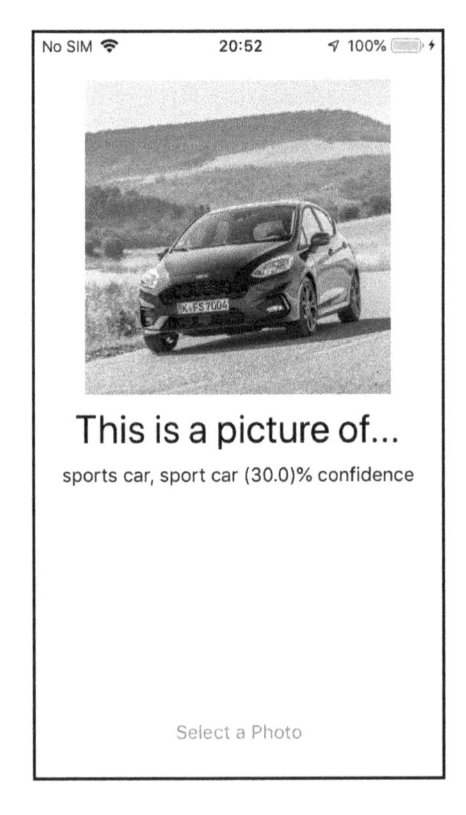

In the next section, you will learn how to train models for purposes that are specific to you and your apps by using CreateML.

Training your own models with CreateML

As part of Xcode 10 and Apple's latest version of macOS, called **Mojave**, they have shipped a new tool that you can use to train your own machine learning models by adding specializations to existing models. This means that you can train your own natural language model that places certain texts in categories that you define. Or, you can train a model that recognizes certain product names or terms in a text that is specific to your application's domain.

If you're building a news app, you might want to train a CoreML model that can automatically categorize the articles in the app. You can then use this model to keep track of the articles your users read, and present articles that are most likely to fit their interests on a dedicated page in your app.

In this segment, you will learn how to train natural language models and how you can train an image recognition model based on the Vision framework. In doing so, you will find that creating a large and optimized training set is crucial when you want to train a machine learning model.

In the code bundle for this chapter, you will find a Playground called `CreateML`. This playground contains all the resources used for training the natural language models.

Training a natural language model

The Natural Language framework has excellent features to analyze text with. Bundled with the power of machine learning models, you can perform some powerful operations on text. Apple has spent a lot of time training several models with vast amounts of data to ensure that the natural language framework can detect names, places, and more.

However, sometimes you might want to add your own analysis tools. To facilitate this, Natural Language works well with CoreML and Apple's new CreateML framework. With CreateML, you can easily and quickly create your own machine learning models that you can use in your apps straight away.

You can use several different types of training for a Natural Language model. In this section, you will learn about two different models:

- A text classifier
- A word tagger

The **text classifier** will classify a particular piece of text with a label. This is similar to the sentiment analysis you have implemented in the **TextAnalyzer** sample app. An example of an entry in your training data would look as follows:

```
{
  "text": "We took an exclusive ride in a flying car",
  "label": "Tech"
}
```

This is a sample of a news article headline that belongs in a category labeled *Tech*. When you feed a large number of samples like this to your model, you could end up with a classifier that can apply labels to news articles based on their headlines. Of course, this assumes that the headlines are specific enough and contain enough information to train the classifier properly. In reality, you will find that short sentences like these will not make the best models. The sample Playground contains a JSON file with training data that attempts to separate news articles into two categories; politics and tech. Let's see how the model can be trained so you can then see for yourself how accurate the model is.

The following code trains and stores the custom CoreML model:

```
import CreateML
import Foundation

let trainingData = try! MLDataTable(contentsOf:
Bundle.main.url(forResource: "texts", withExtension: "json")!)
let model = try! MLTextClassifier(trainingData: trainingData,
textColumn: "text", labelColumn: "label")
try! model.write(to: URL(fileURLWithPath:
"/path/to/folder/TextClassifier.mlmodel"))
```

Training the entire model requires only a couple of lines of code. All you need to do is obtain your training data, create the classifier, and save it somewhere on your machine. You can even do some quick testing to see whether your model works well, right inside of the playground. Note that the preceding code uses a `try!` statement. This is done to keep the code sample brief and simple. In your own apps, you should always strive for proper error handling to avoid surprising crashes.

The string that is passed to the `URL(fileURLWithPath:)` initializer, represents the location where your model will be stored. Make sure to specify the full path here, so, for instance, use `/Users/donnywals/Desktop/TextClassifier.mlmodel`, and not `~/Desktop/TextClassifier.mlmodel`.

The following lines of code test two different headlines to see if the model correctly labels them:

```
let techHeadline = try! model.prediction(from: "Snap users drop for
first time, but revenue climbs")
let politicsHeadline = try! model.prediction(from: "Spike Lee says
President Donald Trump is a 'bullhorn' for racism")
```

If you're happy with the results of your model, you can grab the trained model from the place where you saved it, and immediately add it to your Xcode project. From there, you can use the model like you would use any other model.

Let's see another example of a Natural Language model. In this case, the model should label every word in a text to classify it as a certain type of word. For instance, you could train the model to recognize certain brand names, product names, or other words that have special meanings to your app. An example of training data that you could use to train a model like this is the following:

```
{
  "tokens": ["Apple", "announced", "iOS 12", "and", "Xcode 10", "at",
"WWDC 2018"],
  "labels": ["COMPANY", "NONE", "OPERATING_SYSTEM", "NONE",
"SOFTWARE", "NONE", "EVENT"]
}
```

By collecting many samples that include the words that you want to label, your model will be able to not only match tags based on the word itself, but even on the surrounding words. Essentially, the model would be aware of the context in which each word is used to then determine the correct tag. Once you have collected enough sample data, you can train the model in a similar way as the classifier:

```
let labelTrainingData = try! MLDataTable(contentsOf:
Bundle.main.url(forResource: "labels", withExtension: "json")!)
let model = try! MLWordTagger(trainingData: labelTrainingData,
tokenColumn: "tokens", labelColumn: "labels")
try! model.write(to: URL(fileURLWithPath:
"path/to/folder/TextTagger.mlmodel"))
```

The amount of code to train the model hasn't changed. The only difference is that the previous model was based on the `MLTextClassifier` class, and the current model is based on the `MLWordTagger`. Again, you can immediately use the trained model to make some predictions that you can then use to validate whether the model was trained properly. Providing good data and testing often are the keys to building a great CoreML model.

In addition to text analysis models, CreateML can also help you to train your own image recognition models. Let's see how this works next.

Training a Vision model

In the **ImageAnalyzer** sample app, you saw that picking an image of a certain car would be classified as a sports car with a pretty low confidence score. You can train your own vision model that specializes in recognizing certain cars.

Collecting good training data for image classifiers is tough, because you have to make sure that you gather many pictures of your subjects from all sides and in many different environments. For instance, if all your car images feature cars that are next to trees, or on the road, the model might end up classifying anything with trees or a road to be a car. The only way to obtain a perfect training set is to experiment, tweak, and test.

Training a Vision model works slightly different from training a Natural Language model. You can't use a JSON file to feed your test data to the classifier. So, instead, you should create folders that contain your images where the folder name is the label you want to apply to each image inside that folder. The following screenshot is an example of a training set that contains two kinds of labels:

Once you have collected your set of training data, you can store it anywhere on your computer—for instance, on the desktop. You will then pass the path for your training data to your model training code as follows:

```
import CreateML
import Foundation

let dataUrl = URL(fileURLWithPath: "/path/to/trainingdata")
let source = MLImageClassifier.DataSource.labeledDirectories(at:
dataUrl)
let classifier = try! MLImageClassifier(trainingData: source)
try! classifier.write(toFile: "~/Desktop/CarClassifier.mlmodel")
```

Again, you only need a couple of lines of code to train a model. That's how powerful CreateML is. If you want, you can quickly test your image classifier by dropping the `.mlmodel` file in the **ImageAnalyzer** project, and using that, instead of the `MobileNet` classifier that you used before.

Apart from the simple ways of training models, there are certain parameters that you can pass to the different CreateML classifiers. If you have trouble training your models properly, you could tweak some of the parameters that are used by CreateML. For instance, you could apply more iterations to your training set, so the model gains a deeper understanding of the training data.

As mentioned before in this chapter, machine learning is a subject that could span several books on its own, and even though CreateML makes training models straightforward and simple, it's not easy to train a robust model without any prior machine learning experience.

Summary

In this chapter, you have seen how you can make use of the machine learning capabilities that iOS provides. You saw that adding a machine learning model to your app is extremely simple, since you only have to drag it to Xcode and add it to your target app. You also learned how you can obtain models, and where to look to convert existing models to CoreML models. Creating a machine learning model is not simple, so it's great that Apple has made it so simple to implement machine learning by embedding trained models in your apps.

In addition to CoreML, you also learned about the Vision and Natural Language frameworks. Vision combines the power of CoreML and smart image analysis to create a compelling framework that can perform a massive amount of work on images. Convenient requests, such as facial landmark detection, text analysis, and more are available out of the box without adding any machine learning models to your app. If you do find that you need more power in the form of custom models, you now know how to use CreateML to train, export, and use your own custom, trained CoreML models. You learned that CreateML makes training models simple, but you also learned that the quality of your model is drastically impacted by the quality of your training data.

In the next chapter, you will learn how you can help your users live healthier lives with workouts and activity data.

Questions

1. What is the file format for a CoreML compatible model?

   a) `.coremlmodel`
   b) `.mlmodel`
   c) `.coreml`

2. How do you use a CoreML model in your project?

   a) By adding the model to Xcode and using the model class that is generated for you.
   b) By adding the model to Xcode and training the model on the device.
   c) By sending the CoreML model and the input data to iCloud.

3. Why is it so important the CoreML runs on the end user's device?

   a) It's better for privacy and preserves the user's data plan.
   b) Because the predictions can be calculated faster that way.
   c) Both answers are true.

4. What does the `NaturalLanguage` framework do?

   a) It generates natural text.
   b) It extracts interesting information from text, for instance, names, places, words, and more.
   c) It's used to determine whether a text was written by a human or a machine.

5. What does the `Vision` framework do?

   a) It is used to access the user's camera.
   b) It helps with accessibility for the visually impaired.
   c) It performs complex analysis on images.

6. Which of the following options could negatively impact the quality of a training set for an image classifier?

   a) If the training subject has different colors than the real object.
   b) If the training subject is always shown in similar surroundings.
   c) When the training subject is shown in many different angles. Natural Language

Further reading

- Apple's machine learning website is available at `https://developer.apple.com/machine-learning/`
- The CoreML tools documentation is available at `https://apple.github.io/coremltools/`
- The CreateML documentation can be found at `https://developer.apple.com/documentation/create_ml`
- The Natural Language framework's documentation is available at `https://developer.apple.com/documentation/naturallanguage`
- The Vision framework's documentation is available at `https://developer.apple.com/documentation/vision`

15
Tracking Activity Using HealthKit

More and more people care about their health nowadays. A lot of people try to make sure that they eat the right foods and maintain an active lifestyle. A considerable part of this group uses fitness apps and trackers, such as the Apple Watch, to measure and record their activities. All of this activity data is logged using Apple's HealthKit frameworks.

App developers can integrate HealthKit in their apps to track and record workouts, activities, and many other metrics that are meaningful for monitoring personal health. Because more and more people care about this data, Apple aims to continuously improve the HealthKit framework so that app developers can focus on building great apps, rather than figuring out how to integrate their data with iOS.

This chapter covers the following topics:

- Understanding HealthKit
- Implementing a workout app
- Integrating with iOS

By the end of this chapter, you will know how to build an application that tracks user activity in HealthKit. You will also know how to extract data from HealthKit so you can provide more accurate metrics for the workouts you log to HealthKit.

Understanding HealthKit

HealthKit is a framework that allows developers to interact with a user's health data. This data includes information like a user's weight, BMI, blood type, biological sex, skin type, date of birth, whether they use a wheelchair, and much more. In addition to these metrics, HealthKit also provides access to activity data, such as workouts or the amount of steps the user has walked on a given day.

HealthKit is available on the iPhone and Apple Watch only. The iPad does not have support for HealthKit, so you must take special precautions to determine whether HealthKit is available, by calling `isHealthDataAvailable()` on the `HKHealthStore` class.

Because health data is considered private and thus very sensitive, apps that want access to health data have to specify their reasons for wanting to read and/or write data to HealthKit in the `Info.plist`, under the following keys:

- Privacy: Health Share Usage Description (`NSHealthShareUsageDescription`) for reading data from the HealthKit store
- Privacy: Health Update Usage Description (`NSHealthUpdateUsageDescription`) for writing data to the HealthKit store
- Privacy: Health Records Usage Description (`NSHealthClinicalHealthRecordsShareUsageDescription`) for reading record data from the HealthKit store

In addition to adding the appropriate `Info.plist` entries, you must also enable the HealthKit capability in your app's Capabilities tab. This will configure your app with the required entitlements for accessing the HealthKit store. When you enable the HealthKit capability, you have the option to enable access to health records. Health records are specific records that contain clinical data.

Requesting access to the HealthKit Store

Applications that want access to health-related data must ask permission for every type of data they want to access. If your app requires access to heart rate data, weight data, and activity data, that means that you must request access for each different data type. A single authorization request can ask for access to multiple data types at once, and the user can manually decide which data types can be accessed.

When you want to use certain data types from the HealthKit store, it's essential that you make sure your user understands why you need access to this data and to request access at an appropriate time. If your app has an onboarding flow, you could add a page that outlines the health data your app needs and asks the user for permission once they understand why you need access. The following screenshot shows the access dialog where users can configure the type of data an app is allowed to access:

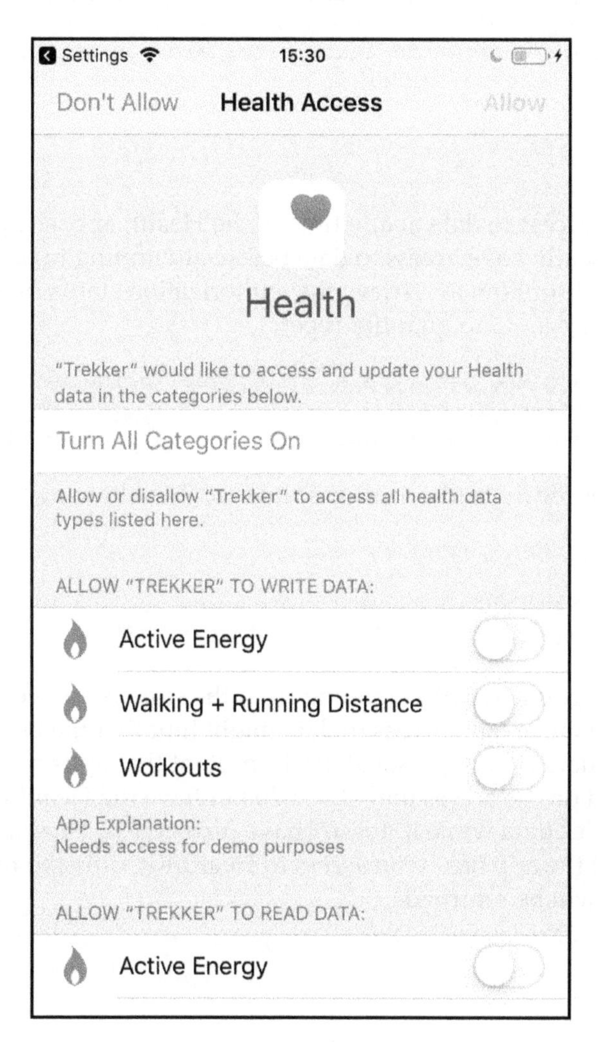

The preceding screenshot shows all the different categories that this app wants to have read and write access for. An authorization request is performed to request access to a set of HealthKit categories. The following code is an example of the request that was used for the screenshot:

```
let objectTypes: Set<HKSampleType> = [HKObjectType.workoutType(),
HKObjectType.quantityType(forIdentifier: .activeEnergyBurned)!,
HKObjectType.quantityType(forIdentifier: .distanceWalkingRunning)!]

healthKitStore.requestAuthorization(toShare: objectTypes, read:
objectTypes) { success, error in
  print(success)
  print(error)
}
```

A user can revoke access to data at any time in the Health app, so apps must make sure that they presently have access to data before attempting to use it. The following code shows a sample of how to retrieve the authorization status for the `.distanceWalkingRunning` quantity type:

```
let objectType = HKObjectType.quantityType(forIdentifier:
.distanceWalkingRunning)!
let status = healthKitStore.authorizationStatus(for: objectType)
```

The possible values for the authorization status are the following:

- `.sharingAuthorized`
- `.sharingDenied`
- `.notDetermined`

Your app can't determine whether reading from the store is allowed. The reasoning for this is that denying an app access to data might hint that the user has something to hide. Since health data is very personal, it's important that apps can't make any assumptions based on the access they do or do not have the HealthKit. This means that if an app queries data while it doesn't have read access, HealthKit will return an empty result set. If the app has write access to HealthKit, only the results that the app wrote to the store will be returned.

Storing and retrieving data with HealthKit

All data that is stored in the HealthKit store is stored as an `HKSample` subclass, and is associated with a corresponding `HKObjectType` and an `HKObject`. Together, these objects describe what a user did, what value is associated with the sample, and when it occurred. The following code is a sample of adding a short walk of 100 meters to the HealthKit store:

```
let object = HKQuantity(unit: HKUnit.meter(), doubleValue: 100)
let objectType = HKObjectType.quantityType(forIdentifier:
.distanceWalkingRunning)
let distance = HKQuantitySample(type: objectType!, quantity: object,
start: Date(), end: Date())

healthKitStore.save(distance) { success, error in
  print(success)
  print(error)
}
```

To provide greater flexibility and unit conversion, HealthKit uses `HKUnit` objects to represent units like meters, inches, pounds, liters, and more. Because all HealthKit operations are asynchronous, the `save(_:withCompletion:)` method takes a completion callback that is called after the operation has completed.

To retrieve data from the HealthKit store, you use `HKQuery` objects. For instance, to retrieve a user's active energy burned for the past ten days, you would use a query that looks as follows:

```
let predicate = NSPredicate(format: "%K > %@",
HKPredicateKeyPathStartDate, (Date() - 60 * 60 * 24 * 10) as NSDate)
let sortDescriptors = [NSSortDescriptor(key:
HKSampleSortIdentifierStartDate, ascending: false)]
let objectType = HKObjectType.quantityType(forIdentifier:
.activeEnergyBurned)!
let query = HKSampleQuery(sampleType: objectType, predicate:
predicate, limit: Int(HKObjectQueryNoLimit), sortDescriptors:
sortDescriptors]) { query, samples, error in
  print(samples)
  print(error)
  print(query)
}

healthKitStore.execute(query)
```

The preceding sample sets up an `NSPredicate` to filter the results that are returned by the query. HealthKit provides several helpers, including `HKPredicateKeyPathStartDate`, so you don't have to manually type the correct key paths that are used to filter and sort data.

The records that you receive from HealthKit contain metadata about the record itself. For instance, you can read the value that is associated with the record, the start and end dates, and even the source that added the record to the HealthKit store. All data that you retrieve from HealthKit is immutable. This means that you can't modify existing records that have already been saved to HealthKit.

The previous samples for storing and fetching data used metrics that are expressed as `HKQuantitySample` objects. This type of object is for data that is represented as a certain number. Walking distance, calories burned, or even weight are examples of this.

HealthKit also records data in the form of `HKCategorySample`. The value for a category sample is expressed as an enum value. Sleep tracking is an example of a metric that is tracked as a category sample. The user is either in bed, asleep, or awake. You can't add custom values to this type of sample.

Now that you are up to speed with gaining access to the HealthKit store and know how to read and write data, let's implement a small application that tracks a user's walks.

Implementing a workout app

In addition to measuring temporary individual metrics, you can track complete workouts with HealthKit. Workouts can contain data like the number of calories burned during the workout, the workout duration, the number of strokes that were made during a swim, and more. The most convenient way to track workouts is by adding a watchOS version of your app that uses all of the sensors in the Apple Watch to track a workout. Since this book focuses on iOS development, you won't learn how to track workouts using an Apple Watch app. Instead, you will implement a workout app that only uses the iPhone to track a running workout.

In the code bundle for this chapter, you will find a project called `Trekker`. This app is a straightforward run-tracking app, where users can tap a button to start tracking a workout. Their GPS location is tracked, and then entered into HealthKit when the workout is complete. The user's route will also be stored in HealthKit by implementing the `HKWorkoutRouteBuilder` class. This class takes GPS locations as they come in from a location manager and associates them with a workout.

The starter project for Trekker has already been set up with the appropriate permissions to begin implementing the HealthKit workout tracking immediately. The majority of your work should go on the `HealthKitHelper` class. This class is very similar to the `LocationHelper` that you worked on in `Chapter 13`, *Improving Apps With Location Services*.

If you examine the `toggleWorkout()` method in `ViewController.swift`, you can already see how the app is supposed to work. If the health kit helper has all the required permissions and no workout is running, the HealthKit helper is asked to start a workout. At the same time, the location helper will begin tracking the user's location. One part that is still missing from this method is passing the locations received by the location helper on to the HealthKit helper.

When you associate locations with a workout route, it's important that the location data is as accurate as possible. So, before adding the location, you must make sure that the accuracy is good enough to use in a workout route. Add the following implementation for the `locationHelper.beginTrackingLocation(_:)` call:

```
locationHelper.beginTrackingLocation { [weak self] location in
  if location.horizontalAccuracy < 30 && location.verticalAccuracy <
30 {
    self?.healthKitHelper.appendWorkoutData(location)
    self?.tableView.reloadData()
  }
}
```

When a location is added to the HealthKit helper, the table view is also reloaded so the user can see the tracked location data appear as it comes in.

With all the calls to the HealthKit helper in place, it's time to implement the missing methods to make sure the workout was tracked correctly.

There are four methods in `HealthKitHelper.swift` that need to be implemented to track the user's run and associate the relevant location data with it. The first step is to ask the user for permission to read and write their workout and workout route data. You need to ask for both of these permissions since you will store the workout and workout route data separately. Add the following implementation for `requestPermissions()` to ask for the correct permissions:

```
func requestPermissions() {
  guard isHealthKitAvailable
    else { return }

  let objectTypes: Set<HKSampleType> = [HKObjectType.workoutType(),
HKSeriesType.workoutRoute()]
  healthKitStore.requestAuthorization(toShare: objectTypes, read:
objectTypes) { success, error in
    if let error = error {
      print(error.localizedDescription)
    }
  }
}
```

If HealthKit is available on the current device, the user is prompted for access to workouts and workout routes. The next step is to implement `startWorkout()`. Add the following implementation to `HealthKitHelper`:

```
func startWorkout() {
  workoutBuilder = HKWorkoutRouteBuilder(healthStore: healthKitStore,
device: nil)
  workoutStartDate = Date()
}
```

The preceding code creates a new instance of `HKWorkoutRouteBuilder` and associates it with the existing health store. You can optionally provide information about the device that is providing the route data. After creating the route builder, the current date is stored as the start date of the workout.

After creating an instance of `HKWorkoutRouteBuilder`, you can add new location data to it as the workout progresses. This location data is automatically attached to the route builder so it can eventually be connected to a workout and stored in HealthKit. Add the following implementation for `appendWorkoutData(_:)` to add new location data to the builder:

```
func appendWorkoutData(_ location: CLLocation) {
  workoutBuilder?.insertRouteData([location]) { success, error in
    if let error = error {
      print(error.localizedDescription)
```

```
      } else {
        print("Location data added to route")
      }
    }
  }
```

The preceding code calls `insertRouteData(_:completion:)` to add new data to the builder. Since this operation is performed asynchronously, this method takes a completion callback. In this case, the result isn't very relevant, so the result is only printed to the console.

There is one more step left to wrap up the workout tracker. When `endWorkout()` is called, an instance of `HKWorkout` should be created and stored to HealthKit. After saving the workout, the route should be associated with this workout and saved to HealthKit. Calling `finishRoute(with:metadata:completion:)` on the route builder will associate it with the workout and store it in HealthKit.

Add the following implementation for `endWorkout()` to the `HealthKitHelper` class:

```
func endWorkout() {
  guard let builder = workoutBuilder, let startDate = workoutStartDate
    else { return }

  let workout = HKWorkout(activityType: .running, start: startDate,
end: Date())

  healthKitStore.save(workout) { success, error in
    builder.finishRoute(with: workout, metadata: [ : ]) { route, error
in
      if let error = error {
        print(error.localizedDescription)
      } else {
        print("route saved: \(route.debugDescription)")
      }
    }
  }
}
```

This final method implementation wraps up the implementation of your custom workout tracker. Go ahead, launch your app and have a short walk outside to see your walk being logged right into your health app.

Summary

In this chapter, you learned about HealthKit. You learned what it is, how to ask the user for permissions, and how you can read and write data in the HealthKit store. This chapter put a lot of focus on using HealthKit to track a user's activity, but HealthKit is capable of tracking much more data. You can build apps that help users track just about anything related to their health.

When you build an app that accesses a user's health data, make sure that you don't request access to information that you don't need. It's usually better to ask the user for permission to access certain data just before you need it than to ask for all permissions at once. For instance, you might build an app that tracks both sleep and blood sugar, but not necessarily at the same time. You might want to ask permission to access each of these data types separately, so the user fully understands why you need access to their data.

In addition to learning how you read and write HealthKit data, you implemented a straightforward workout app that can track a user while they are walking around outside, and associate that with a workout which is then saved to HealthKit. With the knowledge from this chapter, you should have an excellent starting point from which you can start building your own HealthKit-based apps to help your users live a healthier life.

In the next chapter, you will learn how you can use Siri and the new Siri Shortcuts to make your user's life a little bit easier.

Questions

1. Which devices support HealthKit?

    a) The iPhone and iPad.
    b) The iPhone, iPad, and Apple Watch.
    c) The iPhone and Apple Watch.

2. What's the difference between a quantity sample and a category sample?

   a) Quantity samples have an arbitrary value, while categories use enums to represent their value.
   b) Quantity samples take place over time, and category samples are static.
   c) Category samples have more possibilities than quantity samples.

3. How can you check whether you have read access to a certain health metric?

   a) By requesting the authorization status for that metric.
   b) By making a request. If the results are empty, you don't have access.
   c) To protect the user's privacy, there is no way to be sure.

4. How can you filter data that is returned from a HealthKit query?

   a) With predicates
   b) By using the built-in filters
   c) By specifying the values you want to fetch

5. What permissions do you need to use a workout route builder?

   a) Only for workouts.
   b) Just for the workout route builder.
   c) You need permission for both the builder and the workouts.

6. What do you need to keep in mind when you associate location data with a workout route?

   a) You should not add too much data.
   b) You should make sure the location data is very accurate.
   c) You should make sure to add all location data.

Further reading

- Apple's HealthKit documentation is available at `https://developer.apple.com/documentation/healthkit`

16
Streamlining Experiences with Siri

Over the past few years, Apple has provided developers with more and more opportunities to integrate their apps with iOS. One of these opportunities is in integrating your application with Siri using the `SiriKit` framework.

In this chapter, you will work on an existing application called Hairdressers. This application allows users to book an appointment with their hairdresser for a certain day, and can also message their hairdresser directly. You will learn how you can integrate Siri Intents extensions to allow users to message their hairdresser through Siri. You will also implement the new Siri shortcuts to allow users to quickly schedule their appointments with their regular hairdresser.

In this chapter, you will learn about the following topics:

- Understanding intents
- Implementing an intents extension
- Adding a custom UI to a Siri intents extension
- Integrating Siri Shortcuts in your app

If you want to follow along with the code samples in this chapter, make sure to grab the **Hairdressers** starter project from this book's code bundle. Make sure to examine the existing code, so you understand what's going on inside of the app. The code relies heavily on the knowledge you gained in `Chapter 8`, *Adding Core Data to Your App*, so you might want to keep that chapter nearby while you examine the starter code.

Understanding intents

Siri is a powerful, smart, and ever-learning personal assistant that aims to give natural responses to natural speech input. This means that there is often more than one way to say something to Siri. Some users like to be extremely polite when they talk to Siri, saying please and thank you whenever they ask for something. Other users like to be short and to the point; they simply tell Siri what they want and that's it.

This means that Siri has to be really smart about how it interprets language and how it converts the user's requests to actionable items. Not only does Siri take into account the language used; it's also aware of how a user is using Siri. If a user activates Siri by saying *Hey, Siri!*, Siri will be more vocal and verbose than when a user activates Siri by pressing and holding the home or side buttons because it's more likely if they didn't press the home or side button, that the user is not looking at their device.

To convert a user's spoken requests into actions, Siri uses intents. An **intent** is a way to describe an action. These intents are supported by app-specific vocabularies; this allows you to make sure that your users can use terms that are familiar to them because they are also used in your app.

Siri does not handle an unlimited amount of intents. All of the intents that Siri can handle belong to a predefined set of domains. If you plan to create a Siri extension for your own app, it must fit into one of the predefined domains that Siri understands—unless you provide a custom intent. Currently, Siri handles the following domains out of the box:

- Messaging (*Send a message to Donny that says Hello, World*)
- Calling (*Call my sister*)
- Payments (*Transfer 5 euros to Jeff*)
- Workouts (*Start an outdoor run*)
- Ride booking (*Get me a taxi to the airport*)
- Photo search (*Find me photos from Paris*)
- Notes (*Create a note that says "Don't forget the dishes"*)
- Reminders (*Remind me to do the dishes when I get home*)
- Visual codes (*Show me my ticket for tonight's concert*)

If your app is not in one of these domains, you will need to do a bit more work to integrate your app with Siri, as you will learn in the last section of this chapter. In order to integrate Siri with your app, your app must ask the user permission for it to be used with Siri. It's recommended that you do this when your user first opens your app because this ensures that your app is available to Siri as soon as possible. Make sure that you ask for permission appropriately, though; you don't want to unpleasantly surprise your user with a permission request. You also need to add the *Privacy—Siri Usage Description* (`NSSiriUsageDescription`) key to your `Info.plist` file, and enable the Siri capability in your app's Capabilities tab.

You can only enable the Siri capability for your app if you are a paying member of Apple's Developer Program. Unfortunately, you can't enable or test Siri if you haven't got such a membership.

In order for Siri to work with your app, it's important that a user mentions your app's name to Siri when asking it something related to your app. This is to make sure that apps don't hijack certain words or verbs, which would result in a confusing experience. It doesn't matter exactly how or when the user does this. Siri is all about natural language, and it will even understand if your app name is used as a verb. For example, *MyBankApp some money to John Doe* would be interpreted as a money transfer action with an app named *MyBankApp* to a person named *John Doe.*

If a user would literally ask Siri to perform the preceding action, Siri would not be able to fulfill the request yet. In the case of a money transfer Siri has to know two things:

1. Who should receive the money?
2. How much money should be transferred?

These requirements are captured in an intent, and if Siri is missing information, it will ask the user directly for this information in order to collect everything needed to successfully transfer the money.

Now that you know what an intent is and what it's used for, let's see how you can implement an intents extension so Siri knows how to use your app to send messages.

Implementing an intents extension

There are several ways for apps to integrate with iOS. These integrations are always built using app extensions. An app extension is an isolated section of your app that can be instantiated and used by the system as needed to provide certain functionality for users. In iOS 12, there are 25 different ways for apps to integrate with iOS. Not every app can or should implement every possible extension, but a proper integration with iOS can truly make your app stand out. The following screenshot shows a selection of available extensions in iOS 12:

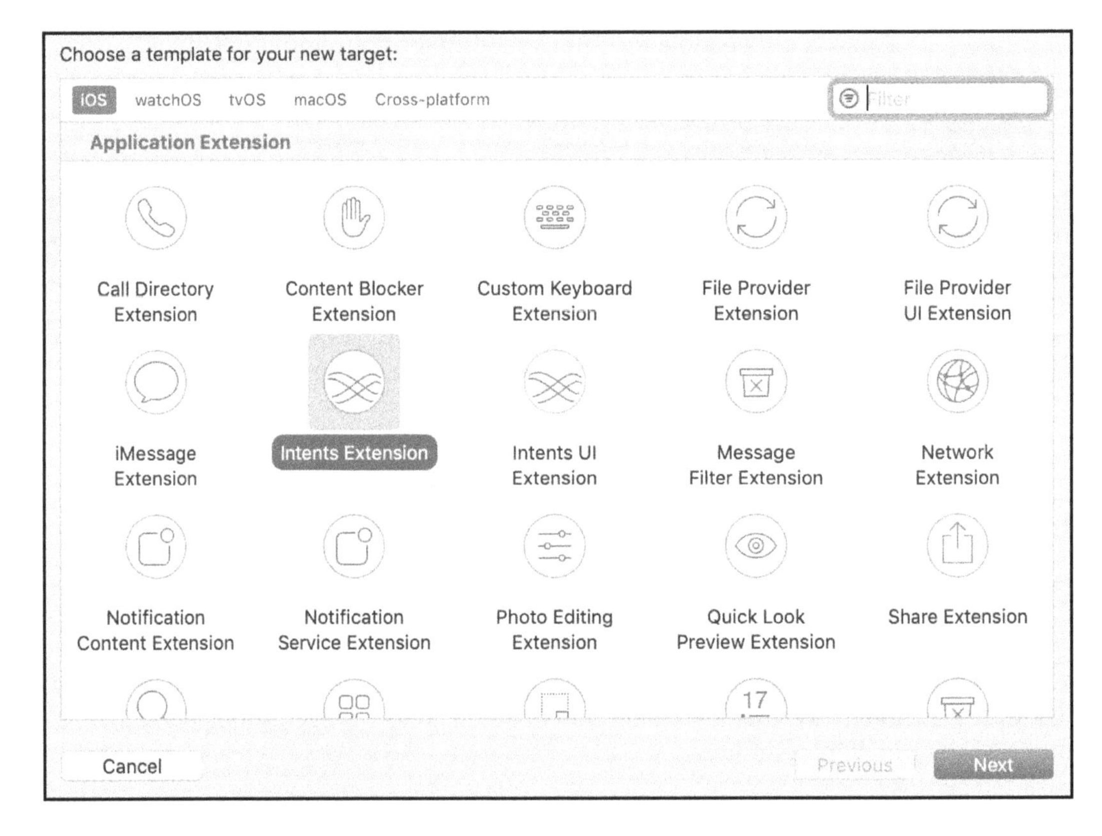

In this chapter, you will implement an **Intents Extension** and an **IntentUI Extension**. These extensions are used to communicate with Siri and to provide a custom UI for the Siri integration. Before you implement your extensions, let's see what extensions are and how they work.

Understanding app extensions

As mentioned before, extensions are used by apps to integrate with iOS. Some extensions have their own UI, such as the **IntentUI Extension**, and others have no UI at all, such as the **Intents Extension**. Every extension has its own purpose and specializes in performing a single task.

When an extension is opened or instantiated, this happens through the extension framework itself. For instance, if you open a photo in the **Photos** app and you tap the share option, you can choose several to choose the selected photo too. If you select one of these apps, its **Share Extension** is launched. This extension then provides its own UI that is used to share the photo in a way that is supported by the share target. The following screenshot is an example of this:

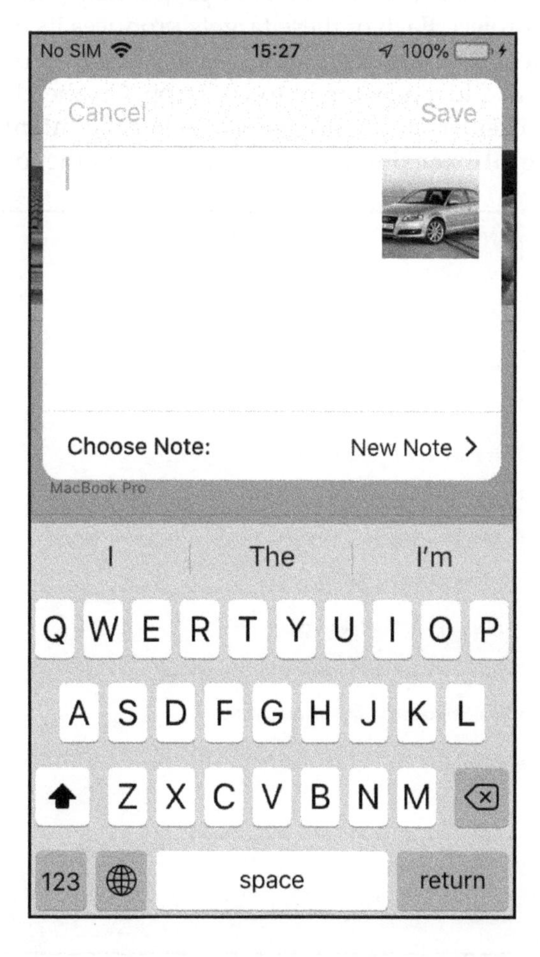

In the preceding screenshot, a photo is shared to the **Notes** app. The Notes app itself does not launch or own the extension. This means that an extension functions completely separately from its host application. Instead, the extensions framework is responsible for managing the extension's life cycle. Because of this, extensions do not work like individual apps. Extensions usually only have a single view controller, and you have to rely on the `UIViewController` life cycle methods to configure your extension. This might seem a little bit counter-intuitive at first because, as you'll find out in a moment, extensions have their own build target and `Info.plist`, which makes them look a lot like a separate app, while they behave more like a view controller.

Extensions are added to an app through the **Project** window. If you look at the sidebar in the center window, you can see your project name and the **Targets** associated with your project. Each of these targets produces its own build product. The app target produces your app itself, for instance. You can use the **+** icon at the bottom of the sidebar to add a new target to your app. Go ahead and add an extension for the Hairdressers app. Make sure to select the **Intents Extension** option and check the **Include UIExtension** checkbox as shown in the following screenshot:

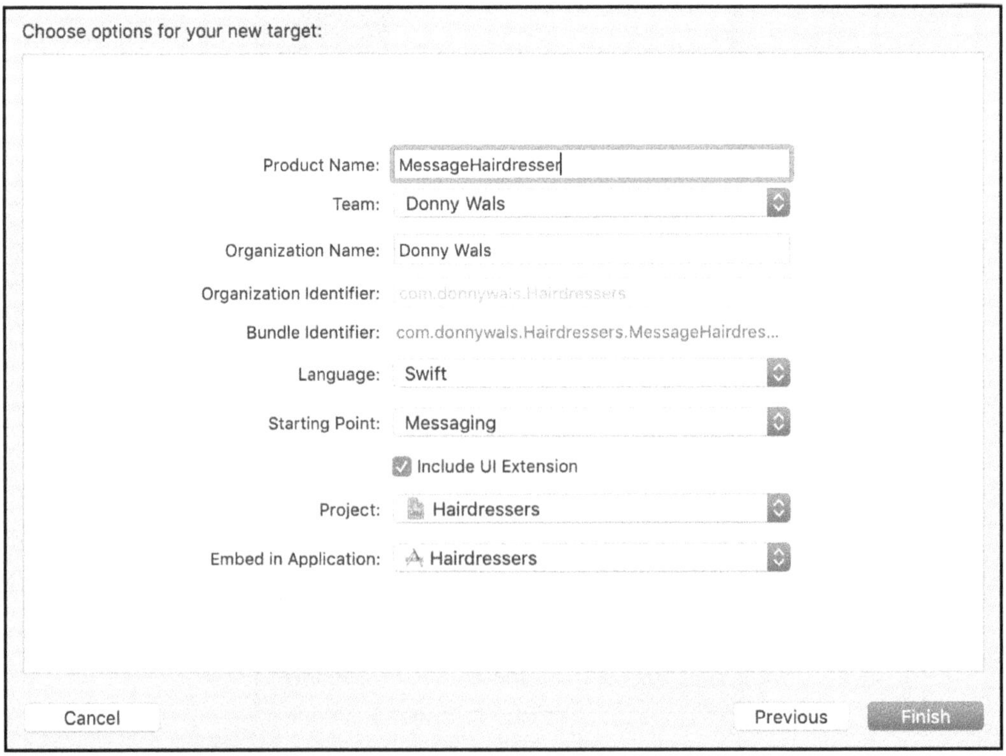

Now that you have an understanding of app extensions, let's see how you can implement your own **Intents Extension** to integrate the Hairdressers app with Siri.

Configuring your extension

Now that you have added your extensions to the Hairdressers project, it's time to configure the app so it can integrate with Siri. Go through the following steps to properly configure your app:

- Make sure you're using a paid developer team, and make sure the app uses this paid team by selecting it in the project settings.
- Enable the Siri capability in the **Capabilities** tab.
- Add the required Privacy key to the app's `Info.plist` file.

After configuring the project, the first step to integrate your app with Siri is to ask the user for permission to do so. Open `AppointmentsViewController.swift` and import the `Intents` framework at the top of the file as follows:

```
import Intents
```

You might have expected all of the Siri-related code to exist in the `SiriKit` framework. Instead, some parts of the Siri integration live in the `Intents` framework, because some other apps, like the **Maps** app on iOS, use intents to determine what action a user wants to perform.

After adding the import, add the following implementation for `viewDidAppear(_:)` to `AppointmentsViewController`:

```
override func viewDidAppear(_ animated: Bool) {
  super.viewDidAppear(animated)

  INPreferences.requestSiriAuthorization { status in
    print(status == .authorized)
  }
}
```

This preceding code asks for permission to use Siri in this app as soon as possible, just like Apple recommends. If permission is already granted, the permissions dialog won't be displayed, so there is no need to check for the current status prior to asking permission. For now, you have done all the required work in the app.

If you open the `MessageHairdresserIntent` folder that Xcode created when you added the extension earlier, you should see that there are two files in there: a file named `IntentHandler.swift`, and the app extension's `Info.plist`. The `IntentHandler` class is responsible for communicating with Siri to resolve, confirm, and handle the intents for which the extension is registered. The `Info.plist` is used to determine which intents can be handled by the extension.

Open the `Info.plist` file and expand the `NSExtension` key. You'll notice that there are two intent-related keys in the file: `IntentsRestrictedWhileLocked` and `IntentsSupported`. The second key contains all of the intents that can be handled by the extension. The `IntentsRestrictedWhileLocked` key specifies which of these supported keys can or can't be used without unlocking the device. SiriKit itself will lock some intents by default. Money transfers, for example, can't be done without unlocking the device, regardless of your extension settings:

NSExtension	Dictionary	(3 items)
▼ NSExtensionAttributes	Dictionary	(2 items)
▼ IntentsRestrictedWhileLocked	Array	(0 items)
▼ IntentsSupported	Array	(3 items)
Item 0	String	INSendMessageIntent
Item 1	String	INSearchForMessagesIntent
Item 2	String	INSetMessageAttributeIntent

The list of intents in **IntentsSupported** is a list of intent class names that your extension is able to handle. Xcode has added a couple of example intents, but this list is not even close to being an exhaustive list of available intents. For a complete list of available intents, you should have a look at the documentation for the `Intents` framework.

The available intents range from starting a workout, to booking a restaurant reservation or requesting that another person transfers money into your bank account. Each of these intents has their own corresponding class that holds all of the properties that are used to describe the intent.

To enable message sending in Hairdressers, only the `INSendMessageIntent` is required, so you can remove the other two intents. If you want to experiment with multiple intents from the get-go, you can keep any intents that you want to play around with, or add more, if you like!

Even though Siri is quite smart when it comes to resolving intents, some apps have their own terminology for specific actions. Custom terms make resolving intents a lot harder for Siri. Luckily, you can help Siri out by adding vocabulary information to Siri.

Adding vocabularies to your app

Siri always makes an effort to understand what a user is trying to do with an app and is usually quite good at this. However, sometimes it's tough for Siri to figure out what's going on or what a user is trying to do. To help Siri figure out what a user might mean, you can provide vocabularies. A **vocabulary** is a set of strings that map to intents, or to parameters for the intent.

There are two ways for you to teach Siri the vocabulary for your app. One way is through a `.plist` file. This approach is mostly used for when your app has a global vocabulary that applies to all users. If your app can't supply all custom vocabularies in a static file, you can provide them dynamically. This is particularly useful if, for instance, you want to teach Siri about contacts in a messaging app.

Adding vocabularies through a .plist file

You already learned that Siri understands when a user wants to do something with your application, even if your user uses your app name as a verb. However, your app name might not be the only thing that's specific to your app. Let's see how this works for a workout app.

If, for example, a user were to tell Siri, *Hey Siri, start an Ultimate Run using RunPotato,* Siri would be able to figure out what RunPotato is since it's the app that is expected to handle the intent. What it won't be able to understand instantly is what it means to *start an Ultimate Run*. This is where a custom vocabulary entry in a `.plist` makes a lot of sense.

The `.plist` that contains the vocabulary should be provided by the application itself; the corresponding extension is not able to do this. You can add a vocabulary of app-specific words, like the workout name above, by adding an extra `.plist` file to the app. To add a new `.plist` file, add a new file to your app target and select the **Property List** file type under the **Resource** header as shown in the following screenshot:

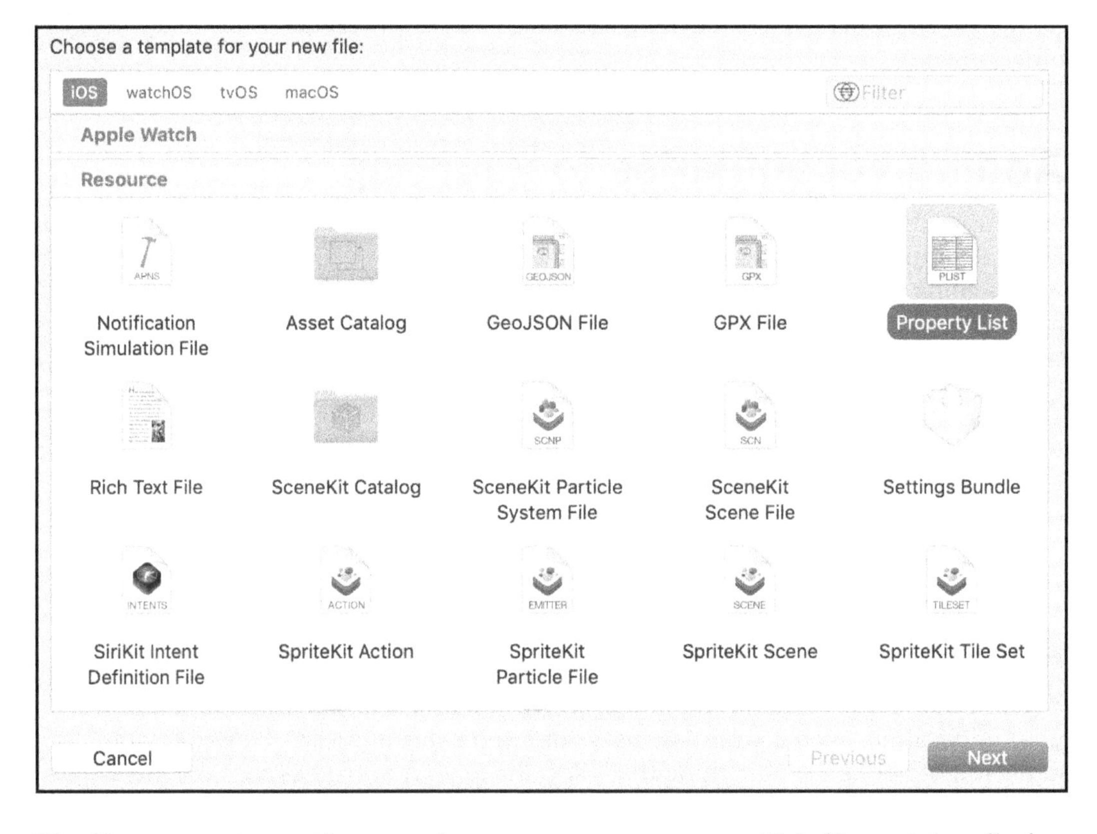

The file you create must be named `AppIntentVocabulary`. This file contains all of the information about an app's custom vocabulary. The vocabulary `.plist` is pretty specific and verbose. This is important because its purpose is to provide a window into the vocabulary your app uses. When you created the `AppIntentVocabulary.plist` file, Xcode added a dictionary at the top level of the file. All custom vocabulary should be added to that dictionary.

To implement the example phrase shown earlier, you need a `Parameter Vocabularies` array. This array will contain `Parameter Vocabulary` items that specify the vocabulary the app uses for workout names. The first item that needs to be defined is a `Parameter Names` array. For the workout app, it would contain a single entry: `INStartWorkoutIntent.workoutName`. In addition to the parameter names, a `Parameter Vocabulary` dictionary must also be added to the entry in the `Parameter Vocabularies` array.

The `Parameter Vocabulary` dictionary has keys for the item identifier, synonyms for this identifier, a pronunciation, and even an example phrase. This is all very verbose, but it provides Siri with all of the information it needs to resolve user input for your app. The following screenshot shows an example of the entry that would be needed to make sure Siri understands what an *Ultimate Run* workout is:

Key	Type	Value
▼ App Intent Vocabulary Property List	Dictionary	(1 item)
▼ Parameter Vocabularies	Array	(1 item)
▼ Item 0 (Parameter Vocabulary)	Dictionary	(2 items)
▼ Parameter Names	Array	(1 item)
Item 0 (Parameter Name)	String	INStartWorkoutIntent.workoutName
▼ Parameter Vocabulary	Array	(1 item)
▼ Item 0 (Parameter Vocabulary)	Dictionary	(2 items)
Vocabulary Item Identifier	String	ultimate run
▼ Vocabulary Item Synonyms	Array	(1 item)
▼ Item 0 (Vocabulary Item Synonym)	Dictionary	(3 items)
▼ Vocabulary Item Examples	Array	(1 item)
Item 0 (Vocabulary Item Example)	String	Start an Ultimate Run with RunPotato
Vocabulary Item Phrase	String	Ultimate Run
Vocabulary Item Pronunciation	String	ultemit run

Teaching Siri new vocabularies at runtime

The second way you can teach Siri about specific content in your app is through the `INVocabulary` class. The `INVocabulary` class is used to teach Siri information that is specific to the user or changes over time.

A great example of user-specific vocabulary is a workout program that your user manually creates inside of your app. Or maybe the user's friends in a messaging app, if those friends don't match the people your user has in their list of contacts on their device.

Updating the vocabularies for an app always occurs in batches. You don't add or remove individual words or phrases for a specific string type. For example, if your user adds a new workout type to a workout application, or if they add new content in a messaging application, you must update the entire vocabulary for the parameter that the vocabulary applies to. Since the Hairdressers app uses its own list of hairdressers, Siri should learn their names so the users can message them. Add the following code to the `application(_:didFinishLaunchingWithOptions:)` method in `AppDelegate`, and don't forget to import the `Intents` framework, as follows:

```
let hairdressers = NSOrderedSet(array:
HairdressersDataSource.hairdressers)
INVocabulary.shared().setVocabularyStrings(hairdressers, of:
.contactName)
```

By supplying the contact names vocabulary when the app starts, it's available to Siri as soon as possible, even before the app's first screen is loaded.

If a user logs out or does something else that makes their custom vocabulary redundant or invalid, you need to make sure you delete the entire user-specific vocabulary by calling `removeAllVocabularyStrings()` on the shared `INVocabulary` class. This method of teaching vocabulary to Siri is not intended for vocabularies that are common to all users. If a vocabulary is the same for all users of your app, you should provide this vocabulary through the `.plist` file that was mentioned earlier. Now that you're completely up to speed regarding intents and vocabularies, let's see how you can start to handle the intents that Siri sends your way.

Handling intents in your extension

Handling intents can be divided into three stages. The first stage is the resolving stage. In this stage, your extension will go back and forth with Siri to figure out the correct parameters for the given intent. In this stage, Siri could ask your app to verify that a certain username exists. Your extension will then have to figure out if the given input is valid, and you'll provide Siri with a response code that tells it whether the parameter is resolved, or maybe requires a little bit more clarification on the user's end.

The second stage is expected to confirm that everything is set up correctly and all requirements for executing the action are met. The third and final stage involves actually acting on the intent and performing the desired action. Let's go through each of stages, one by one.

Resolving the user's input

When you create an intents extension, Xcode creates a single main class for your extension, named `IntentsExtension`. This is the class that serves as an entry point for your extension. It contains a `handler(for:)` method that returns an instance of `Any`. The `Any` type indicates that this method can return virtually anything and the compiler will consider it valid. Whenever you see a method signature like this, you should consider yourself on your own. Being on your own means that the Swift compiler will not help you to validate that you've returned the correct type of object from this method.

In addition to the `handler(for:)` method, Xcode has generated a lot of sample code to show how you could implement an intent handler. After quickly studying this code, you can go ahead and remove it, since you'll write your own implementation for this functionality soon.

The reason the `handler(for:)` method returns `Any` is because this method is supposed to return a handler for every intent that your app supports. If you're handling a send message intent, the handler is expected to conform to the `INSendMessageIntentHandling` protocol. Xcode's default implementation returns `self`, and the `IntentHandler` class conforms to all of the intents the extension handles by default, according to its plist file.

This default approach is not inherently wrong, but if you add an intent to your extension and forget to implement a handler method, you might return an invalid object from the `handler(for:)` method. A cleaner approach is to check the type of intent you're expected to handle, and return an instance of a class that's specialized to handle the intent. This is more maintainable, and will allow for a cleaner implementation of both the intent handler itself, and the `IntentHandler` class.

Replacing Xcode's default implementation ensures that you always return the correct object for every intent that the Hairdressers app supports. Go ahead and replace the `handler(for:)` method with the following implementation:

```
override func handler(for intent: INIntent) -> Any? {
  if intent is INSendMessageIntent {
    return SendMessageIntentHandler()
  }

  return nil
}
```

The `SendMessageIntentHandler` is a class you will define and implement to handle the sending of messages. Create a new `NSObject` subclass, named `SendMessageIntentHandler`, and make it conform to `INSendMessageIntentHandling`. When you create this file, make sure it's added to the **MessageHairdresserIntent** target, and not the Hairdressers app target.

Every intent handler has different required and recommended methods. `INSendMessageIntentHandling` has just one required method: `handle(sendMessage:completion:)`. Other methods are used to confirm and resolve the intent. All of the resolve methods work in similar ways, but are used for different parameters in an intent.

Imagine you're building a messaging app that uses groups to send a message to multiple contacts at once. These groups are defined in your app and Siri wants to resolve a group name. If this occurs, Siri calls the `resolveGroupName(forSendMessage:with:)` method on the intent handler. This method is now expected to resolve the group name and inform Siri about the result by calling the callback it's been passed. Let's see how this works:

```
let supportedGroups = ["neighbors", "coworkers", "developers"]

func resolveGroupName(forSendMessage intent: INSendMessageIntent, with
completion: @escaping (INStringResolutionResult) -> Void) {

  guard let givenGroupName = intent.speakableGroupName else {
    completion(.needsValue())
    return
  }

  let matchingGroups = supportedGroups.filter { group in
    return group.contains(givenGroupName)
  }
```

```
    switch matchingGroups.count {
    case 0:
      completion(.needsValue())
    case 1:
      completion(.success(with: matchingGroups.first!))
    default:
      completion(.disambiguation(with: matchingGroups))
    }
  }
```

To simplify the example a bit, the supported groups are defined as an array. In reality, you would use the given group name as input for a search query in Core Data, your server, or any other place where you might have stored the information about contact groups.

The method itself first makes sure that a group name is present on the intent. If it's not, a callback is used to inform Siri that a group name is required for this app. Note that this might not be desirable for all messaging apps. Actually, most messaging apps will allow users to omit the group name altogether. If this is the case, you'd call the completion handler with a successful result.

If a group name is given, it is used to filter the `supportedGroups` array. Again, most apps would query an actual database at this point. If no results are found, Siri is asked for a value. If a single result is found, the work is done. The code successfully managed to match the intent's group with a group in the app's database and Siri is informed accordingly. If more than one result was found, Siri is asked to disambiguate the results that were found. Siri will then take care of asking the user to specify which one of the provided inputs should be used to send the message to. This could happen if you ask Siri to send a message to a person named *Jeff*, but you have multiple *Jeffs* in your contact list.

In the case of the Hairdressers app, messages are sent to individual hairdressers that are stored in the `Hairdressers.plist` file. The `HairdressersDataSource` helper object can read the data from this `plist` and provides a simple array of hairdresser names. Since this data is currently only part of the Hairdressers app target, you will need to add it to the **MessageHairdresserIntent** target as well.

To do this, select both of the files in the **Project Navigator**, and use the **File Inspector** on the right side of the Xcode window to add these files to the **MessageHairdresserIntent** target, as shown in the following screenshot:

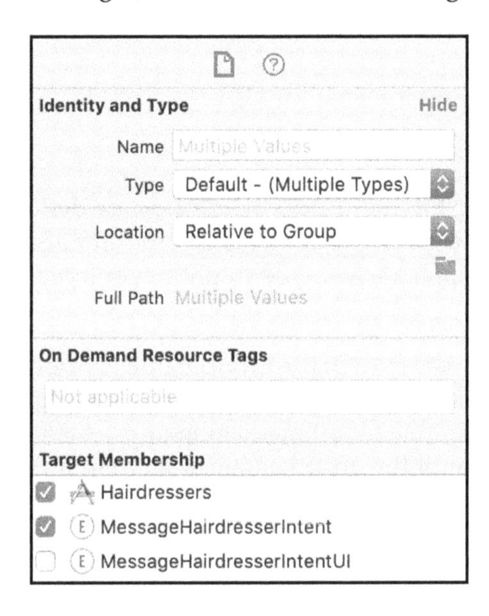

Next, add the following extension to `SendMessageIntentHandler.swift` to implement `resolveRecipients(for:with:)`:

```
extension SendMessageIntentHandler: INSendMessageIntentHandling {
  func resolveRecipients(for intent: INSendMessageIntent, with
completion: @escaping ([INSendMessageRecipientResolutionResult]) ->
Void) {

    guard let recipients = intent.recipients else {
      completion([.needsValue()])
      return
    }

    let results: [INSendMessageRecipientResolutionResult] =
recipients.map { person in
      let matches = HairdressersDataSource.hairdressers.filter {
hairdresser in
        return hairdresser == person.displayName
      }

      switch matches.count {
      case 0: return
INSendMessageRecipientResolutionResult.needsValue()
```

```
      case 1: return
INSendMessageRecipientResolutionResult.success(with: person)
      default: return
INSendMessageRecipientResolutionResult.disambiguation(with: [person])
      }
    }

    completion(results)
  }
}
```

The preceding code should look very familiar, because it's very similar to the code for resolving a group name. The main difference is that a user can choose multiple recipients for the message, so a resolution result is created for each of the recipients that Siri passes to the `resolveRecipients(for:with:)` method. The next stage in handling the intent is to confirm the intent status.

Confirming the intent status

After you've made sure that everything is in place to eventually handle the intent, you must confirm this to Siri. Every intent handler has a `confirm` method. The signature might vary, but there is always some form of confirmation. Refer to the documentation for the intent you are handling to confirm which method you are expected to implement. When you send messages, the confirmation method is `confirm(sendMessage:completion:)`.

You can make the confirmation step as complex as you want. For example, you could check whether a message is too long, contains forbidden content, or virtually anything else. Most commonly, you'll want to make sure that the user is authenticated and allowed to send a message to the recipient.

Again, it's completely up to your best judgment to determine which conditions apply to your extension. The important takeaway for the confirm method is that you're expected to make sure that everything is in place to smoothly perform the action later.

Let's look at an example of a confirmation implementation to explore some of the possible outcomes of the confirmation stage:

```
func confirm(sendMessage intent: INSendMessageIntent, completion:
@escaping (INSendMessageIntentResponse) -> Void) {

  guard let user = User.current(), user.isLoggedIn else {
    completion(INSendMessageIntentResponse(code:
.failureRequiringAppLaunch, userActivity: nil))
```

```
        return
    }

    guard MessagingApi.isAvailable else {
        completion(INSendMessageIntentResponse(code:
.failureMessageServiceNotAvailable, userActivity: nil))
        return
    }

    completion(INSendMessageIntentResponse(code: .ready, userActivity:
nil))
}
```

The preceding implementation checks whether a current user is available, and whether they are logged in or not. If no user exists, the intent handler will inform Siri that the app should be launched so that a user can log in.

Next, the availability of the API that will eventually handle the message sending is checked. Note that the `user` and `api` classes don't exist in the example project, and should be defined by you if you decide to go with this confirmation approach. These classes simply serve as placeholder examples to demonstrate how confirmation of an intent works.

If a user must be taken to your app in order to log in or for any other reason, Siri will automatically create a user activity that's passed to `AppDelegate` in your application. You must implement `application(_:continue:restorationHandler:)` to catch and continue the user activity.

A user activity that's created by Siri has its `interaction` property set. This property contains an `INInteraction` object that reflects the action that the user attempts to complete using Siri. A good implementation of `application(_:continue:restorationHandler:)` will fulfill this interaction as soon as possible inside of the app. It's also possible to create your own user activity if you want to add custom information that Siri doesn't pass on. If you want to do this, you should pass your custom user activity to the `INSendMessageIntentResponse` initializer.

Since the Hairdressers app doesn't use an external API and the user doesn't have to log in, the `confirm(sendMessage:completion:)` method does not have to be implemented. After ensuring that everything is in place, it's time to move on to the third stage of handling an intent: performing the desired action.

Performing the desired action

Once Siri understands what the user wants to do and which parameters to use, and once your app has confirmed that everything is in place to handle the user's request, the time has finally come to execute the requested action. Once this time has come, Siri calls the `handle` method on your intent handler.

Just like the `confirm` method, every intent has their own version of this method, but they all follow a similar pattern. For sending messages, the method signature is `handle(intent:completion:)`, where the `intent` is an instance of `INSendMessageIntent`. The parameters for this method are identical to the ones in the confirmation step. The major difference is that you're now expected to handle the intent, instead of only confirming that the intent is valid.

Once you have handled the intent, you must call the completion handler with an `INSendMessageIntentResponse`. If everything goes well, you're expected to use a `success` response code. If you're unable to process the intent promptly, you must call the completion handler with an `inProgress` status code. Using the `inProgress` status code informs Siri that you're handling the intent, but it's taking a while. If you fail to handle the intent at all, you should pass a `failure` status to the completion handler.

Since the Hairdressers app uses Core Data to store all of its data, you will need to perform some additional steps to share data between the extension and the app. To do this, you must use the **App Groups** capability. **App Groups** associate multiple targets with a single, shared group, which allows them to access shared files or `UserDefaults` instances. To create an app group, you must enable the **App Groups** capability in the project's **Capabilities** tab, and provide a name for your group, as shown in the following screenshot:

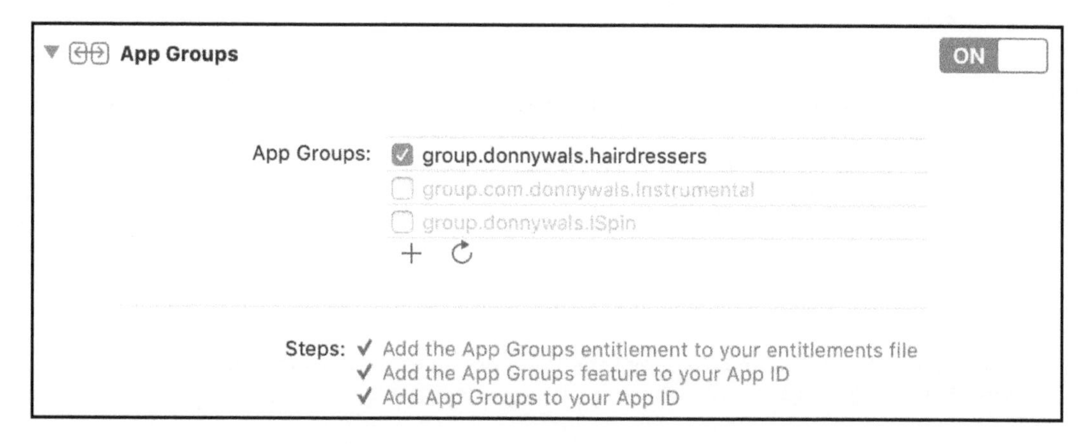

When you have configured the app groups capability for your app, enable it for the **MessageHairdresserIntent** extension as well, and make sure to select the app group you just created to add the extension to that app group.

After doing this, you must configure the app's persistent container to use the app group for storing the database file. Add the following code to the `persistentContainer` computed property in `PersistentHelper`, right before the persistent store is loaded, to configure the persistent container properly:

```
let containerUrl =
FileManager.default.containerURL(forSecurityApplicationGroupIdentifier
: "YOUR.APP.GROUP.NAME")!
let databaseUrl =
containerUrl.appendingPathComponent("Hairdressers.sqlite")
let description = NSPersistentStoreDescription(url: databaseUrl)
container.persistentStoreDescriptions = [description]
```

Also, make sure to add the `PersistentHelper.swift` file, the `NSManagedObjectContext` extension, and the data model in `Hairdressers.xcdatamodeld`, which already exist in the app target, to the extension target.

Now that the app and its Siri extension share the Core Data store, you are ready to take the message the user is trying to send and store it in the Core Data database. Add the following code to the existing extension on `SendMessageIntentHandler`:

```
func handle(intent: INSendMessageIntent, completion: @escaping
(INSendMessageIntentResponse) -> Void) {
  guard let hairdressers = intent.recipients, let content =
intent.content else {
    completion(INSendMessageIntentResponse(code: .failure,
userActivity: nil))
    return
  }

  let moc = PersistentHelper.shared.persistentContainer.viewContext
  moc.persist {
    for hairdresser in hairdressers {
      let message = Message(context: moc)
      message.createdAt = Date()
      message.hairdresser = hairdresser.displayName
      message.content = content
    }

    completion(INSendMessageIntentResponse(code: .success,
userActivity: nil))
```

```
    }
}
```

All the preceding code does is create a new `Message` object for each recipient of the message and stores it in the Core Data database. If no recipients are found, or no message exists, the operation is considered failed. If everything works out well, the operation is completed successfully.

To test your extension, click on your app target in the top of the Xcode window and select your extension. When you attempt to run it, Xcode will ask you for an app to run it with, as shown in the following screenshot. You can pick **Hairdressers** if you want, but any app will do:

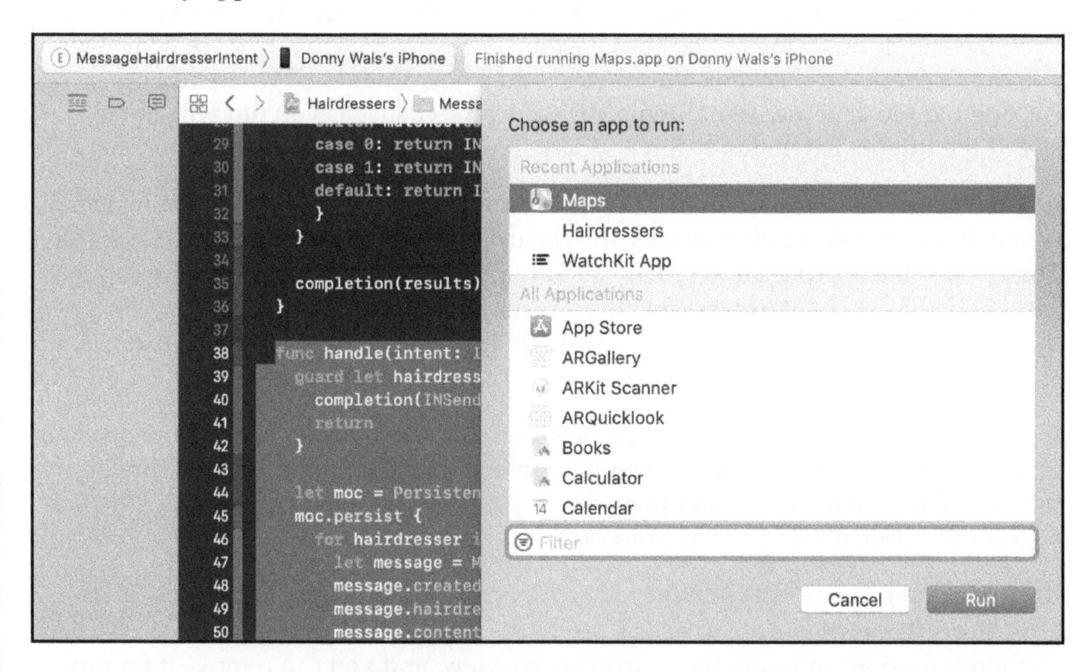

Before testing the extension, make sure that you have run **Hairdressers** at least once to make sure the app has permission to be integrated with Siri.

Even though it's pretty cool that you can now use Siri to send messages, the UI that Siri shows isn't that great. In the next section, you will learn how you can add a custom UI to your Siri intent.

Adding a custom UI to Siri

When your user interacts with Siri, they aren't always looking at their device. But when they are, it's desirable that the experience a user has when using your app through Siri looks and feels a lot like when they're directly interacting with your app. One of the tools to achieve this is using custom vocabularies. You can use a vocabulary to map user- and app-specific terms to Siri's vocabulary, as you have seen in the previous section.

Another way we can customize the Siri experience is through an **Intents UI Extension**. Whenever you add an **Intents Extension** to your project, Xcode asks you if you also want to add a corresponding UI extension. If you select this checkbox, you don't have to do anything to add the UI extension, since it's already there. However, if you didn't check the checkbox, you should add a new **Intents UI Extension** by clicking the + icon in the list of targets for your project, just as you did when you added the **Intents Extension** before.

A custom user interface for an intent works a lot like other UI extensions. When you create the extension, Xcode creates a storyboard, a view controller, and a `.plist` file. The `.plist` file is expected to specify all of the intents that this UI extension can handle. In the case of **Hairdressers**, this is just a single intent; `INSendMessageIntent`. Make sure to update the `Info.plist` for the UI extension accordingly.

If you intend to support multiple intents, it's often a good idea to split your extension up into various extensions, while grouping intents together based on how similar they are. This will make it easier to maintain your code in the long run, and it makes it easier to reason about the way your code and extensions work.

If you open the `IntentViewController` file, you'll find a method named `configureView(for:of:interactiveBehavior:context:completion:)`. This method is used to configure the UI extension's view, and you can remove it for now because you'll implement a slightly simpler version of this method later.

Since the UI extension is used for a messaging feature, it would be nice to show the Hairdressers that will end up receiving the message, and the message that they will receive. Add the following two `@IBOutlet` properties to the `IntentViewController`:

```
@IBOutlet var recipientsLabel: UILabel!
@IBOutlet var messageContentLabel: UILabel!
```

After adding the outlets, add the following method to the `IntentViewController`:

```
func configure(with interaction: INInteraction, context:
INUIHostedViewContext, completion: ((CGSize) -> Void)) {

  guard let messageIntent = interaction.intent as?
INSendMessageIntent,
    let recipients = messageIntent.recipients
    else { return }


  recipientsLabel.text = recipients.map { $0.displayName
}.joined(separator: ", ")
  messageContentLabel.text = messageIntent.content


  let viewWidth = extensionContext?.hostedViewMaximumAllowedSize.width
?? 0
  completion(CGSize(width: viewWidth, height: 100))
}
```

This implementation verifies that a message intent with recipients exists, and then populates the labels you just added accordingly. The code also determines at which width the UI should be displayed. Finally, the completion handler is called with the desired size at which Siri should display the custom UI. A value of 100 should be plenty of room to accommodate the message for now.

Next, open the storyboard file for the UI extension and add two labels. Lay them out as shown in the following screenshot. Don't forget to connect the outlets for these labels as well. You can set the text style for the message label to **Body**, and the recipients label can be set to **Caption** style:

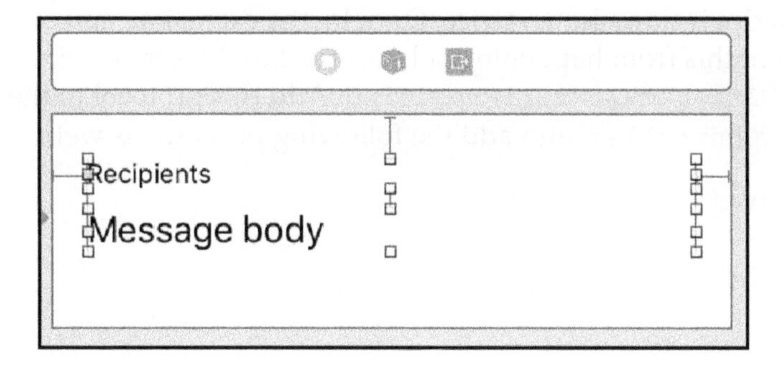

After setting up the UI, run your UI extension instead of the intent extension, and try to send a message to one of the Hairdressers again.

You should see something that looks like the following UI:

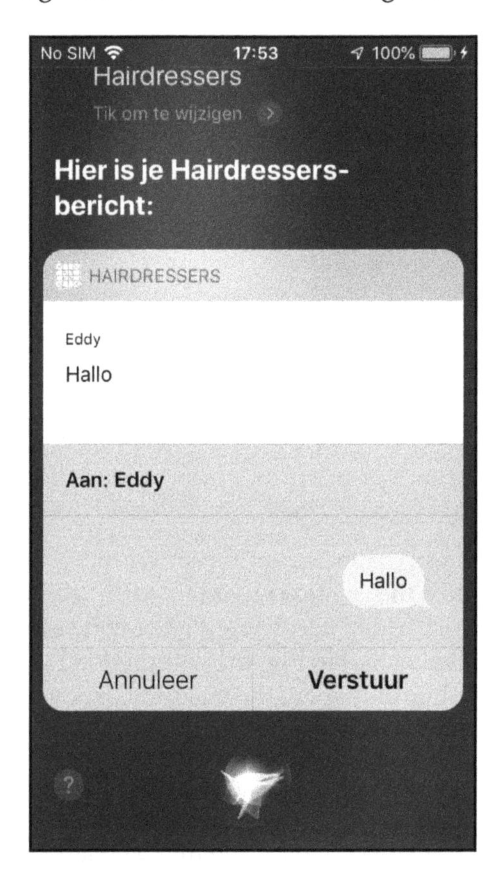

The cool part is that the custom UI appears. There is one problem though. The message transcript is now shown twice: Once by the extension, and once through Siri itself. To prevent this from happening, all you need to do is make our view controller conform to `INUIHostedViewSiriProviding`. Add this protocol to the declaration of `IntentViewController`, and add the following property as well:

```
var displaysMessage = true
```

By doing this, Siri is now aware that you render your own version of the message transcript, so there is no need for Siri to also show the message. Now that you know how to implement a Siri intent and provide a custom UI for it, it's time to take your Siri integration one step further, and have a look at Siri Shortcuts.

Implementing Siri Shortcuts

If you have used iOS for a while, you may have noticed that in places such as the notification center, Siri suggests certain apps you might want to use at a certain time. Siri can do this because it continuously learns about the apps you use, when you use them, and where you are when you use them. By learning your behavior, Siri can make accurate predictions about what you might want to do next.

With the introduction of iOS 12, Apple has announced a new, powerful feature for Siri. This feature is called **Siri Shortcuts**, and it allows app developers to teach Siri about certain actions that a user performs in their apps. Siri then learns about these actions and offers them to the user at the appropriate time.

If you book a taxi home every day at a certain time, and the app you use has implemented Siri Shortcuts, Siri can offer you a shortcut to quickly book your taxi without having to go into the app. This is a great feature for users because it allows them to focus on the things that matter to them, without having to perform repetitive tasks inside of an app.

In this section, you will implement Siri Shortcuts for the Hairdressers app to enable users to quickly book appointments. There are two ways for apps to support shortcuts:

- Through `NSUserActivity` objects
- By donating `INInteraction` objects

First, you will implement the `NSUserActivity` method for adding shortcuts. Then you will define your own custom intent, so you can donate activities through `INInteraction` objects, and handle shortcuts in the background by implementing a second **Intents Extension**.

Implementing shortcuts through NSUserActivity

The simplest way to implement Siri Shortcuts is to use `NSUserActivity` objects. An `NSUserActivity` object contains information about the current activity that a user is performing, and can be used to achieve several goals, one of which is to implement Siri Shortcuts. To find out about other uses of `NSUserActivity`, have a look at `Chapter 21`, *Improved Discoverability with Spotlight and Universal Links.*

Your app must register the identifiers, in the `Info.plist` file, for the user activity objects that it will handle. Add a new entry to the `Info.plist` for **Hairdressers**, and name it **NSUserActivityTypes**. The object type for this key should be an array. In this array, you should add the identifier for the user activity object you are going to use for Siri shortcuts. Usually, this will be a reverse domain name kind of identifier, such as, for example, `com.donnywals.hairdressers.appointment`.

After adding the identifier to your `Info.plist`, you are ready to implement user activities in your app. To nicely encapsulate all logic related to the appointment booking activity, add an extension called `NSUserActivity` to the extensions folder, and implement the following extension in it:

```
extension NSUserActivity {
  static var identifierForAppointment: String {
    return "com.donnywals.hairdressers.appointment"
  }

  static func appointmentActivity() -> NSUserActivity {
    return NSUserActivity(activityType: identifierForAppointment)
  }
}
```

The preceding extension contains the identifier you entered earlier, and a convenience method to create a new user activity that uses the type identifier for an appointment booking activity.

Next, create a folder called `Helpers`, and add a file called `AppointmentShortcutHelper`. You will implement all the logic surrounding shortcuts in this file.

When a user books a new appointment, you should create a new user activity and pass it to Siri Shortcuts. Passing shortcuts to Siri Shortcuts is also called donating a shortcut. Every user activity you donate to Siri needs to have a `title` property, a `userInfo`, a `persistentIdentifier`, and it needs to be marked as `isEligibleForPrediction`.

The following helper method creates a user activity object and should be added to the `AppointmentShortcutHelper`:

```
static func activityForAppointment(_ appointment: Appointment) ->
NSUserActivity {
  let userActivity = NSUserActivity.appointmentActivity()
  let title = "Book an appointment with \(appointment.hairdresser!)
for \(appointment.day!)"

  userActivity.requiredUserInfoKeys = ["hairdresser", "day"]
  userActivity.userInfo = ["hairdresser": appointment.hairdresser!,
"day": appointment.day!]
  userActivity.title = title
  userActivity.isEligibleForPrediction = true
  userActivity.persistentIdentifier = "\(appointment.hairdresser!)-
\(appointment.day!)"
  userActivity.suggestedInvocationPhrase = title

  return userActivity
}
```

The preceding block of code creates a user activity. Note that the sample code uses some force unwrapping with a `!` symbol used to keep the code brief and simple. Make sure to asses whether this is appropriate in your own app, and preferably use safe unwrapping instead. To use and donate the user activity that the preceding snippet creates, add the following code to `AddAppointmentViewController` in the `didTapSave()` method, right before `dismiss(animated:completion:)` is called:

```
let activity =
AppointmentShortcutHelper.activityForAppointment(appointment)
self.userActivity = activity
```

To make an activity the current activity, you need to assign it as the current user activity for the current view controller. If you run the app now, creating a new appointment will donate that appointment to Siri. Normally, Siri will begin showing donations once the user has performed a certain action many times, so Siri is quite sure that it makes sense for your user to see the action offered to them. For debugging purposes, you can go to **Settings** | **Developer**, scroll down a bit, and then enable **Display Recent Shortcuts** and **Display Donations on Lock Screen**, as shown in the following screenshot:

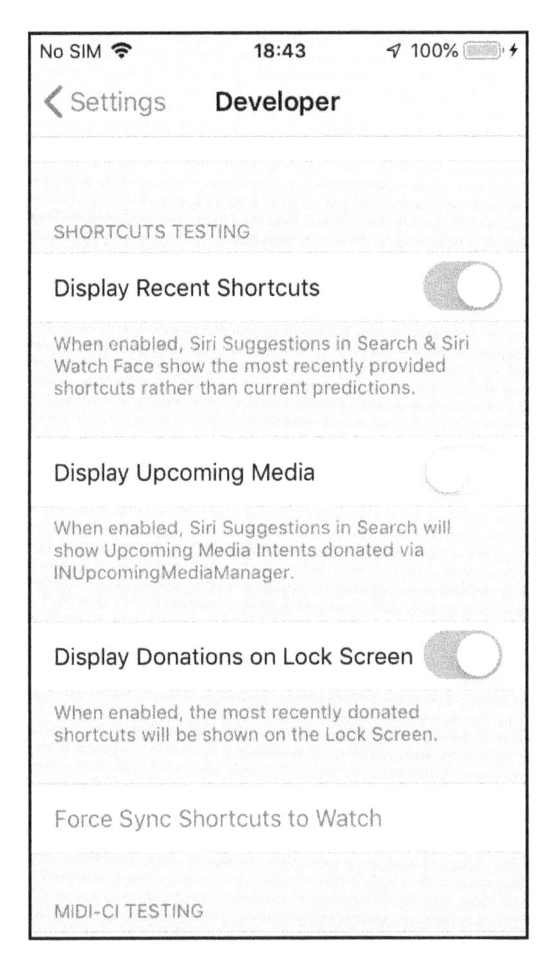

Doing this will immediately show your new donated shortcuts in the notification center and on the lock screen, as shown in the following screenshot:

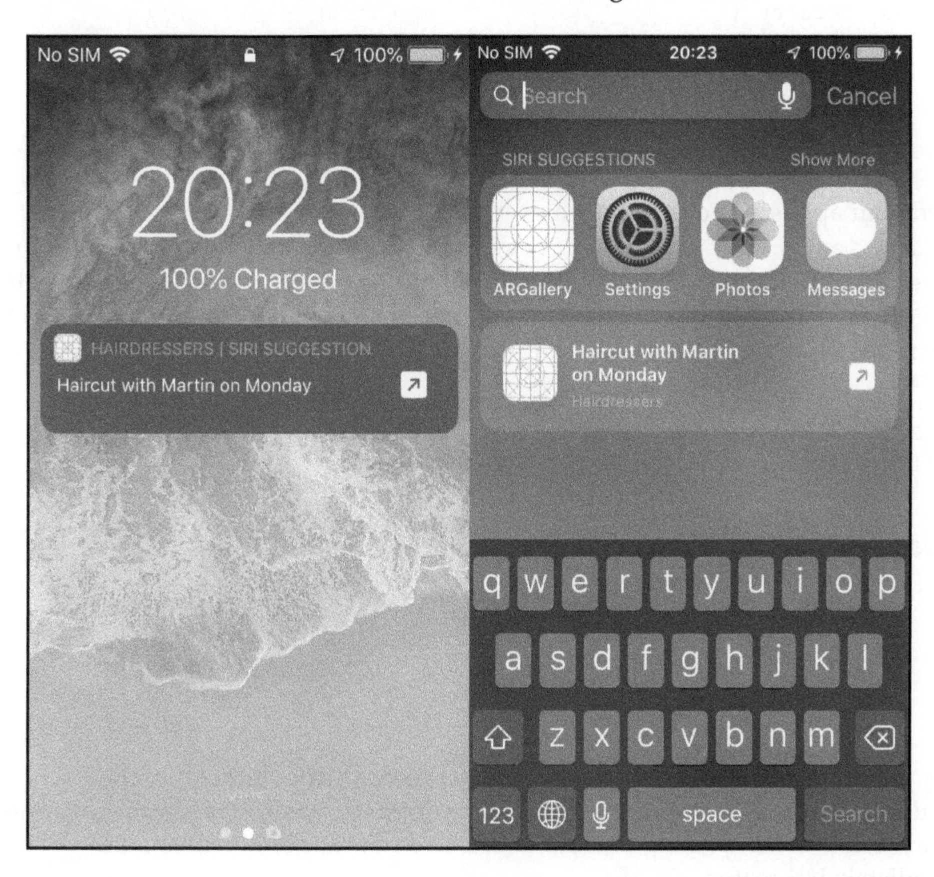

If you only donate your shortcuts to Siri, you're not quite done yet. When a user taps on your shortcut, your application should receive the user activity object and then perform the action the user wants to perform—in this case, booking a repeat appointment with a certain hairdresser.

Add the following helper method to the `AppointmentShortcutHelper` to convert a user activity to an appointment and store it in Core Data:

```
static func storeAppointmentForActivity(_ userActivity:
NSUserActivity) {
  guard let userInfo = userActivity.userInfo,
    let hairdresser = userInfo["hairdresser"] as? String,
    let day = userInfo["day"] as? String
    else { return }
```

```
    let moc = PersistentHelper.shared.persistentContainer.viewContext

    moc.persist {
      let appointment = Appointment(context: moc)
      appointment.day = day
      appointment.hairdresser = hairdresser
    }
  }
```

When your app is asked to handle a user activity, the `application(_:continue:restorationHandler:)` method is called on your `AppDelegate`. Add the following implementation for this method to check whether the received user activity is related to booking an appointment, and pass it on to the helper method to store a new appointment:

```
func application(_ application: UIApplication, continue userActivity:
NSUserActivity, restorationHandler: @escaping
([UIUserActivityRestoring]?) -> Void) -> Bool {

  guard userActivity.userActivityType ==
NSUserActivity.identifierForAppointment
    else { return false }

  AppointmentShortcutHelper.storeAppointmentForActivity(userActivity)

  return true
}
```

After you implement this code, go ahead and book an appointment in the Hairdressers app. You should see your appointment appear in the list of suggested shortcuts, and when you tap on your shortcut, a new appointment should immediately be created.

Users can also add their own voice command for the shortcuts you donate. Users can do this by navigating to the following page in the settings app: **Hairdressers** | **Siri & Search** | **My "Hairdressers" Shortcuts**. A user can then tap on a shortcut to record their own phrase, which can then be used to execute the shortcut they chose, as shown in the following screenshot:

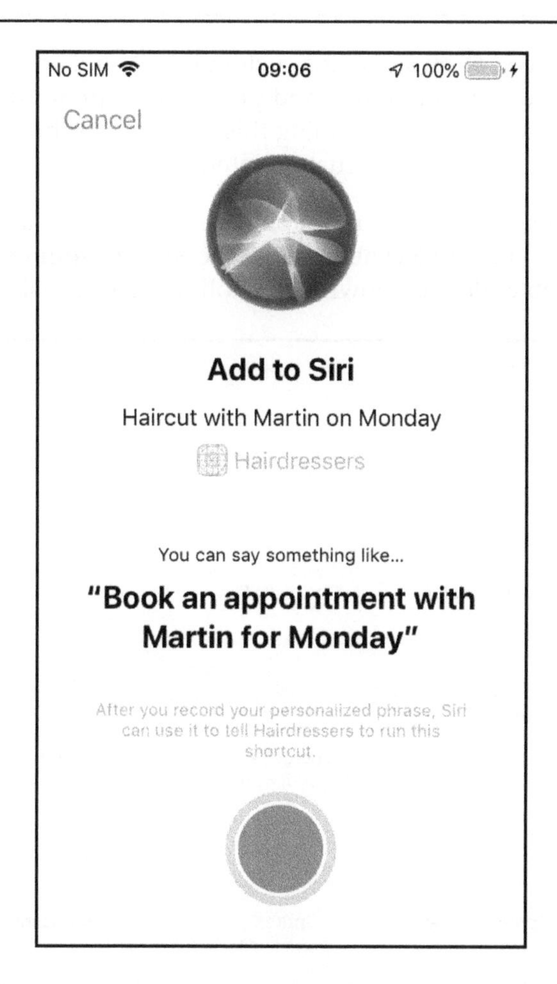

When you donate shortcuts using a user activity, your app will be launched to handle the shortcut. If you want to handle shortcuts in the background, you can create a custom intent and **Intents extension** to achieve this. Let's see how.

Donating shortcuts with INinteractions

Donating shortcuts with user activities is pretty straightforward and powerful. The only downside is that your app has to be launched to handle shortcuts when the user executes them. Sometimes this might be the behavior you need, but for the case of booking an ap

To achieve this, you need to implement a custom intent, and an extension that can handle it. Let's set up the custom intent and prepare the app for donating and potentially handling the intent. After doing this, you will see how to implement an **Intents extension** that handles your custom intent.

The first step to implement your own custom intent is to create an intent definition file. Add a new file to your project and choose the **SiriKit Intent Definition File** type from the list of available files, as shown in the following screenshot:

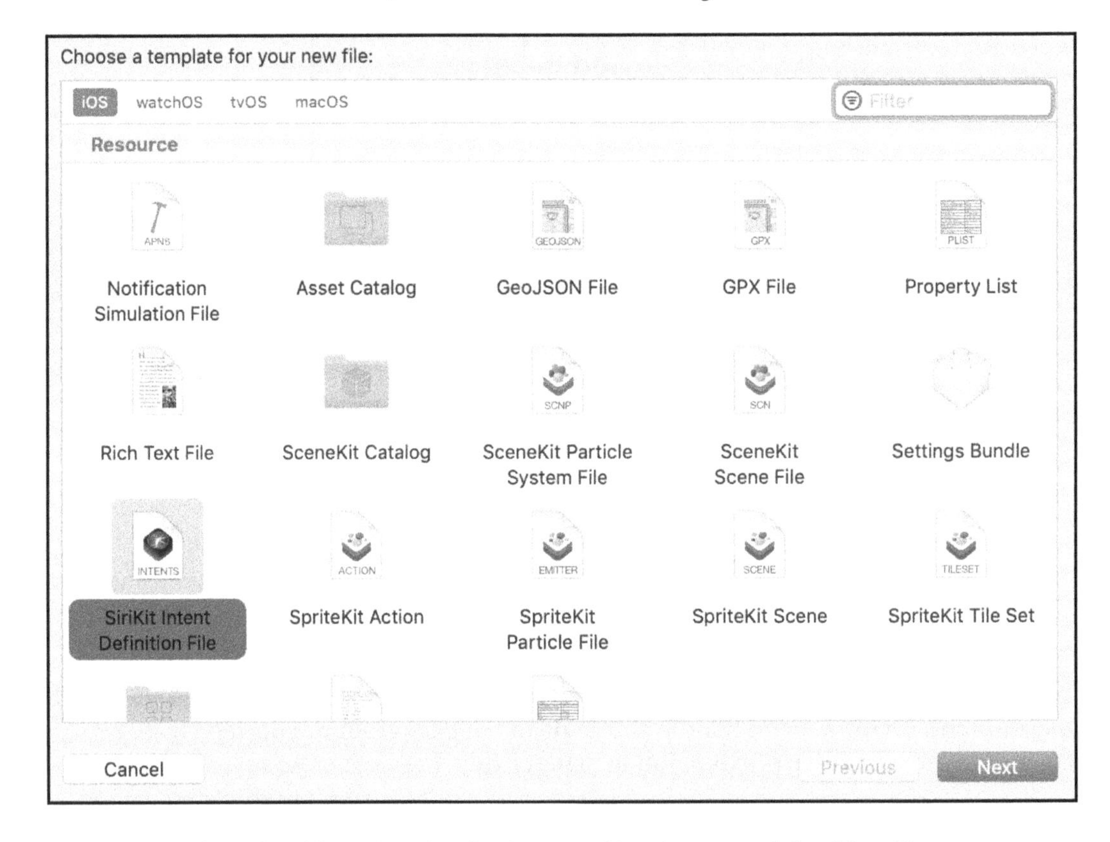

When you select this file in Xcode, the intent editor is opened. In this editor, you can create your custom intents and configure them. Click the + icon in the bottom left corner of the editor to create a new intent. Name your new intent `BookAppointment`.

After creating your intent, you need to choose a category for your intent. There are many available categories to choose from. You should always pick the category that fits your needs as closely as possible. In this case, the **Book** category is a perfect choice. The next step is to give your intent a title and a description, as shown in the following screenshot:

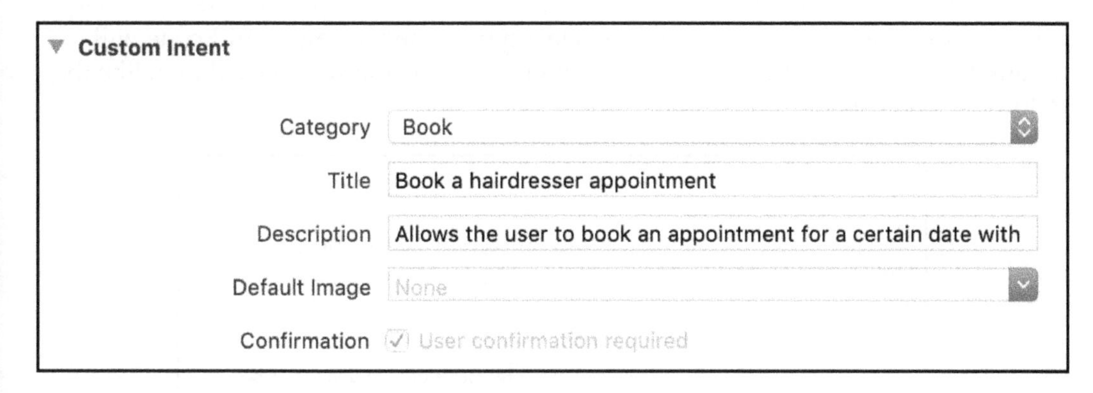

The parameters for your intent are configured in the **Parameters** section. The appointment booking intent should have two parameters:

- `hairdresser` (String)
- `day` (String)

These parameters should appear in Xcode as follows:

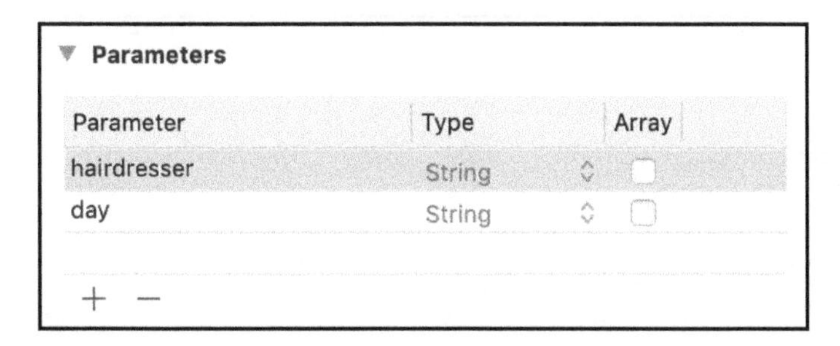

Xcode will generate several subclasses for your custom intent after you have configured it. The parameters you specify in the **Parameters** section will be available as properties on your custom intent.

Now that you have the parameters prepared, you need to configure the shortcut that you want to donate. Every shortcut you base on the intent you just created will have a unique combination of parameters. These parameters can be used to generate a dynamic title for your shortcut.

Add a parameter combination for the `hairdresser` and `day` parameters, and create a title that uses these parameters. You can start typing the message, and Xcode will automatically suggest the parameters for you. The final shortcut should look like the following screenshot:

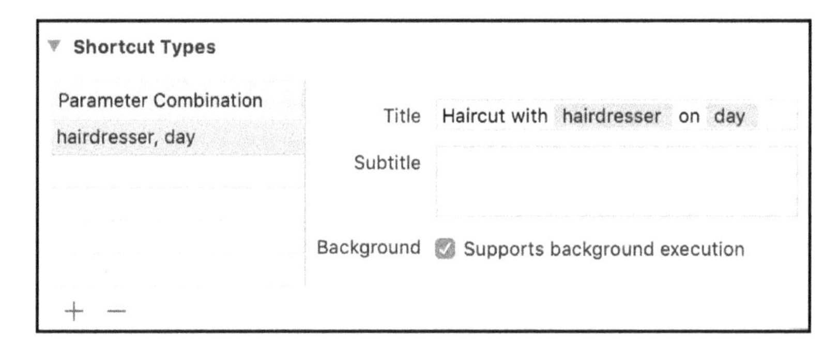

After configuring the intent and shortcut, you can set up response parameters. You can use these parameters to create a helpful response that confirms information for the user. In this case, you can add a `hairdresser` property and use that to verify a user's appointment, as shown in the following screenshot:

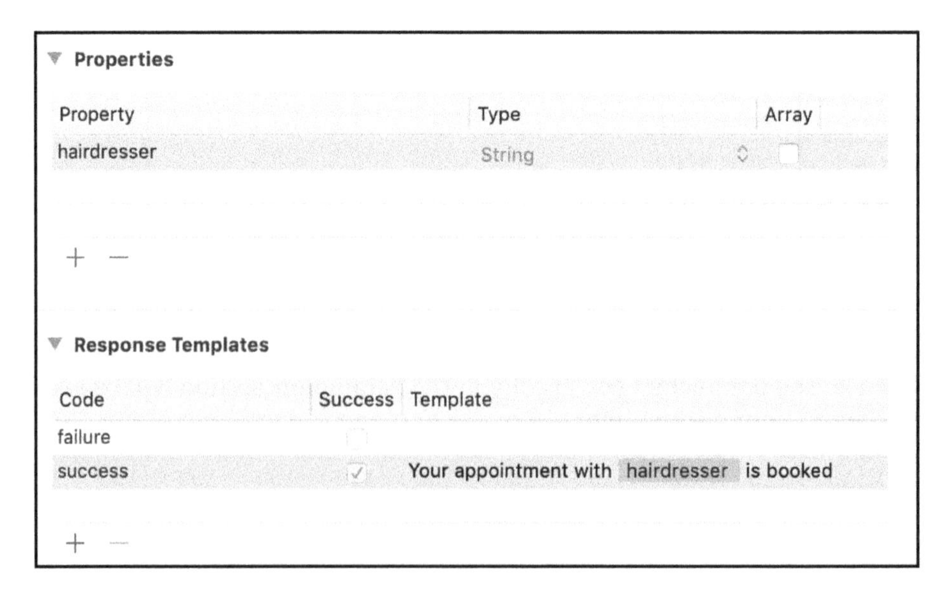

Now that the custom intent is set up, you can update the `AppointmentShortcutHelper` to donate an `INInteraction`, which contains the custom intent, to Siri. Add the following implementation for this helper to `AppointmentShortcutHelper`:

```
static func donateInteractionForAppointment(_ appointment:
Appointment) {
  let intent = BookAppointmentIntent()
  intent.hairdresser = appointment.hairdresser
  intent.day = appointment.day
  let title = "Book an appointment with \(appointment.hairdresser!)
for \(appointment.day!)"
  intent.suggestedInvocationPhrase = title

  let interaction = INInteraction(intent: intent, response: nil)
  interaction.donate(completion: nil)
}
```

Xcode has generated a `BookAppointmentIntent` class based on your custom intent definition. As mentioned before, this intent has the properties that you have defined in the editor, so that you can properly configure your intent.

After configuring the intent, it is wrapped in an `INInteraction` object. To donate the shortcut to Siri, you must call `donate(completion:)` on the interaction object. Next, replace the user activity-based donation code in `AddAppointmentViewController` with the following line:

```
AppointmentShortcutHelper.donateInteractionForAppointment(appointment)
`
```

If your application does happen to be launched by Siri to handle the intent, you should do so in the `application(_:continue:restorationHandler:)` method in `AppDelegate`. The user activity type for your custom intent will be the name of the intent you created. So, if you created an intent called `BookAppointment`, that will be the user activity type that Siri will use. Update the implementation of `application(_:continue:restorationHandler:)` as follows:

```
func application(_ application: UIApplication, continue userActivity:
NSUserActivity, restorationHandler: @escaping
([UIUserActivityRestoring]?) -> Void) -> Bool {

  var allowedActivities = [NSUserActivity.identifierForAppointment,
"BookAppointment"]

  guard allowedActivities.contains(userActivity.activityType)
    else { return false }
```

```
    AppointmentShortcutHelper.storeAppointmentForActivity(userActivity)
    return true
  }
```

This wraps up the configuration of your app. The next step is to create an **Intents Extension** that can handle the new custom intent. Since you have already done this before, the processes won't be explained step by step. Instead, make sure you do all of the following:

1. Create an `Intents Extension`.
2. Enable **App Groups** for the extension.
3. Add the intent definition, managed object context extension, data helpers, and Core Data model to the new extension's target.
4. Configure the extension to support the `BookAppointment` intent.
5. Remove the boilerplate code from the `IntentHandler` class.

Since the `IntentHandler` for the custom intent is only meant to handle a single intent, add the following implementation for the `IntentHandler` class:

```
class IntentHandler: INExtension, BookAppointmentIntentHandling {

  override func handler(for intent: INIntent) -> Any {
    return self
  }

  func handle(intent: BookAppointmentIntent, completion: @escaping
(BookAppointmentIntentResponse) -> Void) {
    guard let hairdresser = intent.hairdresser, let day = intent.day
else {
      completion(BookAppointmentIntentResponse(code: .failure,
userActivity: nil))
      return
    }

    let moc = PersistentHelper.shared.persistentContainer.viewContext

    moc.persist {
      let appointment = Appointment(context: moc)
      appointment.day = day
      appointment.hairdresser = hairdresser

      let response = BookAppointmentIntentResponse(code: .success,
userActivity: nil)
      response.hairdresser = hairdresser
      completion(response)
    }
```

```
    }
}
```

If you book a new appointment in the Hairdressers app and invoke the Siri Shortcut later, the **Intents extension** will be used to handle your intent rather than having to open the app. This is great, because it allows the user to go on with what they were doing, without having to look at your app for confirmation.

Now that your extension is in place, go ahead and play with your Siri Shortcuts a bit. The work for this chapter is completed—you have created an application that is tightly integrated with Siri in order to streamline the user's experience with your app.

Summary

This chapter has taught you a lot about how you can integrate your app with Siri. First, you learned how to handle certain predefined intents, such as sending messages by adding an **Intents Extension**. Then you implemented a custom UI for your extension to enhance the Siri experience.

Next, you learned how to enhance the user's experience by donating Siri Shortcuts. Shortcuts integrate deeply with iOS, which means that Siri picks up and shows the shortcuts added by you at the most appropriate time for the user. You also learned that users can record their own phrases for shortcuts, so they can trigger a shortcut by simply asking Siri.

The next chapter will teach you how you can integrate media like video, audio, and the camera in your app. So make sure to skip on to the next chapter when you're done playing with your Siri integration.

Questions

1. How does Siri understand a user's request?

    a) It has templates for the user's requests.
    b) Siri uses natural language processing.
    c) Siri uses the current time to determine what a user means.

2. How are app extensions launched?

    a) App extensions are launched by the app that owns the extension.
    b) App extensions are launched by the currently active app.
    c) App extensions are launched by the extensions framework.

3. How can you share data between an app and an extension?

    a) By adding the required files to both targets.
    b) By implementing app groups.
    c) Both of the above.

4. What are the three stages of handling a Siri intent?

    a) Resolve, Confirm, Handle.
    b) Confirm, Resolve, Handle.
    c) Handle, Resolve, Confirm.

5. How do you make sure Siri does not render its default UI?

    a) By providing your own UI.
    b) By conforming the `IntentViewController` to the `INUIHostedViewSiriProviding` protocol, and setting `displaysMessage` to `true`.
    c) By setting `showsDefaultUI` to `false` in the UI extension's `Info.plist`.

6. How can you donate actions to Siri?

    a) Through user activities and interaction objects
    b) With the Siri Shortcuts framework
    c) By adding them to a Core Data database

7. How can a user add their own voice commands for Siri Shortcuts?

    a) Through the settings app.
    b) The app must provide its own audio recording.
    c) Siri automatically asks this when an intent is donated.

Further reading

- The Intents documentation is available at `https://developer.apple.com/documentation/sirikit/inextension`
- The SiriKit documentation is available at `https://developer.apple.com/documentation/sirikit`

17
Using Media in Your App

lot of the apps that people use every day make use of media in some way. Some apps show photos and videos in a user's feed. Other apps focus on playing audio or video, while there are also apps that allow users to record media and share it with their peers. You can probably name at least a couple of very well known apps that make use of such media in one way or the other.

Because media has such a significant presence in people's daily life, it's good to know how you can integrate media into your own apps. iOS has excellent support for media playback and offers several different ways to create and consume different types of media. Some ways provide less flexibility, but are more straightforward to implement. Others are more complex, but provide significant power to you as a developer.

In this chapter, you will learn about several ways to play and record media on iOS. You will learn how to play and record video, play audio, take pictures, and you'll even learn how to apply filters to images with Apple's Core Image framework. This chapter covers the following topics:

- Playing audio and video
- Recording video and taking pictures
- Manipulating photos with Core Image

By the end of this chapter, you will have a great foundation that you can build on to create engaging experiences for your users, allowing them to not only view content, but also to create their own content in your app.

Playing audio and video

To make playing audio and video files as simple and straightforward as can be, Apple has created the `AVFoundation` framework. This framework contains a lot of helper classes that provide very low-level control over how iOS plays audio and video files. You can use `AVFoundation` to build a rich, custom media player with as many features as you need for your purpose.

If you're looking for a simpler way to integrate media into your app, the `AVKit` framework might be what you need. `AVKit` contains several helpers that build upon the `AVFoundation` components to provide an excellent default player that supports many features like subtitles, Airplay, and more.

In this section, you will learn how to implement a simple video player with the `AVPlayerViewController` from the `AVKit` framework. You will also implement a more complex audio player with `AVFoundation` components that plays audio in the background and displays, on the lock screen, the audio track currently being played.

To follow along with the samples, you should open the `MediaPlayback_start` project in this chapter's code bundle. The starter app contains a straightforward interface with a tab bar and two pages. You will implement a video player on one page, and the audio player on the other page. The audio page comes with some predefined controls and actions that you will implement later.

Creating a simple video player

The first thing you need to do to implement a video player is to obtain a video file. You can use any video that is encoded in the `h.264` format. A good sample video is the **Big Buck Bunny** sample movie that was created by the Blender Foundation. You can find this video at the following URL: `http://bbb3d.renderfarming.net/download.html`. If you want to use this video to practice with, make sure to download the 2D version of the video.

As stated before, you will implement the video player using the `AVPlayerViewController`. This view controller provides a convenient wrapper around several components from `AVFoundation`, and also provides default video controls, so you don't have to build your entire video player from scratch, as you will do for the audio player later.

The `AVPlayerViewController` is highly configurable, which means that you can choose whether the player supports Airplay, shows playback controls, whether it should be full screen when a video plays, and more. For a complete list of configurable options, you can refer to Apple's `AVPlayerViewController` documentation.

Once you have found your test video, you should add it to the **MediaPlayback** project and ensure that the video is added to the app target. After doing this, open `VideoViewController.swift` and add the following line to import `AVKit`:

```
import AVKit
```

You should also add a property to `VideoViewController` to hold on to your video player instance. Add the following line to the `VideoViewController` class to do this:

```
let playerController = AVPlayerViewController()
```

Since the `AVPlayerViewController` is a `UIViewController` subclass, you should add it to the `VideoViewController` as a child view controller. Doing this will make sure that `VideoViewController` forwards any view controller lifecycle events, such as `viewDidLoad()`, along with any changes in trait collections and more to the video player. To do this, add the following code to the `viewDidLoad()` method in `VideoViewController`:

```
// 1
addChild(playerController)
playerController.didMove(toParent: self)

// 2
view.addSubview(playerController.view)

let playerView = playerController.view!
playerView.translatesAutoresizingMaskIntoConstraints = false

NSLayoutConstraint.activate([
  playerView.widthAnchor.constraint(equalTo: view.widthAnchor,
constant: -20),
  playerView.heightAnchor.constraint(equalTo: playerView.widthAnchor,
multiplier: 9/16),
  playerView.centerXAnchor.constraint(equalTo: view.centerXAnchor),
  playerView.centerYAnchor.constraint(equalTo: view.centerYAnchor)])
```

The previous code snippet adds the video player to the video view controller as a child view controller. When you add a view controller as a child view controller, you must always call `didMove(toParent:)` on the child controller to make sure it knows that it has been added as a child view controller to another view controller. After adding the video player as a child view controller, the video player's view is added as a subview for the video view controller, and some constraints are set up to position the player view.

This is all you need to do to create an instance of the video player and make it appear in your view controller. The last step is to obtain a reference to your video file, create an `AVPlayer` that has a reference to the video file, and assign it to the player. Add the following code to do this:

```
let url = Bundle.main.url(forResource: "samplevideo", withExtension:
"mp4")!
playerController.player = AVPlayer(url: url)
```

The preceding code looks for a video file called `samplevideo.mp4` and obtains a URL for that file. It then creates an instance of `AVPlayer` that points to that video file, and assigns it to the video player. The `AVPlayer` object is responsible for actually playing the video file. The `AVPlayerViewController` instance uses the `AVPlayer` instance to play the video, and manages the actual playback of the video internally.

If you run your app after adding the player this way, you will find that the video plays perfectly well, and that you have access to all the controls you might need. This is a great demonstration of how simple it is to add basic media integration to your app. The next step is a little more complex. You will directly use an `AVAudioPlayer` instance to play an audio file that is controlled through several custom media controls. The player will even play audio in the background and integrate with the lock screen to show information about the current file. In other words, you will build a simple audio player that does everything a user would expect it to do.

Creating an audio player

Before you can implement your audio player, you will need to obtain some `.mp3` files you wish to use in your player. If you don't have any audio files on your computer, you can get some files from The Free Music Archive website, available at `https://freemusicarchive.org/about`, to obtain a couple of free songs that you would like to use for playback. Make sure to add them to the **MediaPlayer** Xcode project and ensure that they are included in the app target.

You will build the audio player using the following steps:

1. Implementing the necessary controls to start and stop the player, and navigate to the next and previous songs.
2. Implementing the time scrubber.
3. Reading the file's metadata and show it to the user.

The user interface, outlets, and actions are already set up, so make sure to familiarize yourself with the existing code before following along with the implementation of the audio player.

Implementing basic audio controls

Before you implement the audio player code, you will need to do a little bit of housekeeping. To be able to play audio, you need a list of the files that the player will play. In addition to this list, you also need to keep track of what song the user is currently playing, so you can determine the next and previous songs. Lastly, you also need to have the audio player itself. Instead of using a pre-built component, you will build your own audio player using an `AVAudioPlayer` object. The `AVAudioPlayer` is perfect for implementing a simple audio player that plays a couple of local `.mp3` files. It offers some convenient helper methods to easily adjust the player's volume, seek to a specific timestamp in the song, and more.

Define the following properties in `AudioViewController.swift`:

```
let files = ["one", "two", "three"]
var currentTrack = 0
var audioPlayer: AVAudioPlayer!
```

Make sure to replace the `files` array with the filenames that you use for your own audio files. The `audioPlayer` does not have a value yet at this point. You will set up the audio player next.

Before you can play audio, you need to obtain a reference to a media file and provide this reference to an `AVAudioPlayer` object. Any time you want to load a new media file, you will have to create a new instance of the audio player, since you can't change the current file once a file is playing. Add the following helper method to the `AudioViewController` to load the current track and create an `AVAudioPlayer` instance:

```
func loadTrack() {
  let url = Bundle.main.url(forResource: files[currentTrack],
withExtension: "mp3")!
```

```
    audioPlayer = try! AVAudioPlayer(contentsOf: url)
    audioPlayer.delegate = self
}
```

This method reads the file name for the current track and retrieves the local URL for it. This URL is then used to create and set the `audioPlayer` property on the `AudioViewController`. The view controller is also assigned as the delegate for the audio player. You won't implement any of the delegate methods just yet, but you can add the following extension to make `AudioViewController` conform to the `AVAudioPlayerDelegate` protocol to ensure your code compiles:

```
extension AudioViewController: AVAudioPlayerDelegate {

}
```

You will implement one of the `AVAudioPlayerDelegate` methods when you add support for navigating to the next and previous tracks.

Add the following two methods to the audio view controller to add support for playing and pausing the current audio file:

```
func startPlayback() {
  audioPlayer.play()
  playPause.setTitle("Pause", for: .normal)
}

func pausePlayback() {
  audioPlayer.pause()
  playPause.setTitle("Play", for: .normal)
}
```

These methods are relatively straightforward. They call the audio player's `play()` and `pause()` methods and update the button's label, so it reflects the current player state. Add the following implementation for `playPauseTapped()` so the play and pause methods get called when the user taps the play/pause button:

```
@IBAction func playPauseTapped() {
  if audioPlayer.isPlaying {
    pausePlayback()
  } else {
    startPlayback()
  }
}
```

If you run the app now, you can tap the play/pause button to start and stop the currently playing file. Make sure your device is not in silent mode, because the audio for your app is muted when the device is in silent mode. You will learn how to fix this when you implement the ability to play audio in the background. The next step is to add support for playing the next and previous tracks. Add the following two implementations to `AudioViewController` to do this:

```
@IBAction func nextTapped() {
  currentTrack += 1
  if currentTrack >= files.count {
    currentTrack = 0
  }
  loadTrack()
  audioPlayer.play()
}

@IBAction func previousTapped() {
  currentTrack -= 1
  if currentTrack < 0 {
    currentTrack = files.count - 1
  }
  loadTrack()
  audioPlayer.play()
}
```

The preceding code adjusts the current track index, loads the new track, and immediately plays it. Note that every time the user taps on the next or previous button, a fresh audio player has to be created by calling `loadTrack()`. If you run the app now, you can play audio, pause it, and skip to the next or previous tracks.

When you allow a full song to play, it won't yet advance to the next song afterward. To implement this, you need to add an implementation for the `audioPlayerDidFinishPlaying(_:successfully:)` method from `AVAudioPlayerDelegate`. Add the following implementation to call `nextTapped()`, so the next song automatically plays when the current song finishes:

```
func audioPlayerDidFinishPlaying(_ player: AVAudioPlayer, successfully
flag: Bool) {
  nextTapped()
}
```

Now that the first features are implemented, the next step is to implement the time scrubber that shows the current song's progress and allows the user to adjust the playhead's position.

Implementing the time scrubber

The user interface for the audio player app already contains a scrubber that is hooked up to the following three actions in the view controller:

- `sliderDragStart()`
- `sliderDragEnd()`
- `sliderChanged()`

When an audio file is playing, the scrubber should automatically update to reflect the current position in the song. However, when a user starts dragging the scrubber, it should not update its position until the user has chosen the scrubber's new position. When the user is done dragging the scrubber, it should adjust itself based on the song's progress again. Any time the value for the slider changes, the audio player should adjust the playhead, so the song's progress matches that of the scrubber.

Unfortunately, the `AVAudioPlayer` object does not expose any delegate method to observe the progress of the current audio file. To update the scrubber regularly, you can implement a timer that updates the scrubber to the audio player's current position every second. Add the following property to the `AudioViewController`, so you can hold on to the timer after you have created it:

```
var timer: Timer?
```

Also, add the following two methods to `AudioViewController` as a convenient way to start the timer when the user starts dragging the scrubber, or when a file starts playing, and stop it when a user starts dragging the scrubber or to preserve resources when the playback is paused:

```
func startTimer() {
  timer = Timer.scheduledTimer(withTimeInterval: 1, repeats: true) {
[unowned self] timer in
    self.slider.value = Float(self.audioPlayer.currentTime /
self.audioPlayer.duration)
  }
}

func stopTimer() {
  timer?.invalidate()
}
```

Add a call to `startTimer()` in the `startPlayback()` method and a call to `stopTimer()` in the `stopPlayback()` method. If you run the app after doing this, the scrubber will immediately begin updating its position when a song starts playing. However, scrubbing does not work yet. Add the following implementations for the scrubber actions to enable manual scrubbing:

```
@IBAction func sliderDragStart() {
  stopTimer()
}

@IBAction func sliderDragEnd() {
  startTimer()
}

@IBAction func sliderChanged() {
  audioPlayer.currentTime = Double(slider.value) *
audioPlayer.duration
}
```

The preceding methods are relatively simple, but they provide a very powerful feature that immediately makes your homemade audio player feel like an audio player you might use every day. The final step for implementing the audio player's functionality is to display metadata about the current song.

Displaying song metadata

Most `.mp3` files contain metadata in the form of `ID3` tags. These metadata tags are used by applications such as iTunes to extract information about a song and display it to the user, as well as to categorize a music library or filter it. You can gain access to an audio file's metadata through code, by loading the audio file into an `AVPlayerItem` object and extracting the metadata for its internal `AVAsset` instance. An `AVAsset` object contains information about a media item, such as its type, location, and more. When you load a file using an `AVPlayerItem` object, it will automatically create a corresponding `AVAsset` object for you.

A single asset can contain loads of metadata in the metadata dictionary. Luckily, Apple has captured all of the valid `ID3` metadata tags in the `AVMetadataIdentifier` object, so once you have extracted the metadata for an `AVAsset`, you can loop over all of its metadata to filter out the data you need. The following method does this, and sets the extracted values on the `titleLabel` variable of `AudioViewController`, as shown here:

```
func showMetadataForURL(_ url: URL) {
  let mediaItem = AVPlayerItem(url: url)
  let metadata = mediaItem.asset.metadata
  var information = [String]()

  for item in metadata {
    guard let identifier = item.identifier
      else { continue }

    switch identifier {
    case .id3MetadataTitleDescription, .id3MetadataBand:
      information.append(item.value?.description ?? "")
    default:
      break
    }
  }

  let trackTitle = information.joined(separator: " - ")
  titleLabel.text = trackTitle
}
```

Make sure to add a call to this method from `loadTrack()`, and pass the audio file's URL that you obtain in `loadTrack()` to `showMetadataForURL(_:)`. If you run your app now, your basic functionality should be all there. The metadata should be shown correctly, the scrubber should work, and you should be able to skip songs or pause the playback.

Even though your media player seems to be pretty much done at this point, did you notice that the music pauses when you send the app to the background? To make your app feel more like a real audio player, you should implement background audio playback and make sure the currently playing song is presented on the user's lock screen, similar to how the native music app for iOS works. This is precisely the functionality you will add next.

Playing media in the background

On iOS, playing audio in the background requires special permissions that you can enable in your app's **Capabilities** tab. If you enable the **Background modes** capability, you can select the **Audio, Airplay**, and **Picture in Picture** option to make your app eligible for playing audio in the background. The following screenshot shows the enabled capability for playing audio in the background:

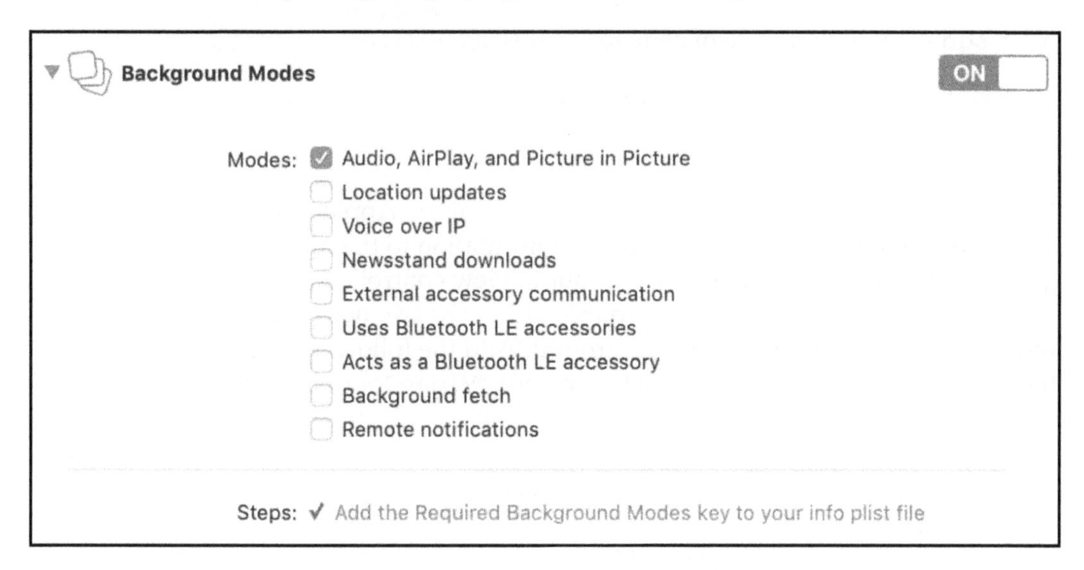

If you want to add proper support for background audio playback, there are three features you need to implement:

- Set up an audio session, so audio continues playing in the background.
- Submit metadata to the "now playing" info center.
- Respond to playback actions from remote sources, such as the lock screen.

You can set up the audio session for your app with just two lines of code. When you create an audio session, iOS will treat the audio played by your app slightly differently. For instance, your songs will play even if the device is set to silent. It also makes sure your audio is played when your app is in the background, if you have the proper capabilities set up. Add the following code to `viewDidLoad()` to set up an audio session for the app:

```
try? AVAudioSession.sharedInstance().setCategory(.playback, mode:
.default, options: [.allowAirPlay])
try? AVAudioSession.sharedInstance().setActive(true, options: [])
```

The second feature to add is to supply information about the currently playing track. All information about the currently playing media file should be passed to the `MPNowPlayingInfoCenter` object. This object is part of the `MediaPlayer` framework, and is responsible for showing the user information about the currently playing media file on the lock screen and in the command center. Before you pass information to the "now playing" info center, make sure to import the `MediaPlayer` framework at the top of the `AudioViewController.swift` file.

Next, add the following line of code to `viewDidLoad()`:

```
NotificationCenter.default.addObserver(self, selector:
#selector(updateNowPlaying), name:
UIApplication.didEnterBackgroundNotification, object: nil)
```

In the documentation for `MPNowPlayingInfoCenter`, Apple states that you should always pass the most recent now playing information to the info center when the app goes to the background. To do this, the audio view controller should listen to the `UIApplication.didEnterBackgroundNotification` notification, so it can respond to the app going to the background. Add the following implementation for the `updateNowPlaying()` method to the `AudioVideoController`:

```
@objc func updateNowPlaying() {
  var nowPlayingInfo = [String: Any]()
  nowPlayingInfo[MPMediaItemPropertyTitle] = titleLabel.text ??
"untitled"
  nowPlayingInfo[MPNowPlayingInfoPropertyElapsedPlaybackTime] =
audioPlayer.currentTime
  nowPlayingInfo[MPMediaItemPropertyPlaybackDuration] =
audioPlayer.duration

  MPNowPlayingInfoCenter.default().nowPlayingInfo = nowPlayingInfo
}
```

The preceding code configures a dictionary with metadata about the currently playing file, and passes it to the "now playing" info center. This method is called automatically when the app goes to the background, but you should also update the "now playing" information when a new song begins playing. Add a call to `updateNowPlaying()` in the `loadTrack()` method to make sure the "now playing" information is updated whenever a new track is loaded.

The next and final step is to respond to remote commands. When the user taps the play/pause button, next button, or previous button on the lock screen, this is sent to your app as a remote command. You should explicitly define the handlers that should be called by iOS when a remote command occurs. Add the following method to `AudioViewController` to add support for remote commands:

```
func configureRemoteCommands() {
  let commandCenter = MPRemoteCommandCenter.shared()

  commandCenter.playCommand.addTarget { [unowned self] event in
    guard self.audioPlayer.isPlaying == false
      else { return .commandFailed }

    self.startPlayback()
    return .success
  }

  commandCenter.pauseCommand.addTarget { [unowned self] event in
    guard self.audioPlayer.isPlaying
      else { return .commandFailed }

    self.pausePlayback()
    return .success
  }

  commandCenter.nextTrackCommand.addTarget { [unowned self] event in
    self.nextTapped()
    return .success
  }

  commandCenter.previousTrackCommand.addTarget { [unowned self] event
in
    self.previousTapped()
    return .success
  }

  UIApplication.shared.beginReceivingRemoteControlEvents()
}
```

The preceding code obtains a reference to the remote command center and registers several handlers. It also calls `beginReceivingRemoteControlEvents()` on the application object to make sure it receives remote commands. Add a call to `configureRemoteCommands()` in `viewDidLoad()`, to make sure that the app begins receiving remote commands as soon as the audio player is configured.

Try to run your app and send it to the background. You should be able to control media playback from both the control center and the lock screen. The visible metadata should correctly update when you skip to the next or previous song, and the scrubber should accurately represent the current position of playback in the song.

At this point, you have implemented a reasonably complete audio player that has pretty sophisticated behaviors. The next step in your exploration of media on iOS is to discover how you can take pictures and record video.

Recording video and taking pictures

In addition to playing existing media, you can also make apps that allow users to create their own content. In this section, you will learn how you can use a built-in component to enable users to take a picture. You will also learn how you can use a raw video feed to record a video. If you want to follow along with the samples in this section, make sure to grab the starter project for **Captured** from this chapter's code bundle.

The starter project contains a couple of view controllers and some connected outlets and actions. Note that there is a `UIViewController` extension in the project, too. This extension includes a helper method that makes displaying an alert to the user a little bit simpler. This extension will be used to show an alert that informs the user when their photo or video is stored to the camera roll.

Since a user's camera and photo library are considered very privacy-sensitive, you need to make sure that you add the following privacy-related keys to the app's `Info.plist`:

- **Privacy - Camera Usage Description**: this property is required in order to access the camera so you can take pictures and record video.
- **Privacy - Microphone Usage Description**: you must add this property so that your videos record audio, as well as images.
- **Privacy - Photo Library Additions Usage Description**: this property allows you to write photos to the user's photo library.

Make sure to provide a good description for the privacy keys, so the user knows why you need access to their camera, microphone, and photo library. The better your description is, the more likely the user is to allow your app to access the associated privacy-sensitive information. After adding the keys, you are ready to see how you can take a picture using the built-in `UIImagePickerController` component of the `UIKit`.

Taking and storing a picture

When you need a user to supply an image, they can do this by either selecting an image from their photo library, or by taking a picture with the camera. The `UIImagePickerController` supports both ways of picking an image. In this sample, you will learn how you can allow users to take an image using the camera. Changing the example to allow users to select an image from their photo library should be trivial, as long as you remember to add the **Privacy - Photo Library Usage Description** key to your `Info.plist`.

Add the following implementation for `viewDidLoad()` to the `ImageViewController` class:

```
override func viewDidLoad() {
  super.viewDidLoad()

  let imagePicker = UIImagePickerController()
  imagePicker.sourceType = .camera
  imagePicker.delegate = self
  present(imagePicker, animated: true, completion: nil)
}
```

The previous implementation creates an instance of the `UIImagePickerController` object, and configures it so that it uses the camera as the image source and presents it to the user. Note that the view controller is set as a delegate for the image picker. When the user has taken a picture, the image picker will notify its delegate about this, so that it can extract the image and use it. In this case, the image should be given the `selectedImage` label in the view controller so it can be shown in the image view, and saved when the user taps on the save button and the `saveImage()` method is called as a result.

Add the following extension to make `ImageViewController` conform to `UIImagePickerControllerDelegate`:

```
extension ImageViewController: UIImagePickerControllerDelegate,
UINavigationControllerDelegate {
  func imagePickerController(_ picker: UIImagePickerController,
didFinishPickingMediaWithInfo info: [UIImagePickerController.InfoKey :
Any]) {
    picker.dismiss(animated: true, completion: nil)

    guard let image = info[.originalImage] as? UIImage
      else { return }

    selectedImage = image
```

```
    }
}
```

Note that this extension also makes the image view controller conform to `UINavigationControllerDelegate`. The `delegate` property on the image picker controller requires all delegates to conform to both `UINavigationControllerDelegate` and `UIImagePickerControllerDelegate`.

When the user has taken a picture with the camera, `imagePickerController(_:didFinishPickingMediaWithInfo)` is called to notify the delegate about the photo that the user took. The first thing that the preceding code does is dismiss the picker, as it's no longer needed. The picture that the user just took is stored in the `info` dictionary as the original image. When the image is extracted from the dictionary, it is set as `selectedImage`.

To store the image, add the following implementation of `saveImage()`:

```
@IBAction func saveImage() {
  guard let image = selectedImage
    else { return }

  UIImageWriteToSavedPhotosAlbum(image, self,
#selector(didSaveImage(_:withError:contextInfo:)), nil)
}

@objc func didSaveImage(_ image: UIImage, withError error: Error?,
contextInfo: UnsafeRawPointer) {
  guard error == nil
    else { return }

  presentAlertWithTitle("Success", message: "Image was saved
succesfully")
}
```

The preceding code calls `UIImageWriteToSavedPhotosAlbum(_:_:_:_)` to store the image in the user's photo library. When the save operation completes, the `didSaveImage(_:withError:contextInfo:)` method will be called. If this method does not receive any errors, then the photo was successfully stored in the photo library and an alert is shown.

Allowing the user to take a picture by implementing `UIImagePickerController` is relatively straightforward, and it's a great way to implement a camera feature in your app without too much effort. Sometimes, you need more advanced access to the camera. In these cases, you can use `AVFoundation` to gain access to the raw video feed from the camera, as you will see next.

Recording and storing video

In the previous section, you used AVFoundation to build a simple audio player app. You will now use AVFoundation again, except instead of playing video or audio, you will now record video and store it in the user's photo library. When using AVFoundation to record a video feed, you do so with an AVCaptureSession object. A capture session is responsible for taking the input from one or more AVCaptureDeviceInput objects, and writing it to an AVCaptureOutput subclass.

The following diagram shows the objects that are involved with recording media through an AVCaptureSession:

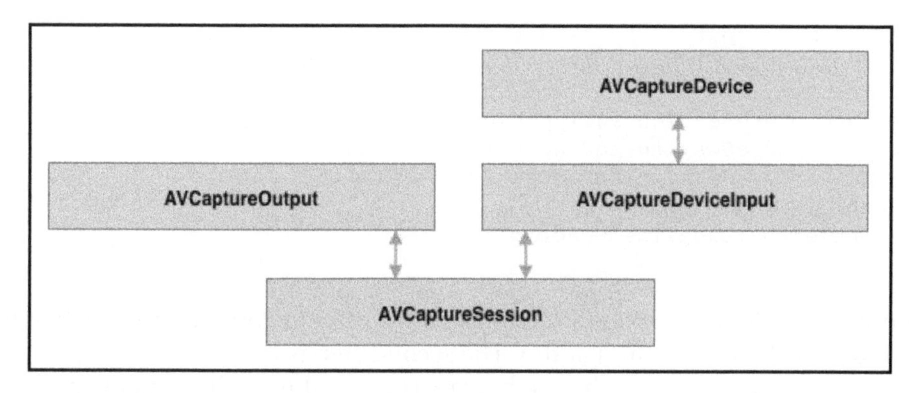

To get started on implementing the video recorder, make sure to import AVFoundation in RecordVideoViewController.swift. Also, add the following properties to the RecordVideoViewController class:

```
let videoCaptureSession = AVCaptureSession()
let videoOutput = AVCaptureMovieFileOutput()

var previewLayer:  AVCaptureVideoPreviewLayer?
```

Most of the preceding properties should look familiar, because they were also shown in the screenshot that outlined the components that are involved with an AVCaptureSession. Note that AVCaptureMovieFileOutput is a subclass of AVCaptureOutput, specialized in capturing video. The preview layer will be used to render the video feed at runtime and present it to the user, so they can see what they are capturing with the camera.

The next step is to set up the `AVCaptureDevice` objects for the camera and microphone and associate them with the `AVCaptureSession`. Add the following code to the `viewDidLoad()` method:

```
// 1
guard let camera = AVCaptureDevice.default(.builtInWideAngleCamera,
for: .video, position: .back),
  let microphone = AVCaptureDevice.default(.builtInMicrophone, for:
.audio, position: .unspecified)
  else { return }

// 2
do {
  let cameraInput = try AVCaptureDeviceInput(device: camera)
  let microphoneInput = try AVCaptureDeviceInput(device: microphone)

  videoCaptureSession.addInput(cameraInput)
  videoCaptureSession.addInput(microphoneInput)
  videoCaptureSession.addOutput(videoOutput)
} catch {
  print(error.localizedDescription)
}
```

The preceding code first obtains a reference to the camera and microphone that will be used to record the video and audio. The second step is to create the `AVCaptureDeviceInput` objects that are associated with the camera and microphone and associate them with the capture session. The video output is also added to the video capture session. If you examine the screenshot that you saw earlier and compare it with the preceding code snippet, you will find that all four components are present in this implementation.

The next step is to provide the user with a view that shows the current camera feed, so they can see what they are recording. Add the following code to `viewDidLoad()` after the capture session setup code:

```
previewLayer = AVCaptureVideoPreviewLayer(session:
videoCaptureSession)
previewLayer?.videoGravity = .resizeAspectFill
videoView.layer.addSublayer(previewLayer!)

videoCaptureSession.startRunning()
```

The preceding code sets up the preview layer and associates it with the video capture session. The preview layer will directly use the capture session to render the camera feed. The capture session is then started. This does not mean that the recording session starts; rather, only that the capture session will begin processing the data from its camera and microphone inputs.

The preview layer is added to the view at this point, but it doesn't cover the video view yet. Add the following implementation for `viewDidLayoutSubviews()` to `RecordVideoViewController` to set the preview layer's size and position, so it matches the size and position of `videoView`:

```
override func viewDidLayoutSubviews() {
  super.viewDidLayoutSubviews()

  previewLayer?.bounds.size = videoView.frame.size
  previewLayer?.position = CGPoint(x: videoView.frame.midX, y:
videoView.frame.size.height / 2)
}
```

Running the app now will already show you the camera feed. However, tapping the record button doesn't work yet, because you haven't yet implemented the `startStopRecording()` method. Add the following implementation for this method:

```
@IBAction func startStopRecording() {
  // 1
  if videoOutput.isRecording {
    videoOutput.stopRecording()
  } else {
    // 2
    guard let path = FileManager.default.urls(for: .documentDirectory,
in: .userDomainMask).first
      else { return }

    let fileUrl = path.appendingPathComponent("recording.mov")

    // 3
    try? FileManager.default.removeItem(at: fileUrl)

    // 4
    videoOutput.startRecording(to: fileUrl, recordingDelegate: self)
  }
}
```

Let's go over the preceding snippet step by step to see what exactly is going on:

1. First, the `isRecording` property for the video output is checked. If a recording is currently active, the recording should be stopped.
2. If no recording is currently active, a new path is created to store the video temporarily.
3. Since the video output can't overwrite an existing file, the `FileManager` object should attempt to remove any existing files at the temporary video file path.
4. The video output will start recording to the temporary file. The view controller itself is passed as a delegate to be notified when the recording has begun and is stopped.

Since `RecordVideoViewController` does not conform to `AVCaptureFileOutputRecordingDelegate` yet, you should add the following extension to add conformance to `AVCaptureFileOutputRecordingDelegate`:

```
extension RecordVideoViewController:
AVCaptureFileOutputRecordingDelegate {
  // 1
  func fileOutput(_ output: AVCaptureFileOutput, didStartRecordingTo
fileURL: URL, from connections: [AVCaptureConnection]) {
    startStopButton.setTitle("Stop Recording", for: .normal)
  }

  // 2
  func fileOutput(_ output: AVCaptureFileOutput, didFinishRecordingTo
outputFileURL: URL, from connections: [AVCaptureConnection], error:
Error?) {
    guard error == nil
      else { return }

    UISaveVideoAtPathToSavedPhotosAlbum(outputFileURL.path, self,
#selector(didSaveVideo(at:withError:contextInfo:)), nil)
  }

  // 3
  @objc func didSaveVideo(at path: String, withError error: Error?,
contextInfo: UnsafeRawPointer?) {
    guard error == nil
      else { return }

    presentAlertWithTitle("Success", message: "Video was saved
succesfully")
    startStopButton.setTitle("Start Recording", for: .normal)
```

```
    }
}
```

The preceding extension contains three methods. The first is a delegate method, called when the video output has begun recording. When the recording has started, the title of the `startStopButton` button is updated to reflect the current state. The second method is also a delegate method. This method is called when the recording has completed. If no errors occur, the video is stored at the temporary location you set up earlier. `UISaveVideoAtPathToSavedPhotosAlbum(_:_:_:_:)` is then called, to move the video from the temporary location to the user's photo library. This method is very similar to the `UIImageWriteToSavedPhotosAlbum(_:_:_:_:)` method that you used to store a picture. The third and final method in the extension is called when the video is stored in the user's photo library. When the video has been successfully stored, an alert is shown, and the title of the `startStopButton` button is updated again.

You can now run the app and record some videos! Even though you have done a lot of manual work by implementing the video recording logic directly with an `AVCaptureSession`, most of the hard work is done inside of the `AVFoundation` framework. One final media-related feature to explore is applying visual filters to images using `Core Image`.

Manipulating photos with Core Image

In this chapter, you have already seen that iOS has powerful capabilities for recording and playing media. In this section, you will learn how you can manipulate images with `Core Image`. The `Core Image` framework provides many different filters that you can use to process both images and video. You will expand on the photo-taking capabilities that you implemented in the **Captured** app so users can grayscale and crop images.

Every `Core Image` filter you apply to images is an instance of the `CIFilter` class. You can create instances of filters as follows:

```
let filter = CIFilter(name: "CIPhotoEffectNoir")
```

The name parameter in the filter's initializer is expected to be a string that refers to a specific filter. You can refer to Apple's documentation on Core Image and the Core Image Filter Reference guide to see an overview of all the filters that you can use in your apps.

Every filter has a certain set of parameters that you need to set on the `CIFilter` instance to use the filter. For instance, a grayscale filter requires you to provide an input image. Other filters might take an intensity, location, or other properties. The best way to see how you can apply a filter to an image is through an example. Add the following implementation for `applyGrayScale()` to `ImageViewController.swift` to implement a grayscale filter:

```
@IBAction func applyGrayScale() {
  // 1
  guard let cgImage = selectedImage?.cgImage,
    // 2
    let initialOrientation = selectedImage?.imageOrientation,
    // 3
    let filter = CIFilter(name: "CIPhotoEffectNoir")
    else { return }

  // 4
  let sourceImage = CIImage(cgImage: cgImage)
  filter.setValue(sourceImage, forKey: kCIInputImageKey)

  // 5
  let context = CIContext(options: nil)
  guard let outputImage = filter.outputImage,
    let cgImageOut = context.createCGImage(outputImage, from:
outputImage.extent)
    else { return }

  // 6
  selectedImage = UIImage(cgImage: cgImageOut, scale: 1, orientation:
initialOrientation)
}
```

The preceding code has a lot of small, interesting details, highlighted with numbered comments. Let's go over the comments one by one to see how the grayscale filter is applied:

1. The `UIImage` instance that is stored in `selectedImage` is converted to a `CGImage` instance. Strictly speaking, this conversion isn't required, but it does make applying other filters to the `UIImage` instance later a bit easier.
2. One downside of using `CGImage`, instead of `UIImage`, is that the orientation information that is stored in the image is lost. To make sure the final image maintains its orientation, the initial orientation is stored.
3. This step creates an instance of the grayscale filter.

4. Since Core Image does not directly support `CGImage` instances, the `CGImage` instance is converted to a `CIImage` instance that can be used with Core Image. The `CIImage` instance is then assigned as the input image for the grayscale filter, by calling `setValue(_:forKey:)` on the filter.
5. The fifth step extracts the new image from the filter, and uses a `CIContext` object to export the `CIImage` output to a `CGImage` instance.
6. The sixth and final step is to create a new `UIImage` instance, based on the `CGImage` output. The initial orientation is passed to the new `UIImage` instance to make sure it has the same orientation as the original image.

Even though there are a lot of steps involved, and you need to convert between different image types quite a bit, applying the filter is relatively simple. Most of the preceding code takes care of switching between image types, while the filter itself is set up in just a couple of lines. Try running the app now and taking a picture. The initial picture will be in full color. After you apply the grayscale filter, the image is automatically replaced with a grayscale version of the image, as shown in the following screenshot:

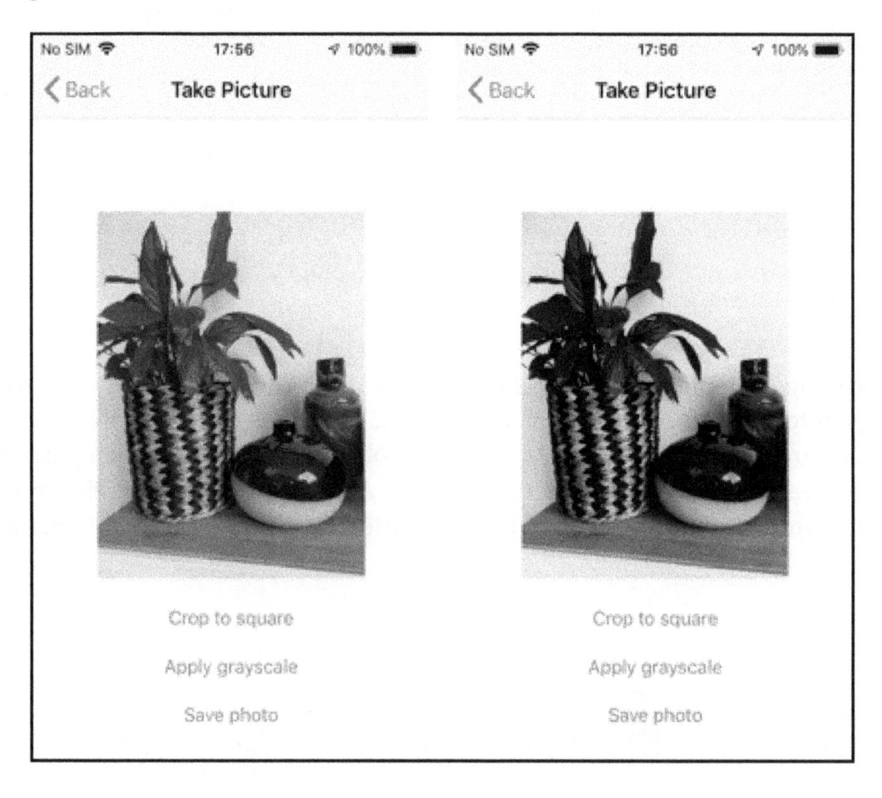

The next filter you will implement is a crop filter. The crop filter will crop the image so that it's a square, rather than a portrait or landscape picture. The process for implementing the crop filter is mostly the same as the grayscale filter, except for the values that need to be passed to the crop filter. Add the following implementation for `cropSquare()` to implement the crop filter:

```
@IBAction func cropSquare() {
  let context = CIContext(options: nil)

  guard let cgImage = selectedImage?.cgImage,
    let initialOrientation = selectedImage?.imageOrientation,
    let filter = CIFilter(name: "CICrop")
    else { return }

  let size = CGFloat(min(cgImage.width, cgImage.height))
  let center = CGPoint(x: cgImage.width / 2, y: cgImage.height / 2)
  let origin = CGPoint(x: center.x - size / 2, y: center.y - size / 2)
  let cropRect = CGRect(origin: origin, size: CGSize(width: size,
height: size))

  let sourceImage = CIImage(cgImage: cgImage)
  filter.setValue(sourceImage, forKey: kCIInputImageKey)
  filter.setValue(CIVector(cgRect: cropRect), forKey:
"inputRectangle")

  guard let outputImage = filter.outputImage,
    let cgImageOut = context.createCGImage(outputImage, from:
outputImage.extent)
    else { return }

  selectedImage = UIImage(cgImage: cgImageOut, scale: 1, orientation:
initialOrientation)
}
```

The preceding code performs several calculations to figure out the best way to crop the image into a square. The `CGRect` instance that specifies the crop coordinates and size, which are then used to create a `CIVector` object. This object is then passed to the filter as the value for the `inputRectangle` key. Apart from specifying the crop values, the process of applying the filter is identical, so the code should look familiar to you.

If you run the app now and tap the crop button, the image will be cropped, as shown in the following screenshot:

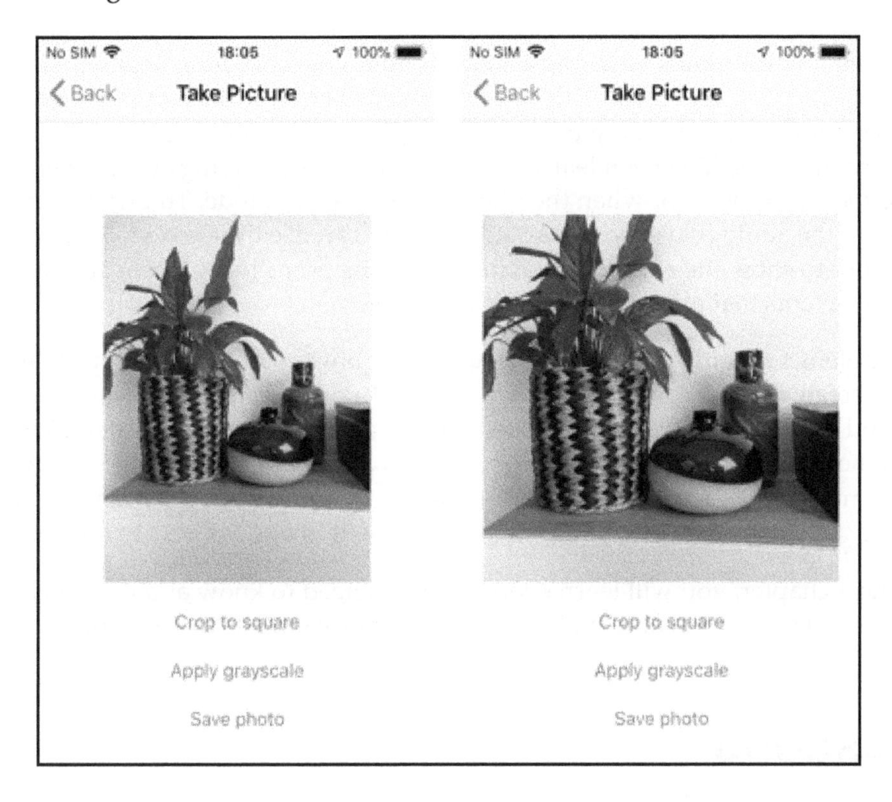

There are many more filters available in Core Image, which you can play with to build pretty advanced filters. You can even apply multiple filters to a single image to create elaborate effects for the pictures in your apps. Because all filters work in very similar ways, it's relatively easy to apply any filter to your images once you understand how the general process of applying a filter works. You can always use the code from the preceding examples if you need a reminder about how to apply Core Image filters.

Summary

In this chapter, you have learned a lot about media in iOS. You saw how you can implement a video player with just a couple of lines of code. After that, you learned how to use `AVFoundation` directly to build an audio player that supports features such as stopping and resuming playback, skipping songs, and scrubbing forward or backward in a song. You even learned how you can keep playing audio when the app goes to the background or when the phone is set to silent mode. To put the finishing touches to the audio player, you learned how you can use the `MediaPlayer` framework to show the currently playing file on the user's lock screen, and to respond to control events that are sent to the app remotely.

After implementing media playback, you learned how you can build apps that help users to create media. You saw that the `UIImagePickerController` provides a quick and simple interface to allow users to take a picture with the camera. You also learned how you can use `AVFoundation` and an `AVCaptureSession` object to implement a custom video recording experience. To wrap it all up, you learned about the `Core Image` framework, and how you can use it to apply filters to images.

In the next chapter, you will learn everything you need to know about notifications and how you can use them to make significant improvements to your apps.

Questions

1. Which framework is used to implement a simple video player?

    a) `AVFoundation`
    b) `MediaPlayer`
    c) `AVKit`

2. Which object is used to play a local audio file?

    a) `AVPlayer`
    b) `AVAudioPlayer`
    c) `AVQueuePlayer`

3. What are remote commands, and where do they come from?

   a) Media playback-related commands that come from the lock screen.
   b) Media playback-related commands that come from the control center.
   c) Media playback-related commands that can come from any external source, including the lock screen, control center, or even the user's headphones.

4. Which object can be used to quickly take a picture?

   a) `UIImagePickerController`
   b) `AVCaptureSession`
   c) `AVCaptureDevice`

5. Which of the following is not needed to record a video with an `AVCaptureSession` object?

   a) A temporary file to store the video
   b) Permission to access the user's photo library
   c) An `AVCaptureFileOutput` object

6. What information tends to get lost when you convert a `UIImage` instance to a `CGImage` instance?

   a) The orientation
   b) The color profile
   c) The previously applied filters

Further reading

- Big Buck Bunny download page: `http://bbb3d.renderfarming.net/download.html`
- The Free Music Archive: `https://freemusicarchive.org/about`
- Core Image Filter Reference: `https://developer.apple.com/library/archive/documentation/GraphicsImaging/Reference/CoreImageFilterReference`
- Core Image documentation: `https://developer.apple.com/documentation/coreimage`
- AVFoundation documentation: `https://developer.apple.com/av-foundation/`

18
Implementing Rich Notifications

`Chapter 16`, *Streamlining Experiences With Siri*, you were introduced to app extensions by implementing an intent extension for Siri. You learned that app extensions allow you to integrate your app with iOS on a very deep level. Since iOS 10, Apple added extension capabilities for notifications, while also completely revamping how notifications work. In this chapter, you will learn how notifications work on iOS, and how you can implement extensions for notifications to create a rich notification experience.

First, you will learn about notifications in general. You will learn how notifications appear in iOS and how they work. Even though you won't dive straight into the fine details of implementing notifications, you will gain a deep understanding of notifications by looking at the different types of notifications, and the different ways of scheduling notifications that are available. Once you know how notifications work, you will learn how to implement and use them in your app.

By the end of this chapter, you will know everything there is to know about notifications and how they can be used to create a unique, engaging user experience. People receive lots of notifications throughout the day, so providing a great experience that stands out will ensure that people don't get annoyed with your notifications to the point of wanting to disable them entirely.

This chapter is divided into the following topics:

- Gaining a deep understanding of notifications
- Scheduling and handling notifications
- Implementing notification groups
- Implementing notification extensions

Gaining a deep understanding of notifications

If you use iOS devices on a daily basis, you are probably very familiar with notifications. Sometimes you may have been using your device when you received such a notification; other times, the device might have been in your pocket, on your desk, or somewhere else.

Notifications are a perfect way for apps to inform users about information in an app that might be of interest to them. New messages in a messaging app, breaking news events, or friend requests in social apps are just a few examples of great use cases for notifications. From the get-go, it's important that you're aware of the fact that a lot of apps are fighting for the user's attention at all times. Notifications are a great way to gain the user's attention, but if you send too many notifications, or only use notifications as a marketing tool, instead of as a way to provide meaningful information, it's very likely that the user is going to disable push notifications for your app at some point.

This is especially true since iOS 12. When a user swipes to ignore a notification, the user will be asked if they would like to adjust the notification settings for the app that sent the notification, or even turn off notifications for the app entirely. This new behavior makes it more important than ever to only send meaningful notifications to users.

The following six forms of notifications can be sent to users:

- Sounds
- Vibrations
- Visual notifications (such as banners and alerts)
- Badges
- Critical notifications
- Provisional

Most apps use a combination of these, and you will learn more about them in a real-world scenario soon. A unique type of notification is the critical notification. A *critical notification* is sent to the user regardless of whether or not the device is in *Do Not Disturb* mode, and plays sound even if the device is muted. Critical notifications can't be implemented by all applications, as they are reserved for apps that have very good use cases for sending potentially disruptive notifications.

Examples of good use cases are health-related notifications or alarm systems. If you plan to send critical notifications to your users, you must ask Apple for permission, so they can grant your app the appropriate entitlements if they decide that your use case is valid.

Sending provisional notifications is a new feature in iOS 12 that allows your app to send notifications silently to your users, without explicit permission beforehand. Asking your user for permission to send them provisional notifications will automatically grant your app permission to send notifications without ever prompting the user. Instead, the user will silently receive notifications from your app, and they will be able to choose whether they want to continue receiving notifications from your app, or deny the notifications permission altogether.

There are two ways for apps to send notifications to their users: local notifications and push notifications. **Local notifications** are scheduled by an app and kept on the device. To trigger these notifications, you use either a location, calendar, or time-based trigger.

Push notifications are pushed to a device through a server and the **Apple Push Notification Service** (**APNS**). Push notifications are always delivered by APNS using best-effort delivery. This means that a notification will be sent to the user, and if this fails, APNS may or may not try again. There are no guarantees made about this, so you should never assume that your notification did or did not end up on the user's device.

Push notifications can send either a user-facing notification, or a silent notification. A user-facing notification will actively notify the user about what happened, often in the shape of a banner or badge icon. A silent notification signals the app that it should be woken up to perform a background fetch. You have seen this mechanism in action in `Chapter 11`, *Syncing Data with CloudKit.*

When a single app has multiple notifications to show to the user, iOS will group them automatically to avoid spamming the user's notification screen with a huge list of notifications from a single app.

The following screenshot shows a sample of grouped notifications:

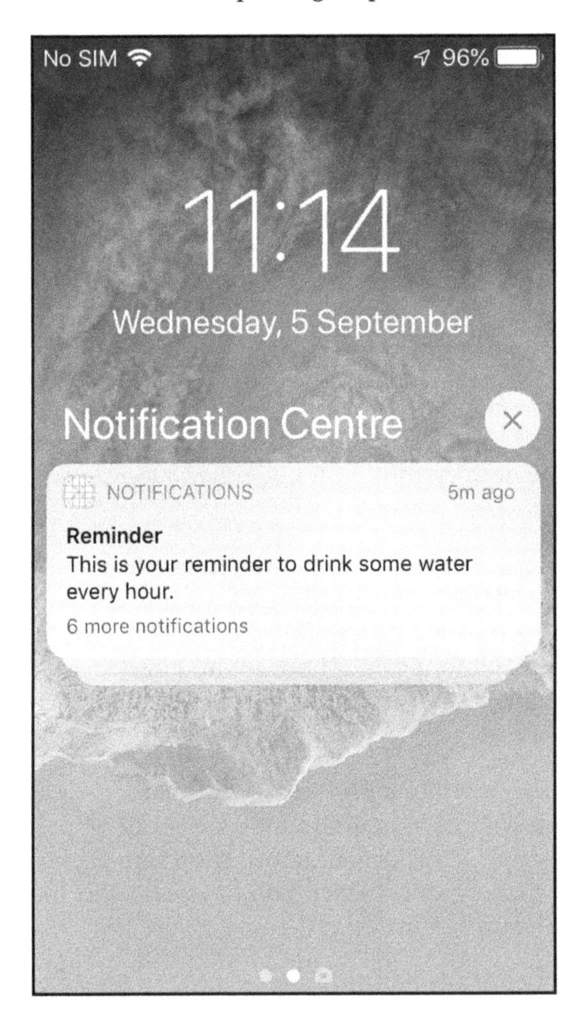

While this type of grouping might be exactly what your app needs, there are cases where you might want to group notifications in a slightly different way. For instance, if you are building a chat application, you might want to group all messages for a single conversation, but show multiple notifications if a user has notifications for several ongoing conversations, as shown in the following screenshot:

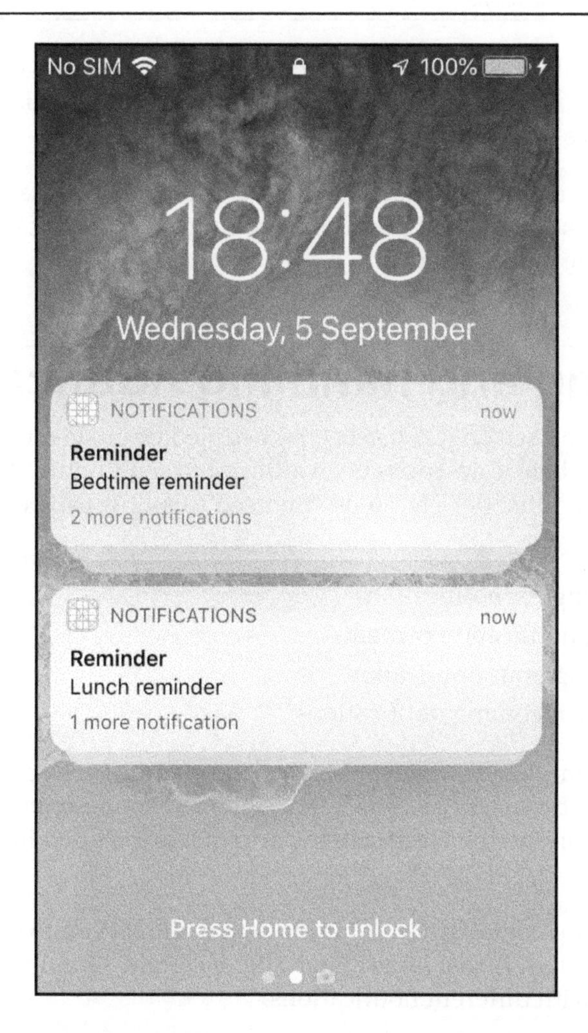

When you schedule or send a notification, you can specify a thread identifier for that notification. If the thread identifier is set for a notification, iOS will group your notifications based on this identifier, rather than grouping all notifications for a single application together.

In order to complement the notifications you send to your user, you can implement **notification extensions**. These extensions add a whole new dimension to notifications, especially for devices that support 3D-Touch. There are two types of extensions you can add to a notification—service extensions and content extensions.

Service extensions allow you to modify or augment the contents of a notification before it is shown to the user. **Content extensions** allow you to display a custom notification interface when the user 3D-Touches a notification.

You will learn more about notification grouping and extensions later. Before that, you will learn how to ask the user for permission to send them notifications, and how to schedule and handle notifications.

Scheduling and handling notifications

Every notification you see on iOS has been scheduled or pushed to the user in one way or another. But before an app is even allowed to send notifications, there are a couple of things you must do. The steps you need to follow can be roughly divided as follows:

1. Registering for notifications
2. Creating notification contents
3. Scheduling your notification
4. Handling incoming notifications

Not every step is completely identical for local and remote notifications. When there are differences, and these will be very minor, they will be described in the relevant section. You will work on the **Notifications** app that is included in the code bundle for this chapter.

The Notifications app contains a user interface that allows you to schedule different notifications for reminders that will trigger after a certain amount of time, so you can easily experiment with different notifications.

Registering for notifications

Any app that wants to present notifications has to ask the user for permission before they do so. This process is called registering for notifications. It's important that you make sure your user understands why you're asking their permission to send notifications. After all, you're interrupting your user while they're trying to achieve something in your app, and the permission alert prevents them from doing so. It's often a great idea to hold off on asking for permission to send notifications until the user does something that can be improved specifically by turning on notifications.

In the Notifications example app, a user can add reminders for certain actions that they need to perform. The best time to ask a user for permission to send them notifications is right after they create a new reminder. This will make it clear to the user that you want to send notifications to inform the user about the reminder they just added.

When you ask for permission to send notifications, you can specify the kinds of notification that you want to send the user. You can pick from the following types:

- Alerts
- Badges
- CarPlay
- Sounds
- Critical
- Provisional

The user can go into your app's page in the *Settings* app at any time to change these settings. You can read these settings with the `UserNotifications` framework. By reading the current notification settings, you can inform your users about the current notification status, and ask them to turn notifications on if needed.

When you ask for a user's permission to send notifications and they deny permission to do so, or they disable notifications at a later stage, it's often a good idea to notify the user that they have notifications turned off, and then provide an easy way to turn notifications back on.

Reading the current notification settings and asking for permission to send notifications is always done through an `UNUserNotificationCenter` instance. You will typically obtain an instance of `UNUserNotificationCenter` through the static `current()` method.

`UNUserNotificationCenter` is not to be confused with `NotificationCenter`. `UNUserNotificationCenter` is responsible for notifications such as push and local notifications, while `NotificationCenter` is used to send notifications internally inside your app. The user never sees or notices these notifications.

As you have learned before, it's good practice to extract certain parts of your code into a dedicated helper object. Create a new helper folder in the **Notifications** project and add a new Swift file, called `NotificationsHelper.swift`, to the group. Add the following implementation code to your new helper file:

```
import UIKit
import UserNotifications

struct NotificationsHelper {
  let notificationCenter = UNUserNotificationCenter.current()
}
```

The best moment to ask a user for permission to send them push notifications for this app is when they have just added a reminder. At this point, the user will understand exactly why they see the permission dialog, because they just performed an action that is closely related to the push notifications they will receive.

Add the following method to the notifications helper to ask the user for permission to send them notifications:

```
func requestNotificationPermissions(_ completion: @escaping (Bool) ->
()) {
  notificationCenter.requestAuthorization(options: [.badge, .sound,
.alert]) { permissionGranted, error in
    completion(permissionGranted)
  }
}
```

The preceding method calls the notification center's `requestAuthorization(options:completion:)`, and calls the completion closure that was passed to `requestNotificationPermissions(_:)` with the result of the permission prompt.

Before you can use the helper in the add notifications view controller to ask the user for notification permissions when they add a new reminder, you need to create an instance of the helper in `AddNotificationViewController`. Add the following property to this class:

```
let notificationsHelper = NotificationsHelper()
```

Also, add the following code to the end of the `createReminder(_:withComponents:inContext:)` method to ask the user for notification permissions after creating a new reminder:

```
DispatchQueue.main.async { [weak self] in
  self?.notificationsHelper.requestNotificationPermissions { result in
    DispatchQueue.main.async {
      self?.enableNotificationsButton.isHidden = result
    }
  }
}
```

Since the code inside of the `createReminder(_:withComponents:inContext:)` method is executed on a managed object context's queue, it's good practice to ensure that the notification prompt will be presented using the main thread, by wrapping the code in a `DispatchQueue.main.async` call. To learn more about how dispatch queues work and how you can use them, refer to `Chapter 25`, *Offloading Tasks with Operations and GCD.*

In addition to asking for notification permissions when the user adds a new reminder, permission should also be asked when the user enables the periodic drink water notification. Add the following implementation for `drinkWaterNotificationToggled(sender:)`:

```
@IBAction func drinkWaterNotificationToggled(sender: UISwitch) {
  if sender.isOn {
    notificationsHelper.requestNotificationPermissions { [weak self]
result in
      DispatchQueue.main.async {
        self?.enableNotificationsButton.isHidden = result
      }
    }
  }
}
```

The add notification view contains a button that prompts the user to enable notifications for the app. This button should only be visible if the user has denied the notification permissions dialog, or if the user has turned off notifications for the app in their settings. Add the following method to the notifications helper to determine whether the user has turned off notifications for the app:

```
func userHasDisabledNotifications(_ completion: @escaping (Bool) ->
()) {
  notificationCenter.getNotificationSettings { settings in
    completion(settings.authorizationStatus == .denied)
  }
}
```

The preceding code retrieves the current notification settings, and checks whether the current status is `.disabled`. There are a total of four possible authorization statuses: `.authorized`, `.denied`, `.notDetermined`, and `.provisional`. The following list describes what each status means:

- `notDetermined` means that the app hasn't asked for permissions before.
- `authorized` means that the app has asked for permissions and is allowed to send notifications to the user.
- `denied` means that the app has asked for permissions to send notifications to the user, and the user denied this permission, because they either declined the permissions prompt, or they turned off notifications in their settings.
- `provisional` means that the app is allowed to deliver quiet notifications to the app on a provisional basis.

It's a good idea to read the current notification settings whenever the add notification view appears, so the button that takes the user to the settings page is always shown and hidden correctly, based on the current notification permissions. Add the following implementation for `viewWillAppear(_:)` to the `AddNotificationViewController` class:

```
notificationsHelper.userHasDisabledNotifications { [weak self]
notificationsDisabled in
  DispatchQueue.main.async {
    self?.enableNotificationsButton.isHidden = !notificationsDisabled
  }
}
```

Before you test the app, add the following implementation for the `openSettingTapped()` method.

Add the following implementation for the `enableNotificationsTapped` action:

```
@IBAction func openSettingsTapped() {
  let settingsUrl = URL(string: UIApplication.openSettingsURLString)
  UIApplication.shared.open(settingsUrl!, options: [:],
completionHandler: nil)
}
```

The preceding implementation will take the user to the *Settings* app, so they can easily enable notifications for the Notifications app. If you test the app now, you can add a new reminder, and the app will ask you for permission to send notifications. If permission is denied, the *enable* button will appear at the bottom of the view. When you tap this button, you will be taken directly to the setting page, where you can again enable notifications.

The preceding code registers the app for local notifications. If you want to register the app for push notifications, you must enable the **Push Notifications** capability in the **Capabilities** tab. Note that you can only enable this capability if you're on Apple's paid developer program. Free developer accounts cannot enable this capability.

After enabling the push notifications capability, you need to add a couple of lines of code to make sure that you properly register the current device with APNS. To do this, you need to call a method on `UIApplication` that registers the current device on APNS. After calling this method, a delegate method in `AppDelegate` is called with the current device token that should be uploaded to your remote server, so you can send messages to the device through the device token.

Apple makes no guarantees about if and when a device's push token can change, so you should always register for remote notifications whenever the app launches. For the Notifications app, it makes sense to register for remote notifications as soon as the app enters the foreground, or when the user has just granted the app permission to send push notifications.

Add the following lines of code to the `requestNotificationPermissions(_:)` method in the notification helper, right after calling the completion handler:

```
DispatchQueue.main.async {
  UIApplication.shared.registerForRemoteNotifications()
}
```

Note that you must always call `UIApplication.shared.registerForRemoteNotifications()` on the main thread. Since the callback for `requestNotificationPermissions(_:)` is not guaranteed to be called on the main thread, you should manually ensure that you register for remote notifications on the main thread.

Next, implement the following methods in `AppDelegate`:

```
func applicationDidBecomeActive(_ application: UIApplication) {
  UIApplication.shared.registerForRemoteNotifications()
}
```

```
func application(_ application: UIApplication,
didRegisterForRemoteNotificationsWithDeviceToken deviceToken: Data) {
  print("received device token: \(deviceToken)")
}

func application(_ application: UIApplication,
didFailToRegisterForRemoteNotificationsWithError error: Error) {
  print("Did not register for remote notifications:
\(error.localizedDescription)")
}
```

For demo purposes, the device token is printed to the console. In a real app, you should upload the device token to your server instead. You're all set to send notifications to your users now. Next, let's look at how you can create and configure notifications in your app.

Creating notification contents

Notifications typically consist of a few pieces of content. First of all, every notification has a title, a subtitle, and a message body. Not all fields have to be present, but you'll often want to provide a bit more information than just a piece of body text. You can also add media attachments to notifications. These attachments are used to provide a preview of a photo, video, or audio file that is stored locally on the device, and are great for providing better-looking, richer notifications to your user.

The biggest gap between local and remote notifications exists in the way you create them. Push notifications are always created on the server and are pushed to the device in a JSON format. Local notifications are created on the device using the `UNNotification` class. Let's look at each notification type separately.

Creating push notifications

If you ignore the complexities of setting up and hosting your own server, push notifications are actually pretty simple to create. Push notifications are always delivered using a JSON payload that your own server sends to APNS. APNS will then take care of sending your notification to the device that's intended to receive it. Let's look at a simple example of a JSON payload that could be used to inform users about a newsworthy event:

```
{
  "aps": {
    "alert": {
      "title" : "Breaking news!",
      "body" : "iOS 12 was released"
    },
    "badge" : 1
  },
  "item_id" : "1"
}
```

This simple notification contains only a title and a body. The number `1` is added to the app icon to indicate the number of news items available. You'll notice that the notification content is placed in the `aps` dictionary. The `aps` dictionary contains the push notification's content. The `item_id` is not part of the `aps` dictionary on purpose; it's a custom property that can be read in the app so it can retrieve data about the news item from a local database, or fetch it from a backend using the ID.

If you want to send a silent push notification to trigger a background data refresh action, like CloudKit does, you should add the `content-available` key to the `aps` dictionary, specifying a value of `1`. If you want to use a custom sound for your notification, you can add a `sound` key to `aps`. The value of this key should point to an audio file that's embedded in your app. Finally, a notification can be part of a category of notifications. If this is the case, add a `category` key to the `aps` dictionary, where the value is the category identifier. For an overview of all of the possible keys that can be added to the `aps` dictionary, have a look at Apple's Payload Key Reference documentation.

Creating local notifications

Local notifications are created and stored locally on the device. They are very similar to push notifications in terms of the content they can contain. The way that content is created is quite different though. For instance, you can't create a local notification for content that doesn't exist yet.

In the context of a news app, this means that a server could push a news item to the application when needed. If you create the notification locally, you don't know what tomorrow's big news item will be, so you can't schedule a notification with content for this news item yet. This is an important distinction between local and push notifications that you should keep in mind when choosing whether you should create a notification locally, or if you should leave it up to a server to create the notification.

The following code snippet demonstrates the creation of a notification for a reminder. You should add this snippet to the `NotificationsHelper`:

```
func createNotificationContentForReminder(_ reminder: Reminder) ->
UNMutableNotificationContent {
  let content = UNMutableNotificationContent()

  content.title = "Reminder"
  content.body = reminder.title ?? ""
  content.badge = 1

  content.userInfo = ["reminder-uuid":
reminder.identifier?.uuidString]

  return content
}
```

This preceding code contains data that is quite similar to the JSON you saw previously. The main difference is that you create an instance of `UNMutableNotificationContent` and assign values to it, instead of defining the entire notification using JSON. A lot of the options that are available in the push example are also available for local notifications. You can add custom properties, such as the unique identifier for a reminder, to a `UNNotification` object's `userInfo` dictionary. You can also assign a custom sound, attachment, and more to local notifications, just as you can for push notifications.

Now that you know how to create notification contents, let's see how you can schedule them.

Scheduling your notification

In order to deliver local notifications to your users, you'll need to schedule them. This is another difference between local and push notifications. Push notifications do not need to be scheduled, because they are pushed through the server and delivered as soon as possible. When you schedule a local notification, you make use of a notification trigger. There are several notification triggers that you can use to send a notification:

- Location-based
- Calendar-based
- Time-based

Before you implement a proper notification to send reminders to a user, you will create a notification that is scheduled to fire a couple of seconds after creating it. You will then learn how to send a repeating notification that uses a time interval. Lastly, you will learn about calendar-based and location-based triggers.

Scheduling a timed notification

A timed notification is the simplest notification to schedule, which makes it a great candidate to explore first. Timed notifications are scheduled with an instance of `UNTimeIntervalNotificationTrigger`. The initializer for this class takes a time interval and a boolean value to indicate whether the notification should be sent to the user repeatedly. Add the following method to the `NotificationHelper`:

```
func scheduleTimedNotificationWithContent(_ content:
UNNotificationContent) {
  let trigger = UNTimeIntervalNotificationTrigger(timeInterval: 10,
repeats: false)
  let request = UNNotificationRequest(identifier: UUID().uuidString,
content: content, trigger: trigger)
  notificationCenter.add(request) { error in
    if let error = error {
      print(error.localizedDescription)
    }
  }
}
```

The preceding method creates a `trigger` for the notification. In this case, the notification is set up to fire after 10 seconds have passed. Next, a notification request is created. Every notification request should have an identifier. In this case, a very simple identifier is used to identify the timed notification. After creating the notification, it is added to the notification with a completion handler.

The `identifier` attribute for a notification should be unique for the content of the notification. The system uses this identifier to avoid sending multiple notifications with the same content. If you send multiple notifications with the same identifier, the last notification you sent will be shown to the user. If you want to display multiple notifications in the Notification Center, you should make sure that every notification you schedule has a unique identifier.

If you want to send a repeating notification to a user, you must set the `repeats` property for the time interval trigger to `true`.

The minimum time interval for repeating notification is 60 seconds. Scheduling a repeating `trigger` with an interval under 60 seconds will crash your application.

To test the interval-based notification, you can add a call to `scheduleTimedNotificationWithContent(_:)` in `AddNotificationViewController` in the `addBedtimeNotification()` or `addLunchTimeNotification()` methods, after the following line:

```
try? self.persistentContainer.viewContext.save()
```

The following code can be added to schedule the timed notification:

```
let content =
notificationsHelper.createNotificationContentForReminder(reminder)
    notificationsHelper.scheduleTimedNotificationWithContent(content)
```

If you run the app now, and schedule a reminder before going to your device's home screen or the lock screen, a notification about the reminder should appear after about 10 seconds. The following screenshot shows an example of a lunchtime reminder:

The next notification you will explore is the calendar based notification.

Scheduling a calendar-based notification

Calendar-based notifications make use of `DateComponents` to determine when they should be fired. The only difference compared to the time-interval scheduling is in the `trigger` that is used to determine when the notification should be displayed.

Go right ahead and add the following method to `NotificationsHelper` to create a notification that is fired every hour, on the hour:

```
func scheduleHourlyWaterReminder() {
  let content = UNMutableNotificationContent()

  content.title = "Reminder"
  content.body = "This is your reminder to drink some water every
hour."

  var hourComponents = DateComponents()
  hourComponents.minute = 0

  let trigger = UNCalendarNotificationTrigger(dateMatching:
hourComponents, repeats: true)
  let request = UNNotificationRequest(identifier: "water-
notification", content: content, trigger: trigger)

  notificationCenter.add(request) { error in
    if let error = error {
      print(error.localizedDescription)
    }
  }
}
```

The preceding code sets up a `trigger` that fires every hour when the value of the `minutes` component is `0`. Since this notification should trigger every hour, the `repeats` argument is set to true.

Since the user can choose to turn off the water reminder, you should also implement a method to remove the repeating notification. Add the following method to the notifications helper to do this:

```
func unscheduleHourlyWaterReminder() {
notificationCenter.removePendingNotificationRequests(withIdentifiers:
["water-notification"])
}
```

Next, update `drinkWaterNotificationToggled(sender:)` in the add notifications view controller, so that `scheduleHourlyWaterReminder()` is called when the switch is on and `unscheduleHourlyWaterReminder()` is called when the switch is off.

Calendar-based notifications provide a powerful way for you to schedule both recurring and one-time-only notifications that are tied to a specific moment in time that's easier to express in days, hours, weeks, or months, than in a time interval.

Scheduling a location-based notification

The last type of notification you can schedule is a location-based notification. To set up a location-based trigger, you use Core Location to set up a region to which the trigger should belong.

A location-based trigger is actually very similar to implementing geofencing. If you need a refresher on geofences, skip back to `Chapter 13`, *Improving Apps With Location Services*. The main difference between a location-based notification and a geofence is that when a geofence triggers, the user is not necessarily informed about this. If a location-based notification triggers, the intent is to inform the user about it so they see a notification. Let's look at an example of how a location-based notification is set up:

```
let coords = CLLocationCoordinate2D(latitude: 52.373095, longitude:
4.8909129)
let region = CLCircularRegion(center: coords, radius: 1000,
identifier: "quote-notification-region")
region.notifyOnExit = false
region.notifyOnEntry = true
let trigger = UNLocationNotificationTrigger(region: region, repeats:
false)
```

The preceding snippet sets up a location-based notification, using a circular region with a radius of `1000` meters. The `notifyOnExit` and `notifyOnEntry` properties are used to specify whether the user should be notified when they enter the region, exit it, or when either event occurs.

Similar to the other notification triggers, you can choose whether the notification should be fired repeatedly, or just once.

This wraps up the exploration of all the ways for you to schedule notifications. Let's explore the last piece of the puzzle of notifications: handling incoming notifications and notification actions.

Handling notifications

When a user receives a notification, it's typically displayed on the lock screen and in the Notification Center. The user can see the contents of the notification, and they have the option to tap the notification to launch the app that the notification belongs to.

In addition to tapping on a notification, users can swipe a notification to make extra buttons appear, which enable them to quickly perform an action that is related to the notification; for instance, accepting a calendar invite or quickly replying to a message. In this section, you will explore the following aspects of notification handling:

- Handling notifications in your app
- Managing pending and delivered notifications
- Adding actions to notifications

Properly implementing the preceding aspects can greatly enhance the value of the notifications that you send to your users.

Handling notifications in your app

Incoming notifications are most often received when the user is not using your app. This means that your app is not directly aware of the incoming notification, and doesn't get to handle it until the user taps the notification to open your application.

When your application launches due to a tapped notification, the `UNUserNotificationCenterDelegate` delegate protocol that belongs to the `UNUserNotificationCenter` object is called. This delegate is responsible for handling the incoming notification whether the app is currently in the foreground, background, or not active at all, perhaps because it's been killed by the system or force-closed by the user.

You are expected to assign a `UNUserNotificationCenterDelegate` delegate before the app has finished launching. This means that you should set the delegate as soon as possible, so `application(_:didFinishLaunchingWithOptions:)` is a great candidate for setting the delegate.

When the user has tapped a notification, or selects one of the custom actions that is attached to the notification, the `userNotificationCenter(_:didReceive:withCompletionHandler:)` method is called on the notification center delegate. If the app is in the foreground and a notification is received, `userNotificationCenter(_:willPresent:withCompletionHandler:)` is called right before the notification is shown to the user. You can use this method to determine whether the notification should be shown to the user. Sometimes, it might not make sense to present a certain notification to the user if the app is already running. For instance, you wouldn't show a notification for an incoming message if the user is already looking at the new message.

Let's see what the notification center delegate methods look like when they are implemented. First, import the `UserNotifications` framework in `AppDelegate` and add the following code to implementation for `application(_:didFinishLaunchWithOptions:)` to set the app delegate as a delegate for the `UNUserNotificationCenter`:

```
func application(_ application: UIApplication,
didFinishLaunchingWithOptions launchOptions:
[UIApplication.LaunchOptionsKey: Any]?) -> Bool {

  // existing code...

  UNUserNotificationCenter.current().delegate = self

  return true
}
```

Next, add the following extension to `AppDelegate.swift` to make the app delegate conform to `UNUserNotificationCenterDelegate`:

```
extension AppDelegate: UNUserNotificationCenterDelegate {
  func userNotificationCenter(_ center: UNUserNotificationCenter,
didReceive response: UNNotificationResponse, withCompletionHandler
completionHandler: @escaping () -> Void) {
    let notification = response.notification
    let action = response.actionIdentifier

    let notificationTitle = notification.request.content.title
    let customAttributes = notification.request.content.userInfo

    completionHandler()
  }
```

```
    func userNotificationCenter(_ center: UNUserNotificationCenter,
willPresent notification: UNNotification, withCompletionHandler
completionHandler: @escaping (UNNotificationPresentationOptions) ->
Void) {
        completionHandler([.alert, .sound])
    }
}
```

Both delegate methods have access to the original notifications and the notification contents. You can use this information to determine exactly which actions should be taken by your app to properly handle the notification. When a notification is tapped, `userNotificationCenter(_:didReceive:withCompletionHandler:)` is called. This method receives a `UNNotificationResponse` object, which provides information about the selected action, in addition to the notification information.

If your app receives a notification and it's in the foreground, `userNotificationCenter(_:willPresent:withCompletionHandler:)` allows you to determine what should happen with the notification. This method is passed a closure that you need to call to inform iOS about what should happen with the notification. If you want to handle the notification yourself, you can simply call the closure without any arguments. In the preceding snippet, the desired behavior is to display all notification alerts to the user and play the corresponding sounds.

Because you can access the original notification, you can determine whether the notification should be shown for each incoming notification individually, depending on the notification content, category, or custom attributes.

Managing pending and delivered notifications

When you learned about scheduling notifications, you saw that each notification is assigned a unique identifier. If a new notification is added with an identifier that has already been used, the new notification will overwrite the existing notification, limiting the amount of similar and duplicate notifications in the Notification Center.

This overwrite behavior applies to both delivered and pending notifications. This means that you can replace a notification that the user has already received by scheduling a second notification with the same identifier. This is especially useful if the new notification makes the old notification obsolete. For instance, consider a situation when you send push messages to update a user about the current score for a sports match. If you push each score update with the same identifier, the user will always see a single notification with the most recent score.

You can also remove delivered notifications by calling `removeDeliveredNotifications(withIdentifiers:)` on the notification center in your app. If your app allows the user to select the types of notification they want to receive, you could use this feature to remove all notifications of a certain type. That way, the user won't see the notifications anymore.

To update notifications that have been sent by your server, you push a new notification with the `apns-collapse-id` header set on the request that your server sends to APNS. If this header is present, the new notification will overwrite any existing notifications that have been pushed with the same `apns-collapse-id`.

Updating your notifications, instead of pushing new ones all the time, is a great way to avoid needless clutter in the Notification Center. More importantly, there won't be any duplicate, redundant, or outdated messages fighting for the user's attention.

Let's see how you can further improve the notification experience through custom actions.

Adding actions to notifications

When you implemented the code to handle incoming notifications, you implemented a method called `userNotificationCenter(_:didReceive:withCompletionHandler:)`. This method contains information about the way the user chose to handle the notification. The could have tapped on the notification itself, which is the default action, or maybe they chose one of the custom actions that are associated with the notification by swiping it. Another way for users to make custom notifications appear is to 3D-Touch on it.

There are three main kinds of actions supported by notifications. The first is a background action. Background actions dismiss the notification and wake the app in the background, so it can take appropriate measures for the selected action. An example of this is accepting a calendar invite through an action. If a user accepts or declines an invite inside the notification, they don't have to be taken to the app, because the response can be handled in the background just fine.

The second type of action is a foreground action. A foreground action takes the user into the app that sent the notification. It's up to the app to handle the selected action in a way that makes sense.

Lastly, you can add text input actions to your notifications. These types of actions are basically background actions, except they take user input. A text input action is great if you're building a messaging app.

All notification actions are associated with a notification category. These categories must be registered on the Notification Center at app launch. Again, the perfect place to set this up is `application(_:didFinishLaunchingWithOptions:)`. Add the following code to this method to configure a notification action that allows a user to mark one of their reminders as completed:

```
let completeAction = UNNotificationAction(identifier: "complete-
reminder", title: "Complete", options: [])

let reminderCategory = UNNotificationCategory(identifier: "reminder",
actions: [completeAction], intentIdentifiers: [], options: [])

UNUserNotificationCenter.current().setNotificationCategories([reminder
Category])
```

The preceding code creates a custom action that is attached to the `reminder` category. The default behavior for a notification is to handle actions in the background. If you want an action to take the user into your app, you need to add the `.foreground` option to the `options` parameter for the action. Lastly, the new category is added to the Notification Center.

Associating a category with a notification is done through the `categoryIdentifier` property of `UNNotificationContent`. Add the following line of code to the `createNotificationContentForReminder(_:)` method in `NotificationsHelper.swift` to add the reminder category to the notifications that are scheduled by the app:

```
content.categoryIdentifier = "reminder"
```

If you 3D-Touch the notifications that is sent by the app now, you're presented with a custom action:

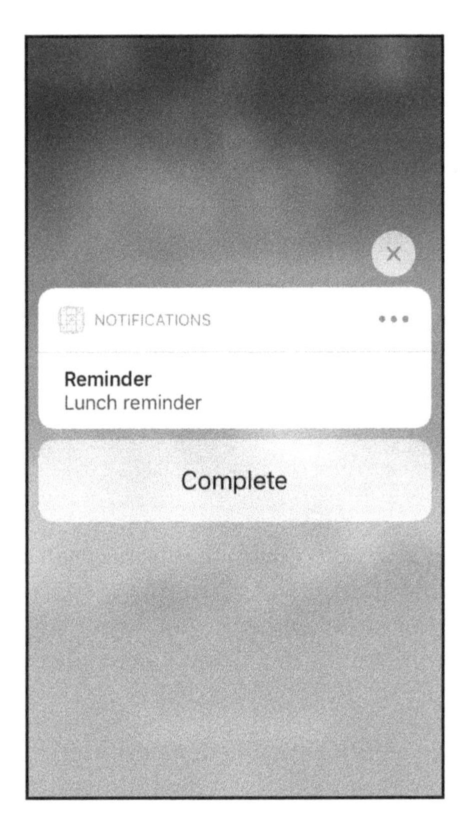

When the user taps on this action, `userNotificationCenter(_:didReceive:)` is called to notify your app about the fact that the user tapped on your custom action. Add the following implementation for this method to `AppDelegate.swift` to mark a reminder as completed when the user selects the **Complete** action:

```
func userNotificationCenter(_ center: UNUserNotificationCenter,
didReceive response: UNNotificationResponse, withCompletionHandler
completionHandler: @escaping () -> Void) {

  let userInfo = response.notification.request.content.userInfo

  guard response.actionIdentifier == "complete-reminder",
    let identifier = userInfo["reminder-uuid"] as? String else {
      completionHandler()
      return
  }
```

```
    let fetchRequest: NSFetchRequest<Reminder> = Reminder.fetchRequest()
    fetchRequest.predicate = NSPredicate(format: "identifier == %@",
  identifier)

    let moc = PersistentHelper.persistentContainer.viewContext
    guard let results = try? moc.fetch(fetchRequest),
      let reminder = results.first else {
        completionHandler()
        return
    }

    moc.perform {
      reminder.isCompleted = true
      try! moc.save()

      completionHandler()
    }
  }
```

The preceding code extracts all relevant information from the notification response and uses it to find a reminder that matches the identifier that was added to the notification's `userInfo` dictionary. If a reminder was found, it's marked as completed and the managed object context is saved to persist the new reminder status.

Notification actions are a great way to add quick interactions to your notifications, and they are already a huge improvement over simply sending a plain notification. Another way to make great improvements to the way you send notifications to your users is to implement the new notification grouping that was introduced in iOS 12.

Implementing grouped notifications

A new feature for notifications in iOS 12 is the possibility to group notifications. iOS will group notifications from an application together by default, showing the user a stack of notifications with the most recent one on top, instead of a long list of mixed notifications from all applications, which is hard to scan through.

In this section, you will learn exactly how you can implement grouped notifications in your own apps.

Grouping notifications based on thread identifiers

Grouped notifications are a huge improvement for the notification system on iOS, which many users have eagerly been waiting for. However, if all notifications for a single app are grouped together, important information might get lost. If a messaging app stacks all notifications in a single group, it's easy to miss an important message if it's hidden somewhere in the group.

For this reason, Apple allows developers to create custom notification groups based on special thread identifiers. The most obvious use case is to group all notifications from a certain conversation in a messaging app together. However, there are other useful implementations of grouped notifications that you can come up with.

You could use thread identifiers to separate certain important, actionable notifications from a group of informative notifications. Doing this will make the important notifications stand out to the user, without cluttering the user's Notification Center with several less important informative messages.

To begin using grouped notifications, all you have to do is assign a thread identifier to a notification content object. For the Notifications app, it would be interesting to group notifications for bedtime and lunchtime notifications together. You can do this by assigning a thread identifier to the notification content object that is created by the notifications helper object. Update the implementation for `addLunchTimeNotification()` as follows to assign a thread identifier to a lunchtime notification:

```
@IBAction func addLunchTimeNotification() {
  persistentContainer.viewContext.perform { [unowned self] in
    // existing implementation ...

    let content =
self.notificationsHelper.createNotificationContentForReminder(reminder
)

    content.threadIdentifier  = "lunchtime-thread"
self.notificationsHelper.scheduleTimedNotificationWithContent(content)
  }
}
```

Only a single line of code was added to implement a thread identifier. Try adding a thread identifier to the bedtime notification as well, and then run the app. Try scheduling a couple of bedtime and lunchtime notifications to see how iOS groups the notifications based on the thread identifier.

Providing a custom summary message for your notification group

When iOS displays a grouped notification, a summary message is shown at the bottom of the notification bubble. By default, this message simply informs the user about the number of notifications that are available. You can customize this message by providing a summary format yourself.

Summary formats are tied to notifications through categories. This means that you can add different messages for different notification categories. A custom summary format is supplied in the form of a format string. Update the `UNNotificationCategoryIdentifier` initialization code in `AppDelegate.swift` to add a custom summary format for the reminder category as follows:

```
let summaryFormat = "%u more reminders for this topic."

let reminderCategory = UNNotificationCategory(identifier: "reminder",
actions: [completeAction], intentIdentifiers: [],
hiddenPreviewsBodyPlaceholder: nil, categorySummaryFormat:
summaryFormat, options: [])
```

The summary format passed to the `UNNotificationCategory` initializer uses the `%u` format specifier to specify where the number of reminders should be displayed. When you run the app and schedule a couple of notifications, the custom summary format should be displayed, as shown in the following screenshot:

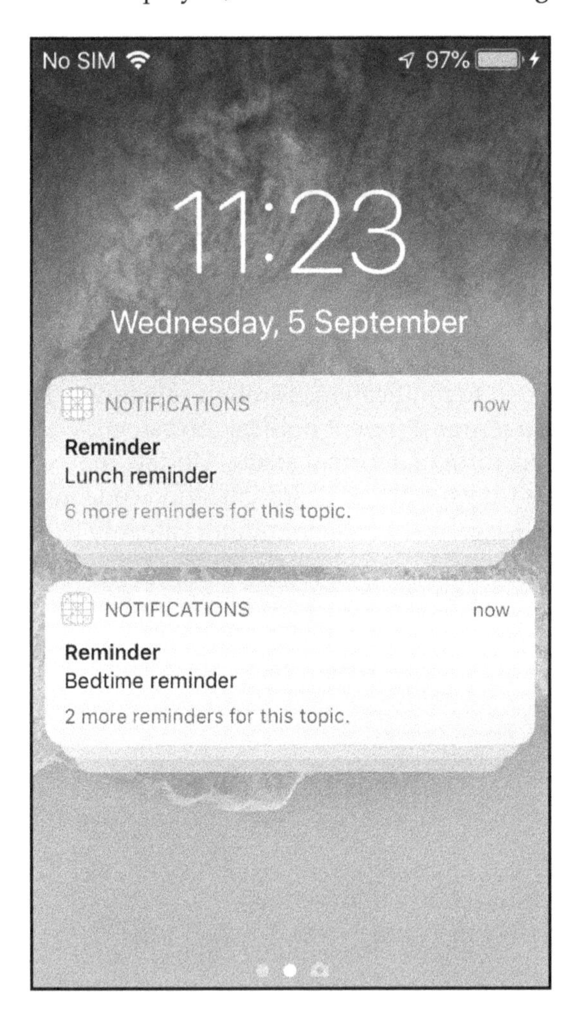

Note that one of the notification groups shows a slightly strange summary message. It says **2 more reminders for this topic**. Note the plural for reminders being used in this summary message. It would be much nicer if the summary format could account for this and use "*1 more reminder for this topic*" instead, if there is only a single extra item in the notification group.

You can achieve this by implementing a localized string for the summary format. To do this, start by defining the summary format string in `AppDelegate` as follows:

```
let summaryFormat = NSString.localizedUserNotificationString(forKey:
"REMINDER_SUMMARY", arguments: nil)
```

The preceding code defines the summary format as a localized string that is specific for notifications. When you define a string like this, iOS will look for the display string in a dedicated file containing all of the strings that you wish to localize for notifications, using the key you have specified.

Localizations for notifications use a special `.stringsdict` localization file that contains information that describes how to localize a string for several quantities. To add a `.stringsdict` file, go to **File | New** and select the `Stringsdict` file from the list of file types. Name your file `Localizable.stringsdict` and make sure it's added to your app target. Next, open the `Localizable.stringsdict` file and make sure its contents resemble the following screenshot:

Key	Type	Value
▼ Strings Dictionary	Dictionary	(1 item)
▼ REMINDER_SUMMARY	Dictionary	(2 items)
NSStringLocalizedFormatKey	String	%#@reminders@
▼ reminders	Dictionary	(4 items)
NSStringFormatSpecTypeKey	String	NSStringPluralRuleType
NSStringFormatValueTypeKey	String	u
one	String	%u more reminder for this topic.
other	String	%u more reminders for this topic.

The strings dictionary contains the key that was used in `AppDelegate` to identify the string that should be used for the summary label. The `NSStringLocalizedFormatKey` string contains a variable that is represented as `%#@reminders@`. This variable is the part of the string that you want to localize. In this case, the variable represents the entire message.

The variable name is then added as a key to the `REMINDER_SUMMARY` dictionary to specify the localization rules. In this case, a rule is added for a single item and for multiple items.

If you want to know more about the `stringsdict` file, make sure to read Apple's documentation on app localization.

Now that you know how to implement grouped notifications, let's see how you can add Notification Extensions to your app.

Implementing notification extensions

Apple has created two extension types for notifications that you can use to take the notification experience up a notch from custom actions. The available extensions are **service extensions** and **content extensions.** These extensions are both very powerful and relatively simple to implement. In this section, you will learn how to add both a service extension and a content extension to the Notifications app.

Adding a service extension to your app

Service extensions are intended to act as middleware for push notifications. A service extension receives a notification before it's displayed to the user. This allows you to manipulate or enrich the notification's content before it's shown to the user.

A service extension is perfect if you're implementing end-to-end encryption, for example. Another great use for a service extension is to download a media attachment from a push notification, save it locally, and add it as a media attachment to the notification contents, because all media attachments that are shown in a notification must be stored locally on the device. This means that a push notification can't really have media attachments, unless a service extension is used to download and store the media locally, before appending it to the notification.

Service extensions can only be implemented for push notifications that have the `mutable-content` property added to their `aps` payload, as shown in the following sample:

```
{
  "aps": {
    "alert": “You have a new message!”,
    "badge": 1,
    "mutable-content": 1
  },
  "custom-encrypted-message": "MyEncryptedMessage"
}
```

When the `mutable-content` property is detected by iOS, your service extension is activated and receives the notification before it's displayed to the user. A service extension is created in the same way as other extensions. You go to the project settings in Xcode, and in the sidebar that shows all targets, you click the **+** icon. In the dialog that appears, you select **Notification Service Extension** and give it a name, and Xcode will provide you with the required boilerplate code. When you add a service extension, a sample extension is added to your project to illustrate what a service extension that updates the notification's body text looks like.

Imagine that the **Notifications** app uses a backend server that sends push notifications when a certain reminder is due. To maintain the user's privacy, you send the following payload to the user:

```
{
  "aps": {
    "alert": "A reminder is due.",
    "badge": 1,
    "mutable-content": 1
  },
  "reminder-uuid": "D82ED5AE-C363-46AF-9B0D-301C2E21C58E"
}
```

The preceding payload contains a very plain message and has the unique identifier for the reminder added as a custom property. You can implement a service extension to read that identifier, and retrieve the appropriate reminder from the Core Data store to show it to the user.

First, add a new **Notification Service Extension** to the Notifications app and name it `ReminderContent`. After doing this, you need to make sure that the extension and the app can both access the Core Data database. Enable the **App Groups** capability for the app and create a new app group identifier. Make sure to enable **App Groups** for the extension as well, and add the extension to the group you just created for the app. Next, make sure the `.xcdatamodeld` and `PersistentHelper.swift` files are added to both the app and the extension target, and then update the code to load the persistent container in `PersistentHelper.swift` as follows:

```
static let persistentContainer: NSPersistentContainer = {
  let container = NSPersistentContainer(name: "Notifications")

  let containerUrl =
FileManager.default.containerURL(forSecurityApplicationGroupIdentifier
: "group.com.donnywals.notifications-app")!
  let databaseUrl =
containerUrl.appendingPathComponent("Notifications.sqlite")
  let description = NSPersistentStoreDescription(url: databaseUrl)
```

```
    container.persistentStoreDescriptions = [description]

    container.loadPersistentStores(completionHandler: {
  (storeDescription, error) in

    })
    return container
  }()
  }
```

The process of sharing a Core Data store, as outlined in the preceding section, is described in more detail in `Chapter 16`, *Streamlining Experiences With Siri*. Once you have set up the shared Core Data store, let's have a look at the service extension example that Xcode has generated for you.

The boilerplate code that Xcode generates for a service extension is rather interesting. Two properties are created: `contentHandler` and `bestAttemptContent`. The properties are initially given a value in `didReceive(_:withContentHandler:)`. This method is called as soon as the extension is expected to handle the notification.

If the extension fails to call the callback handler in a timely manner, the system calls `serviceExtensionTimeWillExpire()`. This is essentially the last chance to quickly come up with content for the notification. If the extension still fails to call the callback, the pushed notification is displayed to the user in its original form. The default version of `serviceExtensionTimeWillExpire()`, generated by Xcode, simply calls the system callback with the notification immediately to ensure a timely response.

Add the following implementation for `didReceive(_:withContentHandler:)` to retrieve the identifier from the push message payload that was shown previously, and create a new notification with the appropriate reminder. Don't forget to import Core Data at the top of `NotificationService.swift`:

```
override func didReceive(_ request: UNNotificationRequest,
withContentHandler contentHandler: @escaping (UNNotificationContent)
-> Void) {
  self.contentHandler = contentHandler
  bestAttemptContent = (request.content.mutableCopy() as?
UNMutableNotificationContent)

  guard let identifier = request.content.userInfo["reminder-uuid"] as?
String else {
    contentHandler(request.content)
    return
  }
```

```
    let predicate = NSPredicate(format: "identifier == %@", identifier)
    let fetchRequest: NSFetchRequest<Reminder> = Reminder.fetchRequest()
    fetchRequest.predicate = predicate

    let moc = PersistentHelper.persistentContainer.viewContext
    guard let results = try? moc.fetch(fetchRequest),
      let reminder = results.first else {
        contentHandler(request.content)
        return
    }

    if let bestAttemptContent = bestAttemptContent {
      bestAttemptContent.title = "Reminder"
      bestAttemptContent.body = reminder.title ?? ""
      bestAttemptContent.categoryIdentifier = "reminder"

      contentHandler(bestAttemptContent)
    }
  }
```

The preceding code extracts data from the pushed notification to find the appropriate reminder to show to the user, and applies this information to the original notification. This means that the server can send a very candid notification that contains no information about the reminder itself, apart from its identifier. It isn't until the notification reaches the device that the notification is transformed and enriched to show the reminder details. Pretty neat, right?

Adding a content extension to your app

The last feature of notifications that you will learn about in this chapter is a feature that can really make your notification experience pop. Content extensions enable developers to take custom notification actions to a whole new level. You already know that 3D-Touching a notification will make custom actions pop up. However, the notification maintains a standard look and feel, which might not be exactly what you want.

Consider receiving an invite for an event. The notification allows you to accept the invite, or decline it. Wouldn't it be great if your calendar popped up inside that notification as well, allowing you to check your calendar without having to actually open your calendar app, before responding to the invite? This is what content extensions are for. When implemented correctly, they can provide users with essential information relevant to the notification that's currently on display.

To demonstrate the possibilities for content extensions, you will use the simple notification you created for the Notifications app, and take it to the next level. Earlier, you scheduled a local notification that simply notified the user about a reminder. The notification itself contained only the notification title and custom actions that allow the user to mark a task as finished.

You will now use a content extension to show more context about the reminder. The content extension will show the user when the notification was due, and whether it is currently pending. In addition to this, you will also update the notification UI when the user taps the **Complete** action, so the UI reflects the current reminder status. If a notification is no longer pending by the time the notification is delivered, the **Complete** action will be replaced with a **Pending** action.

Get started by adding a new extension to your project, and select the **Notification Content Extension** type. Make sure to add your extension to the shared **App Group** you have set up before, and add the Core Data model definition and `PersistentHelper` to the extension target.

When you add a content extension to your project, Xcode generates a small sample implementation for you, which you can use as a reference to see a minimal working implementation for a content extension. Open the `NotificationViewController` class and have a quick look at its content, before removing all generated code and adding the following outlets:

```
@IBOutlet var reminderTitleLabel: UILabel!
@IBOutlet var reminderDueDateLabel: UILabel!
@IBOutlet var reminderStatusLabel: UILabel!
```

Next, add the following implementation for `didReceive(_:)` to set the correct values on the outlets you added:

```
func didReceive(_ notification: UNNotification) {
  guard let identifier =
notification.request.content.userInfo["reminder-uuid"] as? String
    else { return }

  let predicate = NSPredicate(format: "identifier == %@", identifier)
  let fetchRequest: NSFetchRequest<Reminder> = Reminder.fetchRequest()
  fetchRequest.predicate = predicate

  let moc = PersistentHelper.persistentContainer.viewContext
  guard let results = try? moc.fetch(fetchRequest),
    let reminder = results.first
    else { return }
```

```
    let dateFormatter = DateFormatter()
    dateFormatter.dateStyle = .long
    dateFormatter.timeStyle = .medium

    reminderTitleLabel.text = reminder.title
    reminderDueDateLabel.text = dateFormatter.string(from:
  reminder.dueDate!)
    reminderStatusLabel.text = reminder.isCompleted ? "Done" : "Pending"
  }
```

In the storyboard for the content extension, remove the existing label and add three new labels, one for the reminder title, one for the due date, and one for the reminder status. Don't forget to connect the labels to their corresponding outlets, and remove the original outlet that was present in the storyboard. Giving the view controller a height of ±100 points will help you position the labels more easily. You can use stack views and Auto Layout to replicate the following screenshot:

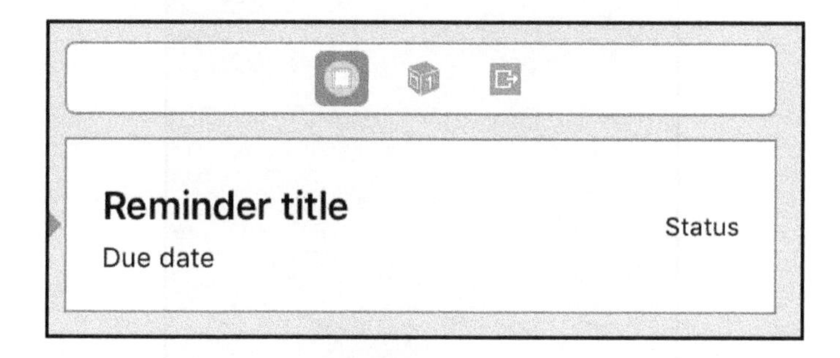

Finally, open the content extension's `Info.plist` file. As mentioned before, content extensions are associated with notifications through categories. The `Info.plist` file is used to specify the category that the current extension should be associated with. Expand the `NSExtension` and `NSExtensionAttributes` properties to find the `UNNotificationExtensionCategory` property. Give this field a value of `reminder`.

You are now ready to take your extension for a spin. Select your extension from the drop-down menu next to the run and stop controls in the top-left corner of Xcode, and run your project. Xcode will ask you for an app to run; pick **Notifications**. Schedule a new reminder in the Notifications app and wait for a notification to appear.

Once the notification appears, the following UI should be shown when you 3D-Touch on the notification:

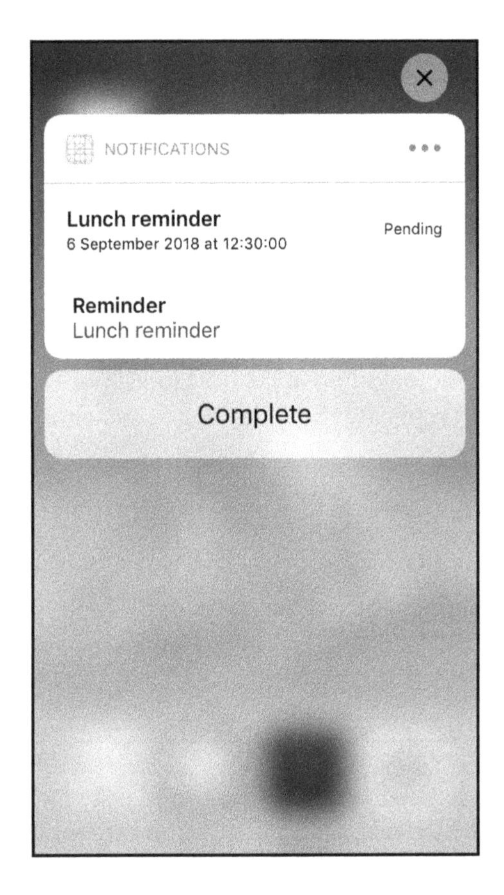

If you've tested this example yourself, you may have noticed that the notification's custom view was too high initially, and that it animated to the correct size for properly fitting your notification contents. The notification determines its final size using `AutoLayout`, so the final size won't be known until the notification content is on screen. The animation doesn't look very good, but luckily, there is a way to minimize the amount of resizing that your content extension has to do.

In the extension's `Info.plist` file, there is a property called `UNNotificationExtensionInitialContentSizeRatio`, which sits right below the notification category property. The default value for this property is `1`, but for your notification, you could probably use a value that's a lot smaller. Try setting this value to `0.2` and run your extension again. It doesn't have to animate nearly as much because the initial height is now just 20% of the extension's width. Much better.

You probably also noticed that the original notification contents are visible below the custom view. You can hide this default content by adding the `UNNotificationExtensionDefaultContentHidden` property to the extension's `Info.plist` file. All the properties you add for your extension should be added at the same level as `UNNotificationExtensionInitialContentSizeRatio` and `UNNotificationExtensionCategory`:

▼ NSExtension	Dictionary	(3 items)
▼ NSExtensionAttributes	Dictionary	(3 items)
UNNotificationExtensionDefault...	Boolean	YES
UNNotificationExtensionCategory	String	reminder
UNNotificationExtensionInitialC...	Number	0.2
NSExtensionMainStoryboard	String	MainInterface
NSExtensionPointIdentifier	String	com.apple.usernotifications.content-extension

This wraps up the interface part of the content extension. In addition to showing a custom UI for a notification, content extensions can respond to actions that the user selects right inside the content extension itself. To do this, you should implement the `didReceive(_:completionHandler:)` delegate method in the extension. Once this method is implemented, the extension becomes responsible for all actions that are chosen by the user. This means that the extension should either handle all the actions, or that it should explicitly pass them on to the host app if the selected action can't be handled within the content extension.

After handling the incoming action, the notification extension determines what should happen next. This is done by calling the completion handler with a `UNNotificationContentExtensionResponseOption`. There are three options to choose from:

- Dismiss the notification and do nothing
- Dismiss the notification and forward the chosen action to the host app
- Keep the extension active so the user can pick more actions or, using Apple's *Messages* app as an example, so that the user can carry on the conversation inside the content extension

Add the following implementation for didReceive(_:completionHandler:) to mark a reminder as completed when a user taps the **Complete** action:

```
func didReceive(_ response: UNNotificationResponse, completionHandler
completion: @escaping (UNNotificationContentExtensionResponseOption)
-> Void) {

  guard let reminder = extractReminderFromResponse(response) else {
    completion(.dismissAndForwardAction)
    return
  }

  if response.actionIdentifier == "complete-reminder" {
    setCompleted(true, forReminder: reminder, completionHandler:
completion)
  } else {
    setCompleted(false, forReminder: reminder, completionHandler:
completion)
  }
}

func setCompleted(_ completed: Bool, forReminder reminder: Reminder,
completionHandler completion: @escaping
(UNNotificationContentExtensionResponseOption) -> Void) {

  reminder.managedObjectContext!.perform {
    reminder.isCompleted = true
    self.reminderStatusLabel.text = reminder.isCompleted ? "Done" :
"Pending"

    try! reminder.managedObjectContext!.save()

    completion(.doNotDismiss)
  }
}

func extractReminderFromResponse(_ response: UNNotificationResponse)
-> Reminder? {
  let userInfo = response.notification.request.content.userInfo

  guard let identifier = userInfo["reminder-uuid"] as? String else {
      return nil
  }

  let fetchRequest: NSFetchRequest<Reminder> = Reminder.fetchRequest()
  fetchRequest.predicate = NSPredicate(format: "identifier == %@",
identifier)
```

```
    let moc = PersistentHelper.persistentContainer.viewContext
    guard let results = try? moc.fetch(fetchRequest),
      let reminder = results.first else {
        return nil
    }

    return reminder
  }
```

The preceding code takes very similar approach to marking the reminder as completed to the one you implemented earlier in `AppDelegate`. Note that the reminder can also be marked as not completed. The reason this is taken into account is that once the user has tapped the **Complete** action, this action should be swapped out for a **Pending** action, in case the user marked the reminder as completed by accident.

Add the following code to replace the action that is shown underneath the content extension:

```
  func setNotificationForReminder(_ reminder: Reminder) {
    let identifier = reminder.isCompleted ? "pending-reminder":
  "complete-reminder"
    let label = reminder.isCompleted ? "Pending" : "Complete"

    let action = UNNotificationAction(identifier: identifier, title:
  label, options: [])
    extensionContext?.notificationActions = [action]
  }
```

The preceding method creates and applies a new action, depending on the current reminder status. Add calls to this method after setting the current reminder's completion status and in the initial `didReceive(_:)` method, so that the correct action is shown in the event that the reminder associated with the notification has already been completed.

As you've seen just in the preceding examples, content extensions are not complex to implement, yet they are amazingly powerful. You can add an entirely new layer of interaction and increase the relevance of notifications by implementing a content extension, and it is strongly recommended that you always consider ways to implement extensions for your notifications to provide the user with rich information about their notification as quickly as possible.

Summary

In this chapter, you explored the amazing world of notifications. Even though notifications are simple in nature and short-lived, they are a vital part of the iOS ecosystem, and users rely on them to provide timely, interesting, and relevant updates about subjects that matter to them, right there and then. You saw how to go from the basics, simply scheduling a notification, to more advanced subjects, such as custom actions and implementing grouped notifications. You saw that service extensions allow you to implement great features, such as true end-to-end encryption, or enrich push notifications with content that is stored on the device.

You then implemented a content extension, and saw how a content extension can take a simple, plain notification, and make it interesting, rich, and more relevant. The example of a calendar that appears with event invites comes to mind immediately, but the possibilities are endless. Proper usage of notification extensions will truly make your app stand out in a positive and engaging way.

The next chapter focuses on another extension: the Today extension!

Questions

1. How are `UNUserNotificationCenter` and `NotificationCenter` different?

    a) One is responsible for showing notifications to users, and the other is a mechanism to send notifications about events inside an app.
    b) One is supported on iOS 9 and below, while the other is supported on iOS 10 and above.
    c) They are the same.

2. What kinds of notifications can you send to a user?

    a) Push notifications and local notifications.
    b) Push notifications and scheduled notifications.
    c) Push notifications.

3. What are provisional notifications?

    a) Special notifications that provide critical information to users.
    b) Notifications that are encrypted.
    c) Notifications that are delivered to the user quietly, so they can preview the notifications your app will send.

4. How does iOS group notifications?

    a) It uses the thread identifier from the notification content and the app that the notification belongs to.
    b) By thread identifier only.
    c) By notification category.

5. What does a Notification Service Extension do?

    a) It implements end-to-end encryption.
    b) It allows you to add custom action handlers.
    c) It allows you to replace a notification's contents before the user sees it.

6. What does a notification content extension do?

    a) It allows you to add custom action handlers.
    b) It allows you to create a custom interface for when a user 3D-Touches a notification.
    c) It implements end-to-end encryption.

Further reading

- Apple's Payload Key Reference documentation: `https://developer.apple.com/documentation/usernotifications/setting_up_a_remote_notification_server/generating_a_remote_notification`
- Apple's documentation on localizing your app: `https://developer.apple.com/library/archive/documentation/MacOSX/Conceptual/BPInternational/LocalizingYourApp/LocalizingYourApp.html`

19
Instant Information with a Today Extension

When a user swipes right on the **Lock Screen** or in the **Notification Center**, they are taken to section called the **Today View**. In this section, users are presented with several widgets that provide useful information at a glance. These widgets are called **Today Extensions**.

In this chapter, you will learn how you can build a **Today Extension** so you can integrate useful information from your app with the **Today View**. A proper integration with the **Today View** can help users see the information they care about at a glance. The basis for the extension that you will build in this chapter is an inspirational quote app called **The Daily Quote**. In its current state, **The Daily Quote** only shows a quote when the user opens the app. You will expand this app by implementing a **Today Extension**.

This chapter covers the following topics:

- Understanding the anatomy of a Today Extension
- Adding a Today Extension to your app

Understanding the anatomy of a Today Extension

The terms **Today Extension** and widget can be used interchangeably; they both refer to a component that is present in iOS's **Today View**. In this chapter, the term *widget* is used throughout.

If you swipe down from the top of the screen to open **Spotlight** and swipe right after that, you're presented with the **Today View**. On the simulator, this view tends to look rather empty, but on your device, there's probably a lot more going on. The following screenshot shows the **Today View** on the simulator with a couple of widgets added to it:

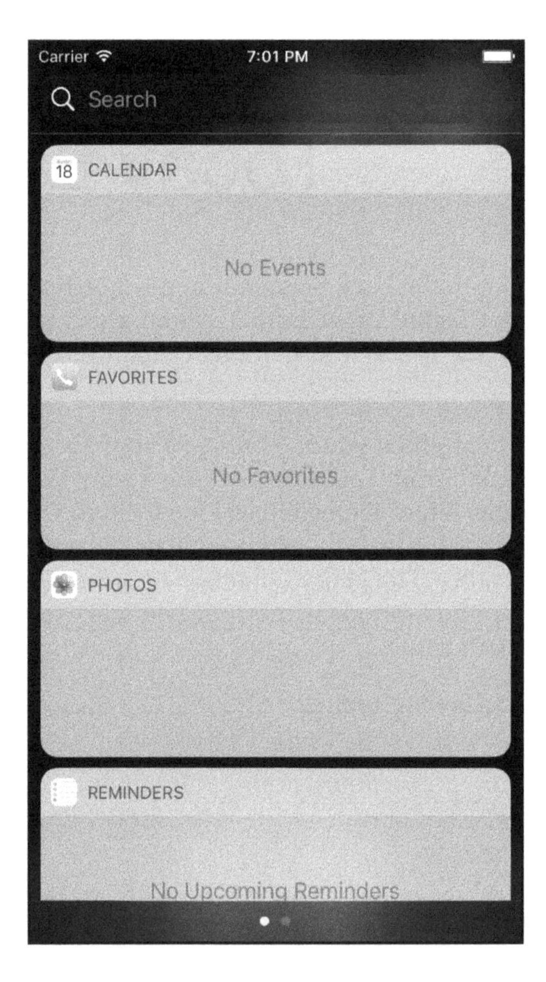

Users can scroll to the bottom of this view and manage their widgets from there. They can add new widgets and remove existing ones. All these widgets have one thing in common: they provide relevant information for the current moment or day. For the **Calendar**, this means showing events that you have planned for today or tomorrow. The **Favorites** widget in the screenshot usually contains contacts that the user interacts with often. Users can even add a widget that suggests apps that the user is likely to use based on the current time and the user's location.

Any widget that you implement for your app should aim to provide your users with relevant, easy-to-scan information. Doing this makes your widget more useful and more likely to be added to a user's Today View.

Before you implement your own Today Extension, let's take a closer look at how a Today Extension works and where a user can find them in iOS.

Finding Today Extensions in iOS

In iOS, there are several places where people can discover and use widgets. For most scenarios, the way the widget looks will be similar to how it looks in the Today View, apart from some very minor modifications. You are already aware that users can find widgets in the Today View, which can be found by swiping right on the lock screen or Notification Center. In addition to these two places, the Today View is also shown on the leftmost screen on the user's Springboard.

One more place where users can find and use Today Extensions is directly on the home screen. When a user 3D-Touches on an app icon for an app that has implemented a Today Extension, the corresponding widget will pop up, as shown in the following screenshot:

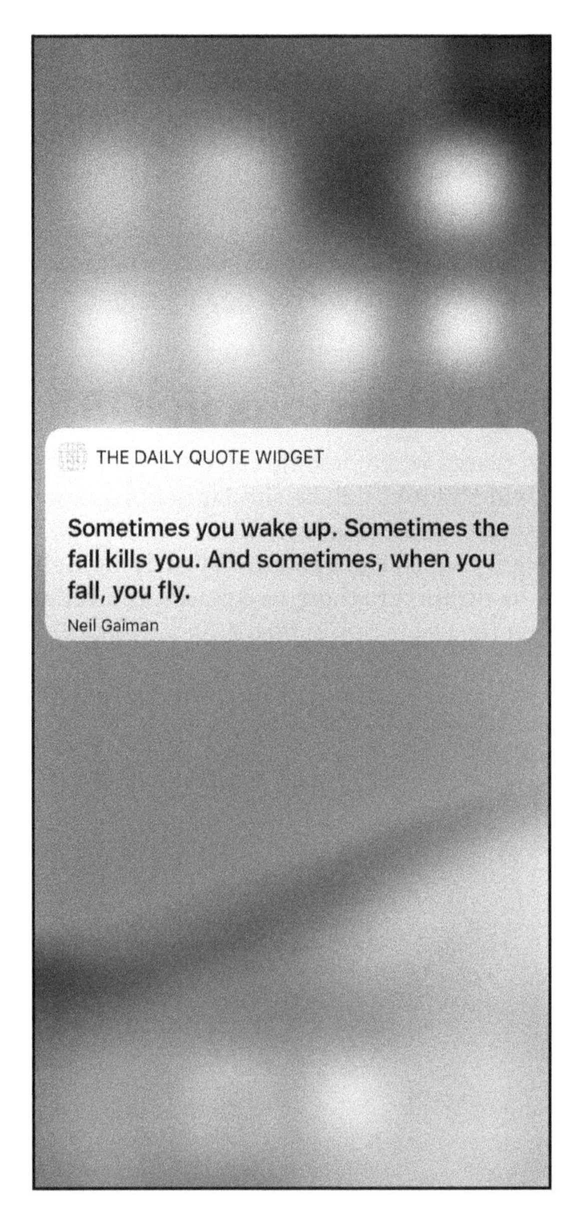

If a user hasn't added the widget for the app that they pressed to the Today View yet, the widget will feature an **Add Widget** button that allows users to quickly add the extension to the Today View so they can access it there in the future.

Understanding Today Extensions

A **Today Extension** isn't much more than a view controller that is embedded inside the notification center's Today View. This means that when your **Today Extension** is initialized, there is no `AppDelegate` involved at all. The extension does not have any application life cycle methods, such as `application(_:didFinishLaunchingWithOptions:)`.

The Today View displays your widget by showing the view controller that you set up for it. This is very similar to how a **Notification Content Extension** works since that extension also has a single view controller to define its functionality.

The Today View is in control of positioning and rendering all of its widgets. This means that you can't set the frame for the widget's view yourself. When you implemented your **Notification Content Extension** in the previous chapter, you could force your notification to be shown at a certain height through AutoLayout. This isn't possible when you implement a **Today Extension**.

Before iOS 10, a widget would resize itself based on its contents or the size that the widget would set as its `preferredContentSize`. Since iOS 10, this behavior has changed, and widgets have a user-defined compact mode and an expanded mode. The compact mode should typically be about 110 points in height, but in reality, this height varies based on the user's text-size preferences.

Because there are no real guarantees regarding the height of your widget can use, it's essential that you only show the most critical information in your widget. Because widgets tend to resize based on the user's text size preferences, it's a good idea to adopt dynamic type in your widget. This makes your widget more accessible by scaling the text inside your widget based on the user's preferences. You'll learn more about this when you implement the widget.

If your widget supports it, the user can switch to the expanded state for your widget. In the expanded state, you have a bit more freedom to set the size of the widget yourself. You will also explore this more when you implement your own widget.

One more key principle of the **Today Extension** is that the widget never communicates with its host app directly. If you have a widget that the user can tap on to open the app, the widget doesn't directly instruct the app to do this. Instead, URLs are used. You can use a custom URL scheme or a Universal Link to achieve this. In several snippets of example code, Apple uses custom URL schemes to redirect users from a widget to an app. **The Daily Quote** app you will extend in this chapter will work similarly.

Now that you're up to speed with **Today Extensions**, it's time to see how you can add your own **Today Extension** to **The Daily Quote**.

Adding a Today Extension to your app

In the code bundle for this chapter, you'll find a project called **The Daily Quote**. It's a straightforward app that displays a different inspirational quote to the user every day. The quotes are hardcoded in the `Quotes.swift` file and the current quote is stored in the `UserDefaults` store. The UI contains just two labels to show the current quote and the person that's quoted.

Even though the project is quite simple, each file contains something interesting. If you select one of the labels in the **Storyboard** and examine its attributes, you'll find that the font for the quote itself is **Title 1** and the font for the quoted person is **Caption 1**. This is different from the default system font that you usually use. When you configure labels with these predefined styles, the text in the label will dynamically adjust based on the user's preferences. Adopting dynamic type such as this is requires only minimal effort and is great from an accessibility standpoint.

If your interface allows it, it's recommended that you make use of accessible type, simply because it will make your app more comfortable to use for all of your users.

If you look at the `QuoteViewController` class, there are only a couple of lines involved in displaying a quote. The reason this code is so simple and concise is that a great deal of preparatory work was done in the `Quote` model file. Go ahead and open that file to see what's going on.

The `Quotes` struct is set up in such a way that it can be used as a simple database. The struct contains several static properties and methods to provide quotes from a predetermined list of quotes.

If you were to build a similar app and put it in the App Store, you'd probably want to download the quotes from a server and store it in Core Data, because pushing an update every time you want to add or remove a couple of quotes is a lot of effort for a simple change. There are only a couple of constant instance properties present on the `Quote` struct:

```
let text: String
let creator: String
```

These properties are the ones that `QuoteViewController` reads and displays to the user.

Furthermore, `UserDefaults` is used to store which quote should be shown for the current day. `UserDefaults` is a simple data store that stores app-related settings. It's essentially a lightweight persistence layer that you can use to store simple objects. **The Daily Quote** makes use of `UserDefaults` to store the date on which the current quote was set as well as the index in the list of `Quote` instances that points to the current quote:

```
static var current: Quote {
  if let lastQuoteDate = userDefaults.object(forKey: "lastQuoteDate")
as? Date {
    if NSCalendar.current.compare(Date(), to: lastQuoteDate,
toGranularity: .day) == .orderedDescending {
      setNewQuote()
    }
  } else {
    setNewQuote()
  }

  guard let quoteIndex = userDefaults.object(forKey: "quoteIndex") as?
Int,
    let quote = Quote.quote(atIndex: quoteIndex)
    else { fatalError("Could not create a quote..") }

  return quote
}

static func setNewQuote() {
  let quoteIndex = Quote.randomIndex
  let date = Date()

  userDefaults.set(date, forKey: "lastQuoteDate")
  userDefaults.set(quoteIndex, forKey: "quoteIndex")
}
```

The preceding snippet illustrates the process of retrieving and storing the current `quote`. First, the code checks whether a `quote` has been set before and if so, it makes sure that the `quote` is at least a day old before generating a new one. Next, the current `quoteIndex` is retrieved from `UserDefaults` and returned as the current `quote`.

If you need a quick refresher on how to use `UserDefaults` in your app, go back to `Chapter 11`, *Syncing Data with CloudKit*.

Don't use `UserDefaults` to store privacy-sensitive data. It's not considered secure; the keychain should be used for this purpose. Also, make sure that you're not storing complex or repetitive data in `UserDefaults`. It's not a database, nor is it optimized for reading and writing many times. Stick to simple, application-specific settings only.

Now that you are familiar with the contents of the starter project, add a new target to your project just like you have done before, but this time, select the **Today Extension**:

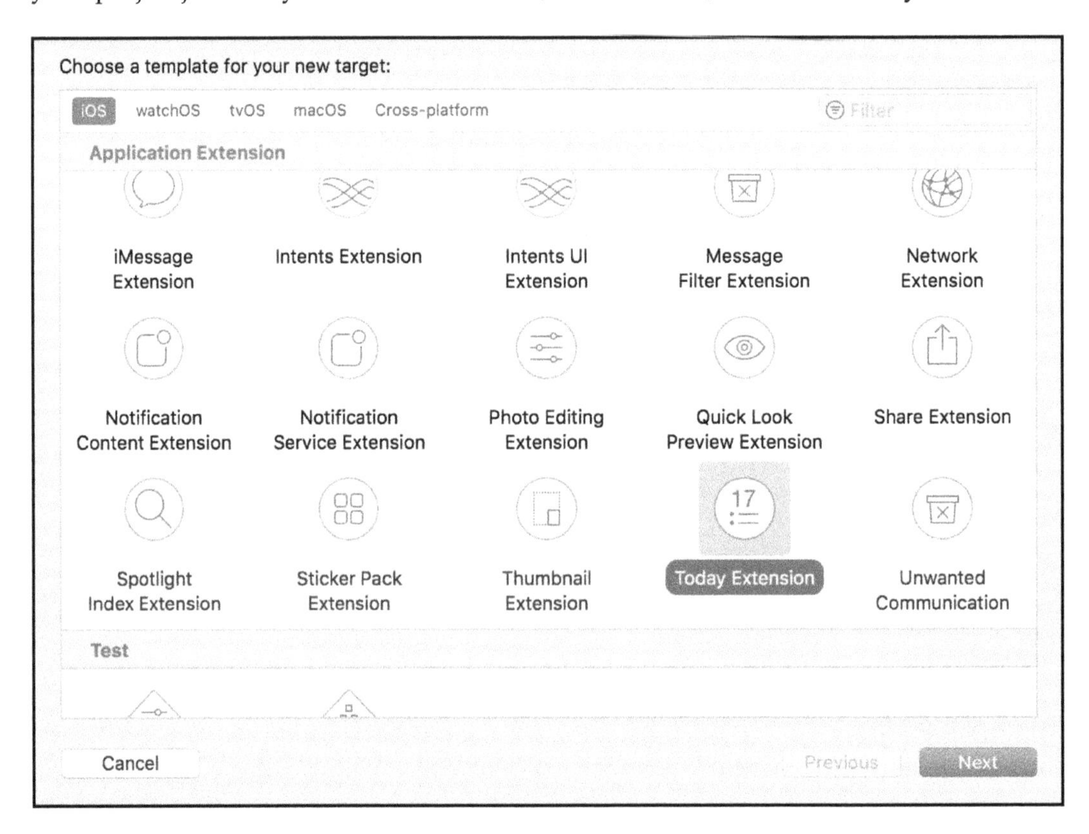

To see the default widget that Xcode generates for you, go ahead and run your extension. The first time you do this, your widget might not show up immediately. Don't worry if this is the case, build and run again, and your app should show up as a widget in the Today View.

Now let's see what kind of files and boilerplate code Xcode has generated for you. In the folder that Xcode created for your extension, you'll find a view controller, a storyboard and an `Info.plist` file. This structure is very similar to the structure that was used for the **Notification Content Extension** you created in the previous chapter.

You can change the size of the view controller in your Storyboard if you like, but note that the display size for the widget is taller than the visible size in the storyboard. This is due to the standard compact widget height that iOS uses for your extension. You can't influence this, so you'll need to keep in mind that you only have limited space for your extension, so you should make sure your layout is flexible enough to look good at the compact size the Today View imposes on the widget.

To give the labels in the extension a little bit of room to breathe, select the view controller (not the view itself) and click on the **Size** inspector. Here you'll see that the simulated size for the view controller is set to free-form and you can set a custom width and height. Leave the width as it is for now and set the height to *110*. This size should be the smallest size at which our widget displays, and it gives you plenty of room to create the interface. Also, delete the default label that has been added to the interface automatically.

Drag a `UILabel` into the view and set its font to **Headline**. Click the **Font** icon in the **Attributes** inspector to change the font and select the **Font** dropdown to find the dynamic text styles. Position the label in the top-left corner using the blue helper lines and adjust its width, so it covers the available space. After doing this, add the following constraints to the label:

- Leading space to Safe Area
- Top space to Safe Area
- Trailing space to Safe Area

Finally, set the number of lines for the label to 3 so the quote doesn't take up more than 3 lines. Now, drag out another label and position it right below the first label. Make sure this label has the same spacing from the superview's left edge and that it has the same with as the other label. Set its font style to **Caption 1**. Also, add the following constraints to this view:

- Leading space to Safe Area
- Trailing space to Safe Area
- Vertical spacing between this label and the label above it

Your layout in **Interface Builder** should resemble the following screenshot:

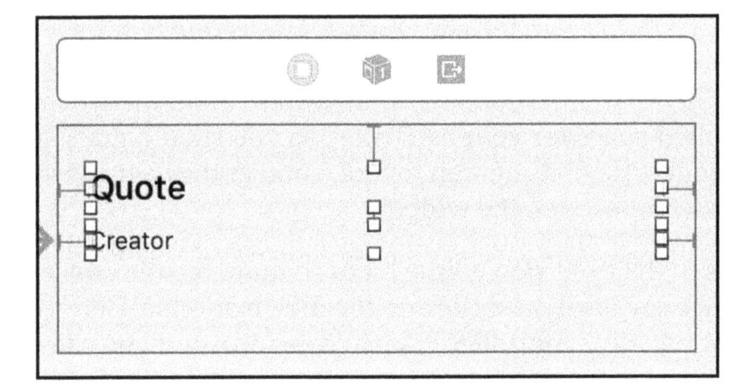

Go ahead and rerun your extension. Your final result should look similar to the one shown in the following screenshot:

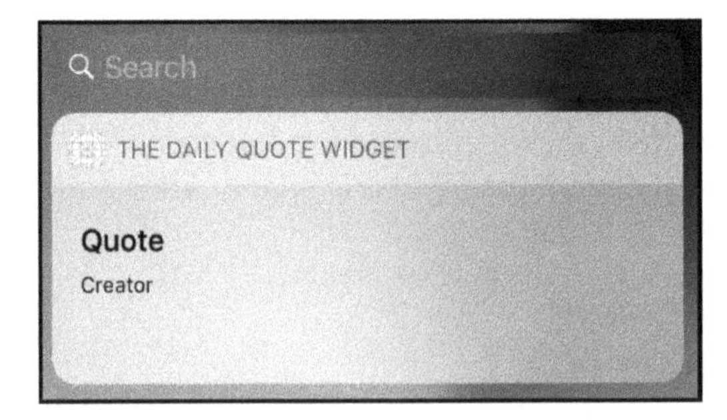

Now that you have set up the layout for your extension, let's some outlets to the widget view controller so you can display the quote of the day inside of the widget:

```
@IBOutlet var quoteLabel: UILabel!
@IBOutlet var quoteCreator: UILabel!
```

Open the widget's Storyboard file and connect the outlets as you've done before. Select the view controller, go to the **Outlet Inspector**, and drag from the **Outlet Inspector** to the corresponding views to connect the outlets to the view.

The next step to build this widget is to load and display the quote of the day in the widget. Since extensions do not communicate with their host apps directly, you will have to use App Groups again to implement a shared `UserDefaults` store. Sharing `UserDefaults` through App Groups will allow both your widget and app to retrieve the current quote index from `UserDefaults`, and ensures that both interfaces always show the same quote.

Create a new App Group for **The Daily Quote** and make sure the app and extension both belong to the same App Group. Also, make sure that you add `Quote.swift` to the extension target as well as the app target.

If you need a quick refresher on what App Groups are and how they work, go back to `Chapter 16`, *Streamlining Experiences with Siri* for more information.

After enabling App Groups and making sure the app and the extension share the `Quote` struct, add the following code to `TodayViewController.swift`:

```
override func viewDidLoad() {
  super.viewDidLoad()

  updateWidget()
}

func updateWidget() {
  let quote = Quote.current
  quoteLabel.text = quote.text
  quoteCreator.text = quote.creator
}
```

If you run the app now, you should see the current quote for the day shown in your widget, as illustrated in the following screenshot:

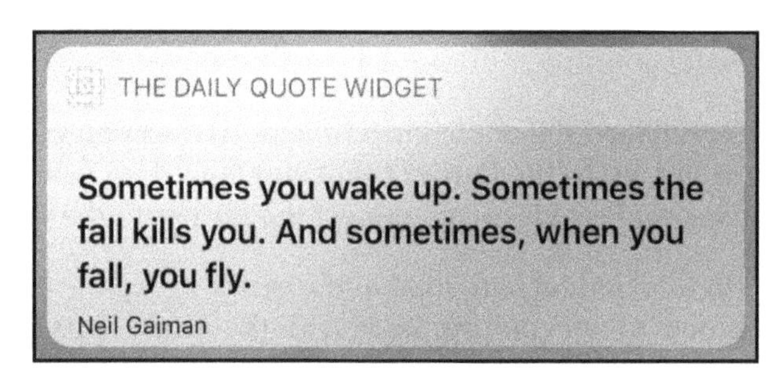

The widget currently loads the quote when its view controller is loaded. There are no guarantees about how and when your widget will be recreated. This means that you can't rely on `viewDidLoad()` to make sure that your widget always contains the most recent information. In the example that Xcode generated when you first created your extension, a method called `widgetPerformUpdate(completionHandler:)` can be found. This method is called regularly by the system to make sure that your widget can update its contents when needed.

The system passes a `completionHandler` to the update method that you must call to let iOS know whether your widget has loaded new data. This allows iOS to optimize the frequency and timing of calls to `widgetPerformUpdate(completionHandler:)` so your widget isn't doing more work than needed, and to make sure it updates when it's most likely to have new data. Add the following implementation for this `widgetPerformUpdate(completionHandler:)` to `TodayViewController.swift`:

```
func widgetPerformUpdate(completionHandler: @escaping (NCUpdateResult)
-> Void) {
  let currentText = quoteLabel.text
  updateWidget()
  let newText = quoteLabel.text

  if currentText == newText {
    completionHandler(NCUpdateResult.noData)
  } else {
    completionHandler(NCUpdateResult.newData)
  }
}
```

The preceding method updates the widget by calling `updateWidget()`, just like `viewDidLoad` does. Before doing this, the current text for the quote is stored. After updating the widget, the new text for the quote is stored. Based on the comparison of these two strings, the callback is informed about the result of the update request.

Some widgets in the Today View have a **Show More** button that can be tapped to expand the widget. You can implement this button in your own widget by setting the `widgetLargestAvailableDisplayMode` property on the notification context to the `.expanded` value. Setting the largest display mode to expanded tells the Today View that your widget needs the ability to show contents in a larger view. The following line of code demonstrates how to set the largest available display mode:

```
extensionContext?.widgetLargestAvailableDisplayMode = .expanded
```

When the user taps the **Show More** button that is rendered after setting the largest display mode, the `widgetActiveDisplayModeDidChange(_:withMaximumSize:)` method is called on your notification extension. In this method, you can set your `preferredContentSize` of `TodayViewController` to the size you would like your widget to be. This should always be smaller than the maximum size that is passed to the method because the Today View will not make your widget any bigger than the maximum size.

The following code is a short sample implementation for `widgetActiveDisplayModeDidChange(_:withMaximumSize:)` that determines the preferred content size based on the new display mode:

```
func widgetActiveDisplayModeDidChange(_ activeDisplayMode:
NCWidgetDisplayMode, withMaximumSize maxSize: CGSize) {

  if activeDisplayMode == .compact {
    preferredContentSize = maxSize
  } else {
    preferredContentSize = CGSize(width: maxSize.width, height: 200)
  }
}
```

With this information, you know everything you need to know to start implementing widgets for your own applications.

Summary

This chapter has shown you everything you need to know about implementing a simple Today Extension that shows a new inspirational quote to your users every day. You learned where users can find and use widgets on iOS and how they work. In addition, you learned how to use App Groups to share a `UserDefaults` store between extensions and apps, which is similar to how you can share a Core Data store in an App Group.

As an exercise, try to change how often a new quote is shown to the user. For instance, try loading a new quote every couple of hours to keep the user entertained.

You also learned how to make sure your widget always has up-to-date data and how you can allow your user to expand a widget, so it can show even more information. As a side note, you also learned that you can adopt accessible and resizable text by using Apple's predefined text styles to improve your widget's accessibility.

In the next chapter, you will learn how to improve an app that you have worked on before by implementing Drag and Drop features for it!

Questions

1. Where can users find Today Extensions?

    a) By Swiping right on the lock screen.
    b) By Swiping right on the lock screen or Notification Center, and on the leftmost screen on the Springboard.
    c) By Swiping right on the lock screen or Notification Center, on the leftmost screen on the Springboard, and 3D-touching an app.

2. At what height are widgets displayed by default?

    a) It depends on the screen width and user settings.
    b) 100 points.
    c) It depends on the size of your contents.

3. How can you allow a user to expand your widget's height?

    a) By setting the preferred content size.
    b) By implementing
    `widgetActiveDisplayModeDidChange(_:withMaximumSize:)`.
    c) By setting the largest available display mode on the extension context to expanded.

4. What is the best place to update your widget contents?

    a) In `viewDidLoad()`.
    b) In `widgetPerformUpdate(completionHandler:)`.
    c) In `viewDidAppear()`.

5. How does iOS know the best moment to reload a widget's contents?

    a) It reloads when the widget is about to appear.
    b) It monitors network requests.
    c) With the completion handler that is passed to
    `widgetPerformUpdate(completionHandler:)`.

Further reading

- Accessibility Programming Guide: `https://developer.apple.com/library/archive/documentation/UserExperience/Conceptual/iPhoneAccessibility/Introduction/Introduction.html`
- Today Extension Programming Guide: `https://developer.apple.com/library/archive/documentation/General/Conceptual/ExtensibilityPG/Today.html`

20
Exchanging Data With Drag And Drop

Until iOS 11 came out, iOS had a pretty terrible drag and drop experience. If you wanted to move a couple of photos from one app on your iPad to another, you would usually have to copy and paste each photo individually. If you compare this experience with the desktop, where you can grab one or multiple items and drag those items from one place to the other, the iOS experience sounds pretty bad.

Since iOS 11, users can finally enjoy a full drag and drop experience on iOS. Implementing drag and drop in your own apps has been made surprisingly simple since all you need to do is implement a couple of required methods from a handful of protocols. This chapter's purpose is to show you how to handle drag and drop in the augmented reality gallery you built in `Chapter 8`, *Using Augmented Reality*. The gallery app will be expanded so that users can add photos from their own photo library, the internet, or any other source to their art library.

This chapter covers the following topics:

- Understanding the drag and drop experience
- Implementing a basic drag and drop functionality
- Customizing the drag and drop experience

Understanding the drag and drop experience

The drag and drop experience on iOS is quite simple to use; pick up an item on the screen, drag it somewhere else, and let it go to make the dragged item appear in the new place. However, iOS didn't have this behavior until iOS 11. And even now, its full range of capabilities is only available on the iPad. Despite this limitation, the drag and drop experience is really powerful on both the iPhone and iPad.

Users can pick up an item from any app, and move it over to any other app that implements drag and drop, as they please. And dragging items is not even limited to just a single item, it's possible for users to pick up multiple items in a single drag session. The items that a user adds to their drag session don't have to be of the same type; this makes the drag and drop experience extremely flexible and fluid. Imagine selecting some text and a picture in Safari and dragging them over to a note you're making. Both the image and the text can be added to a note in just a single gesture.

Unfortunately, apps are not able to handle drag and drop out of the box; you'll need to do a little bit of work to support this feature. At the heart of the drag and drop experience are two protocols and two new interaction classes. Let's briefly go over the protocols and classes before going more in-depth and implementing drag and drop for your augmented reality art gallery.

The first requirement for a drag and drop session is a view that can start or receive a drag session. This ability is added to a view through either `UIDropInteraction` or `UIDragInteraction`. Both interactions are subclasses of the `UIInteraction` base class. `UIInteraction` manifests itself similarly to `UIGestureRecognizer` in the sense that you create an instance of it and attach it to a view. The following figure shows the relationship between a view and the drag and drop interactions:

When adding either `UIDragInteraction` or `UIDropInteraction` to `UIView`, you must also set up a delegate for the interaction you're adding. When you add `UIDropInteraction`, you must set `UIDropInteractionDelegate` on it. When you add `UIDragInteraction`, you must assign `UIDragInteractionDelegate` for it. The following figure illustrates the relationship between the interactions and their delegates:

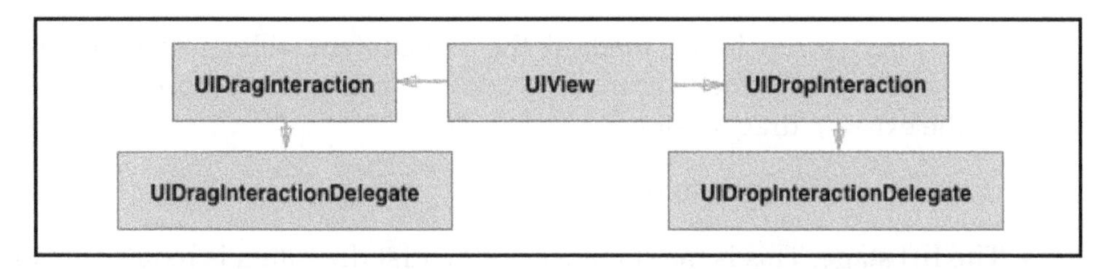

Each protocol is related to a different set of responsibilities, as is suggested by their names. Let's have a look at `UIDragInteractionDelegate` first.

Understanding UIDragInteractionDelegate

Any time the user long-presses an item that has `UIDragInteraction` associated with it, a new drag session is started. This drag session uses the interaction's `UIDragInteractionDelegate` to obtain the data for the item that is being dragged, a drag preview, and more.

`UIDragInteractionDelegate` only has a single required method: `dragInteraction(_:itemsForBeginning:)`. If you implement this method, your app can start supplying data to other apps. The return type for `dragInteraction(_:itemsForBeginning:)` is `[UIDragItem]`. `UIDragItem` is a container for `NSItemProvider`. The drag item is also responsible for providing a preview of the dragged item while it's being dragged. Normally, the selected view will be used, but sometimes you might want to provide a custom view for the preview. If this is the case, you can provide your own preview through `UIDragItem`.

A straightforward example of `dragInteraction(_:itemsForBeginning:)` looks as follows:

```
func dragInteraction(_ interaction: UIDragInteraction,
itemsForBeginning session: UIDragSession) -> [UIDragItem] {
  let text = "Hello, world"
  let provider = NSItemProvider(object: text as NSString)
  let item = UIDragItem(itemProvider: provider)
  return [item]
}
```

As mentioned before, users can add multiple objects to an existing drag session by tapping on them. Whenever the user taps a draggable item while a session is in progress, `dragInteraction(_:itemsForAddingTo:withTouchAt:)` is called on `UIDragInteractionDelegate`.

If you return an empty array in this method, the tapped item will not be added to the drag session. But if you return one or more `UIDragItem` instances, those items will be appended to the existing drag session.

A drag session can be divided into three stages:

- **The lift stage**: This is the stage where the first drag item is requested, and the view that is about to be dragged is animated, so the user sees that they have started a drag session.
- **The dragging stage**: The user is now moving the drag item around.
- **The end stage**: Either a drop has been performed, or the drag session got canceled. If a session is canceled, the dragged item is animated back to its starting position.

You can monitor and respond to each of the stages through `UIDragInteractionDelegate`. To customize the lift animation, you could implement `dragInteraction(_:willAnimateLiftWith:session:)`, for instance. For a full overview of available drag customizations, you should take a look at the documentation for `UIDragInteractionDelegate`; there are a bunch of methods available for you to implement!

Now that you know how to set up your app for dragging, let's see how dropping works.

Understanding UIDropInteractionDelegate

Similar to how `UIDragInteractionDelegate` works, `UIDropInteractionDelegate` is used to respond to different stages in the drag and drop life cycle. Even though `UIDropInteractionDelegate` has no required methods, there are at least two methods that you should implement to support drag and drop. The first method is `dropInteraction(_:sessionDidUpdate:)`.

As soon as the user moves their finger on top of a drop target, the drop target is asked whether it can handle a drop with the contents from the current drop session. To do this, `dropInteraction(_:canHandle:)` is called on the drop target. Assuming the data can be handled, `dropInteraction(_:sessionDidUpdate:)` is called next. You must return an instance of `UIDropProposal` from this method. A drop proposal lets the session know what you'd like to happen if the drop is executed at some point. For instance, you can make a copy proposal or a move proposal. The UI surrounding the contents that are being dragged will update accordingly to let the user know what will happen if they perform the drop.

Now let's say your app can only handle objects of a particular type. You should implement `dropInteraction(_:canHandle:)`. You can use this method to inspect whether the drop session contains items that are relevant for your app to handle. An example of this looks as follows:

```
func dropInteraction(_ interaction: UIDropInteraction, canHandle
session: UIDropSession) -> Bool {
  for item in session.items {
    if item.itemProvider.canLoadObject(ofClass: UIImage.self) {
      return true
    }
  }

  return false
}
```

This example searches for at least one image in the current drop session. You can use `canLoadObject(ofClass:)` on `NSItemProvider` to figure out whether the item provider contains an instance of a particular class, in this case, `UIImage`. If your app restricts drag and drop to the same application, you might not need to implement this method because you know exactly what items in your app are draggable. It is recommended that you always make sure the session can be handled by the drop target.

The second method you should always implement is `dropInteraction(_:performDrop:)`. If you don't implement this method, your app doesn't have a proper implementation of the drop interaction. Once the user drops an item onto a drop target, they expect something to happen. `dropInteraction(_:performDrop:)` is the perfect place to do so.

An example implementation of `dropInteraction(_:performDrop:)` could look as follows:

```
func dropInteraction(_ interaction: UIDropInteraction, performDrop
session: UIDropSession) {
  for item in session.items {
    if item.itemProvider.canLoadObject(ofClass: UIImage.self) {
      item.itemProvider.loadObject(ofClass: UIImage.self) { [weak
self] item, error in
        // handle the item
      }
    }
  }
}
```

The preceding code loops through all items in the drop session. If an item can provide an image, the image is loaded. Note that `loadObject(ofClass:)` uses a callback. This means that the data is loaded asynchronously. The dropped data could be huge and, if it were loaded and processed on the main thread by default, it would make your app freeze for a while. Because Apple made `loadObject(ofClass:)` asynchronous, your app's responsiveness is guaranteed, and your users won't notice any freezes or lag.

Just like a drag session, a drop session typically has three stages:

- The session starts when a user drags content onto a drop target.
- Performing the drop; this is when the user lifts their finger while it's on a drop target.
- Ending the drop; the user has either dragged their finger away from the drop target, or the drop has been performed successfully.

Apple has not defined the stages of the drag and drop experience as you have just learned them. Dividing the experience into the preceding steps is merely intended to help you to grasp the life cycle of drag and drop. If you want to learn everything about the drag and drop life cycle, make sure to check out the documentation for both `UIDropInteractionDelegate` and `UIDragInteractionDelegate`.

Alright, now that you know what drag and drop looks like in theory, let's see what it looks like in practice!

Implementing a basic drag and drop functionality

The previous section explained how drag and drop works from a theoretical point of view. This section focuses on implementing drag and drop in a sample app. First, you will learn how a simple, regular implementation of drag and drop might work. Next, you'll see how Apple has implemented drag and drop for `UICollectionView` and `UITableView`. These two components have received special treatment for drag and drop, making it even easier to implement drag and drop in apps that use these components.

Adding drag and drop to a plain UIView

Before you implement drag and drop in the Augmented Reality gallery, let's see how you can implement a simple version of drag and drop with a simple view and an image. In the code bundle for this chapter, you'll find a sample project named **PlainDragDrop**. Open the starting version for this project and run it on an iPad simulator. You should see the user interface shown in the following screenshot:

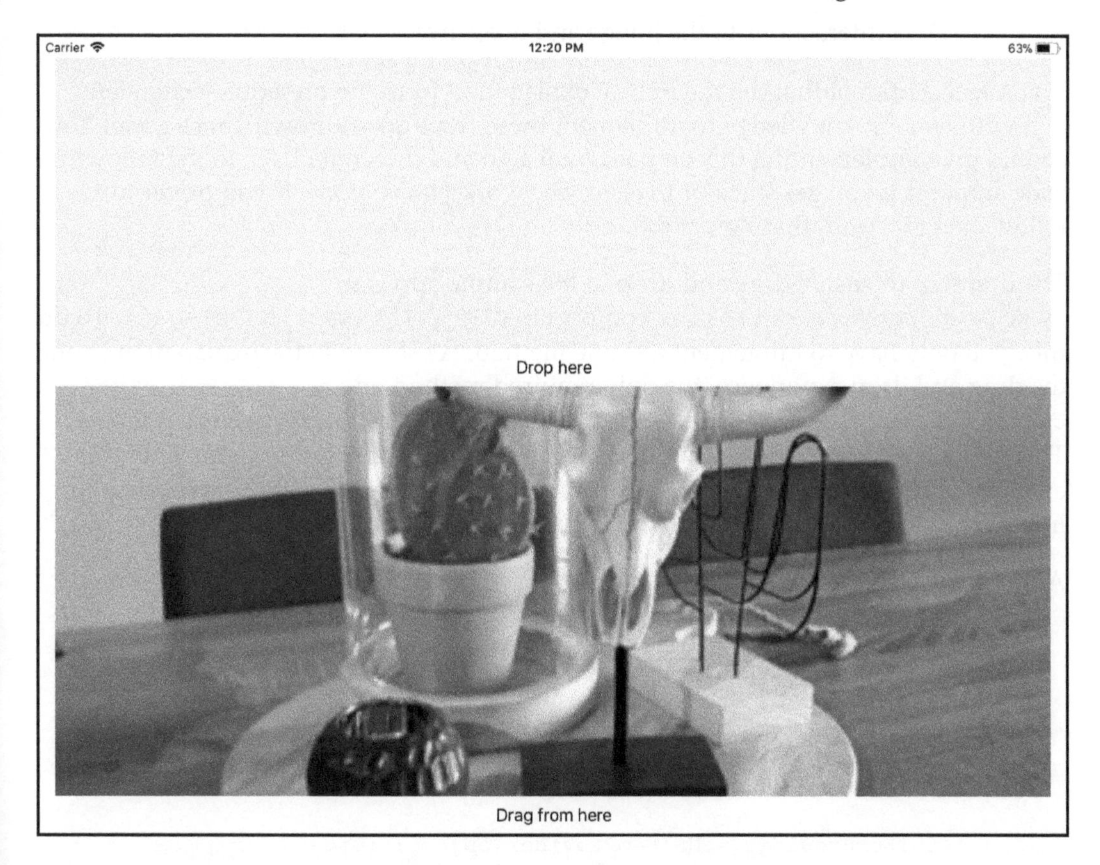

The goal of this example is to allow users to perform the following actions:

- Drag the image to the drop area.
- Drag an external image to the drop area.
- Drag the bottom image to an external app.

While this might sound like you are going to have to do a lot of work, it's quite simple to implement all three features at once. Simply implementing the first feature enables the other two! Quite convenient, right? You are going to implement drag and drop for this example in just three simple steps:

1. Make `ViewController` conform to `UIDragInteractionDelegate`.
2. Make `ViewController` conform to `UIDropInteractionDelegate`.
3. Add interactions to the image and drop area.

If you feel confident that the theoretical explanation from the previous section left you with enough knowledge to implement these steps on your own, that's great! You should give implementing this on your own a go and then refer back to the following code snippets if you get stuck or to check the work you've done. If you prefer to follow along instead, that's great too.

The first step to enable drag and drop in the sample app is to make `DragDropViewController` conform to `UIDragInteractionDelegate`. To do this, you only have to implement a single method. As shown in the explanation about the drag and drop experience, the only required method on `UIDragInteractionDelegate` is `dragInteraction(_:itemsForBeginning:)`. This method requires you to provide a `UIDragItem` for the drag session. Since this example will not support dragging multiple items from the app, there is no need to implement other delegate methods.

Add the following extension to `DragDropViewController.swift` to implement `UIDragInteractionDelegate`:

```
extension DragDropViewController: UIDragInteractionDelegate {
  func dragInteraction(_ interaction: UIDragInteraction,
itemsForBeginning session: UIDragSession) -> [UIDragItem] {
    guard let image = imageView.image
      else { return [] }

    let itemProvider = NSItemProvider(object: image)
    return [UIDragItem(itemProvider: itemProvider)]
  }
}
```

The preceding code should look familiar to you since it closely resembles the example code from the previous section. Now that `DragDropViewConroller` conforms to `UIDragInteractionDelegate`, let's make it conform to `UIDropInteractionDelegate` as well. Add the following extension to `DragDropViewController.swift`:

```
extension DragDropViewController: UIDropInteractionDelegate {
  func dropInteraction(_ interaction: UIDropInteraction,
sessionDidUpdate session: UIDropSession) -> UIDropProposal {
    return UIDropProposal(operation: .copy)
  }

  func dropInteraction(_ interaction: UIDropInteraction, performDrop
session: UIDropSession) {
    guard let itemProvider = session.items.first?.itemProvider,
      itemProvider.canLoadObject(ofClass: UIImage.self)
      else { return }

    itemProvider.loadObject(ofClass: UIImage.self) { [weak self]
loadedItem, error in
      guard let image = loadedItem as? UIImage
        else { return }

      DispatchQueue.main.async {
        self?.dropArea.image = image
      }
    }
  }
}
```

Again, this implementation should look familiar to you. The first method, `dropInteraction(_:sessionDidUpdate:)`, returns a copy proposal since we don't want to move data around. The second method is `dropInteraction(_:performDrop:)`. This method retrieves the image from `itemProvider` that has been created in `UIDragInteractionDelegate` and sets the loaded image as the image for `dropArea`.

The next step is to add the drag and drop interaction objects to the correct views, as follows:

```
override func viewDidLoad() {
  super.viewDidLoad()

  let dragInteraction = UIDragInteraction(delegate: self)
  imageView.addInteraction(dragInteraction)

  let dropInteraction = UIDropInteraction(delegate: self)
  dropArea.addInteraction(dropInteraction)
}
```

Now that all interactions are set up, go ahead and run the app on an iPad. You'll be able to drag the bottom image to the top section, and the image will appear in the top area. If you run the Photos app alongside **PlainDragDrop**, you can drag the bottom image to the photos app, and it will be added to photos. If you drag an image from the Photos app to the top section, the image from Photos will be set as the image for the drop area. Pretty cool stuff! And it was pretty simple to cover all these cases.

Even though the iPhone does not have full drag and drop support, you can still allow a user to drag the image from the top area to the bottom area. Since the iPhone does not support a full drag and drop experience, all drag and drop interaction objects you create are disabled by default. If you want to allow users to use drag and drop in your iPhone app, you must explicitly enable the interaction objects you would like to support. Add the following code to `viewDidLoad` in `DragDropViewController` to enable the drag interaction in the **PlainDragDrop** app:

```
dragInteraction.isEnabled = true
```

Go ahead and test the app on an iPhone; you should be able to drag the top image around. Now that you know how to enable drag and drop for a plain view, let's see how you can add it to a collection view.

Adding drag and drop to a UICollectionView

A lot of iOS apps make extensive use of collections and tables. Therefore, it makes a lot of sense that, whenever Apple introduces a huge feature such as drag and drop, they take a step back and evaluate how the feature should work for collections or tables. Luckily, drag and drop was no exception and Apple truly put some thought into making drag and drop work great.

In this section, you'll implement drag and drop for the collection of images that is at the bottom of the screen for the **ARGallery** app you created in `Chapter 12`, *Using Augmented Reality*. You will implement the following features that use drag and drop for `UICollectionView`:

- Dragging photos from the collection into the AR viewport.
- Reordering items in the collection view.
- Adding items from external sources to the collection view.

As a bonus, you will make sure that the first two features also work on an iPhone since these features are only used inside the app so they can be used on iPhone. Since Apple has tailored drag and drop to work perfectly with `UICollectionView`, the basic concepts for drag and drop still apply; you only have to use slightly different protocols. For instance, instead of implementing `UIDragInteractionDelegate`, you implement `UICollectionViewDragDelegate`.

The first feature, dragging photos from the collection of images to the AR Gallery, is implemented similarly to the drag and drop experience you implemented before. You will implement the relevant protocols first, and then you will implement the interactions. The code bundle for this chapter contains a slightly modified version of the **ARGallery** that you built before. The modifications allow you to focus on implementing drag and drop instead of having to make minor adjustments to the existing code.

Since you should be familiar with the dropping implementation already, add the following extension to `ViewController.swift` in `AugmentedRealityGallery`:

```
extension ViewController: UIDropInteractionDelegate {
  func dropInteraction(_ interaction: UIDropInteraction,
sessionDidUpdate session: UIDropSession) -> UIDropProposal {
    return UIDropProposal(operation: .copy)
  }

  func dropInteraction(_ interaction: UIDropInteraction, performDrop
session: UIDropSession) {
    guard let itemProvider = session.items.first?.itemProvider,
      itemProvider.canLoadObject(ofClass: UIImage.self)
      else { return }

    itemProvider.loadObject(ofClass: UIImage.self) { [weak self] item
,error in
      guard let image = item as? UIImage
        else { return }

      DispatchQueue.main.async {
        self?.addImageToGallery(image)
      }
    }
  }
}
```

Nothing crazy happens in this snippet. In fact, it's so similar to the code you have seen already that you should be able to understand what happens in this snippet on your own. The next step is to implement the `UICollectionViewDragDelegate` protocol. Add the following extension to `ViewController.swift`:

```
extension ViewController: UICollectionViewDragDelegate {
  func collectionView(_ collectionView: UICollectionView,
itemsForBeginning session: UIDragSession, at indexPath: IndexPath) ->
[UIDragItem] {
    let image = UIImage(named: images[indexPath.row])!
    let itemProvider = NSItemProvider(object: image)

    return [UIDragItem(itemProvider: itemProvider)]
  }
}
```

The preceding implementation serves the same purpose as `UIDragInteractionDelegate`. The main difference is that you have access to the `IndexPath` of the item that was selected for dragging. You can use `IndexPath` to obtain the dragged image and create a `UIDragItem` for it. Let's set up the interaction objects for this part of the app now. Add the following lines of code to `viewDidLoad()`:

```
collectionView.dragDelegate = self
collectionView.dragInteractionEnabled = true

let dropInteraction = UIDropInteraction(delegate: self)
arKitScene.addInteraction(dropInteraction)
```

If you build and run the app now, you can drag photos into the AR gallery from the collection view at the bottom. If you run the app on an iPad, you are even able to drag images from external apps into the AR Gallery! This is quite awesome, but you're not done yet. Let's allow users to drag items from external apps into the collection, so they have easy access to it. And while we're at it, let's implement the reordering of the collection using drag and drop as well.

To implement the reordering of items in the collection view and to allow users to add external images to their collection, you must implement `UICollectionViewDropDelegate`. As you will see soon, it is possible to distinguish between a drop session that originated from within the app or outside the app. This information can be used to determine whether the user wants to reorder the collection or add an item to it. Add the following extension for `ViewController`:

```
extension ViewController: UICollectionViewDropDelegate {
  func collectionView(_ collectionView: UICollectionView,
dropSessionDidUpdate session: UIDropSession, withDestinationIndexPath
destinationIndexPath: IndexPath?) -> UICollectionViewDropProposal {

    if session.localDragSession != nil {
      return UICollectionViewDropProposal(operation: .move, intent:
.insertAtDestinationIndexPath)
    }

    return UICollectionViewDropProposal(operation: .copy, intent:
.insertAtDestinationIndexPath)
  }
}
```

The preceding snippet implements `collectionView(_:dropSessionDidUpdate:withDestinationIndexPath:)`. This delegate method is pretty similar to `dropInteraction(_:sessionDidUpdate:)` except you also have access to the destination index path that the user is currently hovering their finger over. By checking whether there is a `localDragSession` on a `UIDropSession`, you can detect whether the user wants to reorder the collection or whether they are adding an item from an external source. By specifying an intent on the drop proposal, `CollectionView` knows how it should update its interface to visualize the action that is taken when the user performs the drop. Speaking of performing drops, add the following code to the `UICollectionViewDropDelegate` extension you just added:

```
unc collectionView(_ collectionView: UICollectionView, performDropWith
coordinator: UICollectionViewDropCoordinator) {
  switch coordinator.proposal.operation {
  case .copy:
    performCopy(forCollectionView: collectionView, with: coordinator)
  case .move:
    performMove(forCollectionView: collectionView, with: coordinator)
  default:
    return
  }
}
```

This method is relatively simple. Depending on the proposal that is used, a different method is called. You will add the copy and move methods soon, but first, let's talk a little bit about `UICollectionViewDropCoordinator`. `UICollectionViewDropCoordinator` contains information about the items that are being dragged, the animations that should be performed, the drop proposal, and of course the drop session. When performing a drop, you make use of the coordinator to request the drag items, but also to make sure the collection view properly updates its view.

The first method you will implement is `performMove(forCollectionView:with:)` since it's the simpler of the two remaining methods to implement. Add the following snippet to the `UICollectionViewDropDelegate` extension:

```
func performMove(forCollectionView collectionView: UICollectionView,
with coordinator: UICollectionViewDropCoordinator) {
  let destinationIndexPath = coordinator.destinationIndexPath ??
IndexPath(item: 0, section: 0)

  guard let item = coordinator.items.first,
    let sourceIndexPath = item.sourceIndexPath
    else { return }

  let image = images.remove(at: sourceIndexPath.row)
  images.insert(image, at: destinationIndexPath.row)

  collectionView.performBatchUpdates({
    collectionView.deleteItems(at: [sourceIndexPath])
    collectionView.insertItems(at: [destinationIndexPath])
  })

  coordinator.drop(item.dragItem, toItemAt: destinationIndexPath)
}
```

The preceding snippet uses the coordinator to retrieve the first item in the drag session. The app doesn't support dragging multiple items at once, so this is alright. Next, the item's source index path is used to remove the image that should be moved from the array of images. The destination index path is then used to add the image back into the array of images at its new location. This is done to make sure the data source is updated before updating the collection view. After the collection view is updated, `drop(_:toItemAt:)` is called on the coordinator to animate the drop action.

The final method you need to implement is `performCopy(forCollectionView:with:)`. Add the following code to the `UICollectionViewDropDelegate` extension:

```
func performCopy(forCollectionView collectionView: UICollectionView,
with coordinator: UICollectionViewDropCoordinator) {
  let destinationIndexPath = coordinator.destinationIndexPath ??
IndexPath(item: 0, section: 0)

  for item in coordinator.items {
      let dragItem = item.dragItem
      guard dragItem.itemProvider.canLoadObject(ofClass: UIImage.self)
else { continue }

      let placeholder =
UICollectionViewDropPlaceholder(insertionIndexPath:
destinationIndexPath, reuseIdentifier: "GalleryCollectionItem")
      let placeholderContext = coordinator.drop(dragItem, to:
placeholder)

      dragItem.itemProvider.loadObject(ofClass: UIImage.self) { [weak
self] item, error in
          DispatchQueue.main.async {
              guard let image = item as? UIImage else {
                  placeholderContext.deletePlaceholder()
                  return
              }

              placeholderContext.commitInsertion { indexPath in
                  self?.images.insert(image, at: indexPath.row)
              }
          }
      }
  }
}
```

Take a close look at `UICollectionViewDropPlaceholder` in this snippet. This class was introduced in iOS 11, and it is used to add temporary items to `CollectionView`. Because it might take a little while to load data from an item provider, you need a mechanism to update the UI while you're loading data. This is the goal of using a placeholder. When you call `drop(_:to:)` on `coordinator`, you receive a placeholder context. You use this context to either remove the placeholder if loading data from the item provider failed, or to commit the insertion if it succeeds. Once it has succeeded and you commit the insertion, you must make sure to update the collection's data source by adding the image to the image array. Otherwise, your app could crash due to data-source inconsistencies.

Since a placeholder is not part of your `CollectionView` data source, it is essential that you proceed with caution if you have a placeholder present in your `CollectionView`. For instance, your placeholder will be gone when you reload `CollectionView` before committing or removing the placeholder.

Lastly, add the following line to the `viewDidLoad` of `ViewController` to set the collection view's drop delegate:

```
collectionView.dropDelegate = self
```

At this point, you should be able to create a very nice implementation of drag and drop in your apps. However, there is more to learn on the topic since you can customize many aspects of how drag and drop works for your app.

Customizing the drag and drop experience

Sometimes you will find yourself working on an app where the default implementations don't work for you. For instance, any application's drop delegate can propose a move action instead of a copy action. However, it's possible that you don't want to support this. You can restrict the allowed proposals by implementing `dragInteraction(_:sessionAllowsMoveOperation:)`. If you only want to allow copy operations, you can return `false` from this method. Another restriction you can enable through a delegate method is `dragInteraction(_:sessionIsRestrictedToDraggingApplication:)`. If you return true from this method, users won't be able to drag content from your app to another app.

Other methods on both the drag and the drop delegates are related to monitoring the state of the drag or drop session. Imagine that your app supports move proposals. When the user decides to move one or more objects from your app to another, you'll need to update the user interface accordingly once the drop is finished. You can implement `dragInteraction(_:sessionDidTransferItems:)` to implement this or perform any other cleanup that you might want to do after a successful drop session. For instance, you could show an alert to let the user know that data was transferred successfully.

In addition to logic, you can also provide custom previews and perform animations alongside the animations performed by the drag and drop sessions. All methods for this are defined in the `UIDropInteractionDelegate` and `UIDragInteractionDelegate` protocols as optional methods. In many cases, you won't need to implement many customizations for drag and drop. If you do find yourself wanting to customize part of the drag and drop experience, the chances are that a quick glance at the available methods on the delegate protocols will guide you in the right direction.

Summary

This chapter showed you how to implement a smooth drag and drop experience for your users. Drag and drop has limited functionality on the iPhone, but this doesn't stop it from being a powerful feature that can be used to support the dragging of contents within an application or to reorder a `UICollectionView`. Collections and tables have received special treatment because they have their own delegate methods and enable you to easily access the cell that is selected by the user for dragging.

While drag and drop might seem complicated at first glance, Apple did a great job of containing this complexity in a couple of relatively simple delegate methods that you can implement. A simple drag and drop implementation only requires you to implement fewer than a handful of methods, which is quite impressive for such a powerful feature!

In the next chapter, you will perform some extra work on the **MustC** app that you have worked on in previous chapters. You will enhance it by integrating it with iOS' powerful Spotlight search database.

Questions

1. What devices support the full drag and drop experience?

    a) Only iPhone.
    b) Only iPad.
    c) Both iPhone and iPad.

2. What kind of object do you have to provide to start a drag session?

   a) A `UIDragItem` object.
   b) A `UIDragInteraction` object.
   c) An `NSItemProvider` object.

3. How can you check whether your app can handle a certain drop action?

   a) By checking the type of object that is associated with the drop action.
   b) By limiting the type of objects that can be used in your drag session.
   c) You receive a special error type if the app failed to handle the drop.

4. How is the drag and drop experience limited on iPhone?

   a) You can only start drag sessions on iPhone.
   b) Your app can only implement drop interactions on iPhone.
   c) Drag and drop is limited to the currently-open app.

5. What do you have to do to support drag and drop on an iPhone?

   a) You must manually enable the drag and drop interactions.
   b) You need a special capability.
   c) You need to add an `Info.plist` entry.

6. How are the drag and drop delegate protocols for collection views different from the regular drag and drop delegate protocols?

   a) They are the same, only with different names.
   b) The collection view versions of the delegate methods receive an index path.
   c) They are called at different times than the regular delegate methods.

Further reading

- The Drag and Drop developer documentation: `https://developer.apple.com/documentation/uikit/drag_and_drop`

21
Improved Discoverability with Spotlight and Universal Links

Many users of macOS and iOS love one feature in particular: Spotlight search. Spotlight keeps an index of all files, contacts, contents, and more on your Mac, and it's been in iOS too for a while now. With iOS 9, Apple made it possible for developers to index the contents of their apps, enabling users to discover app contents right from the Spotlight search they know and love. Ever since opening up Spotlight on iOS to developers, Apple has been pushing to make Spotlight better, more relevant, more impactful, and more helpful to users. In this chapter, we'll explore what Spotlight can do for your apps and how you can make use of the `CoreSpotlight` APIs to provide your users with a fantastic search experience that helps them find and navigate your app.

This chapter covers the following topics:

- Understanding Spotlight search
- Adding your app contents to the Spotlight index
- Handling search-result selection
- Adhering to Spotlight best practices
- Increasing your app's visibility with Universal Links

Let's not waste any time and jump right into Spotlight!

Understanding Spotlight search

If you have a Mac, which you most likely do, you have probably used the blazingly fast Spotlight search feature on it. Furthermore, if you love this feature on your Mac, you have probably used it on iOS as well. Spotlight is a highly optimized search engine that enables you to find content throughout your device.

If you haven't used Spotlight, try swiping down on your home screen on iOS and use the search box. Or, on your Mac, press *command* + space to open the Spotlight search dialog.

Spotlight has been around since iOS 3. However, it wasn't until iOS 9 that Apple opened up Spotlight to developers. This means that starting with iOS 9, Spotlight indexes everything from web results, apps, app store results, and any content you add to the Spotlight index yourself. Spotlight presents search results with an image, a title, and some extra information, and sometimes extra buttons are shown to make a phone call or to start turn-by-turn navigation to an address.

The results that are displayed by Spotlight are a mix of public or online content and private results. If your app has indexed items that are publicly available, Apple will make them visible to all users who have your app installed as long as enough people interact with this particular searchable item.

A great example of public content for almost any app is the main navigation items in an app. Imagine you have an app that has a couple of tabs in `UITabBarController`. Any user that downloads your app and opens it can immediately access each tab if they want to. This means that these tabs are public; any user of your app can access these tabs.

Your app can index these tabs using `CoreSpotlight` and mark them as publicly available. When you mark an item as public, it will not be globally visible right away. Behind the scenes, `CoreSpotlight` indexes the item as it usually would when indexing a private record. Whenever a user selects this particular item from a list of Spotlight search results, it sends an irreversible hash of the item to Apple's servers. If Apple receives the same hash over and over again, a certain threshold will be reached. Once this threshold is reached, the indexed item is pushed to all people who have your app installed. If your indexed item has a public web resource associated with it, your publicly searchable item will even be shown to people who don't have your app installed because the content is available on the web. If a user selects this item but doesn't have your app installed, they will be taken to your website instead to view the requested content there.

The fact that public results in Spotlight are kept private for a while helps to make sure that everything your users see when they use Spotlight is relevant. Imagine that any app developer could mark any content as public and every time a developer does this, the item is pushed directly to all users. This would pollute Spotlight, and the entire feature would become virtually worthless. Spotlight's goal is to provide relevant and interesting results. The threshold for interactions makes sure that enough people consider the displayed item to be appropriate for their search query, which means that lousy indexing or spam doesn't have a chance to thrive.

Another reason this threshold exists is to protect your user's data. It's perfectly fine to add personal and private data to Spotlight. The standard mail app does this so users can search their email messages using Spotlight. These email messages should not be publicly indexed so they shouldn't be marked as such. Mistakes happen though, and in case you do end up marking private data as public by accident, there is no way this specific item will show up for other users because the public result threshold for the item will never be reached since only one user has interacted with the Spotlight search result.

Excellent Spotlight visibility can boost usage of your apps, especially since users don't even need to have your app installed to see results for your app as long as there is a web resource available for your indexed contents. Apple has worked hard to ensure that Spotlight is fast and provides a seamless experience for your users. This means that users won't notice whether a result comes straight from their device or whether they see a result because it was publicly indexed.

You might have noticed that Safari sometimes shows results for in-app content. These results come from Spotlight. Whenever you add a web resource to an item you index, Safari will show this resource if it matches a user's query in the search bar. In this chapter, you will learn all about optimally indexing your app's contents; so, if you have an app that has a website that mirrors its contents, you'll know how to maximize your presence in search results for both Spotlight and Safari.

Adding your app contents to the Spotlight index

If you have ever worked on a website, you must have heard something about **Search Engine Optimization** (**SEO**). More importantly, you will know that any website you create and publish is indexed by several search engines. All you have to do is make sure that you write semantic and structured HTML markup and any web spider will understand what your website is about and what parts of it are more important.

Search engines, such as Google, have indexed billions of web pages based on their contents and semantic markup.

Apps tend to be a little less neatly structured, and crawling them is a lot harder, if not impossible. There is no formal way to figure out what content is on screen and what this content means. Also, more importantly, a lot of content you'd want to index is only available to users who have logged in or created content of their own.

This is why Apple decided that the developers themselves probably know their app's contents best and should be in charge about how, when, and why a particular content is indexed. Even though this does put a little bit of manual burden on the developers, it gives them a considerable advantage over the automatic indexing that's done on the web. Since developers are in control, they can decide precisely which content matters most to specific users. As you'll soon see, you can index content based on the screens a user visits, which means that you will index those pages that your user may want to visit again.

Even more important than being in control is the ability to index private contents safely. The web is limited to indexing public content. If you use online email software to check your inbox or if you have an online project-management tool, you must rely on the internal search functions inside the web page for these tools. You won't find your emails or projects through a regular search query. With Spotlight indexing, your users can do just that: search through the content they own on their device. Indexing private content is secure because the data is not made available to other apps and you can't accidentally make one user's private data visible to other users due to the public indexing threshold mentioned earlier.

So how exactly do developers take control? That's the question that's answered next. You will learn about the following methods that Apple came up with to index app contents:

- `NSUserActivity`
- `CSSearchableItem`
- Universal Links

Indexing your app through user activity

As part of a feature set called **Continuity**, Apple launched Handoff in iOS 8. **Handoff** allows users to start an activity on one device and then continue it on another. In iOS 9, Apple introduced the ability to index these user activities in Spotlight.

There are several advantages to this because apps that support Handoff hardly need to do anything to support Spotlight indexing and vice versa. So, even though this chapter is all about search and Spotlight, you will also learn something about enabling Handoff for your app.

The philosophy behind user activities is that whenever your user interacts with your app, you create an instance of `NSUserActivity`. For Spotlight, these activities revolve solely around viewing content. Any time your user looks at a piece of content in your app is an excellent time to create a user activity and have Spotlight index it. After the user activity is added to the index, the user will be able to find it through Spotlight; when they tap on it, your app can take the user straight to the relevant section of the app, allowing the user to resume their activity.

In the previous chapters, you worked on an app called **MustC**. The **MustC** app collects data about people and their favorite movies. The app also keeps track of a rating for each of the movies. This content is great to index in Spotlight, so let's add some indexing to it.

A few additions were made since the last time you worked on this app. The app now contains a tab bar. There is a tab for the list of family members, and there is one that lists all of the movies that are added to the app. Selecting a movie will display a list of family members that have added this movie to their favorites.

An app such as **MustC** is a great candidate for indexing. You can add the separate family members, the tabs from the navigation bar, and the movies to Spotlight to make them searchable from anywhere within iOS. You will start off with the most straightforward content to track. You will add the family members and movie tabs to Spotlight when a user visits them.

To do this, you will use user activities, so you should create an activity whenever a user opens one of the two tabs. There are two places where you could implement the user activities:

- `viewDidAppear()`
- `viewDidLoad()`

If you create the user activity in `viewDidAppear`, the user activity is pushed every time the user switches tabs or navigates back to the root view controller from another view controller. Even though it doesn't cost much to index a user activity, it seems like it's overkill to index a tab every single time a user sees it.

Even though this train of thought isn't wrong, the best way to go about indexing tabs in Spotlight is actually to create the activity in the `viewDidAppear()` method. If you put this logic in `viewDidLoad()`, it will only be executed once even though the idea of user activities is that they describe each action a user performs. In this case, inserting the same activity over and over again is the desired behavior because it accurately reflects what the user is doing inside the app.

When your app implements user activities, you should always specify them by adding an `NSUserActivityTypes` array to your app's `Info.plist` that contains the user activities your app will handle. Add this key to `Info.plist` for **MustC** and add a single item name, com.familymovies.openTab, to the activity types array.

Let's take a look at some code that indexes the family members tab. You should add it to `FamilyMembersViewController`:

```
override func viewDidAppear(_ animated: Bool) {
  super.viewDidAppear(animated)
  let userActivity = NSUserActivity(activityType:
"com.familymovies.openTab")
  userActivity.title = "Family Members"
  userActivity.isEligibleForSearch = true
  userActivity.isEligibleForPublicIndexing = true
  self.userActivity = userActivity
  self.userActivity?.becomeCurrent()
}
```

The preceding code shows how to create a straightforward user activity. The activity you just created only has a title because there isn't much else to associate with it. The most important thing to note is the `isEligibleForSearch` property. This property tells the system that the user activity that is about to be set as the current activity should be indexed for searching. Other similar properties are `isEligibleForHandoff` and `isEligibleForPublicIndexing`. You used the `isEligibleForPrediction` property in `Chapter 16`, *Streamlining Experiences with Siri*.

It's a great idea for the activity you just created to be eligible for public indexing, so that property is set to `true`. Doing this will make the activity show up in the search results of a lot more people, given that enough people interact with it. Marking an activity as eligible for handoff allows a user to continue an activity they started on one device on another device. Since this app only works on iOS and you don't need to take multiple users with multiple devices into account, you don't have to set this property to `true`.

Finally, the user activity you just created is set as the current activity. This makes sure that iOS registers the activity and adds it to the Spotlight index. It won't be made available publicly right away due to the threshold of people that interact with this activity that must be reached before Apple will push it to all users. Even though it won't be indexed publicly, it will appear in search results locally for the current user.

If you build and run your application, the family members tab should be the first tab to appear. This means that a user activity for that view is created and indexed immediately. After opening the app, go ahead and open Spotlight by swiping down on the home screen and perform a search for the word **family**.

You'll notice that the activity you just added is listed under the **MustC** header in the Spotlight search results, as shown in the following screenshot:

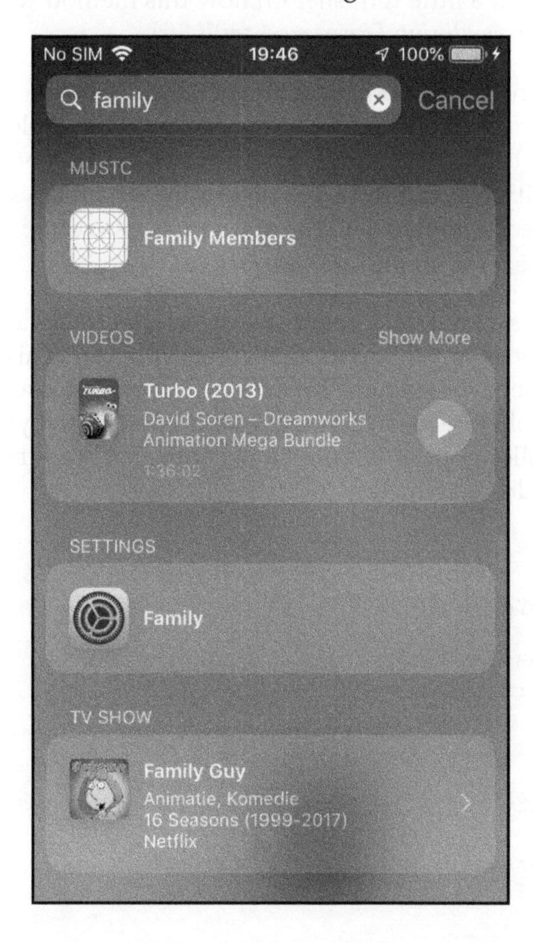

Pretty neat, right? You were able to add a simple entry in Spotlight's search index with very minimal effort. You should be able to add a similar implementation to the **MustC** app to have it index the `Movies` page. Go ahead and add a modified version of the snippet you added to `FamilyMembersViewController` to `MoviesListViewController`.

Now that both tabs show up in Spotlight, how do you make sure that the correct tab is opened when the user selects a result from Spotlight? The answer lies in one of the `AppDelegate` methods. When your app is brought to the foreground because a user selected your app as a Spotlight search result, the `application(_:continueUserActivity:restorationHandler:)` method is called. This is the same method that gets called when a user executes one of their Siri Shortcuts, so if you need a little refresher on how this method works, make sure to go back to `Chapter 16`, *Streamlining Experiences with Siri.*

Currently, there are only two Spotlight entries that the user can select. They either want to see the family members tab or the movies tab. The implementation of `application(_:continueUserActivity:restorationHandler:)` should inspect the user activity that it received to determine which tab should be displayed. Once this is established, a reference to the app's `UITabBarController` should be obtained and the correct tab should become active.

Since each tab contains a navigation controller, you should always make sure to pop the relevant navigation controller to its root view controller. This is needed because the user might have been looking at a detail view controller, and the app would show this detail view controller instead of the root view controller if you don't ensure that the navigation controller pops to its root view controller. Add the following code to `AppDelegate` to add this functionality:

```
func application(_ application: UIApplication, continue userActivity:
NSUserActivity, restorationHandler: @escaping
([UIUserActivityRestoring]?) -> Void) -> Bool {
  // 1
  guard let tabBar = window?.rootViewController as? UITabBarController
    else { return false }

  // 2
  let tabIndex: Int?
  if userActivity.title == "Family Members" {
    tabIndex = 0
  } else if userActivity.title == "Movies" {
    tabIndex = 1
  } else {
    tabIndex = nil
```

```
  }

  guard let index = tabIndex
    else { return false }

  // 3
  guard let navVC = tabBar.viewControllers?[index] as?
UINavigationController
    else { return false }

  // 4
  navVC.popToRootViewController(animated: false)
  tabBar.selectedIndex = index

  return true
}
```

The preceding code first obtains a reference to the tab bar controller. After doing this, the second step is to determine the correct tab index for the selected user activity. Then, the navigation view controller associated with the tab bar is grabbed. The fourth and final step is to pop the navigation controller to its root view controller and to make the correct tab bar index the currently-selected tab bar item.

At this point, two of the screens in **MustC** are indexed in Spotlight. While this is great, there are a couple more screens that can be implemented. Doing this will make the implementation for `application(_:continueUserActivity:restorationHandler:)` more complicated than it is now.

Manually creating and resuming user activities for each screen in an app is tedious and involves quite a lot of boilerplate code. You can solve this problem by utilizing an activity factory. A typical pattern in apps is to use a specific helper object called a **Factory**. The sole purpose of a factory is to act as an object that is responsible for creating objects of a particular type. This dramatically reduces boilerplate code and increases maintainability. Create a new file called `IndexingFactory.swift` in the `Helpers` folder and add the following implementation:

```
import Foundation

struct IndexingFactory {
  enum ActivityType: String {
    case openTab = "com.familymovies.openTab"
    case familyMemberDetailView =
"com.familymovies.familyMemberDetailView"
    case movieDetailView = "com.familymovies.movieDetailView"
  }
```

```
    static func activity(withType type: ActivityType, name: String,
  makePublic: Bool) -> NSUserActivity {
      let userActivity = NSUserActivity(activityType: type.rawValue)
      userActivity.title = name
      userActivity.isEligibleForSearch = true
      userActivity.isEligibleForPublicIndexing = makePublic

      return userActivity
    }
  }
```

Note that this object contains an enum with three different activity types. Don't forget to add the two new activity types to the `NSUserActivityTypes` list in the app's `Info.plist`. `IndexingFactory` has a single static method in it. This method takes a couple of configuration arguments and uses these to create and return a new user activity instance. Let's see this factory in action. You can add the following code to the `MoviesViewController` class:

```
override func viewDidAppear(_ animated: Bool) {
  super.viewDidAppear(animated)

  guard let familyMemberName = familyMember?.name
    else { return }

  self.userActivity = IndexingFactory.activity(withType:
.familyMemberDetailView, name: familyMemberName, makePublic: false)
  self.userActivity?.becomeCurrent()
}
```

The preceding snippet is a lot smaller than creating a user activity from scratch in every `viewDidAppear`. Also, if you decide to make changes to the way user activities are created and configured, it will be easier to refactor your code because you will only have to change a single method. Add a comparable implementation of the preceding method to the `MovieDetailViewController` class so movie pages will also be indexed by Spotlight when the user visits them.

This wraps up simple indexing with `NSUserActivity`. Next up, you will learn how to index content with the `CSSearchableItem` class and how you can use it to index content the user hasn't seen yet. You'll also see how you can associate more sophisticated data with your searchable items and how Spotlight handles updating and re-indexing contents.

Since `AppDelegate` can't handle opening user activity types that should take the user straight to a family member or movie, it's a good exercise for you to try implementing this logic yourself. For a correct implementation, you'll need to add a new `find` method to the **Movie** and **FamilyMember** classes, and you'll have to instantiate view controllers straight from a storyboard. If you get stuck implementing this, don't hesitate to look at the finished code for this chapter in the code bundle.

Indexing with CSSearchableItem

Currently, Spotlight indexing works and users can find any content they have seen before. It's even possible to select results from the Spotlight index and have the app open on the correct page. If you've taken on the challenge of implementing the handling of detail pages, your app should be able to handle the continuation of any activity that was indexed. Wouldn't it be cool if we could be a bit more proactive about indexing though? Ideally, the app would index any new `Family Members` or movies as soon as the user adds them.

This is what `CSSearchableItem` is good at. The `CSSearchableItem` class enables you to index content that the user might not have seen before. Indexing `CSSearchableItem` instances is pretty straightforward. The steps involved are similar to how you could index user activities. To index a searchable item, you should create an instance of `CSSearchableItem` and provide it with attributes that describe the indexed item. These attributes are encapsulated in an instance of `CSSearchableItemAttributeSet`.

Containing information in CSSearchableItemAttributeSet

When you create an attributes set, it's important that you populate it correctly since it describes almost all of the important information about your item for Spotlight. You can set a title, content description, a thumbnail image, keywords, even ratings or phone numbers, GPS information, and much, much more. For a full overview of what's possible, refer to the `CSSearchableItemAttributeSet` documentation. Every time you are about to create a new item that can be indexed, you should take a look at the documentation to make sure you don't miss any attributes.

The better use you make of the available attributes, the more effectively your content can be indexed and the higher your app will rank. Therefore, it's worth putting slightly more time and effort into your search attributes because getting it wrong can be a costly mistake, especially considering the available documentation. At a minimum, you should always try to set `title`, `contentDescription`, `thumbnailData`, `rating`, and `keywords`. This isn't always relevant or even possible for the items you're indexing, but whenever possible make sure that you set these attributes.

You may have noticed that the `NSUserActivity` instances you indexed in the app didn't receive any special attributes. You just set a name and some other basic information, but you didn't have to add a description or a rating to any of the indexed objects. If you index user activities in your applications, it's worth noting that user activities can and should have attributes associated with them. All you need to do is set the `contentAttributeSet` property on the user activity. After you implement indexing through `CSSearchableItem`, you should shortly revisit user-activity indexing to make the indexed item richer and also to make sure that `CoreSpotlight` understands that the user activities and searchable items point to the same underlying index in Spotlight.

When you index items through multiple methods, it's inevitable that you run into data duplication. The **MustC** app indexes every visited screen. So, if a user visits the details page of a movie, a user activity for that movie should be created. However, you also want to index movies as they are created by the user. To avoid duplicate results in Spotlight search, `relatedUniqueIdentifier` should be added to the attributes set. Setting this attribute on a user activity makes sure that Spotlight doesn't add duplicate entries for items with the same identifier.

Let's expand `IndexingFactory` with two methods that can generate attribute sets for searchable items. Putting this functionality in `IndexingFactory` as a separate method is a good idea because, if it is set up correctly, these methods can be used to generate attributes for both user activities and searchable items. This avoids code duplication and makes it a lot easier to add or remove properties in the future. Add the following methods to the `IndexingFactory` struct. Don't forget to import `CoreSpotlight` at the top of the file:

```
static func searchableAttributes(forMovie movie: Movie) ->
CSSearchableItemAttributeSet {
  do {
    try movie.managedObjectContext?.obtainPermanentIDs(for: [movie])
  } catch {
    print("could not obtain permanent movie id")
  }
```

```
  let attributes = CSSearchableItemAttributeSet(itemContentType:
ActivityType.movieDetailView.rawValue)
  attributes.title = movie.title
  attributes.contentDescription = "A movie that is favorited by
(movie.familyMembers?.count ?? 0) family members"
  attributes.rating = NSNumber(value: movie.popularity)
  attributes.identifier =
"(movie.objectID.uriRepresentation().absoluteString)"
  attributes.relatedUniqueIdentifier =
"(movie.objectID.uriRepresentation().absoluteString)"

  return attributes
}

static func searchableAttributes(forFamilyMember familyMember:
FamilyMember) -> CSSearchableItemAttributeSet {
  do {
    try familyMember.managedObjectContext?.obtainPermanentIDs(for:
[familyMember])
  } catch {
    print("could not obtain permanent family member id")
  }

  let attributes = CSSearchableItemAttributeSet(itemContentType:
ActivityType.familyMemberDetailView.rawValue)
  attributes.title = familyMember.name
  attributes.identifier =
"(familyMember.objectID.uriRepresentation().absoluteString)"
  attributes.contentDescription = "Family Member with
(familyMember.favoriteMovies?.count ?? 0) listed movies"
  attributes.relatedUniqueIdentifier =
"(familyMember.objectID.uriRepresentation().absoluteString)"

  return attributes
}
```

For both the `Movie` and `FamilyMember` objects, a set of attributes is created. This set meets Apple's recommendations as closely as possible. The objects don't have any thumbnail images or keywords that can be added to the attributes set other than the movie title or the name of a family member. Adding these to the keywords is kind of pointless because the title in itself is essentially a keyword that the item will match on.

To create an attribute set for a movie or family member, you only have to call the method that matches the object you want to index and you're good to go. Beautiful, convenient, and simple.

Adding CSSearchableItem instances to the search index

In the **MustC** application, the goal is to add family members and movies to the search index as soon as the user adds them. A factory method that creates the `CSSearchableItemAttributeSet` instance that describes the item that should be indexed is already provided. However, you can't directly add these to the index. To add information to the search index manually, you need instances of `CSSearchableItem`. Every searchable item requires a unique identifier and a domain identifier.

The unique identifier is used to identify an indexed item. It's essential that you set this value to something that is unique because otherwise, Spotlight will overwrite the entry with something else that has the same identifier or you'll get duplicate entries if you combine user activities and search items like we're doing for **MustC**.

The domain identifier functions as a namespace. Within any given namespace, all entries must be unique and are identified through their unique identifier. These identifiers only have to be unique within their namespace. Think of this as streets and addresses. In a particular area, every street name is unique (domain, namespace). On each street the house number is unique (unique identifier), but the same number can occur on different streets. The domain identifier for your Spotlight entry is not only used to identify entries uniquely, it's also used to perform specific batch actions on the index, such as deleting all indexed items from a particular domain.

The domain identifier, unique identifier, and the attributes together make up a searchable item. The following code adds factory methods to `IndexingFactory` that will make it simple for the app to add items to the search index:

```
enum DomainIdentifier: String {
  case familyMember = "FamilyMember"
  case movie = "Movie"
}

static func searchableItem(forMovie movie: Movie) -> CSSearchableItem
{
  let attributes = searchableAttributes(forMovie: movie)

  return searachbleItem(withIdentifier:
"(movie.objectID.uriRepresentation().absoluteString)", domain: .movie,
attributes: attributes)
}

static func searchableItem(forFamilyMember familyMember: FamilyMember)
```

```
-> CSSearchableItem {
  let attributes = searchableAttributes(forFamilyMember: familyMember)

  return searachbleItem(withIdentifier: "(familyMember.objectID)",
domain: .familyMember, attributes: attributes)
}

private static func searachbleItem(withIdentifier identifier: String,
domain: DomainIdentifier, attributes: CSSearchableItemAttributeSet) ->
CSSearchableItem {
  let item = CSSearchableItem(uniqueIdentifier: identifier,
domainIdentifier: domain.rawValue, attributeSet: attributes)

  return item
}
```

The preceding code defines an enum that contains domain identifiers for items that should be indexed. Note that `searchableItem(withIdentifier:domain:attributes:)` is marked as private. This is done to make sure anybody using the factory helper has to use `searchableItem(forFamilyMember:)` and `searchableItem(forMovie:)` instead. These methods are simpler to use because they only take a family member or a movie and if you use only these methods, you can rest assured that you only insert consistently configured items into the Spotlight index.

Now that everything is set up for indexing new data, let's begin with indexing family members as soon as the user creates them. Add the following code as the implementation for the `.insert` case in `controller(_:didChange:at:for:newIndexPath:)` in `FamilyMembersViewController`. Make sure that you import `CoreSpotlight` at the top of your file:

```
guard let insertIndex = newIndexPath,
  let familyMember = fetchedResultsController?.object(at: insertIndex)
  else { return }

let item = IndexingFactory.searchableItem(forFamilyMember:
familyMember)
CSSearchableIndex.default().indexSearchableItems([item],
completionHandler: nil)

tableView.insertRows(at: [insertIndex], with: .automatic)
```

Due to the factory methods you set up before, new items can be added to the search index with just a few lines of code. To insert an item into the search index, an instance of `CSSearchableIndex` is created, and Spotlight is told to index this item. If needed, a completion-handler can be passed to the index method. This handler is called with an optional error. If the indexing has failed, the error should tell you why Spotlight couldn't index the item, and you can retry indexing the item or take a different action. In the **MustC** app, it is assumed that the indexing succeeds and no error-handling is implemented.

Update the `mangedObjectContextDidChange(notification:)` method in `MoviesViewController` as follows to index new movies:

```
@objc func managedObjectContextDidChange(notification: NSNotification)
{
  guard let userInfo = notification.userInfo
    else { return }

  if let updatedObjects = userInfo[NSUpdatedObjectsKey] as?
Set<FamilyMember>,
    let familyMember = self.familyMember,
    updatedObjects.contains(familyMember) {

    let item = IndexingFactory.searchableItem(forFamilyMember:
familyMember)
    CSSearchableIndex.default().indexSearchableItems([item],
completionHandler: nil)

    tableView.reloadData()
  }

  if let updatedObjects = userInfo[NSUpdatedObjectsKey] as? Set<Movie>
{
    for object in updatedObjects {
      if let familyMember = self.familyMember,
        let familyMembers = object.familyMembers,
        familyMembers.contains(familyMember) {

        let item = IndexingFactory.searchableItem(forMovie: object)
        CSSearchableIndex.default().indexSearchableItems([item],
completionHandler: nil)

        tableView.reloadData()
        break
      }
    }
  }
}
```

Right after the movie is added to the current family member, it is added to the Spotlight index. You might argue that the item is added too early because the movie's rating has not been fetched at that point. This is OK because when the context is saved the second time with the ratings attached, the item is added to the index again automatically due to the save notification that Core Data sends.

If you run the app now, you should be able to go into Spotlight right after you add a family member or movie, and you should immediately be able to find the freshly added content in Spotlight. If you search for a movie you just added, you'll notice that you can see how many family members have added a certain movie to their favorites list and the rating a movie has. More importantly, there should be only a single entry for each movie because a unique identifier is used to avoid duplicate entries.

Now that you have implemented `CSSearchableItem` and `NSUserActivity` indexing, you need to make sure that both of these objects are correctly configured so you can safely combine both indexing methods.

Safely combining indexing methods

Since no unique information is associated with the user activities yet, Spotlight can't figure out that a family member that's indexed through a user activity is the same family member that was already inserted as a searchable item. To ensure that Spotlight understands this, you should add two more factory methods that create an activity item for either a family member or a movie with the correct information associated with them. Add the following methods to `IndexingFactory`:

```
static func activity(forMovie movie: Movie) -> NSUserActivity {
  let activityItem = activity(withType: .movieDetailView, name:
movie.title!, makePublic: false)
  let attributes = searchableAttributes(forMovie: movie)
  attributes.domainIdentifier = DomainIdentifier.movie.rawValue
  activityItem.contentAttributeSet = attributes

  return activityItem
}

static func activity(forFamilyMember familyMember: FamilyMember) ->
NSUserActivity {
  let activityItem = activity(withType: .movieDetailView, name:
familyMember.name!, makePublic: false)
  let attributes = searchableAttributes(forFamilyMember: familyMember)
  attributes.domainIdentifier = DomainIdentifier.familyMember.rawValue
  activityItem.contentAttributeSet = attributes
```

```
        return activityItem
    }
```

The most important lines to take note of are the ones where a `domainIdentifier` is set on the attributes constant. Since iOS 10, developers have been able to associate a `domainIdentifier` with user activities through `contentAttributeSet`. By adding a `domainIdentifier` to the indexed item, searchable items and user activities are unified even more. Update the `viewDidAppear` implementation for `MovieDetailViewController` as follows:

```
override func viewDidAppear(_ animated: Bool) {
    super.viewDidAppear(animated)

    guard let movie = self.movie
        else { return }

    self.userActivity = IndexingFactory.activity(forMovie: movie)
    self.userActivity?.becomeCurrent()
}
```

We also need to update the `viewDidAppear` method in `MoviesViewController`. You should be able to do this on your own; the code will look similar to the preceding snippet, except you're indexing a family member instead of a movie.

Now that all of your app contents are indexed in Spotlight, it's time to discuss some of the methods Spotlight uses to rate your content and the best practices you should keep in mind when you add your app's contents to Spotlight.

Handling searchable item selection

If a user taps on a search result for one of the items you indexed manually, the `application(_:continue:restorationHandler:)` method is called on `AppDelegate`. This is the same method that's used for user activities, but the internal handling is not quite the same.

In the original code to continue user activities, the user activity's title property was used to determine the appropriate action. Since searchable items don't have this property, a different approach will have to be implemented. To decide whether the user activity that is passed to `application(_:continue:restorationHandler:)` is a searchable item or a user activity, you can read the activity type. If the user activity was originally a searchable item, the activity type should be equal to `CSSearchableItemActivityIdentifier`. If you receive a user activity with this activity type, you should take action accordingly.

Update your code in `AppDelegate` as follows and make sure to import `CoreSpotlight`:

```
func application(_ application: UIApplication, continue userActivity:
NSUserActivity, restorationHandler: @escaping
([UIUserActivityRestoring]?) -> Void) -> Bool {

  if let identifier =
userActivity.userInfo?[CSSearchableItemActivityIdentifier] as? String,
userActivity.activityType == CSSearchableItemActionType {

    return handleCoreSpotlightActivity(withIdentifier: identifier)
  }

  // existing implementation
}

func handleCoreSpotlightActivity(withIdentifier identifier: String) ->
Bool {
  guard let url = URL(string: identifier),
    let objectID =
persistentContainer.persistentStoreCoordinator.managedObjectID(forURIR
epresentation: url),
    let object = try?
persistentContainer.viewContext.existingObject(with: objectID)
    else { return false }

  if let movie = object as? Movie {
    return handleOpenMovieDetail(withName: movie.name!)
  }

  if let familyMember = object as? FamilyMember {
    return handleOpenFamilyMemberDetail(withName: familyMember.name!)
  }
  return false
}
```

The updated version of `application(_:continue:restorationHandler:)` checks whether it received a searchable item. If it did, a special method is called. This method uses the persistent store to convert the string identifier to a managed object ID which is then used to ask the managed object context for the corresponding object. If all of this succeeds, we attempt to cast the fetched object to either a movie or a family member, and if this succeeds, one of the existing handlers is called. If none of the casts succeed, it was not possible to continue the activity, so the method returns false.

Note that this implementation assumes that you did the exercise that was proposed earlier in the chapter so `handleOpenFamilyMemberDetail(withName:)` and `handleOpenMovieDetail(withName:)` do not exist in the project. Refer to this chapter's code bundle to see a suggested implementation for both of these methods.

Understanding Spotlight best practices and ratings

If you implement Spotlight indexing in your app, it's beneficial for both you and your users if you adhere to best practices. Sticking with the best practices will positively affect your ratings and can ultimately drive more users to your app.

There are two best practices that you should always follow if possible. Doing so will give your app great advantages over other apps that might not stick to best practices. The two best practices you should always try to follow are:

- Adding metadata to your web links
- Registering as an indexing delegate

Adding metadata to your web links

In the next section about Universal Links, you will learn what Universal Links are and how you should implement them in your application. Enabling Universal Links benefits your Spotlight search ranking and visibility because Spotlight can associate web resources that Apple's bots can crawl with the items you add to the Spotlight index. To associate Universal Links and your locally indexed items with each other, you should set the `contentUrl` that matches the web resource for a local item on your search attributes objects.

Doing so allows Apple to display search results for your app as public, and it can even show your Universal Links in Safari's search results. This means that it will be even easier for your users to discover and use your app. More importantly, if you associate a `contentUrl` with an indexed item, your app is also shown to people who don't have your app installed. That fact alone should persuade you to make sure that your app has online resources that mirror the public content in your app.

Once you've decided to add content URLs to your content, it's essential that you add proper, structured metadata to your web pages. If your web page has well-formatted metadata available, Spotlight will index your app's contents even if you don't manually add it. This is because Apple continuously crawls the marketing and support URLs you advertise within the App Store.

For Apple's bot to be able to discover the URLs your app can handle, it checks whether there is a Smart App Banner implemented on your web page. The Smart App Banner is visible to people who visit your website. If they already have your app installed, the banner will prompt the user to view this content in your app. If you've added a URL to the banner that your app can handle, Apple's bot can discover this link and associates it with your app.

You should always make sure that you add useful metadata to your web content. When you manually index content, you can set properties on the indexed item's attributes set to represent your data. On the web, you can't create an attribute set, so you need to embed the metadata in your page's HTML markup. If your web page contains Open Graph metadata, Apple's bots will understand how to read that metadata. Even though Open Graph metadata is a great start, Apple recommends that you implement richer and more fine-grained data for your websites.

To do this, it's recommended that you implement metadata as specified in the schema.org standards. These standards provide definitions for a wide range of different entities. For instance, you can express a product rating, pricing, or a business address through `schema.org` definitions of these entities.

If you've implemented metadata for your app, or if you want to make sure that Spotlight can index your contents, you can use Apple's verification tools for searching. Go to `https://search.developer.apple.com` to paste in your website's URL and verify that everything can be indexed as you expected.

Having a complete set of metadata positively impacts your rating. An obvious reason is that it's better because more information means more matches for a search query. However, a different reason is that, if a user sees a search result that provides a lot of information at a glance, it's easier for them to decide whether a specific result is relevant to them.

This quick assessment of relevancy up front makes sure that whenever a user selects a result for your app, they know for sure that they want to interact with that item. Apple measures this form of engagement to ensure that results that are tapped often and are used longer are pushed to the top.

A topic that is directly related to this is the concept of keyword stuffing. When optimizing for a search, it can be tempting to stuff a lot of keywords into your contents. Doing this will not positively impact your ratings in Spotlight. The reasoning behind this is that keyword-stuffed results will often look like spam. The preview won't be a coherent preview of contents, but it will look as if somebody tried to put as many relevant words together as they could. A result that looks poor is less likely to be tapped by a user and will be pushed down in the rankings eventually.

If users do end up tapping a keyword-stuffed item, it's unlikely that they will find what they were looking for. Often, this means that the user exits your app after just a couple of seconds, and this negatively impacts your result rating.

In other words, add as much metadata as you reasonably can, but ensure that any data you add is relevant, authentic, and exists in your app. A good preview in Spotlight and a matching experience in your app is the best way to for apps to rank highly in Spotlight.

Registering as an indexing delegate

An item you add to the Spotlight index today might change over time. Some items you add to Spotlight could even have an expiration date if you've set the `endDate` property on the item attributes. Therefore, it's important that you register your app as an indexing delegate.

The indexing delegate is used for a single purpose: re-indexing content. If you've added items that expire over time, Spotlight will ask you to index them again to make sure that the expiration date is still correct. The indexing delegate is also called if something goes wrong with the index itself; for example, if all pieces of data are lost or if the search index becomes corrupted.

The indexing delegate has two required methods: `searchableIndex(_:reindexAllSearchableItemsWithAcknowledgementHandler:)` and `searchableIndex(_:reindexSearchableItemsWithIdentifiers:acknowledgementHandler:)`.

The first method is called if the index has been lost and everything should be indexed. It's up to you to figure out how to index your entire app. In the **MustC** app, this would probably mean fetching all family members and indexing them. You could use the `movies` relationship on the family members to loop over all of the movies that are stored in the database and index them accordingly. Alternatively, you could query both entities separately and index them separately.

The second indexing delegate method receives an array of identifiers that need to be indexed and is used to re-index a specific set of items. There are several reasons for spotlight to call this method, for instance, if an item is about to expire.

After you index the required items, you should call the `acknowledgementHandler` that the `delegate` method has received. This will make sure that Spotlight understands that you have successfully executed the tasks you needed to perform. To set the indexing delegate, you set the `indexDelegate` of the search index. For most apps, the searchable index is accessed through `CSSearchableIndex.default()`.

Now that you know everything about adequately implementing Spotlight indexing, let's see how you can step up your indexing game a little bit more by implementing Universal Links.

Increasing your app's visibility with Universal Links

A Universal Link is very similar to a deep link. **Deep links** allow apps to link users straight into a certain section of an application. Before Universal Links, developers had to use a custom URL scheme to create their deep links.

You might have seen a URL with a custom scheme in the past. These URLs are easily recognized and look as follows:

```
mustc://FamilyMember/jack
```

It's evident that this isn't a regular web URL because web URLs start with a scheme of either `http://` or `https://`. An application can register itself as capable of opening URLs with a certain scheme. So, the **MustC** app we've been working on could manifest itself as a handler of `mustc:// URLs`.

However, there are a couple of downsides to this approach. First and foremost, this URL isn't shareable at all. You can't send this URL to any friends that don't have the same app installed. If you were to send this URL to somebody and they didn't have the corresponding app installed, they wouldn't be able to open this URL. This is inconvenient because, for others to access the same content, assuming the content is publicly available on the web, users would have to share a different URL that points to the website. But sharing a link to the website usually means that the content is shown in Safari instead of the app, even if it's installed.

Another problem with custom URL schemes is that any app can register as being capable of opening a certain URL scheme. This means that you could create an application that registers as being capable of opening URLs with any scheme you can come up with and unfortunately, iOS offers the users no control over which application opens what URL scheme.

Universal Links were introduced in iOS 9 to solve all of the problems that exist with custom URL schemes and more. First of all, a Universal Link looks identical to a regular web link. A Universal Link is similar to a regular web link. If you've found a great news article on the web and you share it with somebody who has installed the app that belongs to the news website the link is from, the link will redirect straight to the corresponding app. Safari does not open intermediately; no attempts are made to redirect you from a web page to a custom URL scheme. The user merely is taken from the place where they tap the link, right to the app.

Also, not every app can register as capable of opening a Universal Link. Any app that claims to be able to open a certain Universal Link must be accompanied by the server that hosts the website. This means that, if your app claims to be able to open links from a certain domain, you must own that domain. Apple uses a verification file that you must host on the same domain that your app wants to handle links for, to make sure that your app does not try to open links on behalf of another app or website.

Apart from security benefits, Universal Links also provide a more unified, seamless experience for your users. With Universal Links, Apple didn't just open the door to a better, easier way to link to content inside your app; it also opened up an API that makes it easy for your app and website to share login information securely. Just like tying the links for your app and website together, you can also tie your app to the login credentials stored in Safari for your app. Any user that logs into your website through Safari can automatically be logged into your app.

Now that you're aware of the great features and possibilities of Universal Links, let's see how this all works on the server side.

Preparing your server for Universal Links

Setting up Universal Links on the server side has been made as straightforward as possible by Apple. All you have to do is host a file on your server that can prove the connection between your app and your server. This is done through a file called the `apple-app-site-association` file.

The `apple-app-site-association` file contains a dictionary of information that describes exactly how your app is tied to your site. Let's have a look at an example of an `apple-app-site-association` file. The example illustrates the implementation for a recipe app where users can browse and search recipes:

```
{
  "applinks": {
    "apps": [],
    "details": {
      "6QA73RGQR2.com.donny.recipes": {
        "paths": [
          "/recipes/*/",
          "/search/"
        ]
      }
    }
  }
}
```

Let's go over this configuration file bit by bit. Firstly, some mandatory dictionary keys are set up. The `applinks` key tells Apple that this part of the association file applies to Universal Links. Inside this key, an empty `apps` array is defined. This array does not need to have any content in it for a simple integration. The `details` key is a dictionary that contains the actual configuration.

Your app identifier should be added as a key for the `details` dictionary. The prefix you see in front of the app identifier is your team identifier. You can find your team identifier in the Apple Developer portal.

There is also a `paths` array inside the dictionary that's associated to the bundle identifier key. This array specifies all the paths on our website that your app should handle. Imagine that your website's base URL is `https://www.donnysrecipes.com/`. For this URL, `https://` is the scheme, `www.donnysrecipes.com` is the domain, and anything after the trailing `/` is called the path.

The example configuration handles the `/recipes/*/` and `/search/` paths. The `*` in `/recipes/*/` represents a wildcard value that could be anything. In other words, the config is set up to handle URLs such as `https://www.donnysrecipes.com/recipes/10/`, `https://www.donnysrecipes.com/recipes/20/`, or any other URL that looks similar.

The `/search/` path is a bit more interesting. No wildcard values were specified for this path, yet the app will be able to handle URLs such as `https://www.donnysrecipes.com/search/?q=macaroni`. That final part in the URL, `?q=macaroni`, is called the query string and you don't need to specify that you want to match on that because it's not part of the path.

Once you have created and configured your `apple-app-site-association` file, you need to upload it to your server. It's important that you host the association file in a way that makes it accessible on a URL similar to `https://www.donnysrecipes.com/apple-app-site-association`. In other words, the path to this verification file on your server should be `/apple-app-site-association`.

Now that the server is ready to communicate to Apple that your app can open links for your site, it's time to see how you can set up your app to handle Universal Links.

Handling Universal Links in your app

With all parts in place, you can now set your app up to handle Universal Links. In the previous sections of this chapter, a lot of work was done to handle user activities nicely. The setup you created for this will serve as a foundation for handling Universal Links.

Whenever an application is expected to open a Universal Link, the `application(_:open:options:)` method is called on `AppDelegate`. This method receives the URL it's expected to open. If the URL can't be opened, this method is expected to return `false`. If this method does manage to handle the URL, it should return `true`.

Any application that handles Universal Links must have the **Associated Domains** capability enabled in the Capabilities tab. To do this, go to the **Capabilities** tab in your project settings and allow **Associated Domains**. For every domain that you want to handle Universal Links for, you need to create a new `applinks:` entry. The following screenshot shows an example configuration:

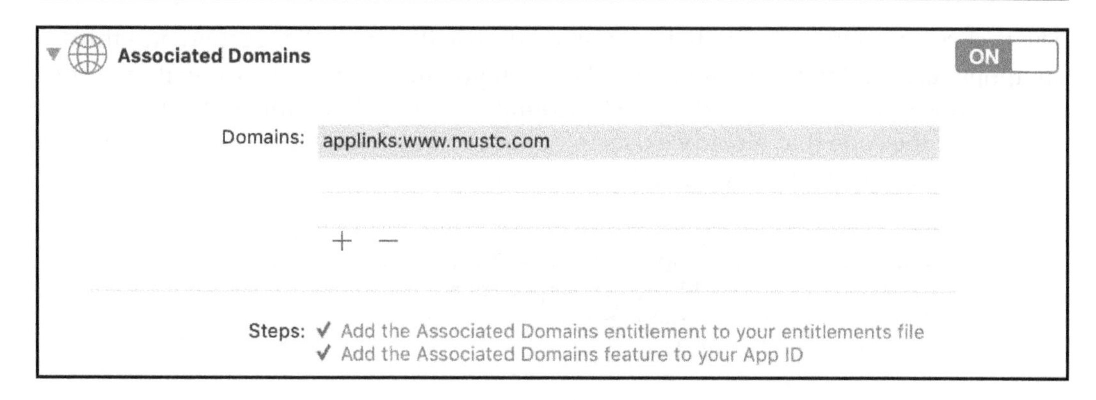

When the `application(_:open:options:)` you are expected to map the URL is passed to this method to a view in the application and take the user to that specific view. This process is called routing, and it's a well-known technique throughout programming languages and platforms.

Let's assume that your application received the following URL:

```
https://www.familymoviesapp.com/familymember/dylan/
```

Just by looking at this URL, you probably have a vague idea of what should happen if an application is asked to open this URL. The user should be taken to a family member screen that shows the detail page for Dylan.

When you receive a URL such as this in `AppDelegate`, three separate parts exist on the URL. One is the `scheme`; this property tells you which URL scheme, for example `https://`, was used to navigate to the app. Usually, the scheme isn't relevant to your app unless you're handling multiple custom URL schemes. Second, there is the host, for example `example.com`. This property describes the domain that the URL belongs to. Again, this is usually not relevant unless your app handles links from multiple hosts. Finally, there is the `pathComponents` property. `pathComponents` is an array of components that are found in the path for the URL. Printing `pathComponents` for the example URL gives the following output:

```
["/", "familymember", "dylan"]
```

The first component can usually be ignored because it's just `/`. The second and third components are a lot more interesting. They tell you more about the route that needs to be resolved. Going back to the **MustC** example, you could handle URLs in that app easily with the code that already exists in `AppDelegate` to navigate to family member and movie pages. The following code shows how:

```
func application(_ app: UIApplication, open URL: URL, options:
  [UIApplication.OpenURLOptionsKey : Any] = [:]) -> Bool {
  let pathComponents = URL.pathComponents
  guard pathComponents.count == 3
    else { return false }

  switch(pathComponents[1], pathComponents[2]) {
  case ("familymember", let name):
    return handleOpenFamilyMemberDetail(withName: name)
  case ("movie", let name):
    return handleOpenMovieDetail(withName: name)
  default:
    return false
  }
}
```

A switch with powerful pattern-matching is used to check whether the second component in the URL points to a family member or a movie and we parse the third component into a variable called `name`. If one of these matches, we call the existing methods; if nothing matches, we return `false` to indicate that we couldn't open the URL.

Earlier, you saw this URL: `https://www.donnysrecipes.com/search/?q=macaroni`. With `pathComponents`, you can easily gain access to the path of the URL. But how do you get the final part of the URL? Well, that's a little harder. You can get the `query` property for a URL, but then you get a single string in the form of `q=macaroni`. What you want is a dictionary where `q` is a key and `macaroni` is a value. The following extension on `URL` implements a naive method to create such a dictionary:

```
extension URL {
  var queryDict: [String: String]? {
    guard let pairs = query?.components(separatedBy: "&")
      else { return nil }

    var dict = [String: String]()

    for pair in pairs {
      let components = pair.components(separatedBy: "=")
```

```
            dict[components[0]] = components[1]
        }

        return dict
    }
}
```

First, the query string is retrieved. Then the string is separated on the & character because multiple key-value pairs in the query string are expected to be separated with that character. The code loops over the resulting array and separates each string on the = character. The first item in the resulting array is used as the dictionary key and the second item is the value. Each key and value are added to a dictionary, and finally, that dictionary is returned. This allows you to get the value for q, shown as follows:

```
URL.queryDict!["q"]
```

Great job! This is all that's needed to implement Universal Links in your app and to enhance discoverability for your users.

Summary

This chapter focused on getting your app indexed with the powerful CoreSpotlight framework. You saw how you can use user activities to index items that the user has already seen and how you can use searchable items to index content that the user might not have seen. You also learned how to make use of unique identifiers to ensure that you don't end up with duplicate entries in the search index. We also took a look at continuing user activities from either a user activity or a searchable item.

After you saw how CoreSpotlight can index your items, you learned about Universal Links and web content. It's important to think about publishing content on the web because it helps your Spotlight indexing tremendously and it enables Apple to show your Spotlight results in Safari. You also learned about the metadata that you should add to your web pages.

After all this Spotlight content, you were introduced to Universal Links. You learned how they relate to your app's web pages, and how you can use them to get bonus points in the Spotlight index and to improve the user's experience with your app.

In the next chapter, you will learn how you can integrate your apps with iMessage through an iMessage app.

Questions

1. Which objects are used to represent an item that you would like to index in Spotlight?

   a) `NSUserActivity` and `CSUserActivity`.
   b) `NSUserActivity` and `CSSearchableItem`.
   c) `CSSearchableItem` and `CSSearchableItemAttributeSet`.

2. What happens when a user selects a Spotlight search result that points to your app?

   a) `AppDelegate` is asked to resume a user activity that represents the item.
   b) `AppDelegate` is asked to handle a URL that represents the item.
   c) `CSSearchableIndexDelegate` is asked to handle the selected item.

3. How do you add an item to the Spotlight index?

   a) By passing a `CSSearchableItem` to `CSSearchableIndex` or setting a user activity as the current activity.
   b) By implementing `CSSearchableIndexDelegate`.
   c) Only by passing a `CSSearchableItem` to `CSSearchableIndex`.

4. Does it make sense to index the same user activity multiple times?

   a) No, the item will only appear once in the Spotlight index.
   b) Yes, the more often an activity is performed in the app, the higher it will rank in the Spotlight search results.
   c) No, Spotlight will punish apps that add duplicate items.

5. What are Smart App Banners?

   a) You can add Smart App Banners to web pages to point visitors to your app.
   b) Metadata objects that you add to your app for better indexing.
   c) Special items that appear in the Spotlight search results.

6. What are Universal Links?

    a) Shareable web links that open your app if it's installed.
    b) Links with a custom URL scheme that point to an app.
    c) Links with a custom URL scheme that point to a website.

7. What happens when your app is expected to open a Universal Link?

    a) `AppDelegate` is asked to continue a user activity.
    b) `AppDelegate` is asked to open the URL.
    c) The link your app should open is passed to the launch arguments dictionary.

Further reading

- Apple's Core Spotlight documentation: `https://developer.apple.com/documentation/corespotlight`
- Allowing Apps and Websites to Link to Your Content: `https://developer.apple.com/documentation/uikit/core_app/allowing_apps_and_websites_to_link_to_your_content`

22
Extending iMessage

As part of Apple's effort to introduce more and more extension points into the iOS ecosystem, iMessage extensions were added in iOS 10. iMessage extensions allow your users to access and share content from your app with their friends right inside the Messages app. This chapter will show you exactly how to create an iMessage extension.

You will learn about two different extension types: Sticker packs and iMessage apps. Note that these extensions are called apps and not just extensions. This is because iMessage extensions behave a lot like apps even though they're not apps. Also, iMessage apps are distributed through their own iMessage App Store. You will learn more about this later in this chapter.

First, you will gain a broad understanding of iMessage apps and what possibilities they give you as a developer. Once this is established, you will deep dive into several aspects of iMessage apps in more detail. To be more specific, the following are the topics that will be covered in this chapter:

- Creating a sticker pack for iMessage
- Implementing custom, interactive iMessage apps
- Understanding sessions, messages, and conversations

By the end of this chapter, you'll understand how iMessage apps work and how you can create an experience for your users that is fun and engaging. More importantly, you'll be able to launch an amazing sticker pack that features cats, dogs, or any other type of sticker you'd like to slap on messages you send and receive.

Understanding iMessage apps

An iMessage app is really an app extension that lives inside iMessage. It's kind of a weird type of extension because it behaves like a *hybrid between an extension and an application*. Extensions typically contain a single view controller and, more importantly, they can't be distributed through the iOS ecosystem without containing an application.

This rule does not apply to iMessage apps. An iMessage app can be distributed directly through the iMessage app store without a containing application. This means that you can build an iMessage app, such as a sticker pack, and distribute it without associating any other app to it. Also, you can add in-app purchases to an iMessage app. In-app purchases are normally unavailable to extensions, which makes iMessage apps behave more like an app than an extension.

Apart from distribution and in-app purchases, iMessage apps behave just like other extensions. They are created on behalf of a host application, Messages in this case, and they have a relatively short lifespan. An iMessage app specializes in sending messages from one user to another. These messages are shaped and created inside your extension and added to the conversation by sending them. An iMessage app can contain or send three different types of messages:

- Interactive messages
- Stickers
- Media content

An interactive message is intended to be tapped by the receiving user so they can interact with it somehow. Stickers are a special kind of image that the user can place right on top of other messages or send as a single image. Media content is distributed in the form of photos, videos, links, and any other type of content that iMessage natively supports. iMessage apps are built using the Messages framework. This framework has been created explicitly for iMessage extensions, and it functions as a gateway from your app into the conversation that your user is currently having.

As mentioned before, iMessage apps can be distributed without a host application. This means that you can distribute your iMessage app independently of any other app. Even though app icons inside of images are not squares, you must still provide an app icon that is square and would otherwise be used on the user's home screen by a containing app. This image is required to display the extension in the settings app and other places in iOS where a square app icon is required.

If you do have a containing app for your iMessage app, users can directly add the iMessage app to their list of available apps through the iMessage App Store without having to download it separately. The simplest app you can build for iMessage is a sticker pack. Let's go ahead and create one so we can start exploring iMessage apps in the simulator.

Creating an iMessage sticker pack

Stickers in iMessage are a fun way to share images with your friends. Stickers can be slapped onto a message as a response to a message you've received or just for fun with no particular reason.

To create a sticker pack, open Xcode and create a new project. One of the available project templates is the **Sticker Pack App**. Select this template and click **Next**. Give your sticker pack a name and click **Next** again to create your project:

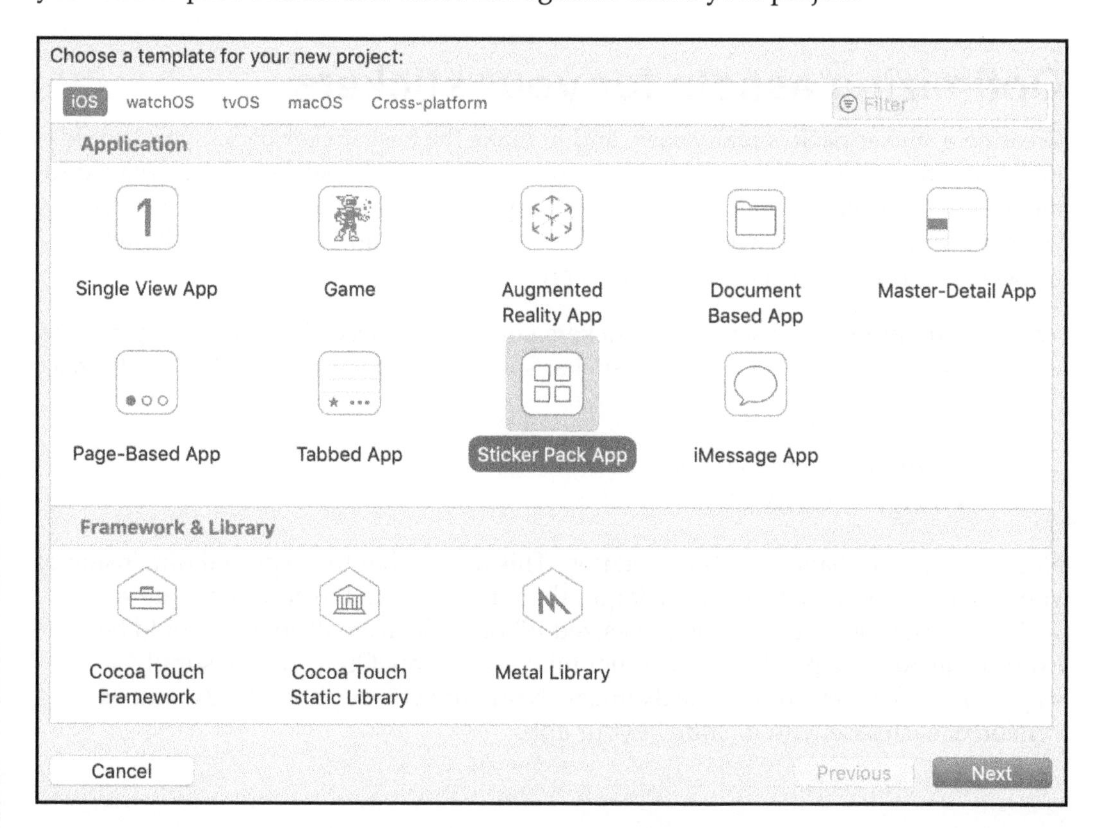

In the generated project, you'll find two folders. One is named `Stickers.xcstickers` and the other is named `Products`. When you create a sticker app, you're only interested in the `stickers` folder. If you open it, you'll find an app icon template and a folder named `Sticker Pack`.

All you need to do to create your sticker pack is to drag images into this `Sticker Pack` folder. After doing this, you can build and run your extension and try it out in the simulator. You'll find that the simulator comes with two active conversations in the Messages app. These are dummy conversations and you'll use them to test your iMessage app.

The simulator cannot send messages outside the simulator so you must use these predefined conversations while testing. Tapping a sticker or dragging it onto an existing speech bubble will send it from one conversation to the other. Go ahead and give that a shot. You can also send text messages and any other type of message you would normally send in the dummy conversation.

Optimizing assets for your stickers

Creating a sticker pack is really easy, and to make the best sticker packs, you should make sure that you properly optimize your image assets. Stickers can be displayed to the user in three sizes: small, medium, and large. To change the size of stickers in your sticker pack, simply click the `Sticker Pack` folder and open the attributes inspector. You can pick the sticker size there.

You should optimize the size of the stickers you provide according to the display size you choose. The following is a list of sticker sizes and their recommended asset sizes:

- **Small**: 100x100 @3x
- **Medium**: 136x136 @3x
- **Large**: 206x206 @3x

Note that all sizes have an @3x annotation. This means that the exported dimension of your images should be three times larger than the size listed to accommodate screens with a 3x resolution, such as the plus-sized iPhones. So, a small sticker should be exported at 300x300 pixels. Unlike other image assets on iOS, you only supply a single image for each sticker, the 3x image. Normally you supply a 1x, 2x, and 3x version of each asset you include in your app.

Sticker packs can contain PNG images, animated PNG images, GIFs, and JPEGs. All images you use must be smaller than 500 kb. It's recommended that you provide all of your assets as either PNG or animated PNG because this image format has superior image quality and supports transparency. Transparent stickers are recommended because they look a lot better when placed on top of other stickers or messages.

Now that you're familiar with sticker packs and testing them in the simulator, let's see how we make our sticker pack a bit more special by customizing the sticker-picking interface a bit.

Creating a custom sticker app

The standard sticker app that you created earlier using the sticker pack template is very plain. The generated iMessage app shows stickers on a white background and that's about it. Sticker apps shouldn't be more complex than this, but wouldn't it be great if you could at least change the background color for a sticker app? This isn't possible if you're using the simple sticker pack template. You can, however, create your own sticker pack app and customize the background. Doing this will allow you to familiarize yourself with the code that's involved in creating an iMessage app, so let's create a custom sticker pack.

In the source code repository for this book, you'll find a project named `CustomStickers`. This project already contains the stickers that will be used in the sample app. The images for these stickers are made available through `openclipart.org` by a user named bocian. Note that the images have not been added to the `Assets.xcassets` folder, but to the `Stickers` folder in `MessagesExtension`. The project was set up like this since the `Assets.xcassets` folder belongs to the containing app, which will be ignored.

In the `MessagesExtension` folder, you'll find a view controller file and a storyboard file. Open the storyboard and remove the default `UILabel` that was placed in the interface. You won't be adding any interface elements through Interface Builder because the interface elements to create a sticker app aren't directly available in Interface Builder.

You will find several boilerplate methods in the `MessagesViewController` class. We'll get into them soon; you can ignore them for now. The `viewDidLoad` method will be used to set up `MSStickerBrowserViewController` to display stickers in.

The `MSStickerBrowserView` instance that is contained inside `MSStickerBrowserViewController` behaves somewhat like `UICollectionView` because it requires a data source to determine how many, and which, stickers to display.

The first step in implementing your own sticker app is to add a property for an instance of `MSStickerBrowserViewController` to `MessagesViewController`:

```
var stickerBrowser = MSStickerBrowserViewController(stickerSize:
.regular)
```

Next, add the following implementation for `viewDidLoad`:

```
override func viewDidLoad() {
  super.viewDidLoad()
  stickerBrowser.willMove(toParent: self)
  addChild(stickerBrowser)
  stickerBrowser.didMove(toParent: self)

  view.addSubview(stickerBrowser.view)
  stickerBrowser.stickerBrowserView.dataSource = self
  stickerBrowser.stickerBrowserView.reloadData()
  stickerBrowser.stickerBrowserView.backgroundColor = UIColor.red
}
```

The preceding snippet should not contain any surprises for you. First, `stickerBrowser` is added as a child view controller of the messages view controller. Then the `stickerBrowser` view is added as a subview of the messages view controller's view. Next, the `dataSource` is set on `stickerBrowserView` and it is told to reload its data. Lastly, the background color for `stickerBrowserView` is set.

If you build your app now, Xcode will complain about `MessagesViewController` not conforming to `MSStickerBrowserViewDataSource`. Add the following extension to `MessagesViewController.swift` to make `MessagesViewController` conform to `MSStickerBrowserViewDataSource`:

```
extension MessagesViewController: MSStickerBrowserViewDataSource {
  func stickerBrowserView(_ stickerBrowserView: MSStickerBrowserView,
stickerAt index: Int) ->; MSSticker {
    return OwlStickerFactory.sticker(forIndex: index)
  }

  func numberOfStickers(in stickerBrowserView: MSStickerBrowserView) ->
Int {
    return OwlStickerFactory.numberOfStickers
```

```
    }
}
```

The first method is expected to return a sticker for a certain index and the second returns the number of stickers in the app. The logic for this is abstracted into a sticker factory. This is done to keep the code in the view controller nice, compact, and to the point. Add a new Swift file to the project and name it `OwlStickerFactory`.

Add the following implementation to this file:

```
import Foundation
import Messages

struct OwlStickerFactory {
  static private let stickerNames = [
    "bike", "books", "bowler", "drunk", "ebook",
    "family", "grill", "normal", "notebook", "party",
    "punk", "rose", "santa", "spring"
  ]

  static var numberOfStickers: Int { return stickerNames.count }

  static func sticker(forIndex index: Int) -> MSSticker {
    let stickerName = stickerNames[index]

    guard let stickerPath = Bundle.main.path(forResource: stickerName,
ofType: "png")
      else { fatalError("Missing sicker with name: \(stickerName)") }
    let stickerUrl = URL(fileURLWithPath: stickerPath)

    guard let sticker = try? MSSticker(contentsOfFileURL: stickerUrl,
localizedDescription: "\(stickerName) owl")
      else { fatalError("Failed to retrieve sticker: \(stickerName)")
}

    return sticker
  }
}
```

Most of the preceding code should speak for itself. There is an array of sticker names and a computed variable that returns the number of stickers for the app. The interesting part of this code is the `sticker(forIndex:)` method. This method retrieves a sticker name from our array of names. Then it retrieves the file path that can be used to retrieve the image file from the application bundle. Finally, it creates a URL with this path to create a sticker.

Note that the `MSSticker` initializer can throw errors so the initialization call is prefixed with `try?`. Also, note that the sticker initializer takes `localizedDescription`. This description is used by screen readers to read out a description of the sticker to users that have certain accessibility features enabled.

When you run your extension now, you should see an interface that looks very similar to the following screenshot:

You can drag stickers from the sticker browser to the messages view. Notice that the stickers are picked and placed with a nice peel animation. Whenever you add an `MSSticker` to the messages view, regardless of whether you use a sticker browser, the Messages framework takes care of the sticker peel, drag, and drop animations for you. This means that you can create a custom interface for your stickers if you'd like.

Keep in mind that most sticker apps will make use of the standard layout and your users might not be too pleased if your app presents them with an unexpected sticker sheet layout. However, apps that aren't about stickers do require a special layer of design and interaction. This is the next topic covered in this chapter.

Implementing custom, interactive iMessage apps

Sticker apps are nice, but they're not particularly interactive. You can build far more interesting and interactive applications for iMessage through the Messages framework. Some of the larger, well-known apps on iOS have been able to implement iMessage extensions that make sharing content from their apps easier. There are even people that have built games in iMessage! The Messages framework allows developers to build a wide range of extensions straight into the Messages app.

You have just seen how you can build sticker packs and how you can create a somewhat customized sticker pack by picking the app template instead of the sticker pack template in Xcode. You haven't gone in-depth into the different life cycle methods that Xcode generates for you when you create a new iMessage app.

To go into more depth with the **Messages** framework, you will build an iMessage app for **The Daily Quote**. You built a widget for this app in `Chapter 19`, *Instant Information with a Today Extension*. First, you will learn what the life cycle for an iMessage app looks like. After that, you will implement your own iMessage app.

Understanding the iMessage app life cycle

An iMessage app lives inside the Messages app, just like you would expect from an extension. As mentioned before, iMessage apps are a special kind of extension, which makes them behave a lot like apps even though they are still extensions at their core.

The main view controller for an iMessage app must always be a subclass of `MSMessagesAppViewController`. You can't have a primary view controller that is not a subclass of this class. When the user navigates to your extension through the iMessage apps drawer, this view controller is added where the keyboard would normally be placed. This size for the extension is called compact mode.

When the Messages framework instantiates your extension, the `willBecomeActive(with:)` method is called, followed by `didBecomeActive(with:)`. These methods are called after `viewDidLoad()` in the view controller life cycle, but before `viewWillAppear(_:)`. When it's time to dismiss your extension, the `viewWillDisappear(_:)` and `viewDidDisappear(_:)` life cycle methods are called. Next, `willResignActive(with:)` and `didResignActive(with:)` are called.

Once the resignation methods are called, the process for your iMessage app is killed shortly thereafter. You do not get any time to do work in the background. This is also true for other extension types. Again, even though Messages extensions behave a lot like apps, they're not.

There are two more methods you should know about for now. These methods are called whenever your iMessage app is transitioning from one display mode to another. The display mode is changed whenever the user expands the iMessage app by dragging the handle that is shown above your extension's view.

You can also trigger this transition from code by calling `requestPresentationStyle(_:)`. The life cycle methods that get called when the extension changes its size are `willTransition(to:)` and `didTransition(to:)`. The first method is called right before the transition occurs, the second is called right after.

Implementing the custom compact view

To start implementing the Messages extension for **The Daily Quote**, a new extension should be added to the project. Pick the **Messages extension** and name it `The Daily Quote Messages`. Enable the **App Groups** capability for this extension and include `Quotes.swift` in the extension target, just like you've done before.

You will use the same view-controller containment technique you used when building the custom sticker pack. This time you will add the view controller that should be displayed to the extension's Storyboard without connecting it to `MessagesViewController`. Before you do this and create the interface, you should implement the code for the compact view. The compact view will feature a quote, the creator of the quote, and a **Share** button.

Create a new `UIViewController` subclass and name it `CompactViewController`. Make sure to add it to the correct target by selecting the extension's folder before creating the new file. The setup for this view controller will be really similar to all of the other view controllers you've created for **The Daily Quote** and its extensions. The view will contain two outlets, one for the quote and one for its creator. In `viewDidLoad()`, the current quote should be fetched and the labels should be populated with the correct values. I trust you to be able to do this on your own. When in doubt, check the other projects. Alternatively, check the source code in the code bundle for this chapter.

When a user taps the Share button, the current quote should be shared as a message in the active conversation. This means that `MessagesViewController` will have to be aware of the share command that occurs in `CompactViewController`. To do this, you can create a `QuoteSelectionDelegate` protocol to which `MessagesViewController` can conform.

The delegate and the share action should be implemented in `CompactViewController`, as shown in the following code snippet:

```
var delegate: QuoteSelectionDelegate?

@IBAction func shareTapped() {
  delegate?.shareQuote(Quote.current)
}
```

The delegate is optional because it can't be set before the view controller is initialized. The tap action simply calls a method on the delegate and passes the current quote along with it. Create a new file to define the protocol and name it `QuoteSelectionDelegate`. The protocol should be implemented as follows:

```
protocol QuoteSelectionDelegate {
  func shareQuote(_ quote: Quote)
}
```

This is a simple protocol with just a single method requirement. Now, let's write all the required code in `MessagesViewController` before implementing the interface for `CompactViewController`. First of all, add the property shown in the following snippet to the message view controller so it can hold on to the compact view controller. Also, update your `viewDidLoad` implementation as follows:

```
var compactViewController: CompactViewController?

override func viewDidLoad() {
  super.viewDidLoad()
```

```
    compactViewController =
  storyboard?.instantiateViewController(withIdentifier:
  "CompactViewController") as? CompactViewController
    compactViewController?.delegate = self
  }
```

The `instantiateViewController(withIdentifier:)` method from `UIStoryboard` is used to obtain an instance of `CompactViewController` from the storyboard.

Add an extension to `MessagesViewController` so it conforms to `QuoteSelectionDelegate`. Add an empty implementation for `shareQuote(_:)` to the extension. You will implement the share functionality later.

Because the user can switch between the compact and expanded state as they please, it's a good idea to abstract the code to show the compact and expanded views into their own methods. Doing this will make your code easier to read, and it avoids code duplication. Add the following method to `MessagesViewController` to show the compact view controller:

```
func showCompactView() {
  guard let compactViewController = self.compactViewController
    else { return }

  compactViewController.willMove(toParent: self)
  addChild(compactViewController)
  compactViewController.didMove(toParent: self)

  view.addSubview(compactViewController.view)
  compactViewController.view.frame = view.frame
}
```

This code is very similar to what you saw when you implemented the sticker pack. The last step to show the compact view is to implement `willBecomeActive(with:)`. `showCompactView()` will be called from this method as follows:

```
override func willBecomeActive(with conversation: MSConversation) {
  if self.presentationStyle == .compact {
    showCompactView()
  }
}
```

Now, open `MainInterface.storyboard` and drag out a view controller. Add two labels and a button to this view controller. Lay them out as shown as shown in the following screenshot:

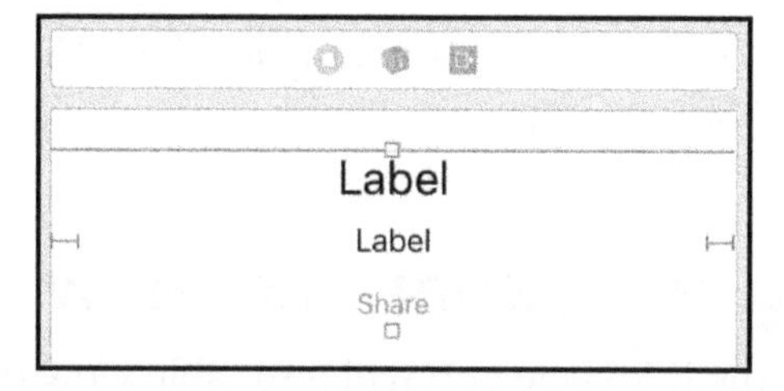

The quote is styled as a `title` and its maximum number of lines is set to `0` lines, so it automatically expands to fit the content. The creator is styled as `caption one`, and it's positioned below the title. The button should be laid out below the caption. Make sure to set `CompactViewController` as both the subclass and the storyboard ID for the view controller you just added. To wrap it up, connect the outlets to the views. Make sure to select **Touch up inside** as the trigger action for `shareTapped()`.

If you build and run your application now, the quote for today should pop up right inside iMessage:

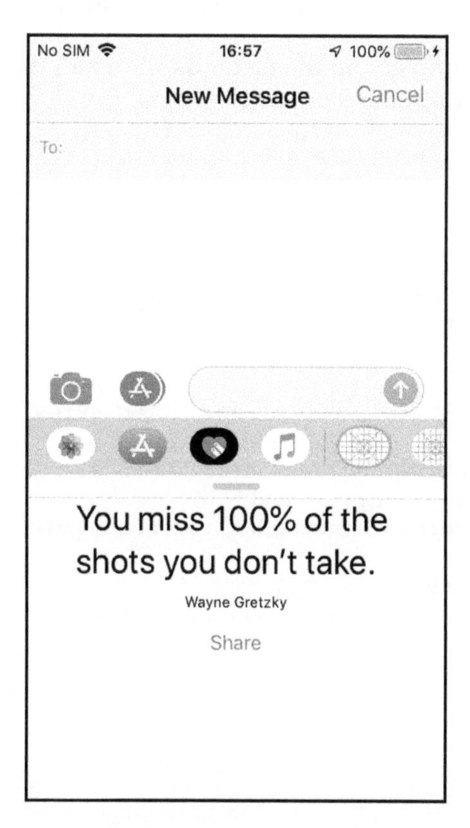

This is all you need to do to implement the compact view. Now let's implement an expanded view for the iMessage app.

Implementing the expanded view

The expanded view for the iMessage app will be a table view that lists all of the quotes in the `Quote` struct. You will use a similar setup to the one you used before by creating a new view controller file and using the delegate protocol to communicate the selection of a quote back to `MessagesViewController`.

First, create a new `UITableViewController` subclass and name it `QuotesTableViewController`. You can remove most of the commented template code; the only methods you should keep are `tableView(_:cellForRowAt:)`, `tableView(_:numberOfRowsInSection:)`, and `numberOfSections(in:)`. In addition to the commented delegate methods, you can remove the `viewDidLoad()` and `didReceiveMemoryWarning()` methods; you don't need them.

For starters, you should implement the methods shown in the following code snippet. These methods provide the table view with its data:

```
override func numberOfSections(in tableView: UITableView) -> Int {
  return 1
}

override func tableView(_ tableView: UITableView,
numberOfRowsInSection section: Int) -> Int {
  return Quote.numberOfQuotes
}

override func tableView(_ tableView: UITableView, cellForRowAt
indexPath: IndexPath) -> UITableViewCell {
  let cell = tableView.dequeueReusableCell(withIdentifier:
"QuoteTableViewCell", for: indexPath)
  let quote = Quote.quote(atIndex: indexPath.row)

  cell.textLabel?.text = quote?.text
  cell.detailTextLabel?.text = quote?.creator
  return cell
}
```

The preceding snippet uses a `numberOfQuotes` property on the `Quote` struct. However, this property is not defined yet. Add it to the `Quote` struct, as shown in the following code snippet:

```
static var numberOfQuotes: Int { return quotes.count }
```

The last thing you should take care of before creating and connecting the interface is the quote-selection delegate and implementing cell selection in `QuotesTableViewController`. The code to do this is pretty straightforward:

```
var delegate: QuoteSelectionDelegate?

override func tableView(_ tableView: UITableView, didSelectRowAt
indexPath: IndexPath) {
  guard let quote = Quote.quote(atIndex: indexPath.row)
    else { return }
  delegate?.shareQuote(quote)
}
```

This leaves you with a complete implementation of the table view in code. Let's create the interface in Interface Builder. Open `MainInterface.storyboard` and drag out a table view controller. Assign `QuotesTableViewController` as its class and Storyboard ID. Also, click the prototype cell and set its style to **Subtitle**. The Identifier for the cell should be set to `QuoteTableViewCell`. That's all you need to do for now. Let's make sure that you display this view controller when the iMessage app is in the expanded presentation mode.

In `MessagesViewController`, update `willBecomeActive(with:)` so it can display both the expanded and compact mode as the initial view, depending on `presentationStyle`:

```
override func willBecomeActive(with conversation: MSConversation) {
  if self.presentationStyle == .compact {
    showCompactView()
  } else if self.presentationStyle == .expanded {
    showExpandedView()
  }
}
```

Since the user can manually change the display mode for the iMessage app, you will need to dynamically switch the view controller that is shown depending on the presentation style. Add the following implementation for `willTransition(to:)`:

```
override func willTransition(to presentationStyle:
MSMessagesAppPresentationStyle) {
  if presentationStyle == .compact {
    showCompactView()
  } else if presentationStyle == .expanded {
    showExpandedView()
  }
}
```

Finally, you should take care of showing the correct view. You should always make sure to remove any existing view controllers before you show a new one. `showExpandedView()` has not been implemented yet, but if you think about what this method should do for a second, you will realize it should look very similar to the `showCompactView()` method. It's a good idea to implement a special cleanup method and refactor `showCompactView()`, so you can reuse it for `showExpandedView()`. First, add the following method that will be used to clean up the displayed view controllers when needed:

```
func cleanupChildViewControllers() {
  for viewController in children {
    viewController.willMove(toParent: nil)
    viewController.removeFromParent()
    viewController.didMove(toParent: nil)

    viewController.view.removeFromSuperview()
  }
}
```

One last thing that needs to be done before implementing the presentation of the expanded view controller is adding a property for this expanded view controller in `MessagesViewController` and updating `viewDidLoad()`, so it creates a new instance of `QuotesTableViewController` that will be used as the expanded view controller. Add the following property and the `viewDidLoad()` implementation to `MessagesViewController`:

```
var expandedViewController: QuotesTableViewController?

override func viewDidLoad() {
  super.viewDidLoad()

  compactViewController =
```

```
storyboard?.instantiateViewController(withIdentifier:
"CompactViewController") as? CompactViewController
  compactViewController?.delegate = self

  expandedViewController =
storyboard?.instantiateViewController(withIdentifier:
"QuotesTableViewController") as? QuotesTableViewController
  expandedViewController?.delegate = self
}
```

Next, let's implement the refactored view controller presentation methods:

```
func showCompactView() {
  guard let compactViewController = self.compactViewController
    else { return }

  showViewController(compactViewController)
}

func showExpandedView() {
  guard let expandedViewController = self.expandedViewController
    else { return }

  showViewController(expandedViewController)
}

func showViewController(_ viewController: UIViewController) {
  cleanupChildViewControllers()

  viewController.willMove(toParent: self)
  addChild(viewController)
  viewController.didMove(toParent: self)

  view.addSubview(viewController.view)
  viewController.view.frame = view.frame
}
```

After doing this, build and run the app to see your extension in action. It flawlessly switches between showing the list of quotes and the single daily quote.

The final step in implementing our iMessage app is to implement the `shareQuote(_:)` method. This method will compose a message that can be shared. Let's have a look at message composition and everything related to it.

Understanding sessions, messages, and conversations

This chapter has mostly focused on user interface-related aspects of iMessage apps. When you create an iMessage app, your app will sooner or later have to send a message. To do this, you use the `MSMessage` and `MSSession` objects. In addition to messages and sessions, the `MSConversation` class represents the conversation context that your extension exists in. These three classes together allow you to send messages, identify recipients in conversations, and even update or collapse existing messages in the messages transcript.

When an iMessage extension is activated, `willBecomeActive(with:)` is called in `MessagesViewController`. This method receives an instance of `MSConversation` that you can use to send messages, stickers, and even attachments. More importantly, the conversation contains unique identifiers for participants in the conversation and the currently selected message.

The `localParticipantIdentifier` and `remoteParticipantIdentifier` properties identify the current user of your app or the person who is sending messages with your app, and the recipients of these messages, respectively. Note that these identifiers are unique to your app so you can't use this identifier to identify users across different iMessage apps. Also, these identifiers can change if the user uninstalls and then reinstalls your app. The participant identifiers are mostly meant to identify users in a group conversation.

The `selectedMessage` property is only set if your extension was launched due to a user tapping on a message that was sent by your extension. This is especially useful if you're building an app that sends multiple messages that build upon each other. In Apple's 2016 WWDC talk, *iMessage Apps and Stickers, Part 2*, they demonstrated an ice-cream designer app. Each user designs a part of the ice cream and `selectedMessage` is used in combination with `MSSession` to collapse old messages so the message history does not get polluted with old pictures of ice creams. You will learn how to use this session in a bit. First, let's have a look at `MSMessage` to see how you can create messages.

Composing a message

The `MSMessage` class encapsulates all the information that is contained inside a message. When you initialize a message, you can initialize it with or without a session. If you instantiate a message with a session, other messages that are attached to this session will be collapsed to only show the message's `summaryText`. The summary is supposed to be a short, clear description of the message so that, even when the message is collapsed, the message still makes sense.

If possible, you should always aim to attach a `URL` and `accessibilityLabel` to your messages, depending on the types of message your app will send. If you're sending plain text messages like **The Daily Quote** will, you don't need an accessibility label since iOS can simply read the text message out loud for users with accessibility needs. The `URL` property is mainly used by platforms that don't support iMessage extensions, such as macOS. Attaching a `URL` makes sure that users can still navigate to your content online. When you're sending plain text, this isn't an issue.

Finally, if you're sending a message that has media attached to it, you should make use of the `MSMessageTemplateLayout` class. `MSMessage` has a layout property that can be set to an instance of `MSMessageTemplateLayout`. The message layout template is highly configurable. You can assign an image or media to it, set a caption, an image title, a sub-caption, and more. Messages will make sure that your layout looks good and is laid out nicely depending on the information you set.

If you set both a media URL and an image on your message layout, the image will be used and the other media items are ignored. Any images you add should be 300x300 @3x. Avoid rendering text on the image as the scaling on different devices might degrade the quality of your image and render the text illegible. Instead, use the image title and image subtitle properties from the message layout object.

An `MSMessage` instance is always intended to either have some form of media associated with it or to be interactive.

Sending a message

Once you have composed a message, you need to attach it to a conversation. To do so, you can use the `activeConversation` property, which is already present on your `MSMessagesAppViewController`. There are several methods on `MSConversation` that will insert a message. You can insert an instance of `MSMessage` if you have composed one. Alternatively, you can insert a sticker or you can simply insert some text.

To share a quote with **The Daily Quote** messages extension, this means that the text from the selected quote should be inserted into the conversation. Add the following implementation for `shareQuote(_:)` to implement the share feature:

```
func shareQuote(_ quote: Quote) {
  guard let conversation = activeConversation
    else { return }

  conversation.insertText("(quote.text) - (quote.creator)",
completionHandler: nil)
}
```

If you run this code and try sharing a quote, two noteworthy things occur:

- The extensions presentation mode does not change to compact after sharing a quote so the user doesn't see that the quote is shared.
- When you change the app to compact mode manually, the quote isn't actually sent yet. Instead, a message is prepared for the user to send manually. Users are always in control of when a certain message is sent. Extensions are only allowed to prepare messages for users so they can decide for themselves whether the message should be sent.

To fix the problem with the extension not switching to compact mode after selecting a quote, you need to add a single line of code to `shareQuote(_:)`. After inserting the quote, add the following line:

```
dismiss()
```

This will dismiss the extension entirely, allowing the user to focus on the quote and send it to the recipient. Alternatively, you could call `requestPresentationStyle(_:)` with a `compact` presentation style to change the current presentation style for your messages extension.

Once the user decides that they want to send the message you've composed on their behalf, they must manually send the message. Once this happens, `didStartSending(_:conversation:)` is called on your messages view controller. You won't be notified when this message was successfully sent to the recipient. This means that you can't make any assumptions in your app about whether the message was delivered to the recipient.

If the user decided to not send your `MSMessage` instance, `didCancelSending(_:)` is called. You can use this method to clean up or do something else in response to the cancel event. Note that both the start and cancel methods are only called when you're working with `MSMessage` instances. If you're sending a simple text, as **The Daily Quote** does, you won't be notified about these events at all.

Summary

In this chapter, you learned everything about iMessage apps. A messages extension is very similar to an app, and it's an exciting, powerful way for you to integrate your apps further into the iOS ecosystem. You now know how you can create a sticker pack to provide your users with a fun, refreshing way to share imagery from your app through standard and custom sticker packs.

Most importantly, you learned that even though iMessage apps can be very complex, they can also be simple. Some iMessage apps will make heavy use of interactive messages, others are merely standard sticker packs, or somewhere in between, such as **The Daily Quote**. The possibilities are endless and you can have a lot of fun with iMessage extensions. Just be aware that during testing you'll only be able to use your extension in the simulator so, unfortunately, you can't surprise your friends by suddenly sending them stickers from a sticker pack you're still developing.

This chapter wraps up the advanced integrations section of the book. The following chapters will focus on ensuring your app is of top-notch quality and submitting it to the App Store.

Questions

1. In what way are iMessage Extensions similar to apps?

   a) They can exist independently and support In-App Purchases.
   b) They can use view controllers.
   c) They are not similar to apps at all.

2. What do you need to create a Sticker Pack?

   a) You need images that can be added to the Sticker Pack template.
   b) You need a simple extension that provides stickers to the sticker browser.
   c) You need an extension that implements the sticker's peel and drop animations.

3. How can you test your Message extensions?

   a) You can use your extensions on a device and send content to your friends.
   b) You must use the predefined conversations in the simulator.
   c) Both a) and b).

4. What object is used to configure a complex message that can be sent through your extension?

   a) The `MSMessageContent` object.
   b) The `MSConversationMedia` object.
   c) The `MSMessage` object.

5. How can you identify the users that participate in a conversation?

   a) By reading their unique, random identifiers.
   b) By retrieving their iCloud accounts.
   c) By their name and account details.

6. How do you know whether a message was delivered to the recipients?

   a) `didStartSending(_:conversation:)` is called on the extension.
   b) `didFinishSending(_:conversation:)` is called on the extension.
   c) There is no way to be sure due to privacy concerns.

Further reading

- iMessage Apps and Stickers, Part 2: `https://developer.apple.com/videos/play/wwdc2016/224/`
- Apple's Messages framework documentation: `https://developer.apple.com/documentation/messages`

23 Ensuring App Quality with Tests

In all chapters so far, the main focus was code that ran as part of an app. The apps you have worked on are small and can easily be tested manually. However, this approach doesn't scale well if your apps become larger. This approach also doesn't scale if you want to verify lots of different user input, lots of screens, convoluted logic, or even if you're going to run tests on many different devices.

Xcode comes with built-in testing tools. These tools allow you to write tests so you can make sure that all of the business logic for your app works as expected. More importantly, you can test that your user interface functions and behaves as intended in many different automated scenarios.

Many developers tend to shy away from testing and postpone it until the end of the project, or don't do it at all. The reason for this is that it's often pretty hard to figure out how to write proper tests. This is especially true if you're just starting out with testing. Lots of developers feel like large parts of the logic their tests validate are so obvious that writing tests for that logic just feels silly. When testing is not approached correctly, they can be more of a burden than a relief by being high-maintenance and not testing the essential areas of code.

This chapter serves as an introduction to writing both logic and interface tests using Xcode and its built-in tools. By the end of this chapter, you should be able to set up a robust suite of tests and understand how you can make use of the tools provided by Xcode to write better code that is testable and reliable.

This chapter covers the following two topics:

- Testing logic with XCTest
- Testing the user interface with XCUITest

Testing logic with XCTest

Even if you haven't written any tests before, you might have thoughts or ideas about it. To start testing code, you don't need to have a computer science degree or spend days studying the absolute best way to test your code. In fact, the chances are that you're already testing your code and you don't even know it.

So, what does it mean to test your code? That's what this section aims to make clear. You will first read about the different types of tests you can write. Then, you'll learn what `XCTest` is and how you can set up a test suite for an app. Finally, you'll learn how to optimally test some actual code and how code can be refactored to make it more testable.

Understanding what it means to test code

When you test your code, you're essentially making sure that certain input produce the desired output. A very basic example of a test would be to make sure that calling a method that increments its input by a given value produces the output you expect.

Any time you launch your application in a simulator or on a device and test whether you can perform specific actions in an app, you are testing your code. Any time you print something to the console to verify that the expected value is printed, you are also testing your code. Once you think about testing this way, a concept that might have sounded hard before actually does not seem as complicated as you may have thought. So if using your app is a way of testing it, then what should you write tests for?

Determining which tests to write

When you start testing, it's often hard to decide what logic you want to test and what logic you don't want to test. Reasons for this could include certain logic being too trivial, too hard, or just not important enough to test. This statement implies that you do not have to test absolutely every line of code in your app and that is intentional. Sometimes it's simply not reasonable to write tests for a certain part of your code. For instance, you don't have to test that `UIKit` behaves as it should; it's Apple's job to make sure that the frameworks they ship are bug-free.

Determining what to test is important, and the longer you defer deciding whether you will add tests for a particular piece of logic, the harder it will be to write tests for it. A simple rule of thumb is that you don't need to test Apple's frameworks. It's safe to assume that Apple makes sure that any code they ship is tested and if it contains bugs, there's not much you can do to fix it anyway. Moreover, you don't want your tests to fail where Apple's tests should have.

What you should at least test is the *call-site* of your methods, structs, and classes. You can think of the call-site as the methods that other objects use to perform tasks. It's a good practice to make anything that's not used by the call-site of your objects private, meaning that outside code can't access that part of the code. More on this later when you learn more about refactoring code to make it more testable.

You should also test code that you might consider too trivial to write tests for. These parts of your code are likely to receive the *too trivial* treatment in other parts of the development process too. This usually causes you and your coworkers to pay less and less attention to this trivial piece of code, and before you know it, a bug gets introduced that might not be spotted until the app is in the App Store. Writing trivial tests for trivial code takes very little time and saves you from minor oversights that could lead to massive complications.

A few simple guidelines that you should follow when you write tests are:

- Test trivial code; this usually requires minimal effort.
- Test the call-site of your objects; these tests will ensure that your public APIs are consistent and work as expected.
- Don't test Apple's frameworks or any other dependencies; doing this is the responsibility of the framework vendor.

Once you've determined what you should test, it's time to start writing the actual tests. However, if you've heard about testing before, you might have heard of terms such as integration tests, unit tests, sanity tests, and a couple of others. The next segment explains a couple of the most important and well-known types of testing.

Choosing the correct test type

When you write tests, it's often a good idea to ask yourself what kind of test you're writing. The kind of test you want to write will typically guide you toward the way your test should be structured and scoped. Having tests that are well-scoped, structured, and focused will ensure that you're building a stable test suite that properly tests your code without unintended side-effects that influence the quality of your tests.

Unit tests

Probably the most well-known type of test is the unit test. A lot of people call virtually any test they write to test their code a unit test, which is probably why this is such a well-known term for testing. Another reason for unit tests being so popular is that it's a very sensible test type.

A unit test is intended to make sure that an isolated object works as expected. This isolated object will usually be a class or struct, but it could just as well be a standalone method. It's important that unit tests do not rely on any other test or object. It's perfectly fine to set up an environment that has all the preconditions you need for your unit test, but none of this setup should be accidental. For instance, you shouldn't accidentally test other objects or depend on the order in which your tests are executed.

When you write a unit test, it's not uncommon to create instances of models that are stored in an array to represent a dummy database or fake REST APIs. Creating such a list of dummy data is done to ensure that a unit test does not fail due to external factors such as a network error. If your test should depend on certain external factors, you are probably writing an *integration test*.

Integration tests

An integration test ensures that a certain part of your code can integrate with other components of the system. Similar to unit tests, an integration test should never rely on other tests. This is important for any test you write. Whenever a test depends on certain preconditions, they must be set up within the test itself. If your test does depend on other tests, this dependency might not be obvious at first, but it can make your tests fail in weird and unexpected ways.

Because no test can depend on another test, integration tests require a little more setup than unit tests. For example, you might want to set up an API helper, fetch some data from the API, and feed it into a database. A test such as this verifies that the API helper can cooperate with the database layer. Both layers should have their separate unit tests to ensure they work in isolation while the integration test ensures that the database and API can work together. There are many other types of test that you can write or learn about, but for now integration tests and unit tests provide an excellent starting point.

Isolating tests

Assumptions are a considerable risk when you're testing. Any time you assume anything about the environment you're testing in, your test is not reliable. If you're just getting into writing tests, it's tempting to make assumptions such as *I'm testing on the simulator and my test user is always logged in so my tests can be written under the assumption that a logged-in user exists*. This assumption makes a lot of sense to a lot of people, but what if one of your tests logs the current user out?

When this happens, a lot of your tests will fail due to assumptions that you made about the test environment. More importantly, these tests might fail even if the code they're testing works flawlessly.

As mentioned before, tests should test a single thing in your app. They should rely on as little outside code as possible, and they should be properly focused. A typical pattern that people use to structure their tests and improve reliability is the 3-As or AAA approach. The name of this pattern is short for Arrange, Act, and Assert. The following is an explanation for each of the As.

Arrange

The arrange step is all about preparation. Make sure a logged-in user exists, populate the (in-memory) database, create instances of your fake API or other helpers. You essentially arrange everything to be in place for your testing environment. Note that this step should not involve too much setup. If you find yourself writing a lot of code in the arrange step, your test might be too broad. Or the code you're testing relies on too many other pieces of code. You can't always avoid this, but if it happens, make sure you consider refactoring your code and test to keep the quality on par with what you're trying to achieve.

Act

In the act step, you set everything for your test in motion. You call methods on the object you're testing, you feed it data, and you manipulate it. This is where you take your code for a proverbial spin. Don't perform too many actions in succession though; too much acting will lead to problems during the next step, assert.

Assert

The final A in the 3-As approach is assert. During the assert step, you make sure that the state of the object you're testing is as you'd expect. Act and assert can be used multiple times in a single test. For instance, you might want to assert that doing something once places the object in a particular state and that doing it again places the object in another state. Or possibly that the state stays the same. Just as with the other two steps, if you're asserting a lot of things, or if you're acting and asserting over and over again in a test, the chances are that your test is too broad. This can't always be avoided, but long tests with a lot of acting and asserting are often an indication of testing too much at once.

Reading about testing can be quite dull, and it tends to get abstract quickly, so let's leave the theory for now. You will set up a test suite for an existing project in Xcode and start writing some tests, so all of the information you've taken in so far becomes a bit more tangible.

Setting up a test suite with XCTest

In this section, you'll work on a test suite for a new app: **MovieTrivia**. You'll find the basic setup for this project in this book's code bundle. If you open the project, there are some view controllers, an `Info.plist` file, and all the other files you would normally expect to find in a project. There's also a JSON file in the project named `TriviaQuestions.json`. This file contains a couple of dummy questions that you can load by uncommenting a bit of code in `LoadTriviaViewController.swift`.

By default, `LoadTriviaViewController.swift` attempts to load questions from a non-existing web server. This is intentional, to demonstrate how one would normally set up a project like this. Since you don't have a web server at your disposal right now, you can swap out the dummy networking code for the JSON file to test this app.

Before you write tests or perform any optimization, you must add a test target to the project. You add a test target in the same way you added extensions before. The only difference is that you select a different type of target. When adding a test target, you should pick the **iOS Unit Testing Bundle** template. The following screenshot shows the correct template you should select:

After adding the target, Xcode adds a new folder to your project. If you choose the default name for the test target, it's called `MovieTriviaTests`. You should add all the tests you write for this project to the test target.

If you think about when you used files in multiple targets with extensions, you might expect that you would need to add all of the files you want to write tests for to both of the targets. Fortunately, this isn't the case. When you write tests, you can import the entire app as a testable target, enabling you to write tests for all of the code in the app target.

If you look inside the `MovieTriviaTests` folder that Xcode created when you added the unit test target, you'll find a single file called `MovieTriviaTests.swift`. This file contains a couple of hints about what tests should look like for your test suite. First of all, note that the test class inherits from `XCTestCase`. All of your test classes should inherit from this `XCTestCase` so they can be identified as a test.

One of the methods you'll find in the test template is the `setUp()` method. This method is executed before every test in the file and helps you to fulfill the first stage of the AAA pattern in testing: Arrange. You use this method to ensure that all of the preconditions for your test are met. You could make sure that your user is logged in or that your database is populated with test data. Of course, the depth of your setup in this method depends on the unit of code for which you're writing a test.

Also, note that there are two methods prefixed with `test` in the test class. These methods are executed as tests, and they are expected to perform the act and assert steps. The majority of work should be performed in these `test` methods. Do note that it's often better to have multiple short test methods rather than a single test method that tests everything. The larger the methods, the harder it will be to maintain and debug your tests.

Finally, you'll find a `tearDown()` method. This method is intended to give you an opportunity to clean up after yourself. When you have inserted dummy data into your database, it's often desirable to remove this data when your tests have completed. This will ensure a clean slate for the next test that runs, and it minimizes the chances that your first test accidentally influences the second test that runs. As mentioned before, tests should never depend on other tests. This means that you also don't want to pollute other tests by leaving traces of previous tests.

Note that `setUp()` and `tearDown()` should be specific to the unit you're testing. This means that you can't put all of your tests in a single class. Separating tests into several classes is a good thing. You should create a test class for every unit of code that you're testing. One test class should typically not test more than a single class or struct in your app. If you're writing an integration test, there might be more than one class involved in the test, but you should still make sure that you're only testing a single thing, which is the integration between the classes involved in the integration you're testing.

Now that you have a test suite in place, let's see how you can write tests for the existing code in the **MovieTrivia** app and how the app can be refactored to be tested appropriately.

Optimizing code for testability

Now that the project has a test target, it's time to start adding some tests to it. Before you add tests, you should determine what to test. Take some time to look at the app and the code and try to think of things to test. Assume that the app is finished and that the trivia questions are loaded from a server.

Some of the things you might have thought of to test are:

- Making sure that we can display the data we load from the network
- Testing that selecting the correct answer triggers the expected code
- Testing that choosing a wrong answer triggers the expected code
- Ensuring that the first question is displayed after we show the last one
- Testing that the question index increments

If you came up with most of the tests on this list, good job. You've successfully identified a lot of good test cases. But how do you test these cases? The project has been made hard to test intentionally, but let's see what tests can be written without refactoring the app right away.

Remove the test class that Xcode has generated for you and create a new one called `LoadQuestionsTest`. Use the following bit of boilerplate code in this file's implementation as a starting point for the tests:

```
import XCTest
@testable import MovieTrivia

typealias JSON = [String: Any]

class LoadQuestionsTest: XCTestCase {
  override func setUp() {
    super.setUp()
  }

  func testLoadQuestions() {

  }
}
```

Note the `@testable import MovieTrivia` line at the top of the file. This line imports the entire app target so you can access it in your tests. Before you implement the test body for `testLoadQuestions`, it's wise to think about what this method should test. If you look at the code in the app target, the trivia questions are loaded in the `viewDidAppear(_:)` method of `LoadTriviaViewController`. Once the questions are loaded, the app moves on to the next screen. An important detail is that the `triviaJSON` property on `LoadTriviaViewController` is set once the questions are loaded.

Based on this information, you could write a test that creates an instance of `LoadTriviaViewController`, makes it appear, so the questions will load, and then waits until `triviaJSON` has a value to verify that the questions were successfully loaded. Writing a test that fits this description would involve many moving parts, way more than you should be comfortable with. **MovieTrivia** uses storyboards, so to obtain an instance of `LoadTriviaViewController`, the storyboard would have to be involved. This means that any changes or mistakes in the user interface would cause the logic test that checks whether data is loaded to fail. This is not desirable because this test should only verify whether it's possible to load data, not whether the user interface updates once the load completes.

This is a great moment to start refactoring some code and make it more testable. The first piece of code that should be revamped for testability is the question-loading code.

Introducing the question loader

To make **MovieTrivia** more testable, you should create a special helper that can load questions. The helper will go to the network and fetch the questions. Once the data is loaded, a callback is called to notify the object that initiated the request about the loaded questions. Because you already know that you're going to write tests for the new helper, you should think of a way to make sure that the helper works with both an offline and an online implementation, so the tests don't have to depend on an internet connection to work.

Because tests should rely on as few outside factors as possible, removing the networking layer from this test would be great. This means that the helper needs to be split into two parts. One part is the helper itself. The other part would be a data-fetcher. The data-fetcher should conform to a protocol that defines the interface that a data-fetcher must have, so you can choose to inject either an online or offline fetcher into the helper.

If the preceding explanation seems a little bit abstract and confusing to you, that's OK. The following code samples will show you the process of separating the different helpers step by step. Add a new Swift file to the application target and call it `QuestionsLoader.swift`. Then add the following implementation to it:

```
typealias QuestionsLoadedCallback = (JSON) -> Void

struct QuestionsLoader {
  func loadQuestions(callback: @escaping QuestionsLoadedCallback) {
    guard let url = URL(string: "http://questions.movietrivia.json")
```

```
        else { return }

      URLSession.shared.dataTask(with: url) { data, response, error in
        guard let data = data,
          let jsonObject = try? JSONSerialization.jsonObject(with: data,
options: []),
          let json = jsonObject as? JSON
          else { return }

        callback(json)
      }
    }
  }
```

This struct defines a method to load questions with a callback. This is already nice and a lot more testable than before. You can now isolate the question loader and test it separated from the rest of the app. A test for the helper in its current state would look like the test shown in the following code snippet:

```
func testLoadQuestions() {
  let questionsLoader = QuestionsLoader()
  let questionsLoadedExpectation = expectation(description: "Expected
the questions to be loaded")
  questionsLoader.loadQuestions { _ in
    questionsLoadedExpectation.fulfill()
  }

  waitForExpectations(timeout: 5, handler: nil)
}
```

The preceding test creates an instance of `QuestionLoader` and sets up an `expectation`. An expectation is used when you expect something to happen in your test eventually. Since `QuestionLoader` loads its questions asynchronously, you can't expect the questions to be loaded by the time this test method is done executing. The callback that's called when the questions are loaded is used to fulfill the expectation in this test. To make sure that the test waits for the expectation to be fulfilled, `waitForExpectations(timeout:handler:)` is called after `loadQuestions(callback:)`. If the expectation isn't fulfilled within the five-second timeout that is specified, the test fails.

Examine this test closely; you should be able to see all of the As that you read about earlier. The first A, arrange, is where the loader and expectation are created. The second A, act, is when `loadQuestions(callback:)` is called. The final A, assert, is inside the callback. This test doesn't validate whether the data passed to the callback is valid, but you'll get to that later.

Separating the loader into its own object is great but is still has one problem. There is no way to configure whether it loads data from a local file or the network. In a production environment, the question loader would load data from the network, which would make the test for the question loader depend on the network as well. This isn't ideal because a test that depends on the network might fail for reasons you can't control.

This can be improved by utilizing some protocol-oriented programming and the dependency-injection pattern. This means that you should define a protocol that defines the public API for a networking layer. Then you should implement a networking object in the app target that conforms to the protocol. `QuestionsLoader` should have a property that holds anything that conforms to the networking protocol. The test target should have its own object that conforms to the networking protocol so you can use that object to provide `QuestionsLoader` with mock data.

By setting the test up like this, you can take the entire networking logic out of the equation and arrange tests in such a way that the networking doesn't matter. The mock networking layer will respond with valid, reliable responses that can be used as test input.

Mocking API responses

It's common practice to mock API responses when you're testing. In this segment, you will implement the mock API that was described before to improve the quality and reliability of the **MovieTrivia** test suite. First, let's define the networking protocol. Create a new file in the app target and name it `TriviaAPIProviding`:

```
typealias QuestionsFetchedCallback = (JSON) -> Void

protocol TriviaAPIProviding {
  func loadTriviaQuestions(callback: @escaping
QuestionsFetchedCallback)
}
```

The protocol only requires a single method. If you want to expand this app later, everything related to the Trivia API must be added to the protocol to make sure that you can create both an online version of your app and an offline version for your tests. Next, create a file named `TriviaAPI` and add the following implementation to it:

```
struct TriviaAPI: TriviaAPIProviding {
  func loadTriviaQuestions(callback: @escaping
QuestionsFetchedCallback) {
```

```
    guard let url = URL(string: "http://questions.movietrivia.json")
      else { return }

    URLSession.shared.dataTask(with: url) { data, response, error in
      guard let data = data,
        let jsonObject = try? JSONSerialization.jsonObject(with: data,
options: []),
        let json = jsonObject as? JSON
        else { return }

      callback(json)
    }
  }
}
```

Lastly, update the `QuestionsLoader` struct with the following implementation:

```
struct QuestionsLoader {
  let apiProvider: TriviaAPIProviding

  func loadQuestions(callback: @escaping QuestionsLoadedCallback) {
    apiProvider.loadTriviaQuestions(callback: callback)
  }
}
```

The question loader now has an `apiProvider` that it uses to load questions. Currently, it delegates any load call over to its API provider, but you'll update this code soon to make sure that it converts the raw JSON data that the API returns to question models.

Update the `viewDidAppear(_:)` method of `LoadTriviaViewController` as shown in the following code snippet. This implementation uses the loader struct instead of directly loading the data inside the view controller:

```
override func viewDidAppear(_ animated: Bool) {
  super.viewDidAppear(animated)

  let apiProvider = TriviaAPI()
  let questionsLoader = QuestionsLoader(apiProvider: apiProvider)
  questionsLoader.loadQuestions { [weak self] json in
    self?.triviaJSON = json
    self?.performSegue(withIdentifier: "TriviaLoadedSegue", sender:
self)
  }
}
```

The preceding code is not only more testable, it's also a lot cleaner. The next step is to create the mock API in the test target so you can use it to provide the question loader with data.

The JSON file in the app target should be removed from the app target and added to the test target. You can leave it in the app folder but make sure to update the **Target Membership**, so the JSON file is only available in the test target. Now add a new Swift file named `MockTriviaAPI` to the test target and add the following code to it:

```
struct MockTriviaAPI: TriviaAPIProviding {
  func loadTriviaQuestions(callback: @escaping
QuestionsFetchedCallback) {

    guard let filename = Bundle(for:
LoadQuestionsTest.self).path(forResource: "TriviaQuestions", ofType:
"json"),
      let triviaString = try? String(contentsOfFile: filename),
      let triviaData = triviaString.data(using: .utf8),
      let jsonObject = try? JSONSerialization.jsonObject(with:
triviaData, options: []),
      let triviaJSON = jsonObject as? JSON
      else { return }

    callback(triviaJSON)
  }
}
```

This code fetches the locally-stored JSON file from the test bundle. To determine the location of the JSON file, one of the test classes is used to retrieve the current bundle. This is not the absolute best way to retrieve a bundle because it relies on an external factor to exist in the test target. However, structs can't be used to look up the current bundle. Luckily, the compiler will throw an error if the class that is used to determine the bundle is removed so the compiler would quickly error and the mistake can be fixed. After loading the file, the callback is called, and the request has been successfully handled. Now update the test in `LoadQuestionsTest`, so it used the mock API as follows:

```
func testLoadQuestions() {
  let mockApi = MockTriviaAPI()
  let questionsLoader = QuestionsLoader(apiProvider: mockApi)
  let questionsLoadedExpectation = expectation(description: "Expected
the questions to be loaded")
  questionsLoader.loadQuestions { _ in
    questionsLoadedExpectation.fulfill()
  }
```

```
        waitForExpectations(timeout: 5, handler: nil)
    }
```

A lot of apps have way more complex interactions than the one you're testing now. When you get to implementing more complex scenarios, the main ideas about how to architect your app and tests remain the same, regardless of application complexity. Protocols can be used to define a common interface for certain objects. Combining this with dependency-injection like you did for `QuestionsLoader` helps to isolate the pieces of your code that you're testing, and it enables you to swap out pieces of code to make sure that you don't rely on external factors if you don't have to.

So far, the test suite is not particularly useful. The only thing that's tested at this point is whether `QuestionsLoader` passes requests on to the `TriviaAPIProviding` object and whether the callbacks are called as expected. Even though this technically qualifies as a test, it's much better also to test whether the loader object can convert the loaded data into question objects that the app can display.

Testing whether `QuestionsLoader` can convert JSON into a `Question` model is a test that's a lot more interesting than just testing whether the callback is called. A refactor such as this might make you wonder whether you should add a new test or modify the existing test.

If you choose to add a new test, your test suite will cover a simple case where you only test that the callback is called and a more complex case that ensures the loader can convert JSON data to models. When you update the existing test, you end up with a test that validates two things. It would make sure that the callback is called but also that the data is converted to models.

While the implications for both choices are similar, the second choice sort off assumes that the callback will be called. You always want to limit your assumptions when writing tests and there's no harm in adding more tests when you add more features. However, if the callback does not get called, none of the tests will work. So in this case, you can work with a single test that makes sure the callback is called and that the loader returns the expected models.

The test you should end up with will have a single expectation and multiple assertions. Writing the test like this makes sure that the expectation for `callback` is fulfilled when the callback is called, and at the same time you can use assertions to ensure that the data that's passed to `callback` is valid and correct.

By making `QuestionsLoader` create instances of a `Question` model rather than using it to return a dictionary of JSON data, it not only makes the test more interesting, it also improves the app code by making it a lot cleaner.

Right now, the app uses a dictionary of JSON data to display questions. If the JSON changes, you would have to update the view controller's code. If the app grows, you might be using the JSON data in multiple places, making the process of updating quite painful and error-prone. This is why it's a much better idea to use the `Codable` protocol to convert raw API responses to `Question` models. Using Codable objects means you can get rid of the JSON dictionaries in the view controllers, which is a vast improvement.

Using models for consistency

Adding a question model to **MovieTrivia** involves quite a bit of refactoring. First, you must define the `Question` model. Create a new Swift file named `Question` and add the following implementation to it:

```
struct Question: Codable {
  let title: String
  let answerA: String
  let answerB: String
  let answerC: String
  let correctAnswer: Int
}
```

If you followed along with `Chapter 10`, *Fetching and Displaying Data from the Network*, this model should look somewhat familiar. The `Question` struct conforms to the `Codable` protocol. Since the dummy JSON data contains a list of questions, you'll want to define a `Codable` object that contains the response as well:

```
struct QuestionsFetchResponse: Codable {
    let questions: [Question]
}
```

Now that the `Question` model and the response container are in place, a couple of changes must be made to the existing code. First of all, `typealias` in the `TriviaAPIProviding` protocol should be modified as follows:

```
typealias QuestionsFetchedCallback = (Data) -> Void
```

Next, update the implementation of the TriviaAPI for the `URLSession` callback in `loadTriviaQuestions(callback:)` as follows:

```
URLSession.shared.dataTask(with: url) { data, response, error in
  guard let data = data
    else { return }
```

```
    callback(data)
  }
```

Also, update `MockTriviaApi` so it executes its callback with data instead of a JSON dictionary:

```
func loadTriviaQuestions(callback: @escaping QuestionsFetchedCallback)
{

  guard let filename = Bundle(for:
LoadQuestionsTest.self).path(forResource: "TriviaQuestions", ofType:
"json"),
    let triviaString = try? String(contentsOfFile: filename),
    let triviaData = triviaString.data(using: .utf8)
    else { return }

  callback(triviaData)
}
```

The `QuestionsLoadedCallback` `typealias` in `QuestionsLoader` should be updated to the following definition:

```
typealias QuestionsLoadedCallback = ([Question]) -> Void
```

And lastly, the implementation for `loadQuestions(callback:)` should be updated as follows:

```
func loadQuestions(callback: @escaping QuestionsLoadedCallback) {
  apiProvider.loadTriviaQuestions { data in
    let decoder = JSONDecoder()
    decoder.keyDecodingStrategy = .convertFromSnakeCase
    guard let questionsResponse = try?
decoder.decode(QuestionsFetchResponse.self, from: data)
      else { return }

    callback(questionsResponse.questions)
  }
}
```

This wraps up the changes for the API. However, there still is some refactoring to be done in the view controllers. Rename the `triviaJSON` property on `LoadTriviaViewController` to the following:

```
var questions: [Question]?
```

Make sure you replace all occurrences of `triviaJSON` with the new questions array. Also, make sure you change the following line in `prepare(for:sender:)`:

```
questionViewController.triviaJSON = triviaJSON
```

Change this line to:

```
questionViewController.questions = questions
```

In `QuestionViewController`, change the type of `questions` to `[Question]` and remove the `triviaJSON` property. At this point, you can clear all of the JSON-related code from the guards in this class. You should be able to do this on your own since the compiler should guide you with errors. If you get stuck, look at the finished project in the code bundle.

By now, you should be able to run the tests, and they should pass. To run your tests, click the Product menu item and select Test. Alternatively, press *Cmd* + *U* to run your tests. The tests run fine, but currently, the test doesn't test whether all of the questions in the JSON data got converted to Question models. To make sure this conversion worked, you can load the JSON file in the test, count the number of questions in the JSON file, and assert that it matches the number of questions in the callback.

Update the `testLoadQuestions()` method as shown in the following code snippet:

```
func testLoadQuestions() {
  let apiProvider = MockTriviaAPI()
  let questionsLoader = QuestionsLoader(apiProvider: apiProvider)
  let questionsLoadedExpectation = expectation(description: "Expected
the questions to be loaded")
  questionsLoader.loadQuestions { questions in
    guard let filename = Bundle(for:
LoadQuestionsTest.self).path(forResource: "TriviaQuestions", ofType:
"json"),
      let triviaString = try? String(contentsOfFile: filename),
      let triviaData = triviaString.data(using: .utf8),
      let jsonObject = try? JSONSerialization.jsonObject(with:
triviaData, options: []),
      let triviaJSON = jsonObject as? JSON,
      let jsonQuestions = triviaJSON["questions"] as? [JSON]
      else { return }

    XCTAssert(questions.count > 0, "More than 0 questions should be
passed to the callback")
    XCTAssert(jsonQuestions.count == questions.count, "Number of
questions in json must match the number of questions in the
callback.")
```

```
        questionsLoadedExpectation.fulfill()
    }

    waitForExpectations(timeout: 5, handler: nil)
}
```

This test loads the dummy JSON file and uses XCTAssert to make sure that more than zero questions were passed to the callback and that the number of questions in the JSON file matches the number of questions that were loaded.

`XCTAssert` takes a Boolean expression and a description. If the assertion fails, the description is shown. Adding good descriptions will help you to quickly figure out which assertion in your test has made your test fail.

This new version of the load-questions test is a small addition to the test suite but has vast consequences. By improving the test suite, you have improved the quality of the app because you can now prove that the question loader correctly transforms JSON into model objects. By adding model objects, you have improved the code in the view controllers as well. Instead of reading raw JSON, you are now reading properties from a model. And lastly, these changes have made your view controllers a lot cleaner.

One more metric that has improved by refactoring your code is the amount of code that is covered by the test suite. You can measure the percentage of code your test suite covers with Xcode's built-in code coverage-tracking. You'll learn how to use this tool next.

Gaining insights through code coverage

Code Coverage is a tool in Xcode that is used to gain insights into how much of your code you are testing with your test suite. It tells you exactly which parts of your code were executed during a test and which parts of your code were not. This is extremely useful because you can take focused action based on the information provided by Code Coverage.

To enable **Code Coverage**, open the scheme editor through the (**Product** | **Scheme**) menu:

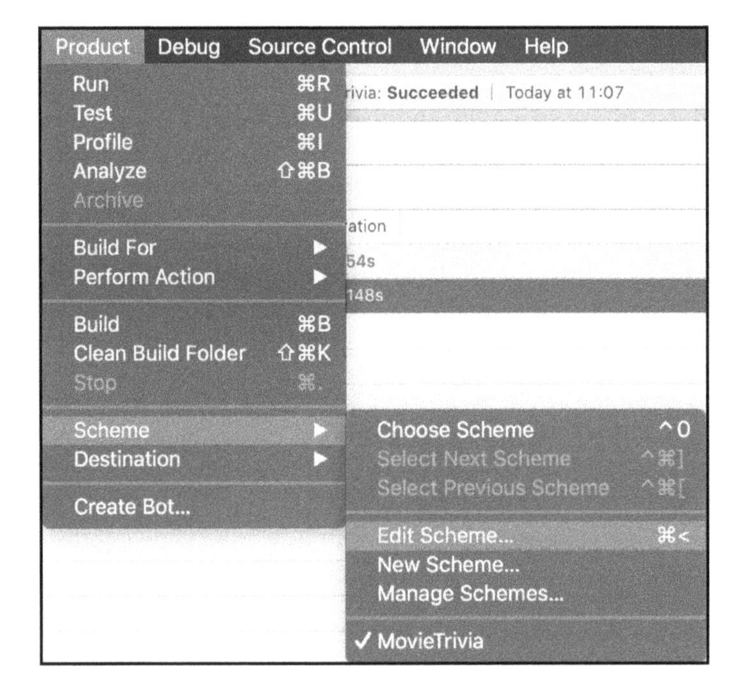

Select the Test action and make sure the **Gather coverage** checkbox on the **Options** tab is checked:

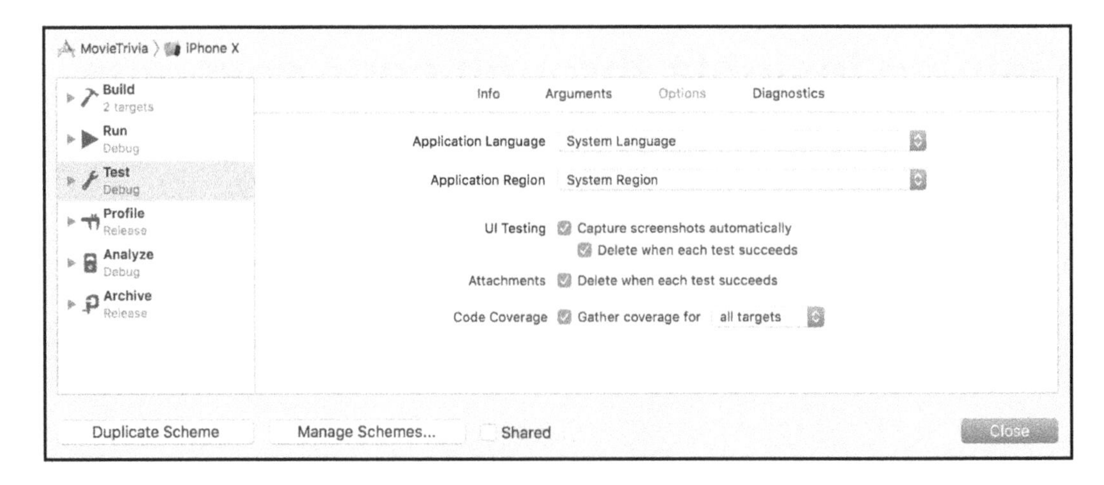

You can also press *Cmd* + < to open the scheme editor quickly.

After doing this, close the scheme editor and run your tests. This time, Xcode will monitor which parts of your code were executed during this test, and which parts weren't. This information can give you some good insights about which parts of your code could use some more testing. To see the coverage data, open the **Report navigator** in the left sidebar in Xcode. The rightmost icon in this sidebar represents the Report navigator:

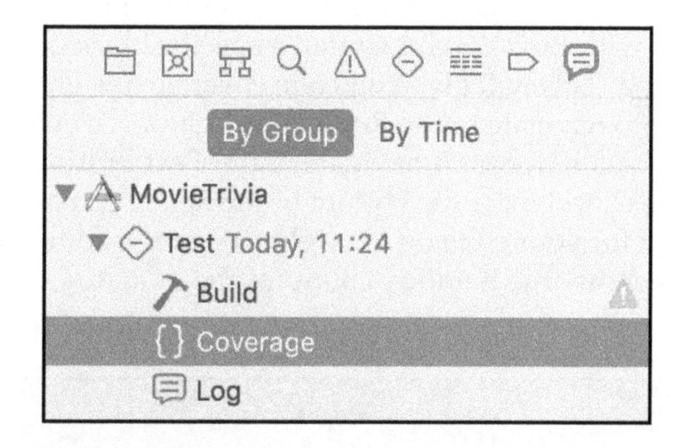

There are several reports listed under your app name. If you select the **Coverage** report, the coverage report will open in the Editor Area in Xcode. You can see all the files in your app and the percentage of code in the file that's covered by your tests. The following screenshot shows Coverage for the **MovieTrivia** app:

Name	Coverage
MovieTrivia.app	37,39%
AppDelegate.swift	100%
LoadTriviaViewController.swift	47,62%
QuestionViewController.swift	0%
QuestionsLoader.swift	100%
TriviaAPI.swift	64,71%

The more a bar is filled, the more lines of code in that file or method were executed during your test. You'll notice that the `AppDelegate.swift` file is covered under the tests even though you haven't written any tests for it. The reason this happens is that the app must launch during the test to act as a host for the test suite. This means that parts of the code in `AppDelegate.swift` are actually executed during the test, and therefore Xcode considers it covered in the tests.

You can see which methods for a specific file were executed by clicking on the triangle next to the class name. This enables you to see exactly which parts of a file are tested and which parts aren't.

One last feature of Code Coverage that's worth mentioning is inline Code coverage. Inline Code coverage will show you how often a specific block of code has been executed during testing. This will give you code coverage insights right next to your code, without having to navigate to the Reports navigator. To enable this feature, open up your Xcode preferences and navigate to the **Text Editing** tab. Check the Show iteration counts checkbox at the bottom of the tab. If you open a file now, you'll see the iteration count for your code on the right side of the editor window. The following screenshot shows the iteration count for the `loadQuestions(callback:)` method:

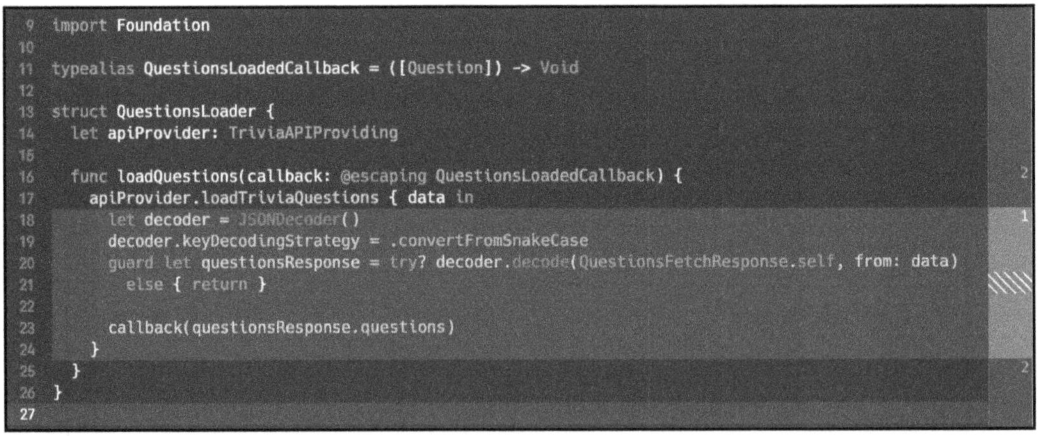

```
9   import Foundation
10
11  typealias QuestionsLoadedCallback = ([Question]) -> Void
12
13  struct QuestionsLoader {
14    let apiProvider: TriviaAPIProviding
15
16    func loadQuestions(callback: @escaping QuestionsLoadedCallback) {          2
17      apiProvider.loadTriviaQuestions { data in
18        let decoder = JSONDecoder()                                            1
19        decoder.keyDecodingStrategy = .convertFromSnakeCase
20        guard let questionsResponse = try? decoder.decode(QuestionsFetchResponse.self, from: data)
21          else { return }
22
23        callback(questionsResponse.questions)
24      }
25    }                                                                          2
26  }
27
```

Even though Code Coverage is a great tool for gaining insights into your tests, you shouldn't let it influence you too much. Regularly check the Code Coverage for your app and look for methods that are untested and are either easy to write tests for, or should be tested because they contain important logic. Code Coverage is also great for discovering parts of your code that should be tested but are hard to test because they're nested deep inside a view controller or otherwise hard to reach.

You should always aim for as much code coverage as possible, but don't push yourself to reach 100%. Doing this will make you jump through all kinds of hoops, and you'll invest way more time in testing than you should. Not all paths in your code have to be tested. However, don't shy away from doing some refactoring. Proper testing helps you to avoid bugs and to structure your code better. Code Coverage is just one extra tool in your tool belt to help identify which parts of your code could benefit from some tests.

If you look at the current state of the coverage in the **MovieTrivia** app, we're doing quite well. Most of the logic in the app is tested. The only parts that are not tested thoroughly are the view controllers. Testing view controllers and navigations flows with `XCTest` can be quite hard and tedious. Luckily, there is one last testing tool that we'll discuss in this chapter: `XCUITest`.

Testing the user interface with XCUITest

Knowing that most of your app logic is covered with tests is great. What's not so great, however, is adding your view controllers to your logic test. Luckily, you can use `XCUITest` to easily record and write tests that focus on the user interface of an app. `XCUITest` uses the accessibility features in iOS to gain access to the user interface of your app. This means that implementing user interface tests forces you to put at least a little bit of effort into accessibility for your applications. The better your app's accessibility is, the easier it will be to write UI Tests for.

`XCUITest` has two great features that we'll look at in greater detail. First of all, UI Tests help you to enhance accessibility for your apps. Secondly, it's easy to get started with UI testing because Xcode can record your tests while you navigate through your app. This can significantly benefit the amount of code that is covered by your test suite since Code Coverage also takes UI Tests into account.

Before we start recording our first UI test, let's have a quick look at accessibility.

Making your app accessible to your tests

One of the lesser thoughts about features in iOS is accessibility. The design teams at Apple work hard to ensure that iOS is accessible for everybody. This includes blind people and people with other disabilities that could somehow affect the user's ability to operate their iOS device.

Just looking at the accessibility settings in the iOS settings app makes it evident that this is a subject that Apple invests a lot of time in. If you're working on an app, Apple expects you to put in the same kind of effort. Doing this will be rewarded by more app downloads and if you're lucky, even a couple of great reviews. In their talk on iOS Accessibility from WWDC 2015, Apple even mentioned that implementing accessibility features can be helpful if you ever want to be featured in the *App Store*. Only the best apps get featured by Apple, and if your app is accessible to all people, that significantly boosts your app's quality.

A common myth surrounding accessibility is that it's hard to implement or that it takes a lot of time. Some people even go so far as to say that it looks ugly or gets in the way of beautiful design. None of this is entirely correct. Sure, making your app accessible requires some effort, but the UIKit framework is very helpful when it comes to accessibility. Using standard components and keeping your user in mind while you design your app will make sure that your app is both accessible and looks good.

So, how does accessibility work on iOS? And how can we make sure our app is accessible? A fun way to experiment with this is to turn on **VoiceOver** on your device. To enable **VoiceOver**, go to the **Accessibility** menu. You'll find several vision-related accessibility settings. **VoiceOver** should be the topmost one. To quickly enable and disable **VoiceOver**, scroll all the way to the bottom of the settings page and select **VoiceOver** as your accessibility shortcut:

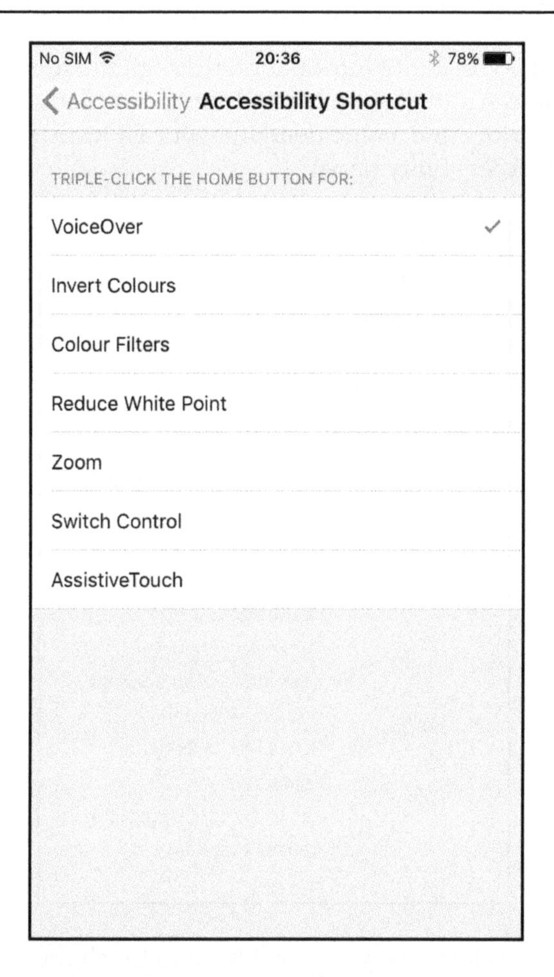

This will allow you to toggle **VoiceOver** off and on by triple-clicking the home button or side button, depending on your device.

After enabling this, run the **MovieTrivia** app on your device and triple-click your home button or side button to enable **VoiceOver**. Swipe around and try to use the app. This is how a person with a visual handicap uses your app. You won't get past the loading screen because the dummy questions aren't loaded, but you should find the splash screen to be pretty accessible, especially considering no special work had to be done to achieve this. UIKit uses great default settings to make sure your app will be accessible by default.

You can set your own accessibility information through the **Identity Inspector** in Interface Builder. You can add custom labels, hints, identifiers, and traits to your interface to aid accessibility and, coincidentally, your UI tests. The following screenshot shows the accessibility panel:

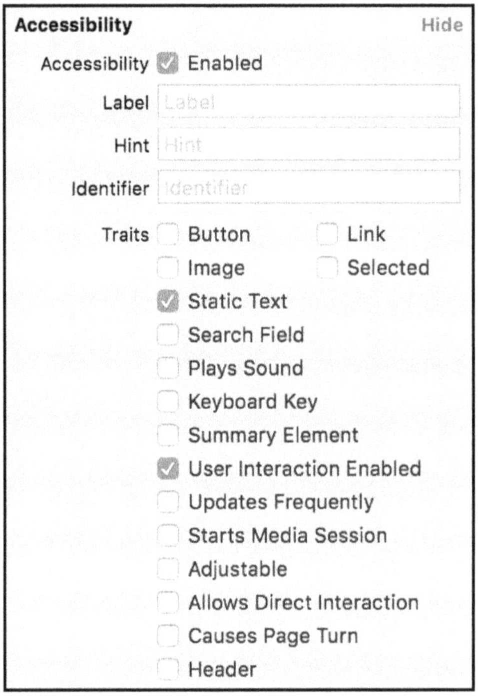

For most UIKit interface elements, you won't have to touch these settings yourself. UIKit will make sure that your objects have sensible defaults that automatically make your app accessible. Now that you have a little bit of background information about accessibility, let's have a look at testing the app's (accessible) UI.

Recording UI tests

Before you can record UI tests, you must add a UI testing target to the project. Follow the same steps as before to add a new testing target, but pick the iOS UI Testing Bundle this time around. If you look inside the newly-created group in your project, the structure for your UI tests looks very similar to the structure for Unit tests.

One significant difference between UI test targets and Unit test targets is that your UI tests do not have access to any code that's inside your app. A UI test can only test the interface of your app and make assertions based on that.

If you open the `MovieTriviaUITest.swift` file, you'll notice the `setUp()` and `tearDown()` methods. Also, all of the tests that must be executed are methods with the test prefix. This is all similar to what you've already seen for `XCUITest`.

One big difference is that the app is launched explicitly in the setup stage. This is because the UI test target is essentially just a different app that can interact with your main app's interface. This limitation is very interesting, and it's also the reason why it's important to make your app accessible.

To start recording a UI test in Xcode, you must start a recording session. If you're editing code in a UI test target, a new interface element is visible in the bottom-left corner of your code editor area:

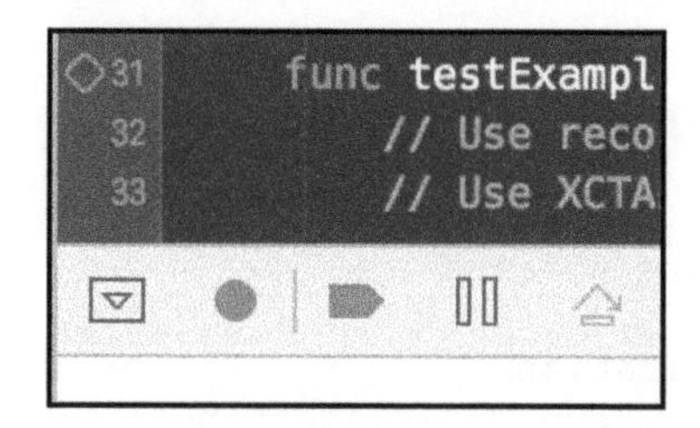

Place your typing cursor inside the `testExample()` method and click the red dot. Your app is launched and anything you do is recorded as a UI test and played back when you run your tests. If you tap on the label and the activity indicator on the loading screen, Xcode produces the following Swift code in the testing method:

```
let app = XCUIApplication()
app.staticTexts["Loading trivia questions..."].tap()
app.otherElements.containing(.activityIndicator, identifier:"In
progress").element.tap()
```

The UI test you recorded is a set of instructions that are sent to the app. In this sample, the test looks for a certain element in the app's UI and calls `tap()` on it. This test doesn't do a lot, so it's not particularly useful. To make the test more useful, we should let the app know that it should run in a special test mode so it can load questions from the JSON file instead of trying to load it from the network. To do this, you can send launch arguments to the app. Launch arguments can be used by the app to enable or disable certain functionalities. You can think of them as variables that are sent to the app when it launches.

Passing launch arguments to your app

To switch the loading of questions from the network to a local file for testing, you can pass your app a launch argument. This launch argument is then read by the app to make sure it loads questions from the JSON file like you did before in the unit tests rather than attempting to load trivia questions from the server.

To prepare for the launch argument and loading the JSON file, make sure you add it to the test target, the app target, and the UI test target. You won't need it in the UI test target just yet, but you will later, so you might as well add it to the UI test target while you're at it.

In order to pass launch arguments to the app, the `setUp()` method in the UI Test class should be modified:

```
override func setUp() {
  super.setUp()

  continueAfterFailure = false

  let app = XCUIApplication()
  app.launchArguments.append("isUITesting")
  app.launch()
}
```

The `XCUIApplication` instance that represents the app has a `launchArguments` property, which is an array of strings. You can add strings to this array before launching the app. These strings can then be extracting inside of the app. Modify the `loadTriviaQuestions(callback:)` method in `TriviaAPI.swift` as shown in the following code snippet:

```
func loadTriviaQuestions(callback: @escaping QuestionsFetchedCallback)
{
  if ProcessInfo.processInfo.arguments.contains("isUITesting") {
```

```
        loadQuestionsFromFile(callback: callback)
        return
      }

      // existing implementation...
    }
```

The preceding code should be inserted above the existing implementation of this method. The snippet checks whether we're UI testing by reading the app's launch arguments. If the UI testing argument is present, we call the `loadQuestionsFromFile(callback:)` method to load the questions from the JSON file instead of loading it from the network.

Note that it's not ideal to perform checks such as the preceding one in your production code. It's often better to wrap configuration such as this in a struct that can be modified easily. You can then use this struct throughout your app instead of directly accessing process info throughout your app. An example of such a configuration could look like this:

```
struct AppConfig {
  var isUITesting: Bool {
    ProcessInfo.processInfo.arguments.contains("isUITesting")
  }
}
```

We won't use this configuration class in this app since it's not needed for an app this small. But for your own apps, you might want to implement a configuration object regardless of app size since it leads to more maintainable code in the long run.

If you build the app right now, you should get a compiler error because `loadQuestionsFromFile(callback:)` is not implemented in the API class yet. Add the following implementation for this method:

```
func loadQuestionsFromFile(callback: @escaping
QuestionsFetchedCallback) {
  guard let filename = Bundle.main.path(forResource:
"TriviaQuestions", ofType: "json"),
    let triviaString = try? String(contentsOfFile: filename),
    let triviaData = triviaString.data(using: .utf8)
    else { return }

  callback(triviaData)
}
```

It's very similar to the question-loading method in the unit tests; the only difference is that it uses a different way to obtain the bundle from which the questions are loaded.

If you run your UI tests now, they will fail. The reason for this is that when the test framework starts looking for the elements it tapped before, they don't exist. This results in a test failure because the test can't tap elements that don't exist.

The test should be adjusted a bit because tapping a loader isn't very useful anyway. It's a lot more useful to make sure that buttons can be tapped and whether the UI updates accordingly. To do this, you can write a UI test that waits for the question and buttons to appear, taps them, and checks whether the UI has updated accordingly. The dummy data will be loaded in this test as well to verify that the correct question is shown and the buttons behave as they should.

Making sure the UI updates as expected

You're going to write two tests to make sure that the trivia game works as expected. The first test will test that the question and answer buttons appear and that they have the correct labels. The second test will make sure that the answers can be tapped and that the UI updates accordingly.

Instead of recording the tests, you'll write them manually. Writing tests manually gives you a bit more control and allows you to do much more than just tapping on elements. Before you do this, you should open the `Main.storyboard` file and give accessibility identifiers to the UI elements. Select the question title and give `UILabel` an identifier of `QuestionTitle`. Select each of the answers and give them the `AnswerA`, `AnswerB`, and `AnswerC` identifiers, respectively. Also, give the next button an accessibility identifier of `NextQuestion`. The following screenshot shows what the question title should look like:

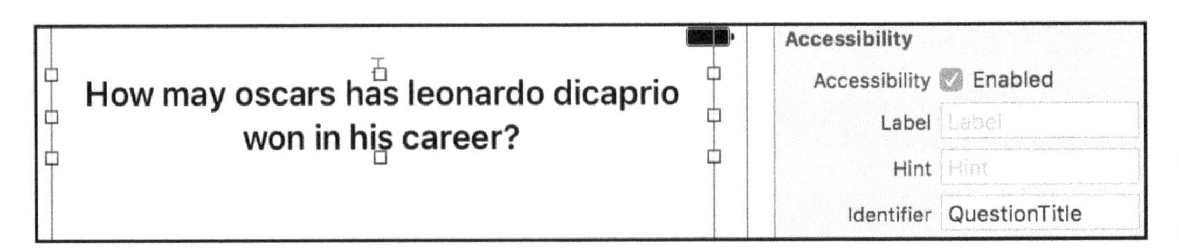

Remove the existing UI test, called `testExample()`, from the `MovieTriviaUITests` class and add the one shown in the following code snippet:

```
func testQuestionAppears() {
  let app = XCUIApplication()

  // 1
  let buttonIdentifiers = ["AnswerA", "AnswerB", "AnswerC"]
  for identifier in buttonIdentifiers {
    let button = app.buttons.matching(identifier: identifier).element

    // 2
    let predicate = NSPredicate(format: "exists == true")
    _ = expectation(for: predicate, evaluatedWith: button, handler:
nil)
  }

  let questionTitle = app.staticTexts.matching(identifier:
"QuestionTitle").element
  let predicate = NSPredicate(format: "exists == true")
  _ = expectation(for: predicate, evaluatedWith: questionTitle,
handler: nil)

  // 3
  waitForExpectations(timeout: 5, handler: nil)
}
```

Each element is selected through its accessibility identifier. You can do this because the `XCUIApplication` instance we create provides easy access to the UI elements. Next, a predicate is created that is used to check whether each element exists and an expectation is created. This expectation will continuously evaluate whether the predicate is true and once it is, the predicate will be fulfilled automatically. Lastly, the UI test will wait for all expectations to be fulfilled.

To make sure the questions are loaded correctly, you should load the JSON file as you did before. Add the following property to the test so you have a place to store the trivia questions:

```
typealias JSON = [String: Any]
var questions: [JSON]?
```

Next, add the following code to the `setUp()` method right after calling `super.setUp()` and before launching the app:

```
guard let filename = Bundle(for:
MovieTriviaUITests.self).path(forResource: "TriviaQuestions", ofType:
"json"),
  let triviaString = try? String(contentsOfFile: filename),
  let triviaData = triviaString.data(using: .utf8),
  let jsonObject = try? JSONSerialization.jsonObject(with: triviaData,
options: []),
  let triviaJSON = jsonObject as? JSON,
  let jsonQuestions = triviaJSON["questions"] as? [JSON]
  else { return }
```

This code should look familiar to you because it's similar to the code you already used to load JSON. To make sure that the correct question is displayed, update the test method as shown here:

```
func testQuestionAppears() {
  // existing implementation...

  waitForExpectations(timeout: 5, handler: nil)

  guard let question = questions?.first
    else { fatalError("Can't continue testing without question
data...") }

  validateQuestionIsDisplayed(question)
}
```

The preceding code calls `validateQuestionIsDisplayed(_:)`, but this method is not implemented yet. Add the following implementation:

```
func validateQuestionIsDisplayed(_ question: JSON) {
  let app = XCUIApplication()
  let questionTitle = app.staticTexts.matching(identifier:
"QuestionTitle").element

  guard let title = question["title"] as? String,
    let answerA = question["answer_a"] as? String,
    let answerB = question["answer_b"] as? String,
    let answerC = question["answer_c"] as? String
    else { fatalError("Can't continue testing without question
data...") }

  XCTAssert(questionTitle.label == title, "Expected question title to
match json data")
```

```
  let buttonA = app.buttons.matching(identifier: "AnswerA").element
  XCTAssert(buttonA.label == answerA, "Expected AnswerA title to match
json data")

  let buttonB = app.buttons.matching(identifier: "AnswerB").element
  XCTAssert(buttonB.label == answerB, "Expected AnswerB title to match
json data")

  let buttonC = app.buttons.matching(identifier: "AnswerC").element
  XCTAssert(buttonC.label == answerC, "Expected AnswerC title to match
json data")
}
```

This code is run after checking that the UI elements exist because it's run after waiting for the expectations we created. The first question is extracted from the JSON data, and all of the relevant labels are then compared to the question data using a reusable method that validates whether a specific question is currently shown.

The second test you should add is intended to check whether the game UI responds as expected. After loading a question, the test will tap on the wrong answers and then makes sure the UI doesn't show the button to go to the next question. Then, the correct answer will be selected, and the test will attempt to navigate to the next question. And of course, the test will then validate that the next question is shown:

```
func testAnswerValidation() {
  let app = XCUIApplication()

  let button = app.buttons.matching(identifier: "AnswerA").element
  let predicate = NSPredicate(format: "exists == true")
  _ = expectation(for: predicate, evaluatedWith: button, handler: nil)
  waitForExpectations(timeout: 5, handler: nil)

  let nextQuestionButton = app.buttons.matching(identifier:
"NextQuestion").element

  guard let question = questions?.first,
    let correctAnswer = question["correct_answer"] as? Int
    else { fatalError("Can't continue testing without question
data...") }

  let buttonIdentifiers = ["AnswerA", "AnswerB", "AnswerC"]
  for (i, identifier) in buttonIdentifiers.enumerated() {
    guard i != correctAnswer
      else { continue }

    app.buttons.matching(identifier: identifier).element.tap()
```

```
    XCTAssert(nextQuestionButton.exists == false, "Next question
button should be hidden")
  }

  app.buttons.matching(identifier:
buttonIdentifiers[correctAnswer]).element.tap()
  XCTAssert(nextQuestionButton.exists == true, "Next question button
should be visible")

  nextQuestionButton.tap()

  guard let nextQuestion = questions?[1]
    else { fatalError("Can't continue testing without question
data...") }

  validateQuestionIsDisplayed(nextQuestion)
  XCTAssert(nextQuestionButton.exists == false, "Next question button
should be hidden")
}
```

The preceding code shows the entire test that validates that the UI responds appropriately to correct and incorrect answers. Tests such as these are quite verbose, but they save you a lot of manual testing.

When you test your UI like this, you can rest assured that your app will at least be somewhat accessible. The beauty in this is that both UI testing and accessibility can significantly improve your app quality and each actively aids the other.

Testing your UI is mostly a matter of looking for elements in the UI, checking their state or availability, and making assertions based on that. In the two tests you written for **MovieTrivia**, we've combined expectations and assertions to test both existing UI elements and elements that might not be on screen yet. Note that your UI tests will always attempt to wait for any animations to complete before the next command is executed. This will make sure that you don't have to write asyncronous expectations for any new UI that is added to the screen with an animation.

Summary

Congratulations! You've made it to the end of this lengthy, information-packed chapter. You should know enough about testing and accessibility right now to begin exploring testing in greater depth than we have in this chapter. No matter how small or big your app, writing automated tests will ensure that your app is of a high quality. More importantly, instead of assuming that something works because it worked before, your automated tests will guarantee that it works because your tests don't pass if you broke your code.

You also learned that writing testable code sometimes requires you to refactor large portions of code. More often than not, these refactoring sessions leave your code in a much better state than before. Code that is easy to test is often cleaner and more robust than code that is hard to test. Now that you know how to cover your app with tests, in the next chapter, we'll look at how you can measure your app's performance using some of the great tools that Xcode provides.

Questions

1. What is a unit test?

   a) A test that checks whether an object can cooperate with other objects.
   b) A test that validates the network connection.
   c) A test that checks an object in isolation.

2. Is it okay to force your tests to execute in a certain order?

   a) Yes.
   b) No.

3. What is the order of the three As?

   a) Act, Assert, Arrange.
   b) Assert, Arrange, Act.
   c) Arrange, Act, Assert.

4. How do protocols help you when writing tests?

    a) You can mock certain objects by conforming them to a protocol.
    b) You can check for traits.
    c) Your code is cleaner with protocols.

5. How can see which code is not tested in your app?

    a) Using Code Coverage.
    b) By checking your unit test classes.
    c) By manually testing your app.

6. What is a launch argument?

    a) A condition under which the app will launch.
    b) A string that is passed to an app when it launches.
    c) A reason to launch an app.

7. How does `XCUITest` identify elements in an app?

    a) It learns where elements are when you record a test.
    b) With launch arguments.
    c) Through accessibility identifiers.

Further reading

- Test-Driven iOS Development with Swift 4 by Dr. Dominik Hauser

24 Discovering Bottlenecks with Instruments

To properly debug and improve your apps, you need to understand what tools are available to you. One of the tools Apple ships as part of Xcode is called Instruments. **Instruments** contains a collection of measurement tools that help you to profile and analyze your app to debug and detect complex problems or performance bottlenecks. For example, Instruments can help you figure out whether your app is suffering from memory leaks. Tracking a memory leak without the right tools is tedious and nearly impossible. A tool such as Instruments can help you track down several possible causes for a memory leak, which can save you a lot of time.

In this chapter, we're going to look at an app named **Mosaic**. This app is a straightforward app that was built to demonstrate how you can profile your app. The app is very slow; the longer it's used and the more the user interacts with the app, the slower the app becomes. You will learn how you can use Instruments to find the problems in **Mosaic** so you can eventually fix them, so the app runs smoothly.

To wrap this chapter up, you'll learn about a feature that is brand new in Xcode 10: Custom Instruments. With **Custom Instruments**, you can add your own measurements and visualizations to Instruments. This enables you to build tailor-made profiling tools that are perfect for your own app or framework.

This chapter is divided into the following segments:

- Exploring the Instruments suite
- Discovering slow code
- Closing memory leaks
- Building your own Instruments

By the end of this chapter, you'll be able to analyze your apps to find potential issues and solve them before they become a problem.

Exploring the Instruments suite

In this book's code bundle, you'll find a project named **Mosaic**. The app is still in the early stages of development, so there is plenty of work that needs to be done to improve it. More specifically, the app doesn't seem to work very well.

Even though the app isn't finished and the code isn't perfect, the app *should* be working fine. None of the features appears to be very performance-heavy, so the app works fine. If you launch the app and click around, you'll notice that the app isn't working fine. It's really slow! Let's see how you can track down the exact cause of the slowness.

If you dig through the source code, you'll immediately see that the app has three main screens:

- A table view
- A collection view
- A detail view

The table view displays a list of 50 items, each of which links to an infinitely-scrolling collection view with a custom layout. The layout for the collection view is custom because it needed to have exactly one pixel of spacing between items. The detail view only shows an image, nothing too crazy.

Based on knowledge from earlier chapters, you should know that a simple app such as **Mosaic** should be able to run smoothly on virtually any iOS device. Collection views and table views are optimized to display huge datasets, so an endlessly-scrolling collection view that keeps adding 50 new rows when you almost reach the end of the list should not be too much of a problem. In other words, this app should be able to run at 60 frames per second with no trouble at all.

However, if you start scrolling through the collection view and move back and forth between screens, and then scroll more and more, you'll find that some collections will stop scrolling smoothly. They will randomly freeze for a little while and then continue scrolling. And the more you scroll through the collections, the worse this issue seems to become.

If you take a look at the memory usage in Xcode's Debug navigator (the sixth icon in the left sidebar), you can tell that something is wrong because memory usage keeps going up while navigating through screens. This in itself is nothing to worry about, temporary memory spikes are normal. When the used memory never gets freed even though you might have expected it to, that's when you should start worrying. Take a look at the following graph:

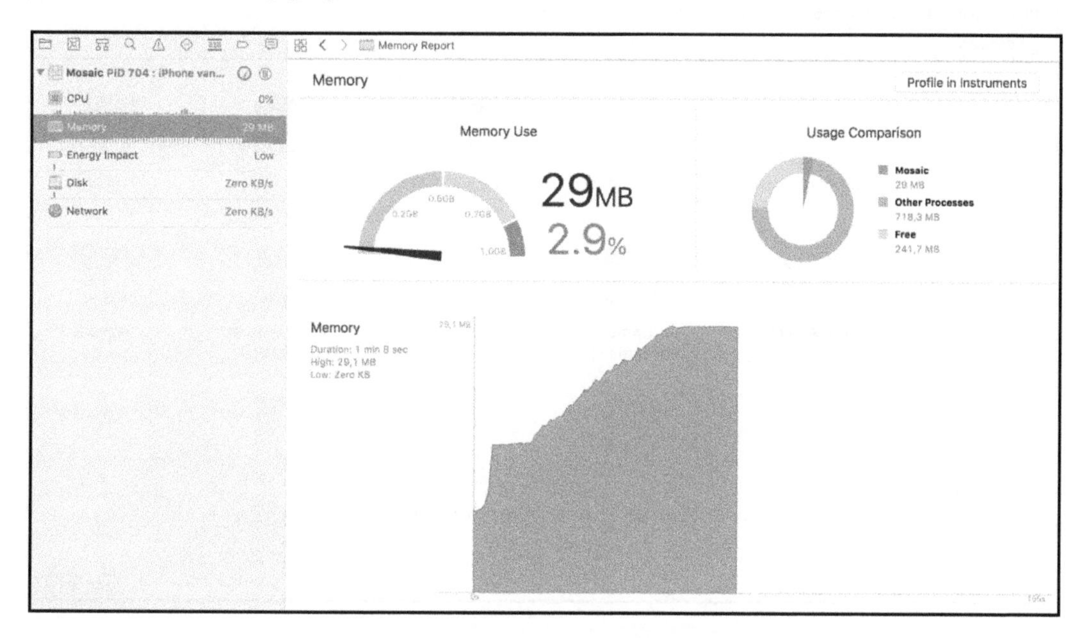

Whenever you debug performance issues or if you're trying to measure performance, make sure to measure on a real device. If you measure using the simulator, your app might run fine even though it might not run fine on a device. This is due to the way the simulator works. It mimics an iOS device, but it's backed up by all the power your Mac has to offer. So, the processor and available memory are those of your Mac, resulting in performance that's often much better than it would be on a device.

For now, make a mental note of the app's rising memory usage. Let's have a look at the Instruments app. To profile your app with Instruments, you can either select **Product** | **Profile** from the Xcode's toolbar or hit *Cmd* + *I* on your keyboard. Xcode will then build your project and launch Instruments so you can select a profiling template:

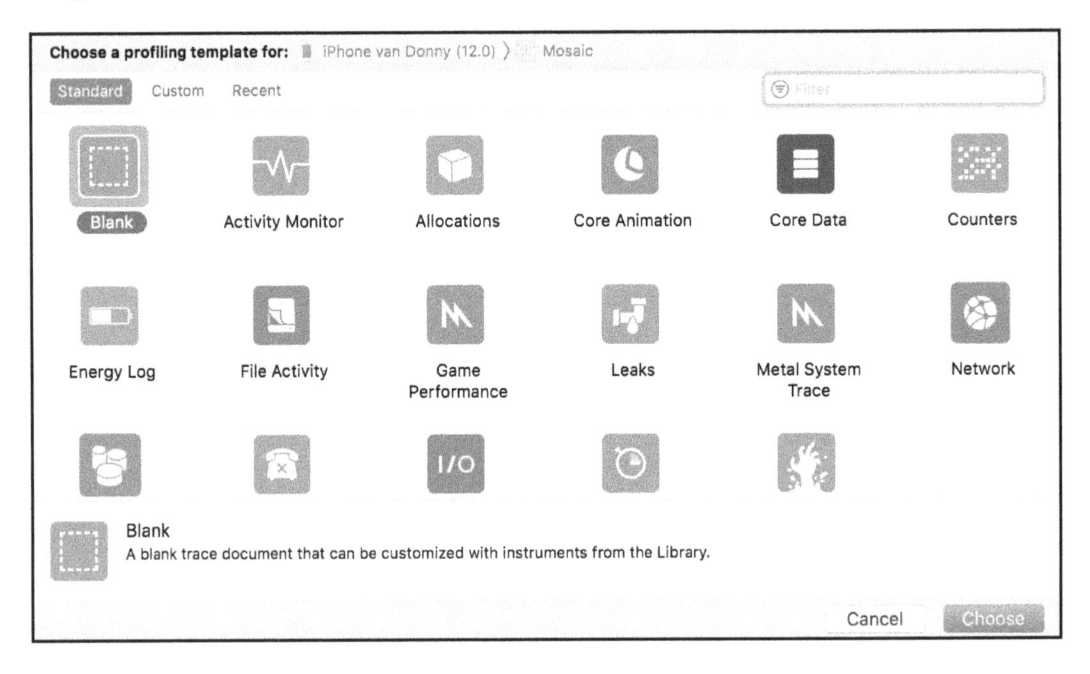

There are many available templates, and most of them are created for a specific case. For instance, the **Network** template will measure network usage. There are also templates available to profile animations, database operations, layout, Metal graphics performance, and more.

If you select a blank template, you're presented with the Instruments interface. The interface contains a record button that is used to start or stop profiling for an app. If you look further to the right side of the top toolbar, there's a plus icon. You can use this icon to add profiling tools for specific metrics yourself. You could use this to create a profiling suite that fits your needs without having to switch between the default templates all of the time. The center of the Instruments interface is the timeline. This is where graphs are drawn, representing the measurements that were made for your app. The bottom section of the screen shows detail about the currently selected metric, as you're about to find out:

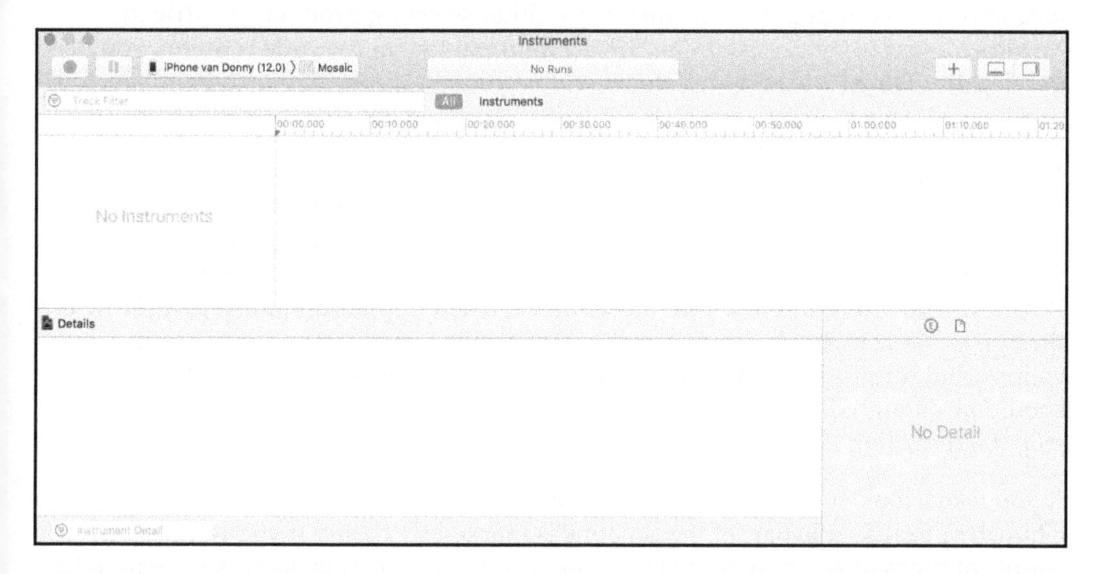

Now that you know a little bit about the interface of the Instruments app, let's make an attempt at profiling the **Mosaic** app. The biggest issue in the app is the choppy scrolling when you scroll down the collection view for several pages. It's clear that the app has a memory issue, as you saw in Xcode, but choppy scrolling occurs even if the app's memory usage is low. This should be a flag that a memory leak probably isn't the problem that is causing the collection views to scroll badly. Something in the code is probably slow, and Instruments can help to figure out which parts of the app's code are taking a long time to complete.

Discovering slow code

Whenever you find that your app is slow or choppy, there's a good chance that something in your code is taking longer than it should, especially if your memory usage appears to be within reasonable range. For instance, if your app uses less than 50 MB, memory is not likely to be an issue for your app, so seeking the problem in your code's performance makes a lot of sense.

To discover where your code is lacking in the performance department, it's a great idea to profile your app in Instruments, by either selecting **Product** | **Profile** in Xcode's toolbar or by pressing *Cmd* + *I*. To figure out what the code is doing, you need to select the **Time Profiler** template when Instruments asks you which template you want to use. This template measures how long certain blocks of code take to run.

To record a profiling session of our app, make sure that a device is connected to your Mac and make sure that it's selected as the device that your app will run on by selecting your iOS device from the list of devices and simulators in the scheme toolbar menu in Xcode. Once you've selected your device, start profiling the app. When Instruments launches, pick the **Time Profiler** template and hit record. Now use the app to navigate to a collection and begin scrolling until the app starts acting choppy and scroll some more. After seeing the app stutter a couple of times, there should be enough data to begin filtering out what's going on. Press the stop button to stop the recording session.

If you take a look at the data Instruments has recorded, you'll notice a graph that has a bunch of peaks. This part of the timeline is marked CPU, and if you hit *Cmd* + a couple of times to zoom in on this timeline, you'll notice that these spikes seem to last longer and longer as the scrolling becomes choppier and choppier. This is a great clue to investigate a bit more. Something must be going on that makes these peaks last longer every time they occur.

In the bottom section of the Instruments window, you'll find an overview of the code in the app that was executed while the recording session was active. The code is separated by thread and since you saw the user interface lagging, you can be pretty sure that something on the main thread is slow. If you drill down into the list of method calls by a couple of levels, you won't find much useful information. It doesn't even look like most of the executed code is part of the app's code, as you can see in the following screenshot:

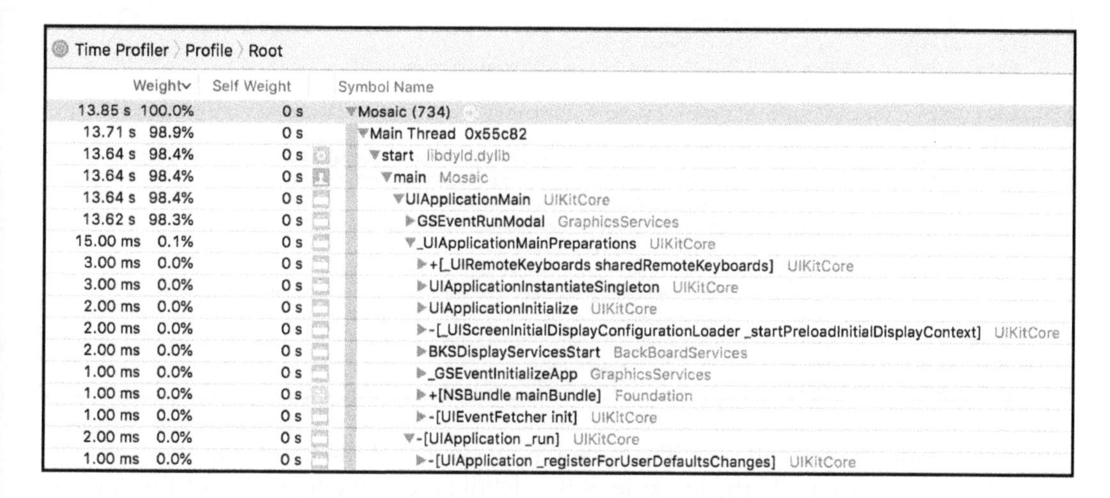

To make sure the code you're interested in is pulled to the surface, you should apply a couple of filtering options. If you click on the **Call Tree** button in the bottom of the window, you can select **Invert Call Tree** and **Hide System Libraries**:

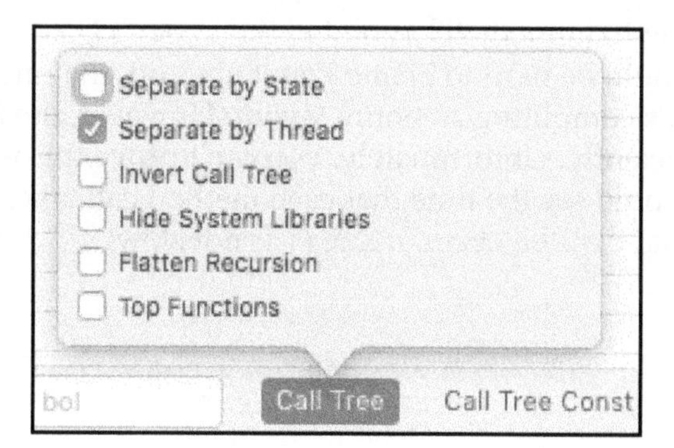

The **Invert Call Tree** option inverts the tree of method calls. The default tree starts at `main` and works its way down to the method calls in your app code. When you invert this tree, your app code will be at the top and works its way down to `main`. By checking **Hide System Libraries**, you make sure that all code shown in the detail view belongs to your app. The call tree is a lot more readable after applying this filter:

Time Profiler 〉 Profile 〉 Root

Weight		Self Weight	Symbol Name
13.77 s	99.3%	0 s	▼Mosaic (734)
13.64 s	98.4%	0 s	▼Main Thread 0x55c82
5.07 s	36.6%	5.07 s	▶ListCollectionViewLayout.createSizeLookup() Mosaic
3.01 s	21.7%	0 s	main Mosaic
2.80 s	20.2%	2.80 s	▶specialized _VariantDictionaryBuffer.nativeUpdateValue(_:forKey:) Mosaic
1.24 s	8.9%	1.24 s	▶protocol witness for Hashable.hash(into:) in conformance CGFloat Mosaic
370.00 ms	2.6%	370.00 ms	▶specialized _VariantDictionaryBuffer.ensureUniqueNativeBuffer(withBucketCount:) Mosaic
258.00 ms	1.8%	0 s	▶ListCollectionViewController.collectionView(_:cellForItemAt:) Mosaic
141.00 ms	1.0%	0 s	▶@objc ListCollectionViewCell.init(coder:) Mosaic
123.00 ms	0.8%	0 s	▶@objc ListCollectionViewController.scrollViewDidScroll(_:) Mosaic
87.00 ms	0.6%	87.00 ms	▶DYLD-STUB$$swift_release Mosaic
85.00 ms	0.6%	85.00 ms	▶ListCollectionViewLayout.prepare() Mosaic
50.00 ms	0.3%	0 s	▶@nonobjc UICollectionViewLayoutAttributes.__allocating_init(forCellWith:) Mosaic
40.00 ms	0.2%	40.00 ms	▶outlined consume of _VariantDictionaryBuffer<A, B><A, B> Mosaic
40.00 ms	0.2%	40.00 ms	▶specialized Dictionary.subscript.setter Mosaic
39.00 ms	0.2%	39.00 ms	▶DYLD-STUB$$swift_retain_n Mosaic
38.00 ms	0.2%	38.00 ms	▶DYLD-STUB$$Hasher.init(_seed:) Mosaic

When you examine the call tree, it's immediately clear that the issue is in the collection view layout. On the left side of the detail area, the percentage of time spent in a certain method is outlined, and you can see that a considerable portion of time is spent in `ListCollectionViewLayout.createSizeLookup()`. This method appears to be much slower than it should be.

It's probably tempting to immediately point at the `createSizeLookup()` method and call it slow. It would be hard to blame you if you did, that method is a pretty rough one because it's computing an entire layout. However, the layout should be computed pretty efficiently. Unfortunately, you can't prove this with the data in Instruments. If you could see the time that each method call takes, you would be able to verify the claim that `createSizeLookup()` is not slow.

What can be proved, though, is that `createSizeLookup()` is called way more often than it should be. If you add a `print` statement at the start of this method, you'll see that there are hundreds of prints as you scroll through the list. If you dig deeper and figure out when `createSizeLookup()` is called exactly, you'll find two places:

- In `prepare()`
- In `rectForThumbAtIndex(_:)`

This is strange because `rectForThumbAtIndex(_:)` is called in a loop inside of the `prepare` method after the size lookup has already been generated.

More importantly, as the number of items in the collection grows, the number of items to loop over grows too. It's very likely that this extraneous call to `createSizeLookup()` is the source of the bug. You should be able to remove the call to `createSizeLookup()` from `rectForThumbAtIndex(_:)` without problems because the lookup is already created before the items are iterated, so it's safe to assume that the lookup exists when `rectForThumbsAtIndex(_:)` is called. Remove the call to `createSizeLookup()` and run your app on the device. Make sure to do plenty of scrolling to verify that the bug was fixed.

To make sure that everything is fixed, you should use Instruments again to see what the Time Profiler report looks like now. Start a new Time Profiler session in Instruments and repeat the steps you performed before. Navigate to a collection and scroll down for a while so the app loads lots of new pages.

Looking at the result of the measurements now, the percentage of time spent in `createSizeLookup()` has dramatically decreased. It has decreased so much that it's not even visible as one of the heaviest methods anymore. The app performs way better now and you have acquired the measurements to prove it.

Now that the scrolling performance issue is solved, you can navigate the app, and everything is working smoothly. But if you take a look at the memory usage in Xcode, the memory usage still goes up every time you navigate to a new item in the table view. This means that the app probably has another problem in the form of a memory leak. Let's see how to discover and fix this leak with Instruments and Xcode.

Closing memory leaks

Usually, if you navigate around in your app, it's normal to see memory usage spike a little. More view controllers on a navigation controller's stack mean that more memory will be consumed by your app. This makes sense. When you navigate back, popping the current view controller off the navigation controller's stack, you would expect the view controller to be deallocated and the memory to be freed up.

The preceding scenario describes how **Mosaic** should work. It's OK if the app uses some more memory if you're deeper in the navigation stack, but this memory should be freed up after the back button is tapped.

In the **Mosaic** app, the memory keeps growing every time you navigate to a new collection view. It doesn't matter if you drill deep into the navigation stack, hit back, or scroll a lot, once memory is allocated it never seems to be deallocated. This is a problem, and you can use Instruments to dig into the app to find out what's wrong. Before you do this, though, let's have a deeper look at memory leaks, how they occur, and what the common causes are.

Understanding what a memory leak is

When your app contains a memory leak, this means that it's using more memory than it should. More specifically, the app fails to release memory that is no longer needed. A couple of scenarios exist where this could occur. Once you're aware of these scenarios, they should be quite easy to spot. If you haven't seen them before or have never even heard of them, it's easy to fall into the trap of having a memory leak in your app.

Preventing objects from using infinite memory

The first type of memory leak is one where an object is allowed to take up an endless amount of memory without any restrictions. A typical example of this is a cache. When you implement a cache that holds on to certain model objects, API responses, or other data that was expensive to obtain in the first place, it's easy to overlook the fact that you have just built yourself a memory leak.

If your user is using your app and your cache object keeps caching more and more data, the device will eventually run out of memory. This is a problem because if the app doesn't free the memory in time, it will be terminated by iOS to make sure that essential processes and other apps don't suffer because of your app's out-of-control memory usage.

Luckily, it's easy to solve issues such as these. The operating system will notify your app through `NotificationCenter` any time it needs you to free up memory. Listening to this notification and purging any cached data you can recreate or reload will prevent your app from hogging memory and, ultimately, it prevents your app from being terminated due to memory reasons.

Here is a straightforward example of an image cache class that purges its cache when memory is tight:

```
class ImageCache: NSObject {
  var cache = [UIImage]()

  override init() {
    super.init()

    NotificationCenter.default.addObserver(self, selector:
#selector(purgeCache), name:
UIApplication.didReceiveMemoryWarningNotification, object: nil)
  }

  deinit {
    NotificationCenter.default.removeObserver(self, name:
UIApplication.didReceiveMemoryWarningNotification, object: nil)
  }

  @objc func purgeCache() {
    cache.removeAll()
  }
}
```

All you need to do is listen for the `UIApplication.didReceiveMemoryWarningNotification` notification and purge any data that you can recreate when needed. Even though it's technically not required, this sample unsubscribes the cache from the memory warning notifications when it's deallocated. Since iOS 9, the OS itself should take care of this but this behavior has proven to be somewhat unpredictable so unsubscribing explicitly avoids potential reference cycles, which is the next type of memory leak to explore.

Avoiding reference cycles

When an object contains references to other objects, you should always be careful to avoid situations where both objects continuously hold a reference to each other. For example, a relationship between a table view and its delegate or data source could become a reference cycle if the relationship isn't managed correctly. Objects can only be deallocated if no objects are referencing them anymore:

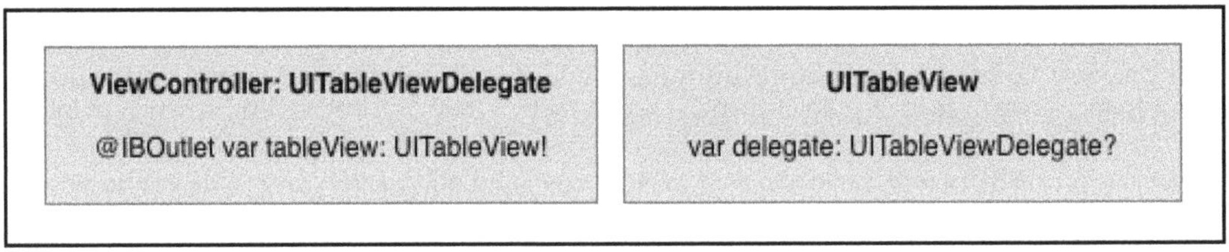

ViewController: UITableViewDelegate	UITableView
@IBOutlet var tableView: UITableView!	var delegate: UITableViewDelegate?

The preceding figure illustrates this. The view controller holds onto `tableView` and `tableView` holds onto its delegate, which is the view controller. This means that neither object can ever be deallocated because for both the view controller and `tableView`, there is always at least one object referencing each at any given time. Of course, Apple has made sure that this doesn't occur in your apps by making sure that `tableView` does not hold onto its delegate forever. You'll see how in just a second.

Another situation where a reference cycle could be created is in a closure. When you implicitly reference `self` in a closure, the compiler complains that you must explicitly refer to `self`. Doing this creates a reference to `self` inside of the closure, potentially resulting in a reference cycle. Throughout this book, you've seen a bunch of closures and they always used a capture list `self` was used inside of the closure:

```
api.fetchData { [weak self] in
  self?.tableView.reloadData()
}
```

The preceding example shows an example of using a capture list. The capture list is the part right before the `in` keyword. The list captures a `weak` reference to `self`, which means that no reference cycle is created between the closure and `self`. If the `api` object stores the closure in a variable and you didn't use a weak reference to `self`, you might have a reference cycle. If the `api` object itself is held onto by another object, you can be pretty sure that a reference cycle is created.

Making the reference `weak` tells the app that the reference to `self` does not add up to the reference count of `self`. This means that if there are only weak references left to an object, it's OK to deallocate it and free the memory. Memory management and reference counts aren't simple. One way to think about this subject is that your app has an internal count of the number of objects that point to another object. For instance, if you create an instance of `UIView` inside of `UIViewController`, the reference count for `UIView` is one. When `UIViewController` is deallocated, the reference count for `UIView` becomes zero because the view controller doesn't use it anymore, meaning that it can be deallocated safely.

If `UIView` has a reference to `UIViewController` as well, both objects will keep each other around because the reference count for each instance won't ever reach zero. This is called a reference cycle. This cycle can be broken by marking at least one of the references as a `weak` reference. Since `weak` references don't contribute to the reference count, they prevent reference cycles from happening. This is how Apple has made sure that `tableView` does not create a reference cycle with its delegate or data source; the references to these objects are marked as `weak`.

As an alternative to making a reference `weak`, you can mark it as `unowned`. While weak is essentially a safe optional value, unowned makes the object implicitly unwrapped. It's often best to take the safe route and mark a captured reference as weak because your app won't crash if the `weak` referenced instance has been deallocated somehow, while it would crash if the reference is `unowned`.

Reference cycles aren't easy to understand, especially if you take into account weak references and reference-counting. It's really easy to get confused about this topic. Luckily, the **Mosaic** app contains a couple of issues with references and retain cycles so you can immediately put your newly acquired knowledge to the test.

Discovering memory leaks

To figure out why the memory usage of **Mosaic** increases every time a new screen is loaded in the app, you should profile the app using the Allocations Instruments template. When you've started a new Allocations profiling session, navigate through the app and you should see the memory usage graph rise consistently. This behavior is typical for a memory leak, so it's time to dig in deeper to figure out what exactly is going on that causes this to happen. Take a look at the following screenshot:

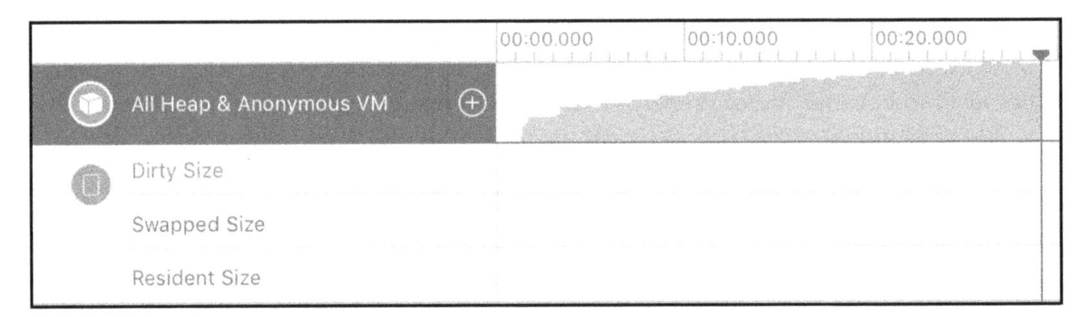

If you look at the detail area in Instruments, there is a lot of information there that does not make a lot of sense. A lot of the objects that are created are not objects that you explicitly wanted to create, which makes it hard to figure out what's causing the memory problem. It's not very likely to be an issue with UIKit, for instance, and even if it is, you can't quite fix that. Luckily, you can use the search dialog in the top-right corner of the detail area to look for objects that do belong to the application code, giving you insight into what's happening with the objects that get created by the application code.

If you type the word **Mosaic** in the search box, you'll find a couple of objects, and they should all look familiar to you. In the detail area, you can now see the number of instances of a particular object that is in memory. Refer to the following screenshot:

Allocations > Statistics > Allocation Summary

Graph	Category	Persistent B...	# Persistent	# Transient	Total Bytes	# Total
	Mosaic.ListCollectionViewCell	134,53 KiB	210	0	134,53 KiB	210
	Mosaic.ListCollectionViewController	12,69 KiB	14	0	12,69 KiB	14
	Mosaic.ListCollectionViewLayout	4,59 KiB	14	0	4,59 KiB	14
	Mosaic.ListViewController	880 Bytes	1	0	880 Bytes	1
	Mosaic.AppDelegate	32 Bytes	1	0	32 Bytes	1

If you look closely, you'll find that there are way more collection view cells, collection view controllers, and collection view layouts present than you'd expect. The app can only show one collection view at a time, and when you navigate back to the list view controller, you would assume the collection view controller to be deallocated since nothing should be referencing it anymore.

When you segue to one of the collection view controllers, it's added to the `viewControllers` array on the navigation controller. This means that the navigation controller keeps a reference to each of the view controllers that are part of the navigation stack. This means that none of the view controllers in that list should be deallocated because the reference count is at least **1**.

When you pop back to the list view controller, the navigation controller removes the collection view controller from its `viewControllers` array. The result of this is that the reference count for the collection view controller is now decremented by one since the navigation controller is not referencing it anymore. This should put the reference count at 0, meaning that the collection view controller can be deallocated and the memory should be freed up.

However, something is preventing this from happening, because the collection view controller remains allocated according to the observations in Instruments. Unfortunately, Instruments does not tell you much more than the information just saw. Objects are sticking around for too long, which means that a reference cycle exists somewhere. This probably means that something is referencing the collection view controller and in turn, the collection view controller is referencing something else.

To figure out what's going on, you should probably start looking for problems in the collection view controller. The collection view controller has a delegate relationship with an object conforming to `ListCollectionDelegate`. It also acts as a delegate for the list collection view cells it displays, and it's also a delegate for the detail view. All the delegate relationships are references to other objects. These delegates could very well be causing the reference cycle. You can use Xcode to visualize all the objects in memory and see how they relate to each other. This means that it's possible to capture the state of the app's memory after browsing to several collection views, and you can see which objects are holding references to other objects. This enables you to visually identify reference cycles instead of blindly guessing.

To visualize your app's memory usage, build and run the app in Xcode and navigate to a couple of collection views. Then open the memory view in the Debug navigator in Xcode and click the **Debug Memory Graph** button in the bottom toolbar of the screen:

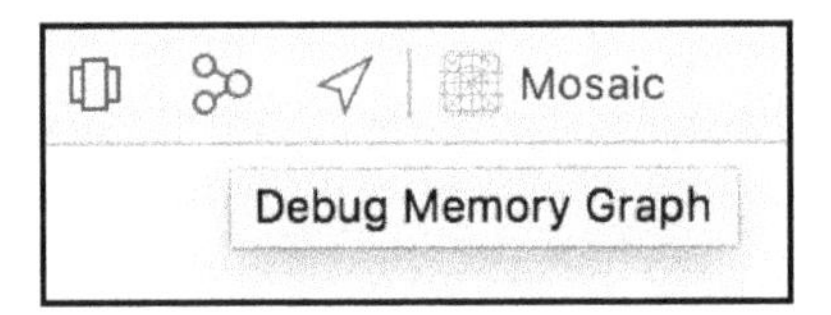

After clicking this button, Xcode will show you a visual representation of all the memory that your app is using. In the toolbar on the left side of the screen, look for `ListCollectionViewCell` and click it. Xcode will show you the relationship between `ListCollectionViewCell` and other objects. At first glance, nothing unusual is happening. There is a list view controller, which holds a reference to a collection view, then there are a couple of other objects, and finally there's the collection view cell:

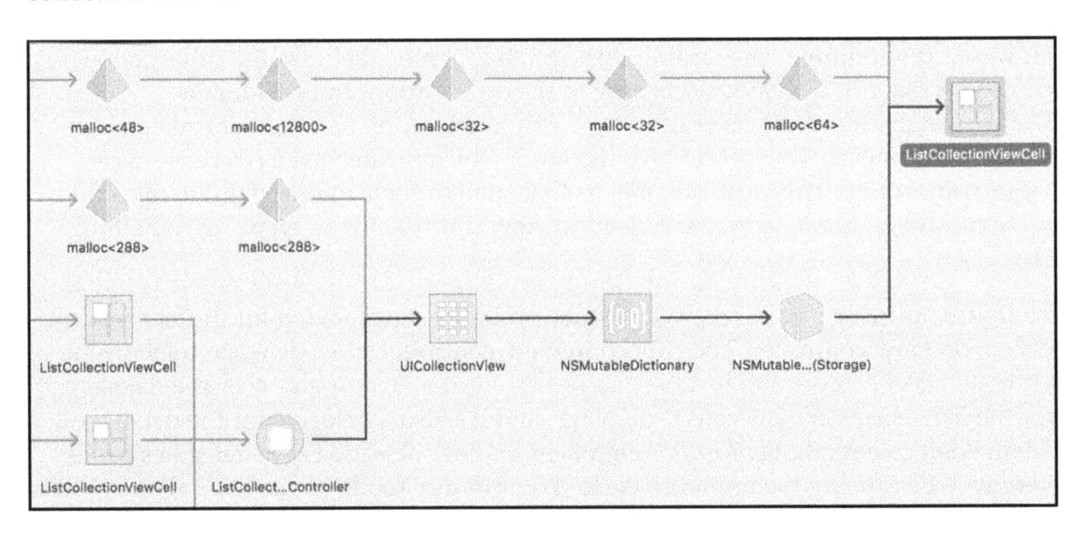

Next, click the collection view controller. When you click it, two different views could be shown to you. One of these two shows a couple of collection view cells pointing to the collection view controller and there should be a navigation controller in the graph as well:

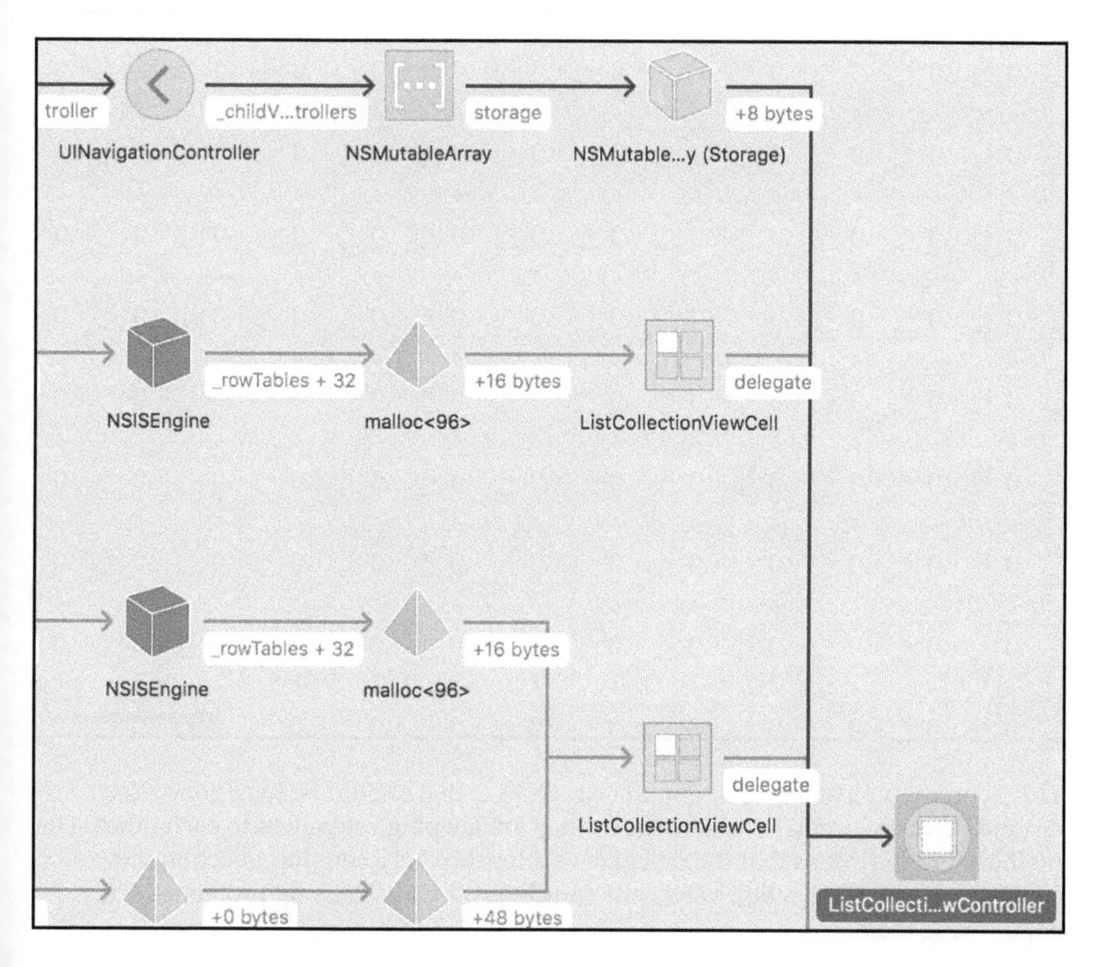

The other situation you might see is slightly different. It also has all the cells pointing to the collection view controller, but the navigation controller is nowhere to be found, as shown in the following screenshot:

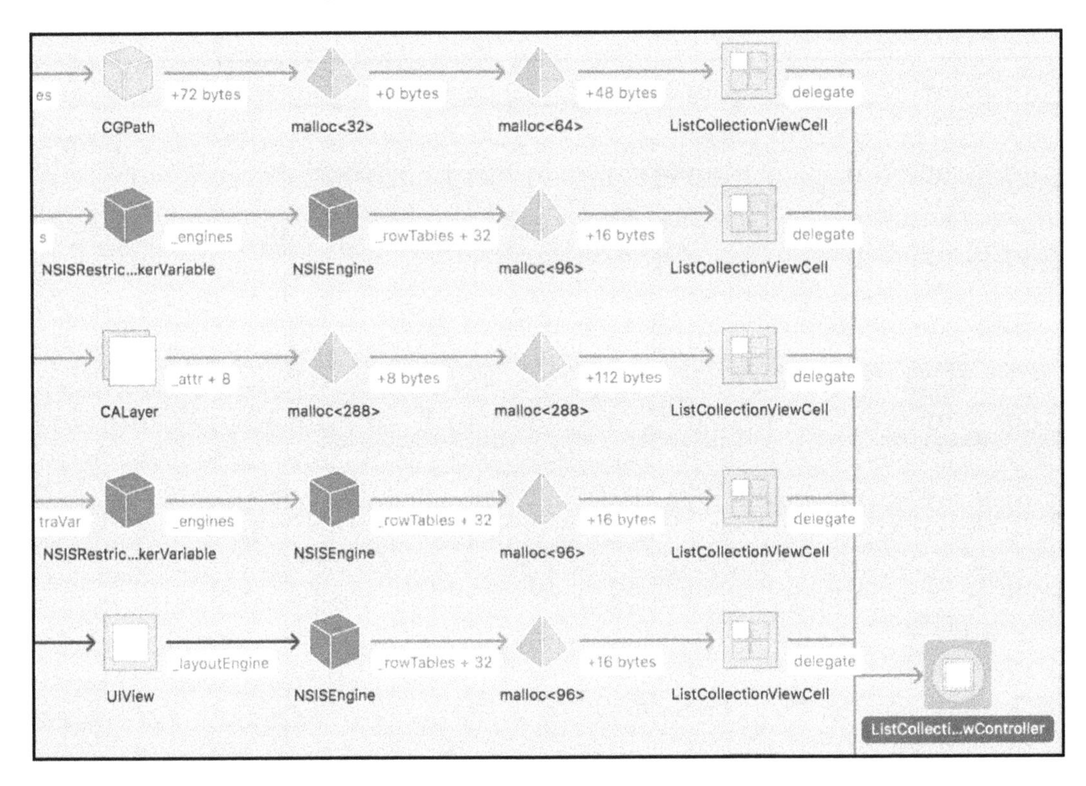

This is pretty typical for a reference cycle. Objects that should be long gone stick around because somewhere, somehow, they are keeping references to each other. The memory graph reveals that the delegate relationship between the collection view controller and the collection view cells could very well be causing problems, you can begin fixing them.

In this case, the fix is pretty simple. You need to make the delegate relationship between the cell and the controller `weak`. Do this by opening `ListCollectionViewCell.swift` and changing the delegate declaration as shown:

```
weak var delegate: CollectionItemDelegate?
```

Marking the delegate as weak breaks the reference cycle because the references aren't strong anymore. This essentially tells our app that it's OK if the referenced object is deallocated. One final adjustment you must make is to constrain `CollectionItemDelegate` to class instances only. Due to the nature of value types, such as structs, you can't mark references to them as `weak`, and since protocols can be adopted by both value and reference types, the protocol must be constrained to class instances only. Update the declaration for the `CollectionItemDelegate` protocol as shown:

```
protocol CollectionItemDelegate: class
```

If you run the app now, you can safely navigate around and both the memory debugger and Instruments will prove that you no longer have any reference cycles in your app. It's possible that Instruments still tells you that your memory usage is growing. Don't worry about this too much; you did your job making sure that your code is safe and that all memory that you don't use anymore can be freed up.

The example of a reference cycle you just saw is a very common one. A common way to avoid reference cycles in delegates is to try to make all of your delegates weak by default. This prevents you from making mistakes that you might not catch right away and it's often the safest route to go. Do note that there are many ways in which a reference cycle can occur, and you should always use the appropriate tools to troubleshoot your memory issues.

Let's wrap up this exploration of Instruments by creating an Instrument of your own.

Creating your own Instrument

The Instruments app comes with some great built-in tools that can be used to profile applications. While this is great, sometimes you might need something more tailor-made. Since Xcode 10, you can create your own Instruments Packages to profile your apps. Custom Instruments can use special logging that you add in your app through the `os_signpost` APIs to keep track of certain processes in your app while collecting interesting data about it. In this section, you will learn how you can add signpost logging to your app as a lightweight mechanism to measure performance in your app. Then, you'll learn how to take this data and funnel it through your own custom Instruments Package.

Adding signpost logging to your app

As a developer, you want to know exactly what your app is doing and how long your app spends on certain parts of your code. It's common to use breakpoints or print statements to do some basic debugging or logging in your app, but keeping track of many print statements in the console can be quite tedious. In iOS 12, you can use signpost logging to flag the beginning and end of a certain task and you can visualize these logs in Instruments using the built-in **os_signpost instrument**. The ossignpost instrument logs all signposts you track and shows them on a timeline. The following image shows an example of what signpost logging in Instruments looks like:

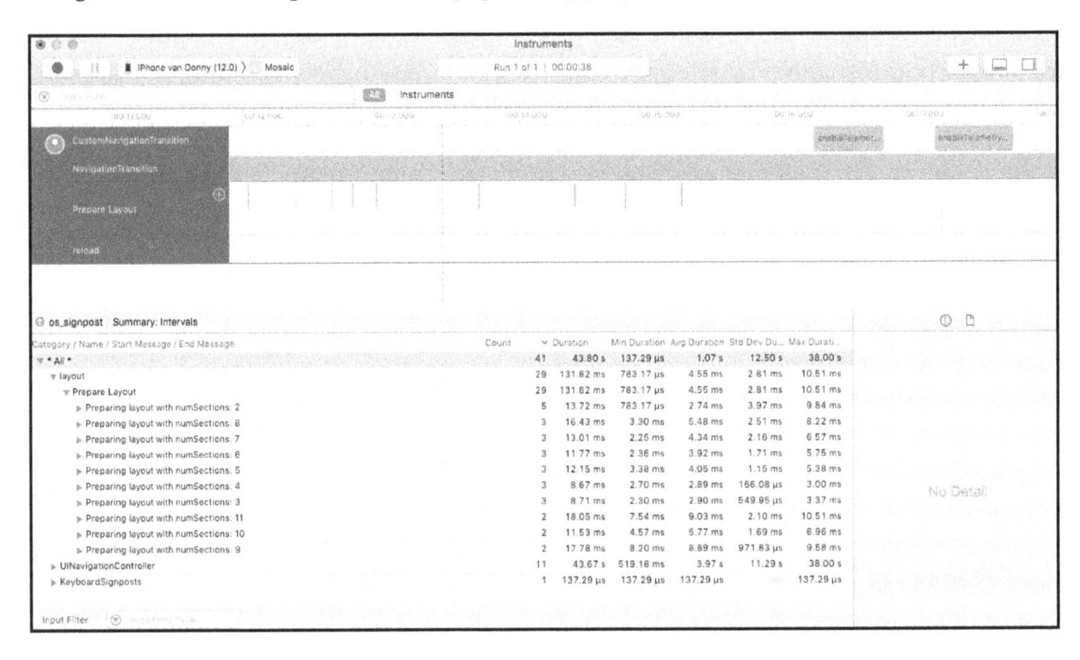

You can see a couple of timelines, one of them is labeled **Prepare Layout**. This timeline is a custom signpost that is sent from the **Mosaic** app. In the detail area, you can see more details about the signposts that were posted. The signposts that are posted by **Mosaic** correspond with the calls to `prepare()` on the custom collection view layout on the collection view page. Using signposts to track the performance of the `prepare()` method is more convenient in certain ways than using the Time Profiler; the Time Profiler provides a lot of detailed information that you might not always want to see right away. In this case, you would be more interested in making sure setting up the layout doesn't take too long. Additionally, you might want to make sure that `prepare()` performs well with many different amounts of sections.

Adding signposts to your app only takes a couple of minutes and they can give you fantastic new insights into what your app is doing. To add signposts to your app, you need an `OSLog` handle. An `OSLog` handle sounds very fancy, but it's just an instance of `OSLog` that is associated with your app. Go ahead and open `ListCollectionViewLayout.swift` in the **Mosaic** project and add the following code to this file:

```
import os.signpost

private let layoutLogger = OSLog(subsystem: "com.donnywals.layout",
category: "layout")
```

The preceding code imports the `os.signpost` framework and creates a log handle that will be used to track the signposts. When you create a signpost, you can flag the signpost as one of three types:

- Begin
- End
- Point of interest

The **begin** and **end** signposts are used to flag the start and end of an operation. You will add these signposts to the `prepare()` method to track the performance of preparing the collection view layout. A **point of interest** signpost is a special kind of signpost that is used to flag interesting moments that can occur in your app. For instance, when a user taps on something or when you reach a certain special moment inside of an operation. To log a signpost, you need to obtain `OSSignpostID`. This object is used to link two begin and end signposts together so it's essential that you use the same signpost ID. To create a signpost ID, you can use one of the following two methods:

```
let one = OSSignpostID(log: layoutLogger, object: self)
let two = OSSignpostID(log: layoutLogger)
```

The first method can only be used if the `object` parameter you pass to the `OSSignpostID` initializer is a class. This way of creating a signpost ID will always return the same for any given object. So if you use this method, every time you use this initializer inside of a particular class instance, the ID will be the same.

The second method generates a random, unique signpost ID every time. So if you're using signposts with a struct, make sure to store the signpost ID so you can reuse it if needed.

Once you have obtained your signpost ID, you can begin logging signpost events. Update the code for `prepare()` in the collection view layout as follows:

```
override func prepare() {
  super.prepare()

  guard let numSections = collectionView?.numberOfSections,
numSections != 0
    else { return }
  // 1
  let id = OSSignpostID(log: layoutLogger, object: self)
  // 2
  os_signpost(.begin, log: layoutLogger, name: "Prepare Layout",
signpostID: id, "Preparing layout with numSections: %{public}@",
"\(numSections)")

  // existing implementation

  // 3
  os_signpost(.end, log: layoutLogger, name: "Prepare Layout",
signpostID: id, "Done preparing layout")
}
```

The first comment signals the creation of a signpost ID. Since the collection view layout is a class, it can be passed to the signpost ID initializer. Once the signpost ID is created, the begin signpost is added. Every signpost is attached to a logger and has a name. The name should be the same for the begin and end signpost so the system understands that they both relate to the same event in your app. The last argument for the signpost begin and end calls is the description of the event. The description is a formatted string so you can add metadata to it through format specifiers. In this case, the `%{public}@` specifier is used to signal that a string will be added to the description. The end signpost does not have any variables associated with it.

If you want to log a point-of-interest signpost, you log it similarly to how the begin and end signposts are logged, except you need a special kind of logger. To create a point-of-interest logger, you need to use `OSLog.Category.pointsOfInterest` as the category for the `OSLog` handle and then associate your signpost with that logger rather than your normal logger.

To see your signposts in Instruments, you need to run your app for profiling like you have done before in this chapter. Instead of selecting one of the predefined templates, select a blank template when Instruments starts up. Use the + icon to find the **os_signpost** instrument:

After locating the **os_signpost** instrument, drag it to the timeline area. You can now begin profiling your app and your signposts should appear. While this is cool, the default view combines your signposts in the detail view. You can see that every time you log *Preparing layout with numSections: 2*, the count for this message is incremented and you can see the minimum, maximum, and average duration. Usually, this is plenty of information, but what if you want to see every individual signpost call in the detail view? You can achieve this by creating your own custom Instruments Package.

Building an Instruments Package

Together with signpost logging, Apple introduced the ability to define custom Instruments Package. An Instruments Package exists as an XML file that describes the type of data that your app wants to track and how to visualize it. Apple uses this approach internally as well, which means that your own Instruments Packages can build upon Apple's own instruments packages. The most interesting type of custom instrument you can create is a signpost instrument. This type of instrument takes the data you log with the `os_signpost` API and you can decide how Instruments should visualize this data.

To create an Instruments Package, add a new target to your app. Instead of selecting the type of target from the list of iOS targets, go to the macOS tab and find the Instruments Package template:

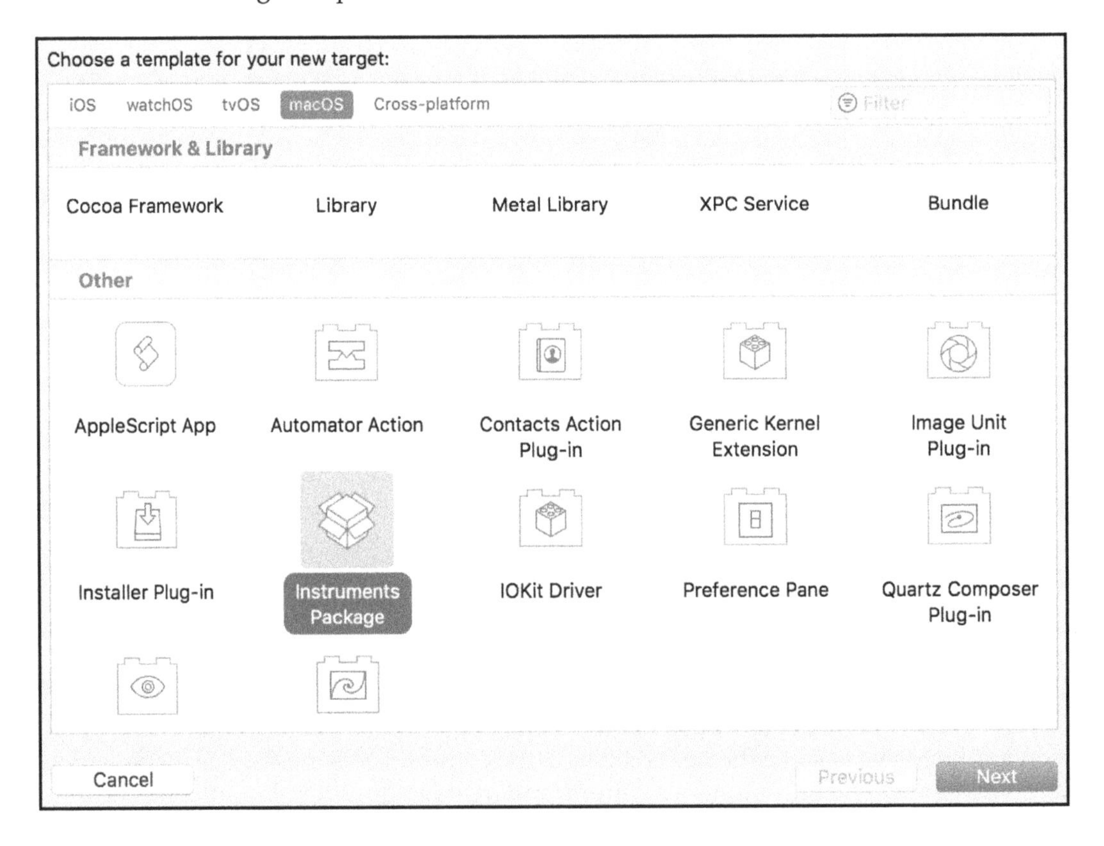

Name the instrument **MosaicLayoutInstrument** if you want to follow along with the steps to create your own instrument. Xcode will create a new folder in your project that has a single file in it. The extension for this file is `instrpkg`. If you open this file, you will find a little bit of XML and a lot if commented-out information to guide you in creating your own instrument. You can remove all of these comments if you want.

Every instrument has roughly three main sections in the XML:

- Metadata about the instrument
- A schema that defines the data that is tracked
- The instrument itself

Depending on your exact needs, you might have multiple schemas or instruments. To track signposts, you need to add an `<os-signpost-interval-schema>` tag to the XML file. If you type `os-signpost-interval-schema` in the XML and hit *Esc*, Xcode will usually offer the option to automatically add a skeleton implementation for the implementation of the tag. The schema for this instrument will only track a single column, the number of sections that were involved in the layout calculation. Add the following implementation for the signpost interval schema:

```
<os-signpost-interval-schema>
  <!-- 1 -->
  <id>signpost-schema</id>
  <title>Layout Signpost Schema</title>

  <!-- 2 -->
  <subsystem>"com.donnywals.layout"</subsystem>
  <category>"layout"</category>
  <name>"Prepare Layout"</name>

  <!-- 3 -->
  <start-pattern>
    <message>"Preparing layout with numSections: "?num-
sections</message>
  </start-pattern>

  <end-pattern>
    <message>"Done preparing layout"</message>
  </end-pattern>

  <!-- 4 -->
  <column>
    <mnemonic>num-sections</mnemonic>
    <title>Number of Sections</title>
    <type>string</type>
```

```
    <expression>?num-sections</expression>
  </column>
</os-signpost-interval-schema>
```

The preceding snippet is defined in four sections:

1. Some metadata about the schema. The ID will be used later to refer to the schema in the instrument.
2. Information about how the information is logged. These fields should match the corresponding values in your app.
3. Patterns for the begin and end messages as they were used in your app. You use the `?num-sections` notation to indicate the location of a variable and to extract it into a new variable that can be used in your schema.
4. The columns that your schema contains. In this case, it's only a single column that uses `num-sections` as its identifier and corresponds with the `?num-sections` variable extracted from the begin message.

After adding the schema, you are ready to create the instrument. Add the following implementation to the XML file, after the schema definition:

```
<instrument>
  <!-- 1 -->
  <id>com.donnywals.layout-instrument</id>
  <title>Layout Instrument</title>
  <category>Behavior</category>
  <purpose>Insight in layout preparation</purpose>
  <icon>Generic</icon>

  <!-- 2 -->
  <create-table>
    <id>layout-table</id>
    <schema-ref>signpost-schema</schema-ref>
  </create-table>

  <!-- 3 -->
  <graph>
    <title>Layout preparation</title>
    <lane>
      <title>Prepare</title>
      <table-ref>layout-table</table-ref>

      <plot-template>
        <instance-by>num-sections</instance-by>
        <value-from>duration</value-from>
      </plot-template>
    </lane>
```

```
    </graph>

    <!-- 4 -->
    <list>
      <title>Layout list</title>
      <table-ref>layout-table</table-ref>
      <column>num-sections</column>
      <column>duration</column>
    </list>
  </instrument>
```

The preceding snippet defines all XML needed for a basic instrument. Again, there are four sections:

1. Metadata about the instrument.
2. Every instrument uses tables that refer to schemas. In this case, a table is created that uses the `signpost-schema` schema for its data.
3. This section defines how the custom instrument should be drawn on the timeline. Every graph uses a table for its underlying data and then extracts the required data from the table. Note that you didn't manually define `duration` on the signpost schema. You get this property for free from the schema.
4. A definition is given for how the instrument should be displayed in the detail area. The format of the detail area is a table, so in this section, you define the detail area in a table-like structure.

This is all the code you need to write a basic custom instrument. To test your instrument, select your Instrument Package target and use your Mac as the target define:

A special version of Instruments will launch. Pick a blank template and look for your custom instruments in the add menu:

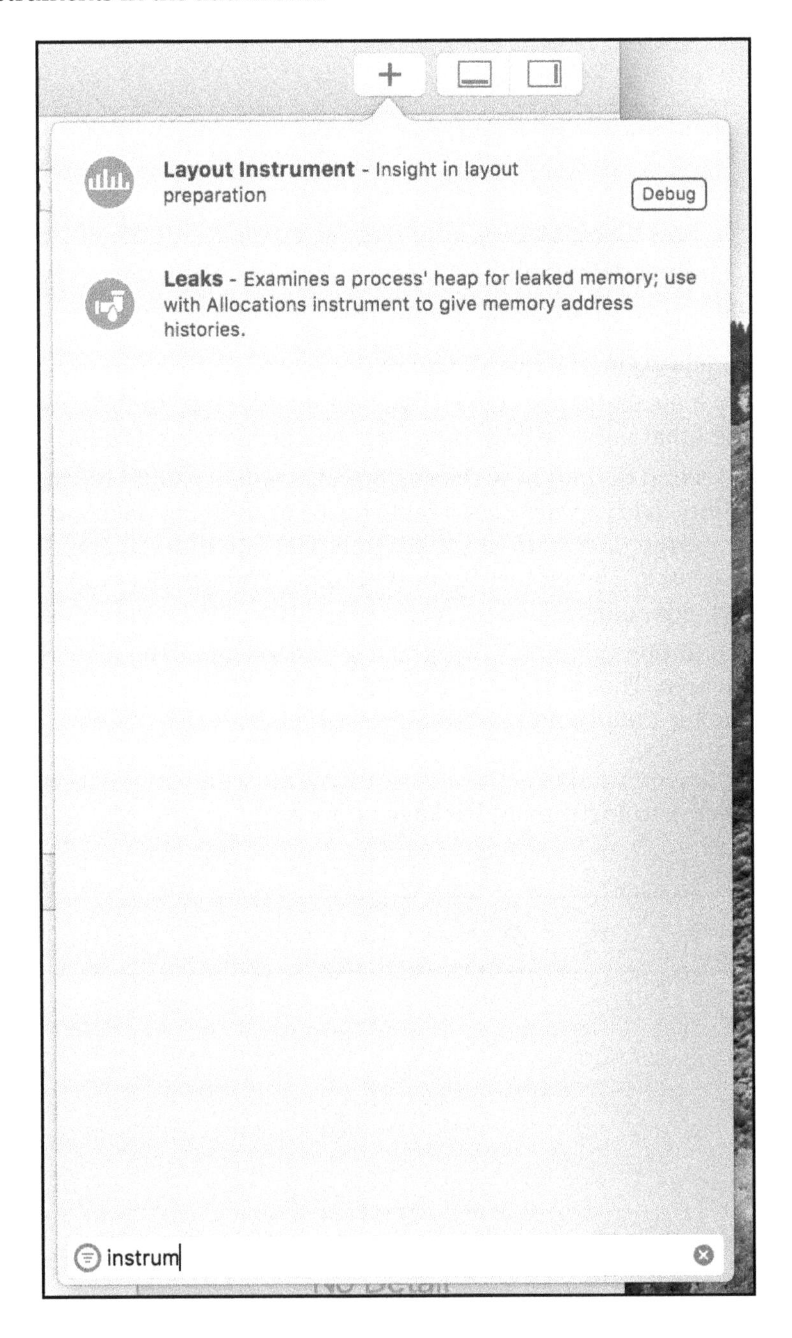

Once you have found your instrument, add it to the timeline area and profile your app. You should see all your signpost data appear as expected. The detail area should now show a single entrance for every time you call `prepare()` so you can easily compare performance between similar calls:

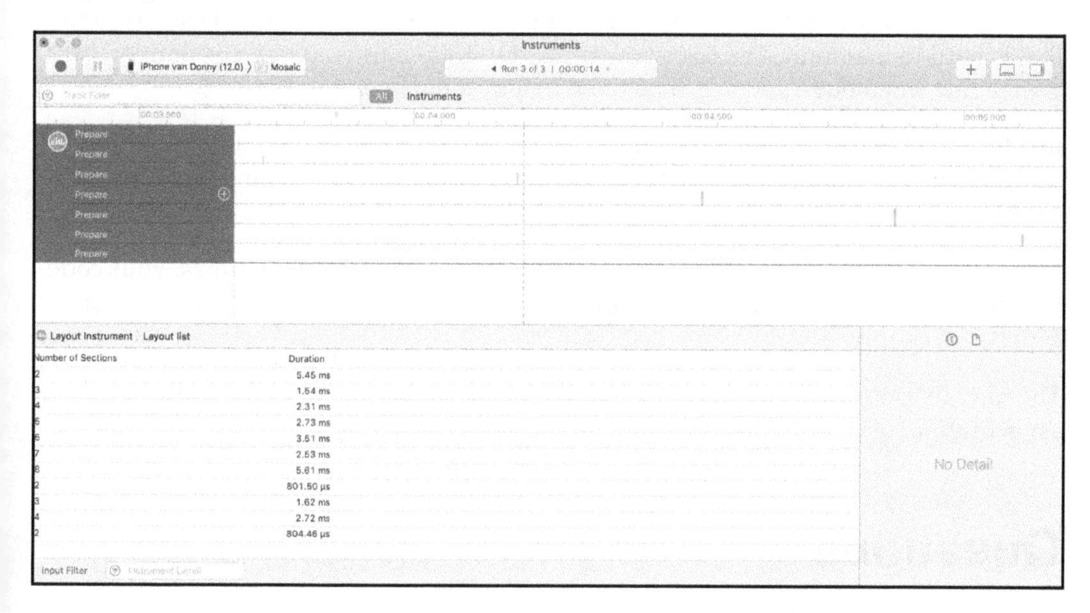

This sample was quite simple but it does an excellent job of showing that you can do powerful things with custom instruments. If you want to learn even more about what kinds of custom instruments you can build and how you can do some custom processing on your signpost data, make sure to catch the *Creating Custom Instruments* talk from WWDC 2018.

Summary

You learned a lot about measuring your app's performance in this chapter. You also learned how to find common issues and how to use Instruments to figure out what's going on behind the scenes for your app. In your day-to-day development cycle, you won't use Instruments or Xcode's memory debugger very often. However, familiarizing yourself with these tools can save you hours of debugging. It can even help you to discover memory leaks or slow code before you ship your app.

Try to audit and measure several aspects of your app while you're developing it, and you can see how the performance of certain aspects of your app improves or degrades over time. This will help you to avoid shipping an app that's full of slow code or memory leaks. However, don't go overboard with optimizing until you encounter actual problems in your app. Prematurely optimizing your code often leads to code that is hard to maintain. The fixes you applied in this chapter are pretty simple; there were a few small bugs that could be fixed in just a few lines of code.

In addition to learning about existing instruments, you learned how to implement signposts in your app to measure your app's performance. You even learned how to create a custom instrument to visualize your signpost calls with.

Unfortunately, fixing issues in your app isn't always this easy. Sometimes, your code is as fast as it can be, yet it still takes too long and it makes scrolling your tables or collections choppy.

The next chapter will show you how to handle such cases by implementing asynchronous programming using operations and GCD.

Questions

1. What does the Time Profiler do?

    a) It shows you when memory leaks occur.
    b) It shows you memory usage over time.
    c) It shows you how much time is spent in certain methods.

2. What does the Allocations instrument do?

    a) It shows you your app's memory usage over time.
    b) It shows you which objects reference another object.
    c) It shows you how much time is spent in certain methods.

3. Which of the following does not necessarily indicate a memory leak?

    a) Growing memory in the memory graph.
    b) Seeing instances of view controllers that were popped off the navigation stack.
    c) When no memory is freed up after your app receives a memory warning.

4. What does Xcode's memory debugger show you?

    a) Corrupt addresses in memory.
    b) The relationship between different objects that are in memory.
    c) Deallocated objects.

5. What is signpost logging?

    a) A low-overhead log to measure your app's performance.
    b) Logs that show the flow of your app.
    c) Logs to discover memory leaks.

6. Which elements does an Instruments Package have?

    a) Metadata, a schema, and the instrument itself.
    b) A schema, import statements for existing instruments, and a graph.
    c) Signposts, tables, and a graph.

Further reading

- Creating Custom Instruments (WWDC 2018): `https://developer.apple.com/videos/play/wwdc2018/410/`
- Measuring Performance Using Logging (WWDC 2018): `https://developer.apple.com/videos/play/wwdc2018/405/`
- Instruments Developer Help: `https://help.apple.com/instruments/developer/mac/10.0/`

25
Offloading Tasks with Operations and GCD

In the previous chapter, you learned how to use Instruments to measure your app's performance. Measuring performance is a vital skill in discovering slow code or memory leaks. You saw that sometimes it's easy to fix slow code and increase performance by fixing a small programming error. However, the fix isn't always that easy.

An example of code that is very hard to make fast is networking code. When you fetch data from the network, you do this asynchronously. If networking code was not asynchronous, the execution of your app would halt until the network request is finished. This means that your app would freeze for a couple of seconds if the network is slow.

Another example of relatively slow code is a data file from the app bundle and decoding the raw data into an image. For small images, this task shouldn't take as long as a networking request, but imagine loading a couple of larger images. Decoding a large amount of data would take long enough to significantly slow down an app's scrolling performance, or render the interface unresponsive for longer than you should be comfortable with.

Making code run asynchronously is not very hard if you do it through dispatch queues. In Objective-C and early versions of Swift, this was mostly known as **GCD** or **Grand Central Dispatch**, and the methods you'd use to interact with GCD were ugly. Nowadays, you use the `DispatchQueue` class, and the methods you interact with are much cleaner.

This chapter will teach you how to make use of dispatch queues to write asynchronous code that can perform slow operations off the main thread, meaning that they don't block the interface while they're running. Once you've got the hang of using dispatch queues, you will abstract some of your asynchronous code into `Operations`, so it's possible to reuse them.

This chapter covers the following topics:

- Writing asynchronous code with dispatch queues
- Creating reusable tasks with operations

By the end of this chapter, you'll be able to enhance your apps by optimizing your code for asynchronous programming. You'll also be able to abstract specific tasks into Operations to make your app more straightforward to understand and maintain.

Writing asynchronous code

In `Chapter 9`, *Syncing Data with CloudKit*, you had an encounter with asynchronous code and multithreading. That chapter didn't go into much detail regarding multithreading and asynchronous code because the subject of threading is rather complex, and it's much more suited to a thorough discussion, which we're explore in this chapter.

If you're unclear on what has already been explained, feel free to go back to `Chapter 9`, *Syncing Data with CloudKit*, to review the information presented there. The biggest takeaway from that chapter is that data fetching is performed on a background thread to avoid blocking the main thread. Once a fetch request is done, a `callback` function is executed, which allows you to use the main thread to update the user interface.

Understanding threads

You've seen the term thread a couple of times now, but you never explored them; you never really learned what a thread is or how a thread works. This section aims to make the subject of threading a lot clearer to you so you can fully understand what a thread is, and why threads are such a vital part of building apps.

A good way to think of a thread is as a stack of instructions. In iOS, your app typically starts off with a single thread – the main thread. This thread is also known as the UI thread. It's called the UI thread because the main thread is where all of the user interface elements are configured, rendered, and pushed to the screen. Anything that is related to the user interface must be executed on the main thread, so if you think of a thread as a stack of instructions that are run one by one, it's easy to see why it's so important that the main thread doesn't get stuck performing a very slow instruction:

The preceding image shows a timeline where all code is executed on the main thread. Notice how the interface can't be updated until the **fetch data** and **parse JSON** instructions are completed. Also, note that fetching data takes a lot longer than displaying the interface or handling a tap. During the fetch data instruction, the app is unable to update any user interface elements or process any gestures or taps. This means that the app is practically frozen until the data is fetched and parsed.

A good, responsive application can't afford to wait for slow instructions. The interface should always respond to user input; if this isn't the case, the app feels slow, buggy, choppy, and just all-around bad.

If an app uses multiple threads, it can run various instruction stacks at the same time. Each *stack* is called a *thread*, and specific instructions can be performed on a different thread to ensure that the main thread remains responsive:

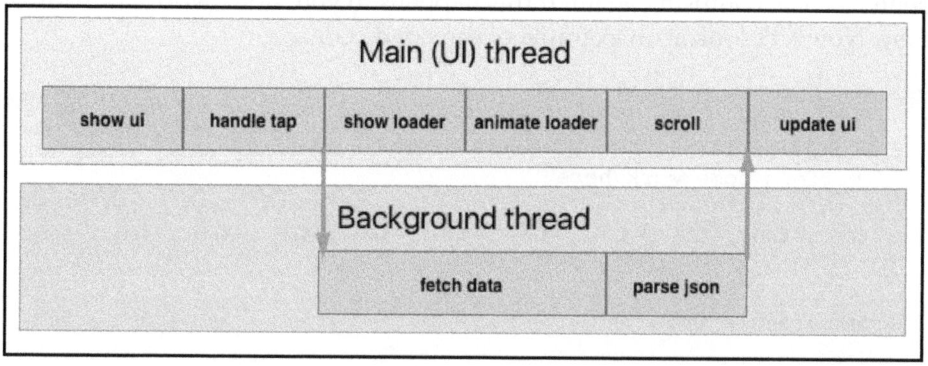

This second figure shows a more desirable scenario. The main thread only handles tasks that are related to the user interface, such as taps, animations, and scrolling. The background thread takes care of the tasks that are not related to the user interface and could potentially take a while to finish. By removing these instructions from the main thread and placing them on a different thread, like iOS does by default for networking, you can ensure that your app remains responsive, even if the network requests take several seconds to finish or never finish at all. Your app can utilize a large number of threads for different tasks.

The number of threads isn't infinite, so make sure that you optimize your code as much as possible to avoid locking up several threads with slow code.

In `Chapter 24`, *Discovering Bottlenecks with Instruments*, you used Instruments to locate slow code in a sample app. The slow code was an instruction on the main thread that took a very long time to complete, resulting in a frozen interface. However, threading would not have solved this particular issue. The slow code you discovered calculated the layout for a collection view. A collection view can't be rendered without calculating the layout first, so this is a scenario where it's essential to make sure that you write optimized code, instead of relying on threads for anything that's slow.

Now that you have a better understanding of threads and how they can be used, let's have a look at how to offload tasks to different threads.

Using dispatch queues in your application

A basic understanding of threads is good enough for you to start using them in your applications. However, once you start using them, there's a good chance that they suddenly become confusing again. If this happens to you, don't worry; threading is not easy. Now let's look at an example of threaded code:

```
var someBoolean = false

DispatchQueue(label: "MutateSomeBoolean").async {
  // perform some work here
  for i in 0..<100 {
      continue
  }

  someBoolean = true
}
```

```
print(someBoolean)
```

The preceding snippet demonstrates how you could mutate a variable after performing a task that is too slow to execute on the main thread. In the preceding code, an instance of `DispatchQueue` is created, and it's given a label. This creates a new queue on which you can execute instructions. The queue represents the background thread from the visualization you looked at earlier.

Then, the `async` method on `DispatchQueue` is called with the closure that should be executed on the freshly created queue. The loop inside of this block is performed on the background thread. In the visualization from before, this would roughly compare to the fetch data and parse JSON instructions. Once the task is done, `someBoolean` is mutated.

The last line in the snippet prints the value of `someBoolean`. What do you think the value of `someBoolean` is at that point? If your answer is `false`, good job! If you thought `true`, you're not alone. A lot of people who start writing multithreaded, asynchronous code don't immediately grasp how it works.

The following image shows what the code does in the background. This should make what happens and why `someBoolean` was `false` when it got printed more obvious:

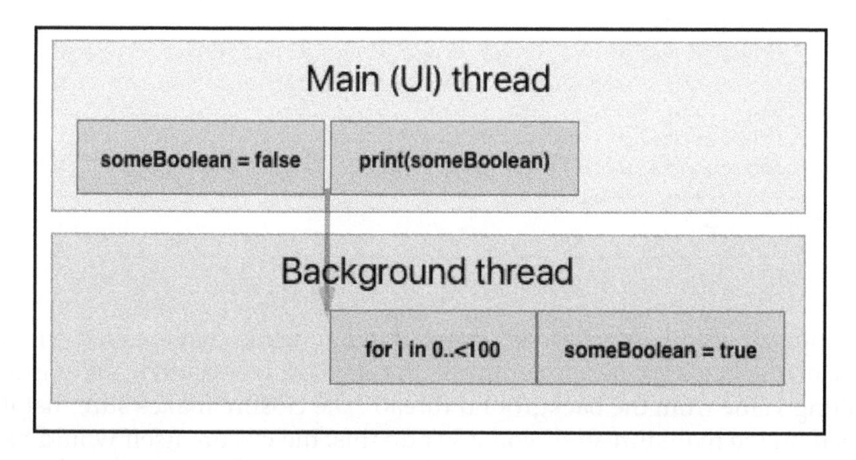

Because this code uses a background thread, the main thread can immediately move to the next instruction. This means that the for loop and the print run simultaneously. In other words, `someBoolean` is printed before it's mutated on the background thread. This is both the beauty and a caveat of using threads. When everything starts running simultaneously, it is hard to keep track of when something completes.

The preceding visualization also exposes a potential problem in the code. A variable was created on the main thread, and then it got captured in the background thread which then mutated the variable. Doing this is not recommended; your code could suffer from unintended side-effects, such as race conditions, where both the main thread and the background thread mutate a value, or worse, you could accidentally try to access a Core Data object on a different thread than the one it was created on. Core Data objects do not support being used on multiple threads so you should always try to make sure that you avoid mutating or accessing objects that are not on the same thread as the one where you access them.

So, how can you mutate `someBoolean` safely and print its value after mutating it? Well, you could use a `callback` closure to achieve this. The following code is a sample of what that would look like:

```
func executeSlowOperation(withCallback callback: @escaping ((Bool) -
> Void)) {
  DispatchQueue(label: "MutateSomeBoolean").async {
    // perform some work here
    for i in 0..<100 {
      continue
    }

    callback(true)
  }
}

executeSlowOperation { result in
  DispatchQueue.main.async {
    someBoolean = result
    print(someBoolean)
  }
}
```

In this snippet, the slow operation is wrapped in a function that is called with a callback closure. Once the task is complete, the callback is executed, and it is passed the resulting value from the background thread. The closure makes sure that its code is executed on the main thread. If you don't do this, the closure itself would have been executed on the background thread. It's important to keep this in mind when calling asynchronous code.

The callback-based approach is excellent if your callback should be executed when a single task is finished. However, there are scenarios where you want to complete some tasks before moving to the next task. You have already used this approach in `Chapter 11`, *Being Proactive with Background Fetch*.

Let's review the heart of the background fetch logic that was used in that chapter:

```
func application(_ application: UIApplication,
performFetchWithCompletionHandler completionHandler: @escaping
(UIBackgroundFetchResult) -> Void) {

  let fetchRequest: NSFetchRequest<Movie> = Movie.fetchRequest()
  let managedObjectContext = persistentContainer.viewContext
  guard let allMovies = try? managedObjectContext.fetch(fetchRequest)
else {
    completionHandler(.failed)
    return
  }

  let queue = DispatchQueue(label: "movieDBQueue")
  let group = DispatchGroup()
  let helper = MovieDBHelper()
  var dataChanged = false

  for movie in allMovies {
    queue.async(group: group) {
      group.enter()
      helper.fetchRating(forMovieId: movie.remoteId) { id,
        popularity in
        guard let popularity = popularity, popularity !=
          movie.popularity else {
            group.leave()
            return
        }

        dataChanged = true

        managedObjectContext.persist {
          movie.popularity = popularity
          group.leave()
        }
      }
    }
  }

  group.notify(queue: DispatchQueue.main) {
    if dataChanged {
      completionHandler(.newData)
    } else {
      completionHandler(.noData)
    }
  }
}
```

When you first saw this code, you were probably able to follow along, but it's unlikely that you were completely aware of how complex this method is. Multiple dispatch queues are used in this snippet. To give you an idea, this code starts off on the main thread. Then, for each movie, a background queue is used to fetch its rating. Once fetching all these ratings is complete, the managed object context's dispatch queue is used to update the movie. Think about all this switching between dispatch queues that is going on for a second. Quite complex, isn't it?

The background fetch method needs to call a completion handler when it is done fetching all the data. However, a lot of different queues are used, and it's kind of hard to tell when all the fetch operations have completed. This is where dispatch groups come in. A dispatch group can hold onto a set of tasks that are executed either serially or in parallel.

When you call `enter()` on a dispatch group, you are also expected to call `leave()` on the group. The enter call tells the group that there is unfinished work in the dispatch group. When you call `leave()`, the task is marked as completed. Once all tasks are completed, the group executes a closure on any thread you desire. In the example, `notify(queue:)` is the method used to perform the completion handler on the main queue.

It's okay if this is a bit daunting or confusing right now. As mentioned before, asynchronous programming and threads are pretty complex topics, and dispatch groups are no different.

The most important takeaways regarding dispatch groups are that you call `enter()` on a group to submit an unfinished task, you call `leave()` to mark the task finished, and you use `notify(queue:)` to execute a closure on the queue passed to this method once all tasks are marked as completed.

The approach you've seen so far makes direct use of closures to perform tasks. This causes your methods to become long and relatively complex since everything is written in line with the rest of your code. You already saw how mixing code that exists on different threads can lead to confusion because it's not very obvious which code belongs on which queue. Also, all this inline code is not particularly reusable. You can't pick up a particular task and execute it on a different queue for instance, because the code is tightly coupled to a specific dispatch queue.

You can use `Operations` to make tasks that are easy to reuse and decoupled from running on a specific queue.

Creating reusable tasks with Operations

You have just learned about `DispatchQueues`, and how they are used to schedule tasks that need to be performed on a different thread. You saw how this speeds up code and how it avoids blocking the main thread. In this section, you will take this one step further. The first reason for this is because asynchronous work is better organized if it is implemented as an isolated object that can be scheduled for execution instead of having several closures that make your code less readable.

The solution to having closures all over your asynchronous code is to use `Operation` instead of a closure. And instead of queueing everything in a dispatch queue, `Operation` instances are queued on `OperationQueue`. `OperationQueue` and `DispatchQueue` are similar, but not quite the same. `OperationQueue` can schedule `Operations` on one or more `DispatchQueues`. This is important because of the way in which `Operations` work.

When using `OperationQueue`, you can execute `Operations` in parallel or serially. It is also possible to specify dependencies for `Operations`. This means that you can make sure that certain `Operations` are completed before the next operation is executed. `OperationQueue` will manage the `DispatchQueues` needed to make everything happen, and it will execute `Operations` in the order in which they become ready to execute. Moreover, code that uses `Operations` will typically be easier to maintain and understand because all work is beautifully encapsulated in isolated building blocks.

The next section will briefly cover some of the basic concepts of `Operations`.

If you're looking to learn more about using `Operations` in interesting and advanced ways, make sure to check out *Apple's Advanced NSOperations talk* from *WWDC 2015*. All code for this talk is presented in Swift 2.0, so you'll need to translate this code to Swift 4.2 yourself, but it's definitely worth a watch.

Using Operations in your apps

Let's take a deep dive into `Operations` and refactor the background fetch code from the **MustC** app to make it use `Operations`. To do this, you will create two `Operation` subclasses: one that fetches data and updates a `movie` object, and one that calls the completion handler once all other operations are complete.

The setup will have a single `OperationQueue` that will execute all of the instances of the fetch `Operation` subclass and a single operation that calls the background fetch completion handler. The completion operation will have all of the fetch operations as its dependencies, so it's automatically executed when all fetch operations are completed.

Whenever you create an `OperationQueue` instance, you can specify the number of concurrent `Operation` instances that can be executed on the queue. If you set this number to one, you have a serial queue that runs all operations one by one in the order in which they become ready to execute.

An operation is considered ready when all preconditions for the operation are met. A great example of this is dependencies. An operation with dependencies is not ready to execute until all of the operations that it depends on are completed. Another example is exclusivity. It's possible to configure your setup to allow only a single operation of a particular type to be running at a time. An operation that is set up like that won't become ready until no other operations of that type are running.

If you set the maximum number of concurrent operations to a higher number, it's not guaranteed that the queue will run that number of operations simultaneously. Imagine setting the maximum amount to 1,000, and you put 2,000 operations in the queue. It's not likely that you will see 1,000 operations running in parallel. The system ultimately decides how many operations it will run at the same time, but it's never more than your maximum value.

As mentioned before, you can improve the **MustC** app by using `OperationQueue`. Get started with this refactor by creating an `OperationQueue` instance. You will add all download operations to this queue. Replace the implementation of `application(_:performBackgroundFetchWithCompletionHandler:)` in `AppDelegate` with the following:

```
func application(_ application: UIApplication,
performFetchWithCompletionHandler completionHandler: @escaping
(UIBackgroundFetchResult) -> Void) {

  let queue = OperationQueue()
```

```
    let fetchRequest: NSFetchRequest<Movie> = Movie.fetchRequest()
    let managedObjectContext = persistentContainer.viewContext
    guard let allMovies = try? managedObjectContext.fetch(fetchRequest)
  else {
      completionHandler(.failed)
      return
    }
  }
```

The preceding code creates an operation queue. This is the queue that will be used to fetch all movies and call the background fetch completion handler once all downloads are done. Before you can add operations to the queue, you must create the appropriate classes to contain your operation. Create a new file and name it `UpdateMovieOperation.swift`.

Every custom operation you create should subclass the `Operation` base class. The `Operation` class implements most of the glue and boilerplate code involved in managing and executing operations and queues. There are a couple of mandatory read-only variables that you need to implement for your operation. Also, you need to make sure these variables are **KVO (Key-Value Observing)** compliant, so the operation queue can observe the various states your operation will go through during its lifetime. Add the following basic implementation for `UpdateMovieOperation`:

```
import Foundation

class UpdateMovieOperation: Operation {
  override var isAsynchronous: Bool { return true }
  override var isExecuting: Bool { return _isExecuting }
  override var isFinished: Bool { return _isFinished }

  private var _isExecuting = false
  private var _isFinished = false
  var didLoadNewData = false

  let movie: Movie

  init(movie: Movie) {
    self.movie = movie
  }
}
```

You'll immediately notice that this implementation overrides a couple of variables. These are the read-only variables that were mentioned earlier. The `isExecuting` and `isFinished` variables return the value of two private variables that will be mutated to reflect the operation's state later.

Furthermore, the operation keeps track of whether new data was loaded, and there are a property and initializer to attach a `Movie` to the operation. So far, this operation isn't very exciting. Add the following methods to your operation so it can do something:

```
override func start() {
  super.start()

  willChangeValue(forKey: #keyPath(isExecuting))
  _isExecuting = true
  didChangeValue(forKey: #keyPath(isExecuting))

  let helper = MovieDBHelper()
  helper.fetchRating(forMovieId: movie.remoteId) { [weak self] id,
    popularity in
    defer {
      self?.finish()
    }

    guard let popularity = popularity,
      let movie = self?.movie,
      popularity != movie.popularity
      else { return }

    self?.didLoadNewData = true

    movie.managedObjectContext?.persist {
      movie.popularity = popularity
    }
  }
}

func finish() {
  willChangeValue(forKey: #keyPath(isFinished))
  _isFinished = true
  didChangeValue(forKey: #keyPath(isFinished))
}
```

Apple's guidelines state that every subclass of `Operation` should override the `start()` method and initiate the operation from there. You'll note that the superclass implementation is called first. This is because the superclass takes care of several under-the-hood tasks that must be performed to make the operation work. Next, `willChangeValue(forKey:)` and `didChangeValue(forKey:)` are called to fire the KVO notifications, so the operation queue knows when the state of the operation has changed or is about to change.

You might notice that the values of `isFinished` and `isExecuted` are not changed after calling `willChangeValue(forKey:)`. This is okay because these methods only tell any observers that reading specific properties after `didChangeValue(forKey:)` will yield a different value than before. Since `isFinished` and `isExecuting` return the value of the private properties that are changed, `isFinished` and `isExecuting` will return different values, as the observer expects.

Next, the code from before is used to fetch and update the movie. A `defer` block is used to call the `finish()` method when the operation is done, regardless of how the network request went. By using `defer` instead of manually calling `finish()` when appropriate, you can't forget to call `finish()` if the code changes at some point. The `finish()` method makes sure that the operation queue is notified about the operation's new status by firing the corresponding KVO notifications.

Another operation that calls the background fetch completion handler should be created. This operation should loop through all of its dependencies, check whether it's a movie update operation, and if it is, it should check whether at least one of these operations has loaded new data. After doing this, the completion handler should be called with the corresponding result, and finally, the operation should finish itself. Create a new file in the `Operations` folder and name it `BackgroundFetchCompletionOperation`. Add the following implementation:

```
import Foundation
import UIKit

class BackgroundFetchCompletionOperation: Operation {
  override var isAsynchronous: Bool { return true }
  override var isExecuting: Bool { return _isExecuting }
  override var isFinished: Bool { return _isFinished }
  var _isExecuting = false
  var _isFinished = false

  let completionHandler: (UIBackgroundFetchResult) -> Void

  init(completionHandler: @escaping (UIBackgroundFetchResult) -> Void)
{
    self.completionHandler = completionHandler

  }

  override func start() {
    super.start()
```

```
    willChangeValue(forKey: #keyPath(isExecuting))
    _isExecuting = true
    didChangeValue(forKey: #keyPath(isExecuting))

    var didLoadNewData = false

    for operation in dependencies {
      guard let updateOperation = operation as? UpdateMovieOperation
else { continue }

      if updateOperation.didLoadNewData {
        didLoadNewData = true
        break
      }
    }

    if didLoadNewData {
      completionHandler(.newData)
    } else {
      completionHandler(.noData)
    }

    willChangeValue(forKey: #keyPath(isFinished))
    _isFinished = true
    didChangeValue(forKey: #keyPath(isFinished)) }
}
```

The implementation for this operation is pretty similar to the movie update operation. The operation should be initialized with the completion handler that was passed to the background fetch method in `AppDelegate`, and it's called after figuring out whether new data was fetched by looping through all movie update operations. Let's see how all this comes together by updating the background fetch logic in `AppDelegate`. Add the following code to the `application(_:performFetchWithCompletionHandler:)` method, right after fetching the movies:

```
let completionOperation =
BackgroundFetchCompletionOperation(completionHandler:
completionHandler)

for movie in allMovies {
  let updateOperation = UpdateMovieOperation(movie: movie)
  completionOperation.addDependency(updateOperation)

  queue.addOperation(updateOperation)
```

```
}

queue.addOperation(completionOperation)
```

This code is a lot more readable than what was there before. First, the completion operation is created. Next, an update operation is created for each movie, this operation is added as a dependency for the completion operation, and it's added to the operation queue. Once all dependencies are set up, the completion operation itself is added to the queue as well. All of the movie update operations will automatically start executing simultaneously, and once they're all done, the completion operation becomes ready to execute. Once this happens, the completion operation will run and the background fetch completion handler will be called.

Even though `Operations` involve a bit more boilerplate code regarding managing execution state, you do end up with code that is a lot easier to read and understand. This chapter only covered setting up dependencies for operation, but if you study *Apple's Advanced NSOperations* video that was mentioned earlier, you'll find that you can do powerful, complex, and amazing things with operations.

Summary

This chapter showed you that asynchronous code can be hard to understand and reason about, especially when a lot of code is running at the same time; it can be easy to lose track of what you're doing. You also learned that `Operations` are a convenient way to reduce complexity in your application, resulting in code that is easier to read, change, and maintain. When an operation depends on multiple other `Operations` to be completed, it can be extremely convenient to use `OperationQueue`, as it dramatically reduces the complexity of the code you write.

It's been mentioned before, but if you intend to make use of operations in your app, make sure to check out Apple's Demonstration of Advanced Operations from WWDC 2015. Operations are capable of far more than you've seen in this chapter, and it's highly recommended to see how Apple uses operations to create rock-solid apps. Once your app is covered by tests, measured with instruments, and improved with asynchronous code and operations, it's probably time that you share your app with others.

The next and final chapter of this book will show you how to set yourself up for deploying your app through TestFlight and the App Store.

Questions

1. What are the downsides of running certain code on the main thread?

   a) It can block the UI.
   b) It's against best practices.
   c) The main thread can only handle UI code.

2. In which circumstances will threading not help your performance?

   a) When you have a lengthy layout-related calculation.
   b) When you load data from the network.
   c) When you load many images from disk.

3. Which GCD feature can you use to execute an action when several other actions are done executing?

   a) `DispatchClub`.
   b) `DispatchGroup`.
   c) `GCD_Group`.

4. Which of the following is not a benefit of operations?

   a) More readable code.
   b) Faster code execution.
   c) Operations can depend on other operations.

5. What's the maximum number of Operations you can run at the same time?

   a) There is no hard limit, iOS decides on the fly.
   b) 4.
   c) 10.

6. What is KVO?

   a) A way for objects to observe properties of another object.
   b) A way to boost the execution speed of threads.
   c) A technique that is used to make operations depend on each other.

7. How do you start executing operations on an operation queue?

   a) By calling `start()` on the queue.
   b) By calling `resume()` on the queue.
   c) By adding operations to it.

Further reading

- Modernizing Grand Central Dispatch Usage: `https://developer.apple.com/videos/play/wwdc2017/706/`
- Advanced NSOperations: `https://developer.apple.com/videos/play/wwdc2015/226/`

26
Submitting Your App to the App Store

Possibly the most exciting part of the development cycle is getting your app into the hands of some real users. The first step in doing so is usually to send out a beta version of your app so you can get feedback and gather some data about how your app is performing before you submit to the App Store and release your app to the world. Once you're satisfied with the results of your beta test, you must submit your app to Apple so they can review it before your app is released to the App Store.

In this chapter, you'll learn everything about packing up your app and submitting it to Apple's App Store Connect portal. Using App Store Connect, you can start beta-testing your app, and you can also submit it to Apple for review. App Store Connect is also used to manage your app's App Store description, keywords, promotional images, and more. This chapter will show you how to fill out all the information for your app correctly. You will go through the following steps in this chapter:

- Adding your application to App Store Connect
- Packaging and uploading your app for beta testing
- Preparing your app for launch

These steps closely resemble the process you'll go through when you're ready to launch your app. Let's get right to it, shall we?

Adding your application to App Store Connect

The first thing you're going to want to do when you're ready to release your app is to register your app with App Store Connect. To access App Store COnnect, you must be enrolled in the Apple Developer program. You can do this through Apple's developer portal at `https://developer.apple.com`. After purchasing your membership, you can log in to your App Store Connect account on `https://appstoreconnect.apple.com` using your Apple ID.

After logging into your App Store Connect account, you are presented with a screen that has a couple of icons on it:

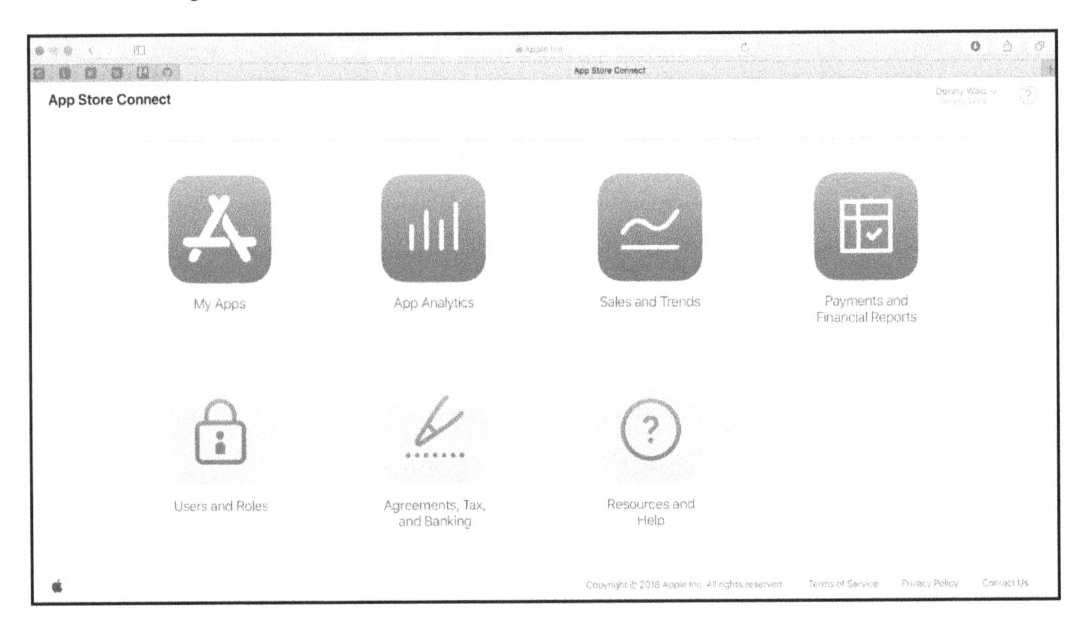

This screen is your portal to manage your App Store presence. From here, you can manage test users, track your app's downloads, monitor app usage, and more. But most importantly, it's where you create, upload, and publish your apps to Apple's beta-distribution program, called **TestFlight**, and once you're done beta testing, to the App Store. Go ahead and peek around a bit; there won't be much to see yet, but it's good to familiarize yourself with the App Store Connect portal.

The first step in getting your app out to your users is to navigate to the **My Apps** section. This section is where you'll find all the apps you have created and where you can add new apps. To add a new app, click the + icon in the top left and click **New App**:

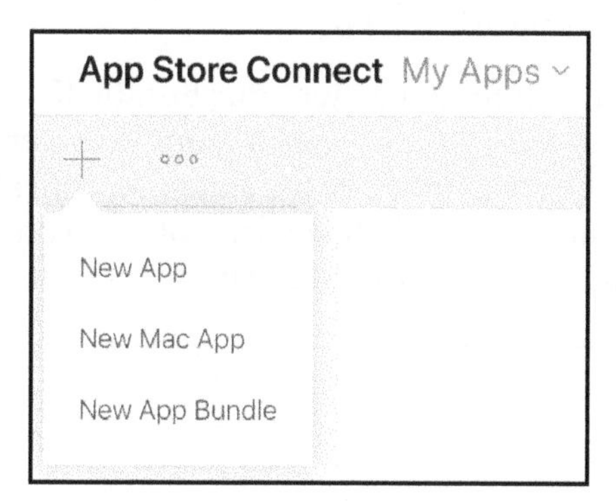

After clicking this, you're presented with a window in which you can fill out all the necessary information about your app. This is where you pick your app's name, select the platform on which it will be released, and a few more properties:

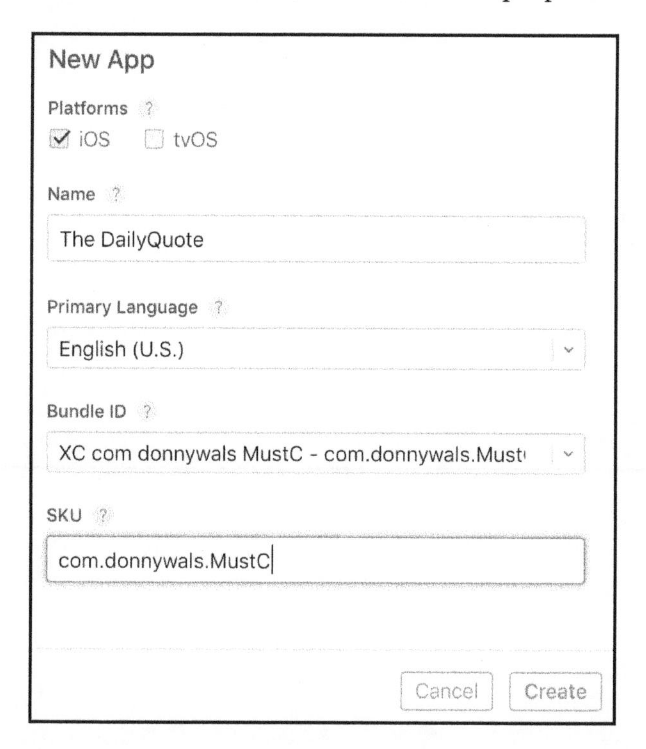

The Bundle ID field is a drop-down menu that contains all of the app IDs that you have registered for your team in Apple's developer portal. If you've been developing with the free-tier developer account up until the last minute, or you haven't used any special features such as push notifications or App Groups, it's possible that your app's Bundle ID is not in the drop-down menu.

If this is the case, you can manually register your Bundle ID in the developer portal. Navigate to `https://developer.apple.com/` in your browser and click the **Account** menu item. You can use the Account page to manage certificates, Bundle IDs, devices, and more. A lot of this is automatically taken care of by Xcode, but you'll occasionally find yourself in this portal. For instance to manually register your app's Bundle ID:

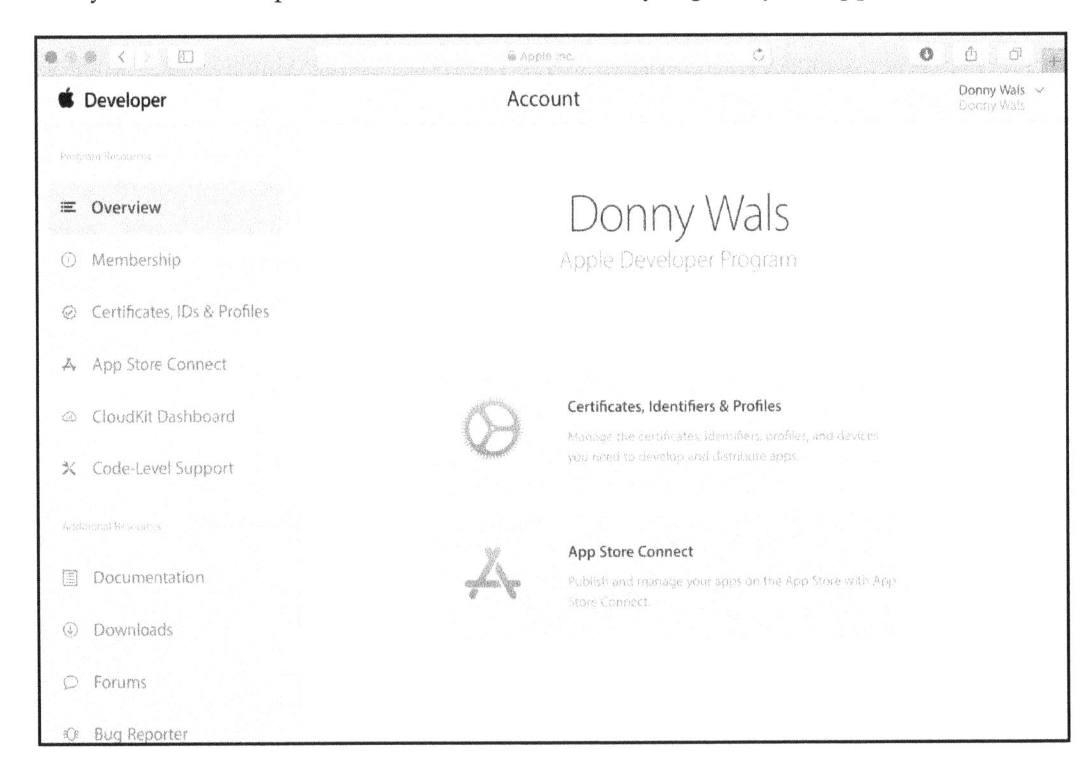

To register your Bundle ID, click on the **Certificates, IDs & Profiles** item on the left-hand side of the page. On the Certificates, IDs & Profiles page, click **App IDs** in the menu on the left-hand side. This will present you with a list of currently registered apps. In the top-right corner, there's a **+** icon. Click this to add a new ID to your profile:

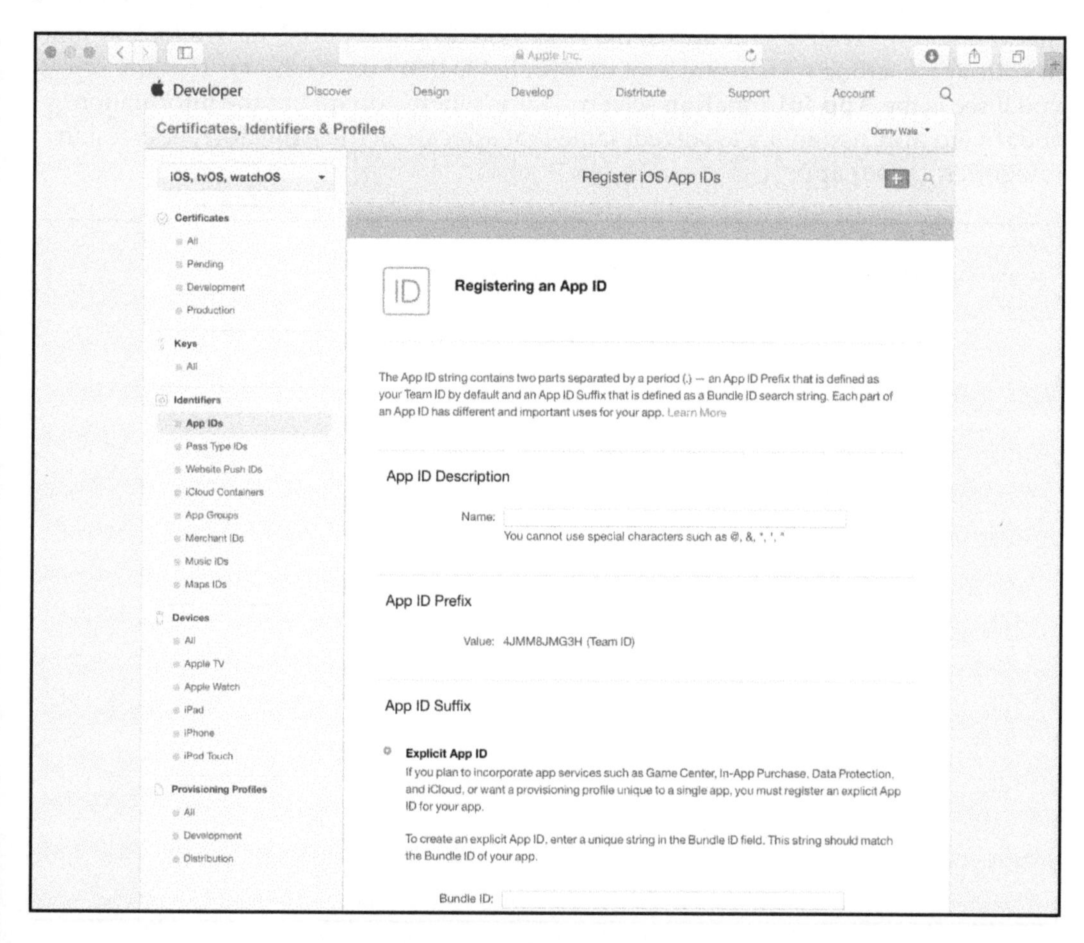

To add your Bundle ID, all you need to do is fill in the form fields. You'll want to use a descriptive name for your app name. It can be the same as the name you have set for your app in Xcode, but it can also be a different name; it doesn't have to match. Make sure to select the **Explicit App ID** field and copy the Bundle ID from your Xcode project. It's important that you perfectly match this. If you don't do this, you'll run into issues later because your app can't be identified if the Bundle ID is incorrect.

Once you've done this, you can scroll all the way down and save your new ID. You don't have to select any capabilities since Xcode will automatically manage this for you when you enable or disable them in the **Capabilities** tab.

After manually registering your Bundle ID, you should be able to move back to App Store Connect, add a new app, and select your app's Bundle ID. After you've done this and you've created your app in the App Store Connect portal, have a look around in your app's settings. There are a lot of form fields that you can fill in. The first screen you'll see is the **App Information** screen. This is where you fill out the information about your app, assign it a localized name that appears in the App Store, and assign categories to your app:

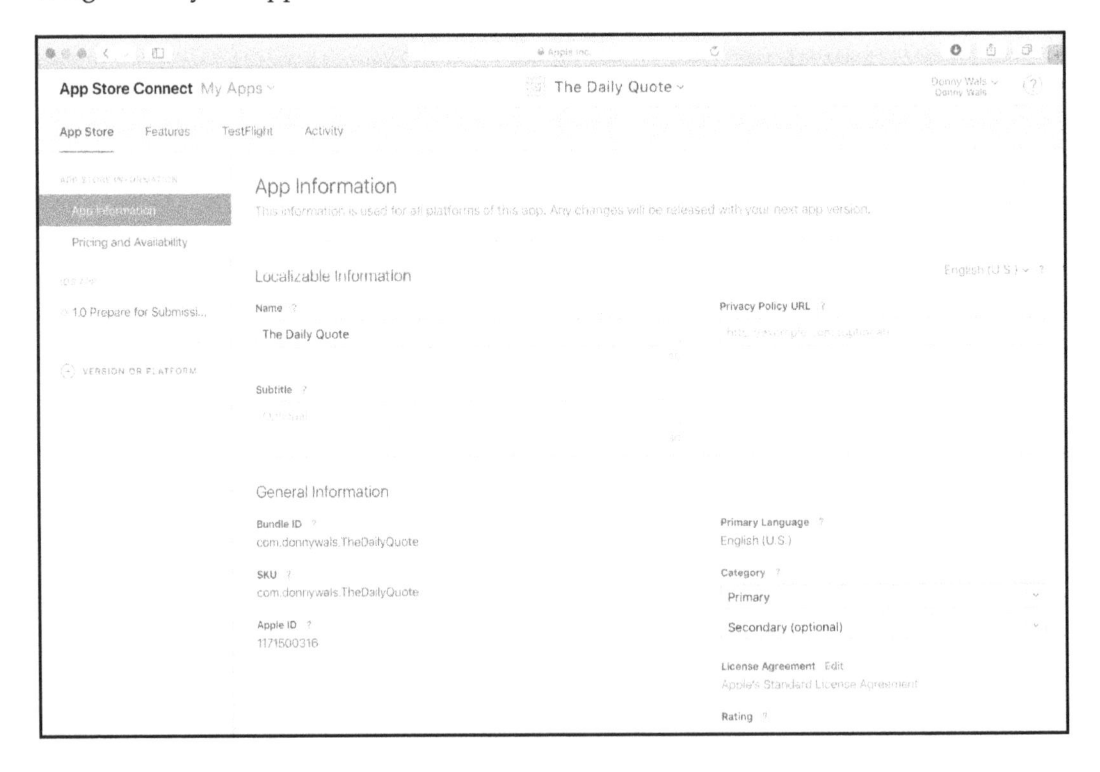

Next, there's the **Pricing and Availability** screen. This is where you decide in which countries your app can be downloaded and also how much it costs. Lastly, there is the **Prepare for Submission** menu item.

Whenever you add a new version of your app, you should fill out the form fields on this screen, and there are quite a lot of them. The **Prepare for Submission** form is used to provide screenshots, keywords, a description for your app, privacy policies, and more. Go ahead and have a look at what's in there. Luckily, everything you have to fill in is pretty straightforward.

Once you have registered your app on App Store Connect, you can upload your app. To do this, you use Xcode to package up your app and send it off to App Store Connect.

Packaging and uploading your app for beta testing

To send your app out to beta testers and eventually real users, you must first archive your app using Xcode. Archiving your app will package up all contents, code, and assets. To archive your app, you must select **Generic iOS Device** from the list of devices your app runs on in Xcode:

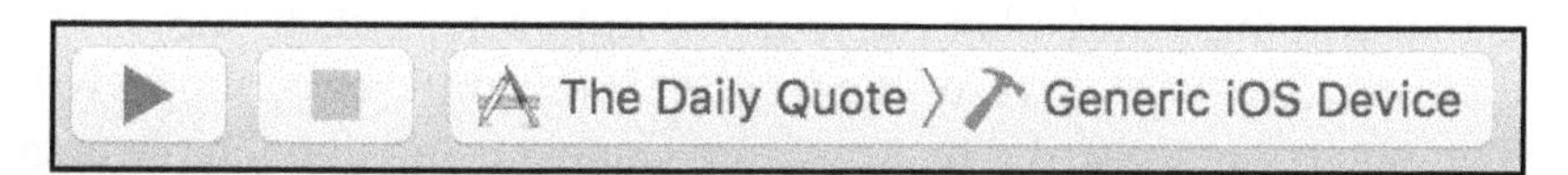

With this build device selected, select **Product** | **Archive** from the top menu in Xcode. When you do this, a build will start that's a lot slower than usual. That's because Xcode is building your app in release mode so it is optimized and can run on any device.

Once the archive is created, Xcode will automatically open the organizer panel for you. In this panel, you get an overview of all apps and archives that you have created:

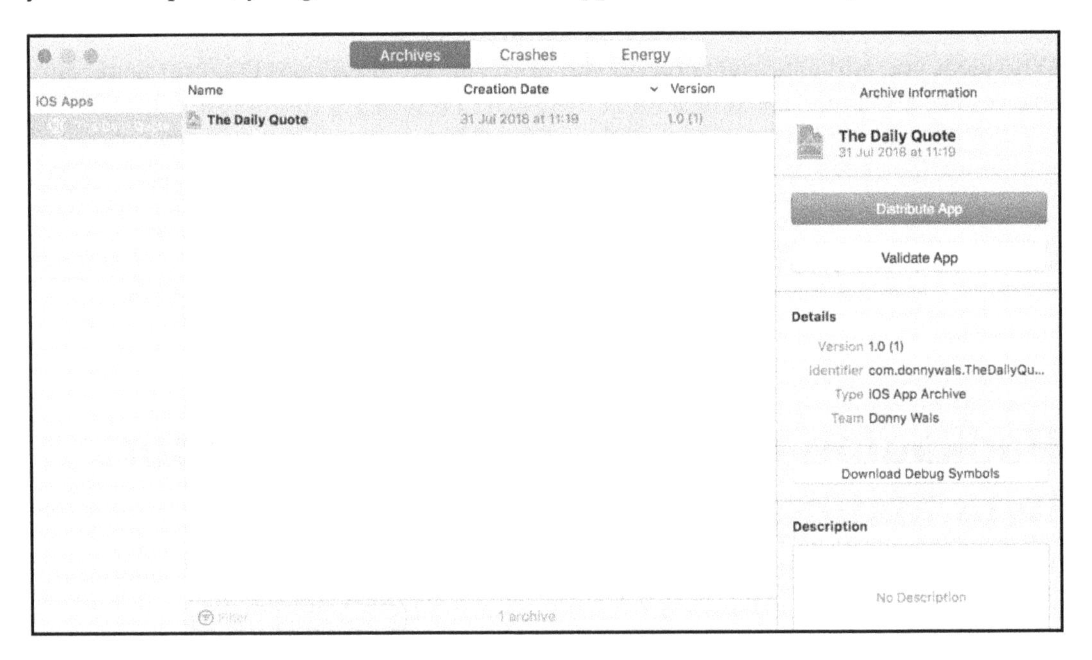

Before you archive your app, you should make sure that your app is ready for release. This means that you must add all of the required app icon assets to the `Images.xcassets` resource. If your app icon set is incomplete, your app will be rejected when you try to upload it to App Store Connect, and you'll have to generate your archive all over again.

When you're ready to upload your build to App Store Connect to send it to your beta testers and eventually your users through the App Store, you should select your latest build and click the **Upload to App Store** button. A popup will appear to guide you through a couple of settings. You should use the default settings most of the time and click **Next**. Finally, your app will be verified based on metadata and uploaded to App Store Connect. If there are errors in your archive, such as missing assets or unsigned code, you'll learn about it through error messages in the upload popup window. When your upload succeeds, you'll also be notified through the popup.

Once you've uploaded your build, you can go to the activity panel for your app in App Store Connect. This is where you'll see your build's current status. If you have just uploaded a build, its status will be **Processing**. Sometimes this step takes a while, but usually no longer than a couple of hours:

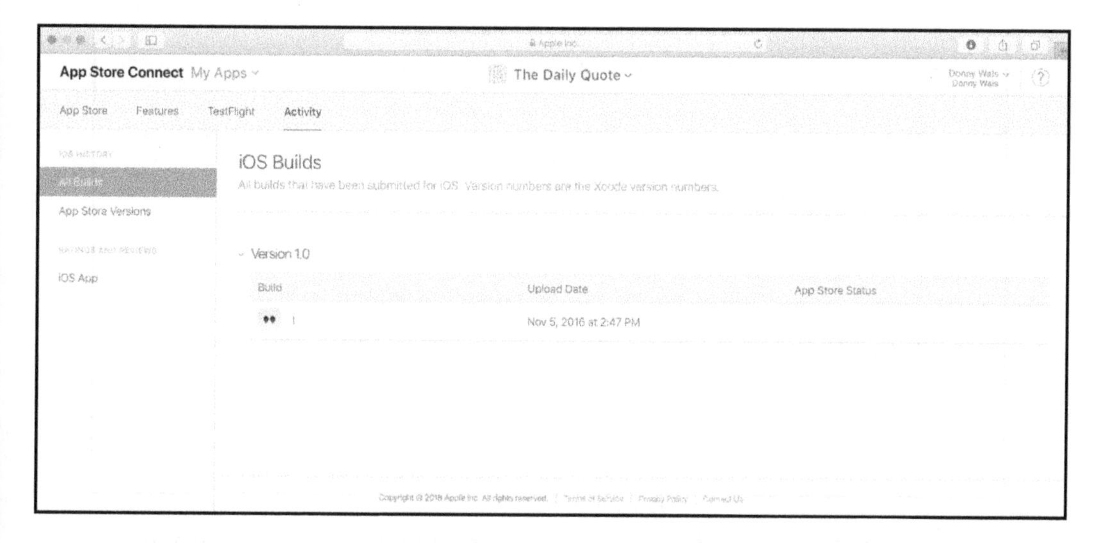

While the app is processing, you can start to prepare your **TestFlight** settings. Select the **TestFlight** menu item and fill in the Test Information form. If you're rolling out your beta test internationally, you might want to provide information in multiple languages, but you don't have to.

Next, select the **Internal Testing** menu item in the sidebar on the right. In this panel, you can select users that are added to your account for beta testing. This type of testing is often the first type of testing you do, and it's mostly intended for testing apps internally with members of your team or with close friends and family. You can add more internal users to your account through the **Users and Roles** section in App Store Connect.

Once your app is processed, you'll receive an email, and you can select which version of your app should be used for internal testing:

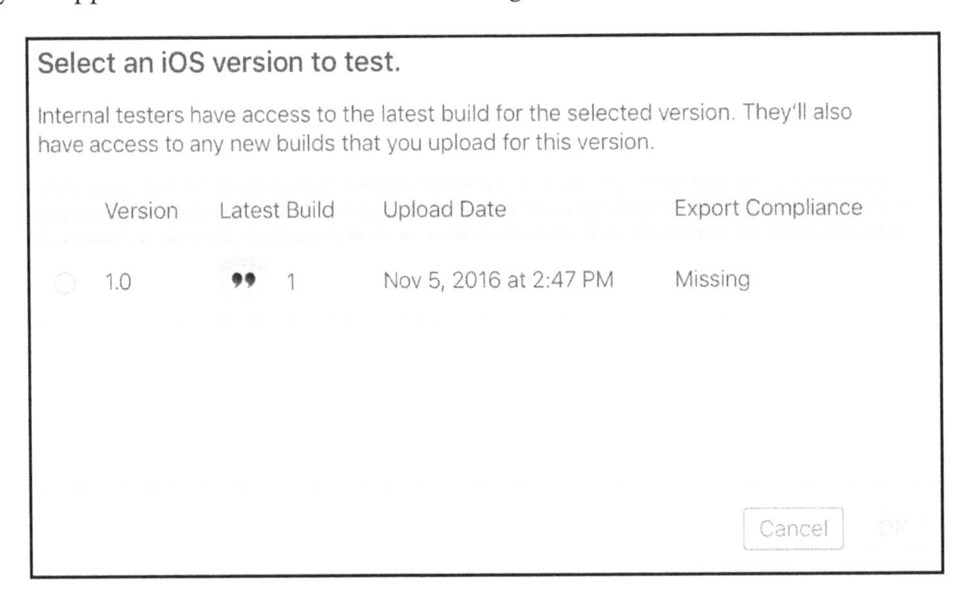

Once you've added testers and you've selected a build to test, you can click the `Start Testing` button to send out a beta invite to your selected testers. They will receive an email that enables them to download your app through the **TestFlight** app for iOS.

Once your internal testers are happy with your app, you can select some external beta testers for your app. **External testers** are typically people who aren't in your team or organization, such as existing users of your app or a selection of people from your app's target audience.

Setting up an external beta test is done identically to how you set up the internal test. You can even use the same build that you used for internal testing for external testing. However, external tests typically require a quick review by Apple before invites can be sent out. These reviews don't take long and passing the beta test review does not mean you'll also pass the App Store review.

This is all you need to know about setting up a beta test through **TestFlight**. When you're happy with the results of your beta test and your app has passed the real-world test, it's time for you to prepare to release your app into the wild through the App Store.

Preparing your app for launch

Moving from beta testing to releasing your app does not require much effort. You use the same version of your app as you've already exported and tested with your users. To be able to submit your app for review by Apple, you have to add more information about your app, and you should set up your App Store presence. The first thing you should do is create a couple of screenshots of your app. You add these screenshots to your App Store page, and they should look as good as possible because potential users will use screenshots to determine whether they want to buy or download your app. The simplest way to create screenshots is to take them on a 5.5-inch iPhone and a 12.9-inch iPad. You can provide screenshots for every type of device that exists, but you must at least provide them for a 5.5-inch iPhone and a 12.9-inch iPad. You can use the **Media Manager** feature in App Store Connect to upload the large-sized media and have it scale down for smaller devices:

After submitting screenshots, you should also fill out a description and keywords for your application. Make sure that your app description is clear, concise, and convincing. You should always try to use as many keywords as you can. Apple uses your keywords to determine whether your app matches search queries from users. Try to come up with synonyms or words you would look for when you'd search for an app that does what your app does.

If your app features an iMessage or Watch app, you should upload screenshots for these apps as well. You can't provide separate keywords or descriptions for these extensions, but they will have their own image galleries in the App Store.

The next step in the submission form is to select your app binary and provide some general information about the app and the person responsible for publishing the app. Usually, you'll want to choose the version of the app you've been beta testing up to the point of release.

Lastly, you must provide some information to Apple about how your app should be reviewed. If your app required a demo account, provide credentials to the reviewer. If your app has been rejected before due to something being unclear, it's usually a good idea to clarify the past misunderstanding in the notes section. This has proven to help for some apps, resulting in accepted reviews at the first try rather than being rejected and providing explanations afterward. When everything is filled out, hit the **Save** button to store all of the information you just entered. Then, if you're satisfied with everything and ready to take the final leap toward releasing your app, press **Submit for Review** to initiate the review process.

Getting your app reviewed by Apple can take from just a single day to a couple of days or even longer. Once you have submitted your app, it's important that you patiently wait until you hear from Apple. Sending them queries about reviewing your app faster or asking them about the current status often yields no results, so there's no point in trying to push Apple for a review.

If you do need to get your app reviewed and released fast and you have a legitimate reason, you can always apply for expedited review. If Apple agrees that a faster review will benefit not just you but also your users, your app could be reviewed in a matter of hours. Note that you should not abuse this. The more often you apply for an expedited review, the less likely Apple is to grant you an exception. Expedited reviews should only be requested in exceptional cases.

Well, time to get some cocoa, coffee, tea, or whatever you prefer to drink. You can now sit back for a while and wait while Apple reviews your app so you can release it to the App Store.

Summary

This final chapter covered preparing to release your app. You learned how to archive and export your app. You saw how to upload it to App Store Connect and how to distribute your app as a beta release. To wrap everything up, you saw how to submit your app for review by Apple to release it to the App Store. Releasing an app is exciting; you don't know how well your app will perform or whether people will enjoy using it. A good beta test will help a lot, you'll be able to spot bugs or usability issues, but there's nothing like having your app in the hands of actual users.

Most developers invest a lot of time and effort into building their apps, and you are one of them. You picked up this book and went from an iOS enthusiast to an iOS master who knows exactly how to build great apps that make use of iOS' newest and greatest features. When you're ready to launch your own app into the App Store, you'll learn how exciting and nerve-racking it can be to wait for Apple to review and hopefully approve your app. Maybe you get rejected on your first try; that's possible. Don't worry too much about it; even the biggest names get rejected sometimes, and often fixing the reason isn't too complicated. Do make sure to read the App Store review guidelines before you submit; these guidelines give a pretty good indication about what you can and can't do in your apps.

Since this is the last chapter in this book, I would like to sincerely thank you for picking up this book and using it as one of the stepping stones toward becoming a master of iOS programming. I hope to have left you with the skills required to venture out on your own, read Apple's documentation, and build amazing applications. Thanks again, and if you've created something cool using this book, please feel free to reach out to me. I would love to see your application.

Questions

No questions for this chapter!

Further reading

- App Store Review Guidelines: `https://developer.apple.com/app-store/review/guidelines/`

Answers

Chapter 1

1. What happens if you don't provide a reason for wanting to access a user's contacts?

    c) The app crashes.

2. What is a reuse-identifier on a table view cell used for?

    c) It is used by the table view to optimize performance.

3. Where does a table view obtain information about the cells it displays from

    b) `UITableViewDataSource`.

4. How does a table view make sure to keep its memory footprint as small as possible?

    b) It reuses cells that were displayed before.

5. What is a placeholder cell called in Interface Builder?

    a) Prototype cell.

6. What is a connection between an Interface Builder item and a variable in code called

    c) `@IBOutlet`.

7. Where is the best place to reset a table view cell?

    c) In `prepareForReuse()`.

Chapter 2

1. `UICollectionView` is very similar to another component on iOS. Which component is this?

    c) `UITableView`.

2. Which class did you have to subclass to create your own collection view layout

    b) `UICollectionViewLayout`.

2. Which technique is applied by collection views to ensure great performance?

    a) Fetching cells a couple of rows ahead of the user's current position.

3. Where in **Interface Builder** can you find the collection view's layout object?

    b) The **Document Outline**.

4. Which feature that can be implemented for collection views is not available for table views?

    c) Horizontal scrolling.

5. What is a gesture-recognizer?

    c) An object that responds to several user interactions, such as tapping, swiping, and pinching.

6. What is the type of the button that you added to the navigation bar to toggle the collection view's editing state?

    c) `UIBarButtonItem`.

Chapter 3

1. Where can you configure the supported Interface Orientations for an iPad in the Hello-Contacts project?

   b) `Info.plist`.

2. How many different size classes exist today?

   a) Two.

3. What object contains information about the current environment your app is running in?

   c) `UITraitCollection`.

4. What is the best way to lay out several objects on top of each other?

   b) Using `UIStackView`.

5. How do you animate an update to layout constraints?

   a) By calling `view.layoutSubviews()` in an animation.

6. What is the correct way to create an instance of a view controller from a storyboard?

   b) `let myViewController = storyboard?.instantiateViewController(withIdentifier:"MyViewController")`.

7. How many methods did you have to implement to support peek and pop?

   b) Two.

Chapter 4

1. Name two ways to implement simple animations in iOS.

    b) `UIViewPropertyAnimator` and `UIView.animate`.

2. What does the damping property do in a spring animation?

    a) It is used for the bounciness of the animation.

3. Name an advantage of `UIViewPropertyAnimator`.

    c) Easier to start, pause, and interact with.

4. What is the difference between implementing `UIViewControllerTransiting` for interactive and non-interactive transitions?

    c) You must implement `interruptibleAnimator(using:)`.

5. How do you prevent a view from showing out-of-bounds contents?

    a) By setting `clipToBounds` to true on the view.

6. Is it possible to add animations to a running property animator?

    b) Yes.

Chapter 5

1. Which statement about reference types is false?

    b) Reference types can always be subclassed.

2. Which statement about value types is true?

    c) Values types can conform to multiple protocols.

3. What does heap allocation mean?

    b) An object allocated on the heap is stored with a variable size in RAM memory.

4. You should always use a struct for your data models. True or false?

   b) False, it depends on whether your models have an identity.

5. What keyword should you add to a function that mutates a property on a struct?

   b) `mutating`.

6. Which type of object is not allocated on the heap?

   c) An enum.

7. Which statement about the stack is true?

   a) The stack has a fixed size.

8. In which of the following cases would a reference type be a good choice?

   a) When you want to create a new button in your app.

9. In which of the following cases would a value type be a good choice?

   b) When writing a networking object.

Chapter 6

1. Which of the following is relevant in Protocol-Oriented Programming?

   c) What an object can do.

2. Which of the following is not a downside of subclassing?

   a) Shared functionality.

3. Protocols can inherit from several other protocols. True or false?

   a) True.

4. An associated type has to be a struct or class. True or false?

   b) False.

5. In which two ways can a struct specify what type an associated type for a protocol is?

    a) By using a typealias.
    b) By using generics on the struct.

6. What does T mean in `genericFunction<T>()`?

    c) T is a placeholder for the generic type.

Chapter 7

No questions.

Chapter 8

1. What does a persistent store coordinator do?

    b) It talks to the persistent store.

2. What is the function of managed object context?

    b) It mediates between managed objects and the persistent store coordinator.

3. How many managed object contexts can you use in an app?

    c) An unlimited amount.

4. What is the correct way to call `save()` on a managed object context?

    c) `try save()`.

5. When should you use a fetched results controller?

    b) Only when you need to fetch data from the database and want to react to changes.

6. Does a fetched results controller always go to the database to fetch data?

    c) Yes, all fetch requests must go to the database.

Chapter 9

1. What object should you use to make web requests?

    a) `URLSession`.

2. How are web requests executed?

    a) Asynchronously.

3. What is a typealias?

    c) An alias for any other type, for instance a callback.

4. What happens if you use managed objects on the wrong thread?

    c) Your app might crash.

5. Which object do you need if you make an HTTP POST request?

    c) `POSTRequest`.

6. What is ATS?

    b) Apple's way of enforcing secure web requests.

Chapter 10

1. For how long can you perform work when your app is woken up for background refresh?

    b) There is no way to be sure.

2. How do you enable background fetch for your app?

    c) Both.

3. Which statement about the minimum refresh interval is true?

    b) iOS will determine when it's the best time wake up your app. It will be no more often that you specify.

4. What did you use a dispatch group for in this chapter?

   b) To be notified when all network calls were done.

5. When should you refactor your code?

   c) Any time you think you can improve something.

6. What do you call the ID that is passed to the movieById case on the `MovieDBHelper.Endpoint` enum?

   a) Associated value.

Chapter 11

1. How does the CloudKit server notify apps about updates?

   b) Silent push notifications.

2. Which object do you send along with a fetch request to CloudKit so you only receive new changes?

   a) A change token.

3. How does your code know whether a certain `CKRecord` is a movie, family member, or something else?

   c) Through `recordType`.

4. What kind of data do you have to store alongside your objects in Core Data to be able to properly sync local data with CloudKit?.

   b) The encoded CloudKit metadata.

5. What's a good place to store server-change tokens?

   c) `UserDefaults`.

6. Why is it smart to import CloudKit data on a background managed object context?

   a) Because you don't know how long the import will take and you don't want to block the main thread.

Chapter 12

1. What is SceneKit typically used for?

   c) 3D games.

2. What is a view in a SpriteKit game called?

   b) A sprite.

3. What unit of measurement do SceneKit and ARKit use?

   c) Centimeters.

4. Which tracking mode tracks both device movement and rotation?

   c) World Tracking.

5. What should you keep in mind when you prepare images for image tracking in ARKit?

   c) The images should have sufficient features.

6. What does ARKit need to accurately track objects?

   c) Surfaces with lots of textures.

Chapter 13

1. What different types of location access exist?

   a) Always and When in use.

2. When should you ideally ask for access to location services?

   c) Once the user understands why you need access to their location.

3. How do you obtain a user's current location?

   b) By calling `startUpdatingLocation()` on the location manager.

4. Which of the following location-tracking techniques is the most battery-friendly?

    c) Visit tracking.

5. What different types of regions can you monitor?

    a) Geofences and iBeacons.

6. Up to how long could it take for region-monitoring to start?

    b) 10 minutes.

7. How can you check whether your app launched due to a significant location change?

    b) By inspecting the app's launch options.

Chapter 14

1. What is the file format for a CoreML-compatible model?

    b) `.mlmodel`.

2. How do you use a CoreML model in your project?

    a) By adding the model to Xcode and using the model class that is generated for you.

3. Why is it so important the CoreML runs on the end user's device?

    a) It's better for privacy and preserves the user's data plan.

4. What does the `NaturalLanguage` framework do?

    b) It extracts interesting information from text, for instance, names, places, and words.

5. What does the `Vision` framework do?

    c) It performs complex analyses on images.

6. Which of the following options could negatively impact the quality of a training set for an image classifier?

    b) If the training subject is always shown in similar surroundings.

Chapter 15

1. Which devices support HealthKit?

    c) The iPhone and Apple Watch.

2. What's the difference between a quantity sample and a category sample?

    a) Quantity samples have an arbitrary value, while categories use enums to represent their value.

3. How can you check whether you have read access to a certain health metric?

    c) To protect the user's privacy, there is no way to be sure.

4. How can you filter data that is returned from a HealthKit query?

    a) With predicates.

5. What permissions do you need to use a workout route builder?

    c) You need permission for both the builder and the workouts.

6. What do you need to keep in mind when you associate location data with a workout route?

    b) You should make sure the location data is very accurate.

Chapter 16

1. How does Siri understand a user's request?

    b) Siri uses natural language processing.

2. How are app extensions launched?

    c) App extensions are launched by the extensions framework.

3. How can you share data between an app and an extension?

   c) Both of the above.

4. What are the three stages of handling a Siri intent?

   a) Resolve, Confirm, Handle.

5. How do you make sure Siri does not render its default UI?

   b) By conforming `IntentViewController` to the `INUIHostedViewSiriProviding` protocol, and setting `displaysMessage` to `true`.

6. How can you donate actions to Siri?

   a) Through user activities and interaction objects.

7. How can a user add their own voice commands for Siri Shortcuts?

   a) Through the settings app.

Chapter 17

1. Which framework is used to implement a simple video player?

   c) `AVKit`.

2. Which object is used to play a local audio file?

   b) `AVAudioPlayer`.

3. What are remote commands, and where do they come from?

   c) Media playback-related commands that can come from any external source, including the lock screen, control center, or even the user's headphones.

4. Which object can be used to quickly take a picture?

   a) `UIImagePickerController`.

5. Which of the following is not needed to record a video with an `AVCaptureSession` object?

    b) Permission to access the user's photo library.

6. What information tends to get lost when you convert a `UIImage` instance to a `CGImage` instance?

    a) The orientation.

Chapter 18

1. How are `UNUserNotificationCenter` and `NotificationCenter` different?

    a) One is responsible for showing notifications to users, and the other is a mechanism to send notifications about events inside an app.

2. What kinds of notifications can you send to a user?

    a) Push notifications and local notifications.

3. What are provisional notifications?

    c) Notifications that are delivered to the user quietly, so they can preview the notifications your app will send.

4. How does iOS group notifications?

    a) It uses the thread identifier from the notification content and the app that the notification belongs to.

5. What does a notification service extension do?

    c) It allows you to replace a notification's contents before the user sees it.

6. What does a notification content extension do?

    b) It allows you to create a custom interface for when a user 3D-touches a notification.

Chapter 19

1. Where can users find Today Extensions?

    c) By Swiping right on the lock screen or Notification Center, on the leftmost screen on the Springboard, and 3D-touching an app.

2. At what height are widgets displayed by default?

    a) It depends on the screen width and user settings.

3. How can you allow a user to expand your widget's height?

    c) By setting the largest available display mode on the extension context to expanded.

4. What is the best place to update your widget contents?

    b) In `widgetPerformUpdate(completionHandler:)`.

5. How does iOS know the best moment to reload a widget's contents?

    c) With the completion handler that is passed to `widgetPerformUpdate(completionHandler:)`.

Chapter 20

1. What devices support the full drag-and-drop experience?

    b) Only iPad.

2. What kind of object do you have to provide to start a drag session?

    a) `UIDragItem`.

3. How can you check whether your app can handle a certain drop action?

    a) By checking the type of object that is associated with the drop action.

4. How is the drag-and-drop experience limited on iPhone?

    c) Drag and drop is limited to the currently-open app.

5. What do you have to do to support drag and drop on an iPhone?

    a) You must manually enable the drag-and-drop interactions.

6. How are drag and drop delegate protocols for collection views different from the regular drag and drop delegate protocols?

    b) The collection view versions of the delegate methods receive an index path.

Chapter 21

1. Which objects are used to represent an item that you would like to index in Spotlight?

    b) `NSUserActivity` and `CSSearchableItem`.

2. What happens when a user selects a Spotlight search result that points to your app?

    a) `AppDelegate` is asked to resume a user activity that represents the item.

3. How do you add an item to the Spotlight index?

    a) By passing `CSSearchableItem` to `CSSearchableIndex` or setting a user activity as the current activity.

4. Does it make sense to index the same user activity multiple times?

    b) Yes, the more often an activity is performed in the app, the higher it will rank in the Spotlight search results.

5. What are Smart App Banners?

    a) You can add Smart App Banners to web pages to point visitors to your app.

6. What are Universal Links?

   a) Universal links are shareable web links that open your app if it's installed.

7. What happens when your app is expected to open a Universal Link?

   b) `AppDelegate` is asked to open the URL.

Chapter 22

1. In what way are Message Extensions similar to apps?

   a) They can exist independently and support In-App Purchases.

2. What do you need to create a Sticker Pack?

   a) You need images that can be added to the Sticker Pack template.

3. How can you test your Message extensions?

   b) You must use the predefined conversations in the simulator.

3. What object is used to configure a complex message that can be sent through your extension?

   c) The `MSMessage` object.

4. How can you identify the users that participate in a conversation?

   a) By reading their unique, random identifiers.

5. How do you know whether a message was delivered to the recipients?

   c) There is no way to be sure due to privacy concerns.

Chapter 23

1. What is a unit test?

   c) A test that checks an object in isolation.

2. Is it okay to force your tests to execute in a certain order?

    b) No.

3. What is the order of the three As?

    c) Arrange, Act, Assert.

4. How do protocols help you when writing tests?

    a) You can mock certain objects by conforming them to a protocol.

5. How can you see which code is not tested in your app?

    a) Using Code Coverage.

6. What is a launch argument?

    b) A string that is passed to an app when it launches.

7. How does `XCUITest` identify elements in an app?

    c) Through accessibility identifiers.

Chapter 24

1. What does the Time Profiler do?

    c) Shows you how much time is spent in certain methods.

2. What does the Allocations instrument do?

    a) Shows you your app's memory usage over time.

3. Which of the following does not necessarily indicate a memory leak?

    a) Growing memory in the memory graph.

4. What does Xcode's memory debugger show you?

    b) The relationship between different objects that are in memory.

5. What are is signpost logging?

    a) A low-overhead log to measure your app's performance.

6. Which elements does an Instruments Package have?

    a) Metadata, a schema, and the instrument itself.

Chapter 25

1. What are the downsides of running certain code on the main thread?

    a) It can block the UI.

2. In which circumstances will threading not help your performance?

    a) When you have a lengthy layout-related calculation.

3. Which GCD feature can you use to execute an action when several other actions are done executing?

    b) `DispatchGroup`.

4. Which of the following is not a benefit of operations?

    b) Faster code execution.

5. What's the maximum number of Operations you can run at the same time?

    a) There is no hard limit, iOS decides on the fly.

6. What is KVO?

    a) A way for objects to observe properties of another object.

7. How do you start executing operations on an operation queue?

    c) By adding operations to it.

Chapter 26

No questions.

Other Books You May Enjoy

If you enjoyed this book, you may be interested in these other books by Packt:

Swift Game Development - Third Edition
Siddharth Shekar

ISBN: 9781788471152

- Deliver powerful graphics, physics, and sound in your game by using SpriteKit and SceneKit
- Set up a scene using the new capabilities of the scene editor and custom classes
- Maximize gameplay with little-known tips and strategies for fun, repeatable action
- Make use of animations, graphics, and particles to polish your game
- Understand the current mobile monetization landscape
- Integrate your game with Game Center
- Develop 2D and 3D Augmented Reality games using Apple's new ARKit framework
- Publish your game to the App Store

Hands-On Full-Stack Development with Swift
Ankur Patel

ISBN: 9781788625241

- Get accustomed to server-side programming as well as the Vapor framework
- Learn how to build a RESTful API
- Make network requests from your app and handle error states when a network request fails
- Deploy your app to Heroku using the CLI command
- Write a test for the Vapor backend
- Create a tvOS version of your shopping list app and explore code-sharing with an iOS platform
- Add registration and authentication so that users can have their own shopping lists

Leave a review - let other readers know what you think

Please share your thoughts on this book with others by leaving a review on the site that you bought it from. If you purchased the book from Amazon, please leave us an honest review on this book's Amazon page. This is vital so that other potential readers can see and use your unbiased opinion to make purchasing decisions, we can understand what our customers think about our products, and our authors can see your feedback on the title that they have worked with Packt to create. It will only take a few minutes of your time, but is valuable to other potential customers, our authors, and Packt. Thank you!

Index

A

B

D

E

F

G

H

I

J

K

L

M

N

O

P

R

S

T

U

V

Lightning Source UK Ltd.
Milton Keynes UK
UKHW031802220419
341426UK00004B/74/P